Undergraduate Texts in Mathematics

Editors:
S. Axler
K. Ribet

Undergraduate Texts in Mathematics

For other titles published in this series, go to
http://www.springer.com/series/666

Lindsay N. Childs

A Concrete Introduction to Higher Algebra

Third edition

 Springer

Lindsay N. Childs
University at Albany
State University of New York
Department of Mathematics
1400 Washington Avenue
Albany, NY 12222
USA
lc802@albany.edu

ISSN: 0172-6056
ISBN: 978-1-4419-2561-9   e-ISBN: 978-0-387-74725-5
DOI: 10.1007/978-0-387-74725-5

Mathematics Subject Classification (2000): 12-01

Printed on acid-free paper

springer.com

Acknowledgements

My thanks to the numerous colleagues at Albany who have used and commented on the book over the years, including Lou Brickman, Ed Davis, Hugh Gordon, Bill Hammond, Ben Jamison, Tom MacGregor, Antun Milas, Malcolm Smiley, Anupam Srivastav, Mark Steinberger, Ted Turner and Don Wilken. Among those elsewhere who have provided errata or otherwise commented on the book, I wish to acknowledge with appreciation Kirby Baker, Hyman Bass, Jamie Bessich, Matthew Bridiga, Ken Brown, Sanford Bucklan, L. H. Chan, Tat-Hung Chan, Steve Chase, Donald Crowe, David Dobbs, Linda Dineen, Richard Ehrenborg, David Ford, Saab Yaqub Hassan, Olav Hjortaas, Chris Jeuell, Irving Kaplansky, Keith Kendig, Barry Loneck, Morris Orzech, Richard Patterson, Kristin Pfabe, Margaret Readdy, Joel Roberts, Michael Rosen, David Saltman, Marcus Schmidmeier, Alan Sprague, Christopher Systsma, Bethany Talbot, Mel Thornton, S. Wang and Bill Waterhouse. I particularly thank David Drasin for his comprehensive reading of the manuscript for the first edition, Ernst Selmer and Frank Gerrish for their extensive comments on the first edition, and Keith Conrad for numerous insightful comments on the material. I wish also to thank the students at Albany who have inspired me to think more deeply about topics in the book, in particular Harold H. Smith III, Daniel Zimmerman and Nicholas Manning.

Finally, my greatest thanks go to Rhonda for her love, understanding and support while I worked on various versions of the book over the past 30+ years.

July 2008

Introduction

This book is an introduction to higher algebra for students with a background of a year of calculus. The first edition of this book emerged from a set of notes written in the 1970's for a sophomore-junior level undergraduate course at the University at Albany entitled "Classical Algebra".

The objective of the course, and the book, is to offer students a highly motivated introduction to the basic concepts of abstract algebra—rings and fields, groups, homomorphisms—by developing the algebraic theory of the familiar examples of integers and polynomials, and introducing the abstract concepts as needed to help illuminate the theory. By building the algebra out of numbers and polynomials, the book takes maximal advantage of the student's prior experience in algebra and arithmetic from secondary school and calculus. The new concepts of abstract algebra arise in a familiar context.

An ultimate goal of the presentation is to reach a substantial result in abstract algebra, namely, the classification of finite fields. But while heading generally towards that goal, motivation is maintained by many applications of the new concepts. The student can see throughout that the concepts of abstract algebra help illuminate more concrete mathematics, as well as lead to substantial theoretical results.

Thus a student who asks, "Why am I learning this?" will find answers usually within a chapter or two.

While our course is called "Classical Algebra" and the book begins with mathematics dating from Euclid (300 BC) and before, the book also includes a substantial amount of mathematics discovered only within the past three generations. Thus this book is explicitly offered as a counterexample to Alan Hammond's (1980) remark that "Mathematics is one of the few subjects that a student can study through high school and even a few years into college without coming into contact with any results invented since 1800."

The extent and variety of applications in the book have led to the book being used in courses rather different than the course for which it was originally designed. The book has been used for courses in Applied Algebra or Applicable Algebra, in Elementary Number Theory, and in Algebra for Teachers. Possible outlines for such courses are described below.

Notes on the Third Edition

The first and second editions of this book were published in 1979 and 1995, a gratifyingly long time ago. The second edition was an extensive revision and expansion (160 pages) of the first edition. This third edition is an extensive revision and somewhat more modest expansion (around 85 pages) of the second edition.

The new edition is organized into seven parts, as follows:

I. Numbers. Chapters 1-5 present the elementary theory of the integers—induction, divisibility, Euclid's Algorithm, unique factorization into prime numbers and congruence modulo m. These ideas and techniques lay the groundwork for the mathematics and applications in the remainder of the book. Section 4D begins a discussion of prime numbers, and Sections 5C and 5D introduce two applications of congruence to error detection—casting out nines and Luhn's formula.

II. Congruence classes and rings. Chapter 6 introduces the ring of congruence classes modulo m, and Chapter 7 introduces some basic concepts of abstract algebra—rings, fields, groups, homomorphisms—that can be used to help identify properties and features of congruence classes. Chapter 8 reviews elementary ideas of matrices and linear equations needed for subsequent applications. Applications include Karatsuba multiplication, the use of modular arithmetic to the design of tournaments and to factoring by trial division, and the use of matrices with entries in the integers modulo m to error-correcting codes (Hamming codes) and to cryptography (the Hill cryptosystem).

III. Congruences and groups. Chapters 9–11 focus on the multiplicative group of units of the integers modulo m. Chapter 9 obtains Fermat's and Euler's Theorems, with two distinctly different proofs; Chapter 11 develops some finite group theory, enough to prove Lagrange's Theorem and to introduce quotient groups. Lagrange's Theorem yields a third proof of Fermat's and Euler's Theorems. Sections 9E and 9F introduce useful techniques for computing modulo m. Chapter 12 presents the Chinese Remainder Theorem, an important result in modular arithmetic. The CRT yields information on the size of groups of units modulo a composite number. Applications include the connection between Euler's Theorem and the period of repeating decimals, and the application of Euler's Theorem to the famous RSA cryptosystem. The RSA cryptosystem motivates the search for methods for identifying possible prime numbers, for factoring large numbers, and for multiplying large numbers. Approaches to all three problems (pseudoprime testing, the Pollard $p - 1$ method, using the CRT for multiplying) are presented in Sections 10B, 10C, 11D, and 12C.

IV. Polynomials. Chapters 13–18 introduce the elementary theory of polynomials in one variable over a field, a theory that closely parallels the theory of the integers presented in Chapters 2–12—divisibility, Euclid's Algorithm, unique factorization, congruence. For integers, prime numbers are the "atoms" that generate all numbers; in a similar way, irreducible polynomials are the atoms for all polynomials. So Chapters 15 and 16 are devoted to studying irreducible polynomials and factorization of polynomials over the rational numbers, the real numbers and the complex numbers (Fundamental Theorem of Algebra). Chapter 17 introduces congruence for polynomials and the Chinese Remainder Theorem, which in turn leads to interpolation and a proof that factoring polynomials over the rational numbers is a finite process. Chapter 18 presents a fast method of multiplying polynomials, based on the Chinese Remainder Theorem and a method for evaluating polynomials known as the Fast Fourier Transform.

V. Primitive Roots. With the availability of D'Alembert's Theorem in Section 14A, Chapters 19–22 return to the study of groups of units of integers modulo m. Chapter 19 introduces enough additional finite group theory (the exponent of an abelian group, finite cyclic groups) to prove the Primitive Root Theorem. Chapter 21 applies the Primitive Root Theorem and quotient groups from Section 11G to determine how to decide whether a number is a square modulo m—the famous Law of Quadratic Reciprocity. These results are applied to cryptography—Diffie-Hellman key exchange, Blum-Goldwasser cryptography and refinements of RSA cryptography; to pseudorandom numbers—the linear congruential and Blum-Blum-Shub methods; and to primality testing and factorization of integers—strong pseudoprimes, Carmichael numbers and Rabin's Theorem, the Pollard rho factoring method.

VI. Finite Fields. Chapters 23–25 continue the theory for polynomials that parallels the theory for integers in Chapters 6–7. The construction of congruence classes for polynomials yields many new fields, fields that can be used to split polynomials defined over subfields into linear factors. The construction yields a complete classification of finite fields in Section 24C. Applications of finite fields include Latin squares (Section 25D) and multiple error-correcting (BCH) codes (Chapter 26).

VII. Factoring Polynomials. Chapters 26 and 27 extend the ideas of Chapters 16 and 17 on factoring polynomials over the rational numbers and integers. Chapter 26 reproves that factoring polynomials over the rationals is a finite process, and then presents methods to make the process more efficient—Berlekamp's factoring algorithm and the Hensel factoring method. Chapter 27 extends the ideas of Sections 24B and C to give a count of irreducible polynomials of every degree over the field of p elements for every prime p, then uses ideas of Section 16B and Chapter 26 to obtain a result of Van der Waerden that in a certain sense, almost all polynomials over the rational numbers are irreducible.

Changes in the new edition not already noted include:

- Material on solving equations in the integers and modulo m has been rewritten and placed in separate sections (3E, 6F).
- The axioms for a field are motivated by looking at how to solve linear equations (7A).
- There is a greater emphasis on homomorphisms of polynomial rings (starting at 13D)
- The section on congruence modulo a polynomial has been rewritten and expanded (17A).
- There is more emphasis on the exponent of an abelian group (Chapter 19)
- The section on finite cyclic groups has been rewritten (19B)

- The material on bounding the roots and factors of polynomials with integer co-efficients has been greatly improved (26 A, B).
- There are new sections on cosets and solutions of equations (11E) and on quotient groups and the "fundamental homomorphism theorem", with applications to groups of units (11G), quadratic reciprocity (21G), and (via Cauchy's Theorem), the cardinality of finite fields (24C).
- There is a new application of Eisenstein's irreducibility criterion to Chebyshev polynomials (16B)
- There is a new proof of a weak form of Rabin's Theorem using subgroups and cosets (20C), and a new proof of quadratic reciprocity (due to G. Rousseau) that uses the Chinese Remainder Theorem and different coset representatives of a quotient group (21C)

In order to keep this edition from getting any larger than it already is, a few sections found in the second edition have been omitted in the third: the connection between Euclid's algorithm and incommensurability, Sturm's Algorithm, knapsack cryptosystems, a proof of Rabin's theorem in its full strength, and Reed-Solomon codes. This last topic was a close call, but I finally decided that to do Reed-Solomon well would stray too far from the main objective of the book.

Prerequisites

The explicit prerequisite consists of precalculus algebra. However, long experience suggests that three or four semesters of college-level mathematics, such as the calculus sequence and a semester of linear algebra, is helpful. Only a few sections of the book use calculus or linear algebra. Elementary row operations show up with the extended Euclidean algorithm in Chapter 3, but otherwise a course can easily be designed to avoid linear algebra (used in 8E, 8F, 11H, 18, and 25). On the other hand, if linear algebra is a prerequisite, then a number of the applications can be done more efficiently.

Designing a course

For an Introduction to Abstract Algebra course that seeks to reach the classification of finite fields, the theoretical core of the course is found in sections 2B, 3A-E, 4A, 5A-B, E-F, 6A-F, 7A, C-D, 9A-C, 11A-C, E-G, 13, 14, 17A, 19A-B, 23, 24A-C.

For our Classical Algebra course, most instructors do chapters 2, 3A-E, 4, 5, 6A-F, 7A,C,D, 9A-D, 10A, 11A-C, most of 12A-D, 13, 14, 16A,B, 17A, 19A-B, plus other topics as time allows.

A course in Number Theory could follows chapters 2, 3, 4, 5, 6, 7A, 9, 12, 13A, 14, 20, 21, 22B and 23ABC (plus handouts on special topics of interest to the instructor).

A course on Applicable Algebra could follow chapters 2, 3, 4, 5, 6, 7ACD, 8E, 9ABC, 10A, 11A-C, E-H, 12ABD, 13, 14, 17ABC, 19AB, 23, 25 (plus supplemental notes on error-correcting codes).

A course emphasizing polynomials (e.g. for future secondary teachers) could do chapters 2, 5A-B, 6C, 7A, C, 13–18, 23–27.

Contents

Part I
Numbers

Chapter 1
Numbers

Mathematics grows through the development and study of new concepts. The history of numbers illustrates this growth.

Ancient cultures began mathematics by counting, keeping tallies. In the ancient Near East an increasingly urban economy the need to keep records, at first by the use of tokens, small clay objects, to correspond to quantities of goods. Sometime around 3100 B.C. ancient accountants began abstracting quantity from the objects being counted, and written numbers were born (see Schmandt-Besserat (1993)).

Once the natural numbers, $1, 2, 3, \ldots$ were available, manipulating them and solving problems involving numbers led to positive fractions, known to the ancient Babylonians (2000 B.C.), who also knew some square roots and cube roots. The classical Greek geometers (400 B.C.) studied positive quantities that could be obtained from natural numbers by the processes of addition, subtraction, multiplication, division, and taking square roots. The number 0 did not come into use, however, until after 300 B.C.; negative numbers first arose around A.D. 600, but became acceptable only in the 1600's (Descartes, in 1637, called them "false"); and complex numbers gradually won acceptance between the early 1500's and 1800. A precise understanding of the real numbers was reached only in the 1870's.

As the domain of numbers broadened, so did mathematics, and its applications. For example, with only the natural numbers available, calculus would be unthinkable.

Since 1800 mathematics has developed many new systems of objects that can be manipulated in the same way as these classical sets of numbers. Just as with classical numbers, once the domain of numbers is expanded, so do the uses that can be made of them.

This book is about these new sets of numbers and some of their uses.

In this chapter we'll set up some notation for classical sets of numbers, and introduce the idea of an equivalence relation, which will be the basis for constructing new sets of numbers in later chapters.

First, notation.

L.N. Childs, *A Concrete Introduction to Higher Algebra*, Undergraduate Texts in Mathematics, © Springer Science+Business Media LLC 2009

$\mathbb{N}$ is the set of natural or counting numbers:

$$1, 2, 3, 4, 5, \ldots,$$

$\mathbb{Z}$ is the set of integers

$$\ldots, -3, -2, -1, 0, 1, 2, 3, 4, \ldots,$$

obtained from $\mathbb{N}$ by including 0 (zero) and the negatives of the natural numbers.

$\mathbb{Q}$ is the set of rational numbers, that is, the set of all fractions a/b, where a and b are integers, and $b \neq 0$. We shall look at $\mathbb{Q}$ in more detail shortly. We think of $\mathbb{Z}$ as being a subset of $\mathbb{Q}$ by identifying the integer a with the fraction $a/1$. $\mathbb{Q}$ is large enough so that every nonzero integer has an inverse, or reciprocal, in $\mathbb{Q}$, as does every nonzero fraction. Thus $\mathbb{Q}$ is an example of a field (see Chapter 7).

$\mathbb{R}$ is the set of real numbers. A useful way to think of $\mathbb{R}$ is as the set of all infinite decimals, or as the set of coordinates on a line (such as the x-axis used in calculus). Any rational number is a real number–the decimal expansion of a fraction a/b is obtained by the familiar process of dividing b into a. So is every quantity that can be constructed by straightedge and compass from a natural number.

The real numbers form a complete Archimedean ordered field.

"Ordered" means that for any two numbers r and s, either $r < s$ or $r = s$ or $r > s$. This property corresponds to the intuitive representation of the real numbers as coordinates on a line, where $r > s$ if the point with coordinate r is to the right of the point with coordinate s.

"Archimedean" means that for every positive real number r there is a natural number n with $n > r$.

"Complete" can be described by the Least Upper Bound axiom: given any non-empty set of real numbers S, if there is an upper bound C for S, that is, a real number C so that for every r in S, $r \leq C$, then there is a least upper bound B for S, that is, an upper bound B so that any upper bound C for S satisfies $B \leq C$. These properties imply the existence of limits, which in turn enables the definition of continuous functions and the Intermediate Value Theorem.

Readers who have had calculus (or even precalculus) should be comfortable about working with the real numbers, and so we will say no more about them here.

$\mathbb{C}$ denotes the complex numbers, which we will describe in Chapter 15D.

The 19th century effort to formulate a satisfactory definition of real numbers led to the idea that sets should be the primitive mathematical objects from which all numbers should be defined.

The general procedure for the construction of new sets of "numbers" is to take a set, call it S, consisting of mathematical objects, such as numbers you are already familiar with, split the set S up ("partition the set S") into a collection of subsets in a suitable way, and then attach names, or labels, to each of the subsets. These subsets then will be elements of a new number system: new "numbers."

We'll use this procedure to construct many new sets of numbers in this book.

We illustrate the idea by defining the set $\mathbb{Q}$ of rational numbers, or fractions.

Rational Numbers. Rational numbers are fractions of integers, numbers like $\frac{6}{9}$ or $\frac{72}{22}$. But they have the curious property, different from integers, that two fractions, like $\frac{6}{9}$ and $\frac{72}{108}$, are really equal:

$$\frac{6}{9} = \frac{72}{108},$$

even though they look different. So any definition of rational numbers must deal with the problem that the same rational number can be expressed as a fraction of two integers in many different ways.

This problem was solved by the ancient Greeks (500-300 B.C.) in their development of the ratio of two numbers. For numbers a and b, the ratio of a to b is the same as the ratio of c to d if $ad = bc$. Thus the ratio of 1 to 2 is the same as the ratio of 3 to 6. The ratio of 2 to $1 + \sqrt{7}$ is the same as the ratio of $\sqrt{7} - 1$ to 3.

From this idea of ratio we can give a set-theoretic definition of the rational numbers as follows.

Take the set S of all ordered pairs (a, b) of integers, where $b \neq 0$. Partition the set S into subsets, by the rule:

two pairs (a, b) and (c, d) are in the same subset if the ratio of a to b is the same as the ratio of c to d, that is, if and only if $ad = bc$ (or, in fractional notation, if $a/b = c/d$ – but using fractional notation assumes that we have defined fractions already!)

We give the name a/b to the set of ordered pairs containing the pair (a, b). Thus the symbol a/b is a *label* for the set of all ordered pairs (m, n) such that the ratio of m to n is the same as the ratio of a to b.

For example, 6/9 is a label for the set of all pairs (m, n) with $6n = 9m$: in set notation,

$$\frac{6}{9} = \{(m, n) | 6n = 9m\},$$

and

$$\frac{72}{108} = \{(m, n) | 72n = 108m\},$$

Since the set $\{(m, n) | 6n = 9m\} = \{(m, n) | 72n = 108m\}$, it follows that $6/9 = 72/108$. In fact, if (c, d) is any ordered pair in the set $\{(m, n) | 6n = 9m\}$, then

$$\{(m, n) | 6n = 9m\} = \{(m, n) | cn = dm\}$$

(intuitively, because if $6/9 = c/d$ then $6/9 = m/n$ if and only if $c/d = m/n$) and so we can name the set $\{(m, n) | 6n = 9m\}$ by c/d.

Thus we define equality of fractions by:

$$\frac{a}{b} = \frac{c}{d} \text{ if and only if } \{(m, n) | an = bm\} = \{(m, n) | cn = dm\}.$$

So $6/9 = 72/108$ because 6/9 and $72/108$ are names for the same set. On the other hand, the fractions 1/2 and 3/4 are not equal, because the sets they are names for,

$$\{(m, n) | n = 2m\} \text{ and } \{(m, n) | 3n = 4m\},$$

are different–in fact, are disjoint: if (m,n) is a pair with $n = 2m$, then $3m \neq 4m$.

To summarize, elements of $\mathbb{Q}$ may be defined as certain subsets of the set of pairs (m,n) of integers with $b \neq 0$. The subset containing the pair (a,b) we label by the fraction a/b; then the subset containing the pair (a,b) is $\{(m,n)|an = bm\}$. Two fractions a/b and c/d are equal if and only if $\{(m,n)|an = bm\} = \{(m,n)|cn = dm\}$, which is the case if and only if $ad = bc$.

Fractions in lowest terms and arithmetic. When we learn fractions, we develop a bias towards using fractions a/b that are reduced, that is, such that a and b have no common factor except 1. Thus we prefer $1/2$ over $5/10$, and $3/5$ over $6/10$. But when adding fractions, the use of nonreduced fractions is unavoidable, because we need to find a common denominator for the fractions we are adding. For example:

$$\frac{1}{6} + \frac{3}{8} = \frac{8}{48} + \frac{18}{48} = \frac{26}{48}.$$

Thus it is often necessary to replace a fraction, for example, a reduced fraction, by a fraction equal to it that is not reduced, in order to do arithmetic.

But there is a nice fact about the arithmetic of fractions: arithmetic operations are not affected by replacing fractions by equal fractions. For example, $1/2 = 3/6$, and $3/5 = 12/20$; multiplying $1/2$ and $3/5$ gives $3/10$; multiplying $3/6$ and $12/20$ gives $36/120$, and the resulting fractions are equal. Similarly $1/6 + 3/8 = 8/48 + 18/48 = 26/48$, and also $1/6 + 3/8 = 4/24 + 9/24 = 13/24$, and the results are the same. Choosing different labels, or representatives, for the set of pairs represented by a fraction, such as choosing $8/48$ instead of $4/24$ for the set

$$\{(m,n)|n = 6m\}$$

does not affect the result of doing arithmetic on these sets. This is an important point that we will need to consider when we define other sets of numbers in later chapters.

Equivalence classes. The basic mathematical strategy at work in the definition just given of the rational numbers $\mathbb{Q}$ is the notion of dividing a set S up into equivalence classes.

An *equivalence relation* is a relation on a set S that satisfies the following three properties:

(R) an element in S is equivalent to itself (reflexive property);

(S) if one element in S is equivalent to a second element in S then the second element is equivalent to the first (symmetry property); and

(T) if one element in S is equivalent to a second element in S, and the second to a third, then the first is equivalent to the third (transitivity property).

If we denote the equivalence relation by $\sim$, then, in symbols, we have:

(R) for all a in S, $a \sim a$;

(S) for all a and b in S, if $a \sim b$ then $b \sim a$; and

(T) for all a, b and c in S, if $a \sim b$ and $b \sim c$, then $a \sim c$.

When a set S has an equivalence relation on it, then the set S is partitioned into subsets, called *equivalence classes*, which are defined by the property that two elements are in the same equivalence class if they are equivalent. The three properties of an equivalence relation listed above imply that if two equivalence classes have any elements at all in common, then they coincide. (See Exercise 3.)

In the case of the rational numbers, we consider the set S of all ordered pairs of numbers (a,b) with $b \neq 0$, and say that the ordered pair (a,b) is equivalent to the ordered pair (c,d), $(a,b) \sim (c,d)$, if $ad = bc$. The relation, $(a,b) \sim (c,d)$ if $ad = bc$, is an equivalence relation on the set S of all ordered pairs of numbers with the second number not 0. (See Exercise 1.) The set S of all ordered pairs (a,b) with $b \neq 0$ is split up, or partitioned, into equivalence classes by that equivalence relation. The fraction a/b is the label for the equivalence class containing the ordered pair (a,b).

The strategy of taking a set of elements, partitioning the set up into a set of equivalence classes by means of some equivalence relation, and then thinking of the equivalence classes, or the labels of the equivalence classes, as new objects, in our case new "numbers", is fundamental in mathematics. All three of the ways of defining the real numbers $\mathbb{R}$ (infinite decimals, Cauchy sequences, Dedekind cuts) use that strategy. It is the strategy we will use to construct new sets of numbers as we proceed in the book.

Exercises.

1. Verify that the relation, two ordered pairs (a,b) and (c,d) are equivalent if $ad = bc$, is an equivalence relation on the set S of all ordered pairs (a,b) of integers with $b \neq 0$.

2. Consider the following relations on $\mathbb{Z}$. In each case, decide if the relation is an equivalence relation. If so, describe the corresponding partition of $\mathbb{Z}$. If not, determine which properties of an equivalence relation fail:
(i) $a \sim b$ if $ab \geq 0$;
(ii) $a \sim b$ if $a - b$ is divisible by 3;
(iii) $a \sim b$ if $ab > 0$;
(iv) $a \sim b$ if $a + b$ is divisible by 3; and
(v) $a \sim b$ if $a \geq b$.

3. Suppose a set S has an equivalence relation on it. Use the properties of an equivalence relation to show that two equivalence classes of S which have any element in common, must be equal. (Two subsets A and B of S are equal if every element of A is an element of B and vice versa.)

Chapter 2
Induction

This chapter describes the method of proof by induction, in several versions. The last section presents the Binomial Theorem.

A. Induction

Induction is the basic method of proof for facts involving natural numbers. It allows us to obtain, in a finite number of steps, proofs of statements about all the numbers in the infinite set $\mathbb{N}$.

Induction comes in various formulations. Here is the best-known version.

Theorem 1 (Induction). *Fix an integer n_0 and let $P(n)$ be a statement which makes sense for every integer $n \geq n_0$. Then $P(n)$ is true for all $n \geq n_0$, if the following two statements are true:*
(a) $P(n_0)$ is true; and
(b) for all $k \geq n_0$, if $P(k)$ is true then $P(k+1)$ is true.

When using induction to prove a theorem, proving (a) is called the *base case*, and proving (b) is called the *induction step*.

You have almost certainly seen this principle used before, perhaps in calculus, in evaluating sums arising in connection with the definite integral.

Here is a simple example.

Example 1. For all $n \geq 1$,

$$1 + 3 + 5 + \ldots + (2n - 1) = n^2.$$

Proof. Let $P(n)$ be the statement

$$1 + 3 + 5 + \ldots + (2n - 1) = n^2,$$

or in words, "the sum of the first n odd numbers is n^2". Thus $P(1)$ is the statement

$$1 = 1^2,$$

$P(2)$ is the statement

$$1 + 3 = 2^2,$$

$P(5)$ is the statement

$$1 + 3 + 5 + 7 + 9 = 5^2,$$

and so on. All of these are clearly true, but just looking at $P(n)$ for many specific values of n does not suffice to prove $P(n)$ for *every* natural number $n \geq 1$. So we let $n_0 = 1$ and use induction to prove $P(n)$ for all $n \geq 1$.

The base case $P(1)$ is true, since $1 = 1^2$.

For the induction step, let k be some unspecified number ≥ 1, and assume that $P(k)$ is true, that is,

$$1 + 3 + \ldots + (2k - 1) = k^2.$$

We want to show that then $P(k+1)$ is true, that is,

$$1 + 3 + \ldots + (2k - 1) + (2k + 1) = (k + 1)^2.$$

To do so, we can add $(2k + 1)$ to both sides of the equation $P(k)$ to get

$$1 + 3 + \ldots + (2k - 1) + (2k + 1) = k^2 + (2k + 1). \tag{2.1}$$

The left side of (2.1) is the left side of the statement $P(k+1)$, and, since $k^2 + 2k + 1 = (k + 1)^2$, the right side of (2.1) is equal to $(k + 1)^2$, the right side of $P(k+1)$. Thus assuming $P(k)$ is true, it follows that $P(k+1)$ is true.

By induction, $P(n)$ is true for all $n \geq 1$.   □

The rationale behind induction is that if the base case (a) and the induction step (b) are true, then for any $n > n_0$, one can prove, in $n - n_0$ logical steps, that $P(n)$ is true. For example, if $P(n)$ is the equation of Example 1, above and we wish to prove that $P(5)$ is true, we can argue logically as follows:

$P(1)$ is true, by the base case.

Since $P(1)$ is true, $P(2)$ is true, by the induction step with $k = 1$;

Since $P(2)$ is true, $P(3)$ is true, by the induction step with $k = 2$;

Since $P(3)$ is true, $P(4)$ is true, by the induction step with $k = 3$;

Since $P(4)$ is true, $P(5)$ is true, by the induction step with $k = 4$.

This same reasoning can be used to show that $P(n)$ is true for any given number n. We simply start with the base case, which says that $P(n_0)$ is true, and then successively infer that $P(n_0 + 1), P(n_0 + 2), \ldots, P(n)$ is true by $n - n_0$ uses of the induction step. The principle of induction simply asserts that given the validity of the base case and of the induction step for all $n \geq n_0$, then for any $n > n_0$, $P(n)$ can be shown true, and therefore *is* true.

Here are some more examples.

Example 2. For all $n \geq 1$, $2^n \geq 1 + n$.

Proof. Here $n_0 = 1$.

The statement

$$P(n) : 2^n \geq 1 + n$$

is clearly true when $n = 1$, so the base case is true.

For the induction step, let k be a number ≥ 1 and assume

$$P(k) : 2^k \geq 1 + k$$

is true. Then multiplying both sides by 2 gives

$$2^k \cdot 2 \geq (1 + k) \cdot 2,$$

so

$$2^{k+1} = 2^k \cdot 2 \geq (1 + k) \cdot 2 = 2 + 2k > (1 + 1) + k = 1 + (k + 1).$$

Thus the statement

$$P(k+1) : 2^{k+1} \geq 1 + (k+1)$$

is true. We've shown that for every $k \geq 1$, the induction step is true. Hence the inequality $P(n)$ is true for all $n \geq 1$ by induction. $\qquad\square$

Example 3. The number 8 divides $3^{2n} - 1$ for all $n \geq 0$. That is, for every $n \geq 0$, $3^{2n} - 1 = 8m$ for some natural number m.

Proof. The statement $P(n)$: 8 divides $3^{2n} - 1$, is true for $n = 0$ since 8 divides $3^0 - 1 = 0$. The induction step involves a little "trick" of subtracting and adding the same quantity. Suppose 8 divides $3^{2k} - 1$. We examine $3^{2(k+1)} - 1$:

$$3^{2(k+1)} - 1 = 3^{2k} \cdot 3^2 - 1$$
$$= 3^{2k} \cdot 3^2 - 3^2 + 3^2 - 1$$
$$= 3^2(3^{2k} - 1) + (3^2 - 1).$$

Since 8 divides $3^{2k} - 1$ and 8 divides $3^2 - 1$, therefore 8 divides $3^2(3^{2k} - 1) + (3^2 - 1) = 3^{2(k+1)} - 1$. Thus the statement $P(n)$ is true for all $n \geq 0$. $\qquad\square$

Example 4. The number $2n^3 - 3n^2 + n + 31 \geq 0$ for all $n \geq -2$.

Proof. Let us set $f(n) = 2n^3 - 3n^2 + n + 31$. Then for each $n \geq -2$, the statement $P(n)$ is the inequality

$$P(n) : f(n) \geq 0.$$

In particular, for the base case, $P(-2)$ is the inequality $f(-2) \geq 0$, which is true because $f(-2) = 1$. For the induction step, suppose that for some $k \geq -2$, the statement $P(k)$ is true, that is, $f(k) > 0$. Then expanding $f(k+1)$ and collecting terms, we find

$$f(k+1) = 2(k+1)^3 - 3(k+1)^2 + (k+1) + 31$$
$$= 2(k^3 + 3k^2 + 3k + 1) - 3(k^2 + 2k + 1) + (k+1) + 31$$
$$= 2k^3 + 6k^2 + 6k + 2 - 3k^2 - 6k - 3 + k + 1 + 31$$
$$= 2k^3 + 3k^2 + k + 31$$
$$= f(k) + 6k^2 \geq f(k) \geq 0.$$

So $P(k+1)$ is true. Thus $P(n)$ is true for all $n \geq -2$, that is, $f(n) \geq 0$ for all $n \geq -2$. □

Example 5. In calculus, after the rules for the derivative of a constant and of x, and the product rule are presented, the rule for the derivative of x^n can be proved by induction:
$$\frac{dx^n}{dx} = nx^{n-1}.$$

Proof. Let $P(n)$ be the statement

$$\frac{dx^n}{dx} = nx^{n-1}.$$

Then $P(0)$ is the statement that the derivative of the constant function 1 is 0, and $P(1)$ is the statement that the derivative of x is 1. To prove $P(n)$ by induction, suppose that for some $k \geq 0$,

$$P(k): \frac{dx^k}{dx} = kx^{k-1}$$

is true. Then consider $\frac{dx^{k+1}}{dx}$. By the product rule, we have

$$\frac{dx^{k+1}}{dx} = \frac{d(x \cdot x^k)}{dx}$$
$$= \frac{dx}{dx} \cdot x^k + x \cdot \frac{dx^k}{dx}$$
$$= x^k + x \cdot kx^{k-1}$$

since we know $P(1)$ is true and we have assumed $P(k)$ is true. Collecting terms, we obtain
$$\frac{dx^{k+1}}{dx} = (k+1)x^k$$

and so $P(k+1)$ is true. Thus by induction,

$$P(n): \frac{dx^n}{dx} = nx^{n-1}$$

is true for all $n \geq 0$.   □

Exercises. In the exercises, n is always an integer.

1. Prove that $1+2+3+\ldots+n = n(n+1)/2$ for all $n \geq 1$.

2. Prove that $1^3 + 2^3 + \ldots + n^3 = [n(n+1)/2]^2$ for all $n \geq 1$.

3. Prove that
$$1 + 2 + 2^2 + \ldots + 2^{n-1} = 2^n - 1$$
for every $n > 1$.

4. Prove that for all $n \geq 1$,
$$1^4 + 2^4 + \ldots + n^4 = \frac{n(n+1)(2n+1)(3n^2+3n-1)}{30}$$

5. Prove that for any real number x and for all numbers $n > 1$,
$$x^n - 1 = (x-1)(x^{n-1} + x^{n-2} + \ldots + x^{n-r} + \ldots + x + 1).$$

6. Using the last exercise, prove that for all $n > 1$,
$$\lim_{r \to 1} \frac{r^n - 1}{r - 1} = n.$$

7. (Askey) Show that $\frac{dx^n}{dx} = nx^{n-1}$ as follows: by the definition of the derivative,
$$\frac{dx^n}{dx} = \lim_{y \to x} \frac{y^n - x^n}{y - x}.$$

Set $y = rx$ and compute the limit using the last exercise.

8. Prove that $n! > 2^n$ for all $n \geq 4$.

9. Prove that $2^{2n} > n^4$ for all $n \geq 4$.

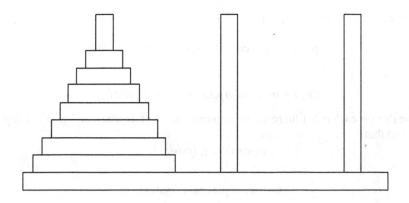

10. Let
$$t_n = \frac{n(n+1)}{2} = 1+2+\ldots+n$$
be the n-th triangular number. Define $t_0 = 0$.
   (i) Show that the odd square number $(2n+1)^2 = 8t_n + 1$ for all $n \geq 1$.
   (ii) Prove that
$$\frac{1}{t_1} + \frac{1}{t_2} + \ldots + \frac{1}{t_n} = 2 - \frac{2}{n+1}.$$
(Hint: observe that $\frac{1}{n(n+1)} = \frac{1}{n} - \frac{1}{n+1}$.)

   (iii) Prove that for all $n \geq 1$,
$$\frac{1}{1} + \frac{1}{9} + \frac{1}{25} + \ldots + \frac{1}{(2n+1)^2} \leq \frac{5}{4} - \frac{1}{4(n+1)}$$
in two ways: directly by induction, and by using (i) and (ii).

11. Let a be a natural number >1. Prove that for all integers $r_0, r_1, \ldots, r_{n-1}$ with $0 \leq r_j < a$,
$$r_0 + r_1 a + r_2 a^2 + \ldots + r_{n-1} a^{n-1} < a^n.$$
When $n = 10$ this says that 10^n is larger than any n-digit number.

12. Let b be a number ≥ 2. Prove that for all $n \geq 1$,
$$(b^n - 1)(b^n - b)(b^n - b^2) \cdot \ldots \cdot (b^n - b^{n-2}) \geq b^{n(n-1)} - b^{n(n-1)-1}.$$

13. Prove that for every $n \geq 1$, 24 divides $16^n - 16$.

14. Prove that for every $n \geq 1$, 5 divides $8^n - 3^n$.

15. Prove that for every $n \geq 1$, 5 divides $3^{4n} - 1$.

16. Prove that for every odd number $n \geq 1$, 9 divides $4^n + 5^n$.

17. Prove that for every $n \geq 0$, 3 divides $2^{2n+1} + 1$.

18. Using the addition formulas
$$\cos(a+b) = \cos(a)\cos(b) - \sin(a)\sin(b)$$
and
$$\sin(a+b) = \sin(a)\cos(b) + \cos(a)\sin(b),$$
prove that for each $n > 1$ there are polynomials $f_n(x)$ of degree n and $g_n(x)$ of degree $n-1$ so that
$$\cos(nx) = f_n(\cos(x))$$
and
$$\sin(nx) = g_n(\cos(x))\sin(x).$$

19. For any real number a, define $a^0 = 1$, and for every number $k \geq 0$, define $a^{k+1} = a^k \cdot a$. Using induction, prove that for all natural numbers m and n, $a^{m+n} = a^m \cdot a^n$.

20. Consider the puzzle called the Tower of Hanoi (attributed to the French mathematician Edouard Lucas, 1883). The puzzle consists of n disks of decreasing diameters placed on a pole. There are two other poles. The problem is to move the entire stack of disks to another pole by moving one disk at a time to any other pole, except that no disk may be placed on top of a smaller disk. Find a formula for the least number of moves needed to move a stack of n disks from one pole to another, and prove the formula by induction.

21. (Neal Hill). Suppose in the Tower of Hanoi, the three poles are in a row, and a disc can only be moved from a pole to an adjacent pole. All other rules apply. How many moves does it take to move a stack of n discs from the leftmost pole to the rightmost pole?

22. Show that for every positive integer n, one of the numbers $n, n+1, n+2, \ldots, 2n$ is the square of an integer.

23. What is wrong with the proof of the following (true) theorem?

Theorem 2. *All new 1922 Ford Model T cars had the same exterior color.*

Proof. The case $n = 1$ is obvious.

Suppose that in any set of n new Model T's, all had the same exterior color. Consider a set of $n + 1$ new Model T's, lined up from left to right.

We may assume by induction that in the set L of the n Model T's to the left all had the same exterior color, and similarly that in the set R of the n Model T's to the right all had the same exterior color. But then evidently all the $n + 1$ Model T's had the same exterior color, for the leftmost and rightmost Model T's had the same exterior color as all the Model T's in between.

By induction, for every number n, in every set of n new Model T's all had the same exterior color. Since the set of all new 1922 Model T's was one such set, the theorem is proved.   □

(Henry Ford was reputed to have said of the Model T, "You can paint it any color, so long as it's black.")

24. Show that for $n \geq 1$,

$$1 + 7 + 13 + \ldots (6n - 5) = 3n^2 - 2n.$$

B. Complete Induction

Complete Induction is a reformulation of induction that is often more convenient to use.

Theorem 3 (Complete Induction). *Let n_0 be a fixed integer and let $P(n)$ be a statement which makes sense for every integer $n \geq n_0$. Then $P(n)$ is true for all integers $n > n_0$, if the following two statements are true:*
(a') (base case) $P(n_0)$ is true, and
(b') (induction step) For all $m > n_0$:
if $P(k)$ is true for all k with $n_0 \leq k < m$, then $P(m)$ is true.

Complete induction appears more complicated than ordinary induction, but in fact it is easier to use. Compare the induction step (b') with the induction step (b) for ordinary induction:
(b) *For all $m > n_0$,*
if $P(m-1)$ is true, then $P(m)$ is true.

In attempting to prove the induction step in a proof by induction, complete induction allows us to assume more than we can with ordinary induction. With complete induction, in order to prove $P(m)$, you may assume that $P(k)$ is true for every k, $n_0 \leq k < m$. In ordinary induction you are allowed only to assume that $P(m-1)$ is true. So complete induction is more flexible than ordinary induction.

For certain kinds of results involving multiplication, ordinary induction is awkward to apply, while complete induction is quite natural. The next example is such a result.

Recall that a natural number n is *prime* if $n \geq 2$ and does not factor into the product of two natural numbers each smaller than n. Also, a number q *divides* a number n, or n *is divisible by* q, if $n = qr$ for some natural number r. Thus 3 divides 12, but 3 does not divide 14.

Proposition 4. *Every natural number $n \geq 2$ is divisible by a prime number.*

Proof. Let $P(n)$ be the statement, "n is divisible by a prime number." Then the base case $P(2)$ is true, because 2 is prime and 2 divides itself.

We'll use complete induction for the induction step. Thus we assume that $P(k)$ is true for all k where $2 \leq k < m$: that is, we assume that every natural number ≥ 2 and $< m$ is divisible by a prime number. Now consider m. If m is prime, then m is divisible by a prime number, namely itself, and $P(m)$ is true. If m is not prime, then m factors as $m = ab$, where $2 \leq a < m$ and also $2 \leq b < m$. Since $2 \leq a < m$, by assumption $P(a)$ is true, that is, a is divisible by a prime. Since a is divisible by a prime, and a divides m, m is divisible by the same prime. So $P(m)$ is true.

Thus $P(n)$ is true for all $n \geq 2$ by complete induction. $\square$

Notice that had we tried to use ordinary induction to prove $P(m)$: "m is divisible by a prime" for all $m \geq 2$, then in the induction step we would have been permitted only to assume that $m-1$ is divisible by a prime, in order to try to prove that m is divisible by a prime. But knowing about factors of $m-1$ is of little direct help in finding factors of m, since no factor of $m-1$ other than 1 can possibly be a factor of m. (Why?) Thus if we wanted to prove Proposition 4 by ordinary induction, we would need to change the statement $P(n)$. See the proof of Theorem 6, below.

If we want to prove something using induction, complete induction will work just as well. For suppose we can prove

For all $k \geq n_0$, if $P(k)$ is true, then $P(k+1)$ is true.

Then we can prove

For all $k \geq n_0$, if $P(m)$ is true for all m with $n_0 \leq m < k$, then $P(k+1)$ is true.

For if we can prove $P(k+1)$ assuming only $P(k)$, then we can prove $P(k+1)$ assuming $P(m)$ for all $n_0 \leq m \leq k$.

Hence:

Theorem 5. *If a statement $P(n)$ can be proved for all $n \geq n_0$ by ordinary induction, it can be proved by complete induction.*

It turns out, however, that the two forms of induction are logically equivalent. We prove

Theorem 6. *If a statement $P(n)$ can be proved for all $n \geq n_0$ by complete induction, it can be proved by ordinary induction.*

Proof. Suppose we know that:
 (a') $P(n_0)$ is true, and
 (b') if $P(k)$ is true for all k, $n_0 \leq k < m$, then $P(m)$ is true.
 Then $P(n)$ is true for all $n \geq n_0$ by complete induction. We show how to prove $P(n)$ for all n by ordinary induction. To do so, we consider a new statement
 $Q(n)$: $P(m)$ is true for all m, $n_0 \leq m \leq n$.
 We prove $Q(n)$ is true for all $n \geq n_0$ by ordinary induction. Note that if $Q(n)$ is true, then $P(n)$ is true.
 For the base case, we need to show:
 (a) $Q(n_0)$ is true.
 But because $Q(n_0)$ is the statement "$P(m)$ is true for all m, $n_0 \leq m \leq n_0$," we have that $Q(n_0)$ is true because by (a), $P(n_0)$ is true.
 For the induction step, we need to show:
 (b) If $Q(m-1)$ is true then $Q(m)$ is true.
 To see this, observe that if $Q(m-1)$ is true, then $P(k)$ is true for all k with $n_0 \leq k \leq m-1$. So since we assumed (b') holds for all $n \geq n_0$, therefore $P(m)$ is true. But then $P(k)$ is true for all k with $n_0 \leq k \leq m$, and so $Q(m)$ is true.
 Thus by ordinary induction, $Q(n)$ is true for all $n \geq n_0$. But if $Q(n)$ is true, then $P(n)$ is true. So $P(n)$ is true for all $n \geq n_0$.   □

This theorem implies that whenever we want to prove a statement about natural numbers, we can use whichever version of induction is most convenient. Henceforth, when we refer to "induction", we mean either version.

Exercises.

25. Prove "For all $n \geq 2$, every number m with $1 < m \leq n$ is divisible by a prime number" by ordinary induction.

26. Prove that any natural number $n \geq 2$ either is prime or factors into a product of primes.

27. Prove that the sum of the interior angles of an n-sided convex polygon is $180 \times (n-2)$.

28. Let $f : \mathbb{N} \to \mathbb{N}$ be a function with the properties that $f(1) = 1$ and for all numbers $n > 1$, $f(n) < n$. Prove that for every n there is some k so that the function $f^{(k)}$, obtained by composing f with itself $k - 1$ times, maps n to 1. (Thus $f^{(1)}(n) = f(n), f^{(2)}(n) = f(f(n)), f^{(3)}(n) = f(f(f(n)))$, etc.)

29. *Russian peasant arithmetic.* Here is a way of multiplying which has been attributed to Russian peasants who could only add, and multiply and divide by 2. In fact this method of multiplying was also used by the ancient Egyptians (2000 B.C.) (see [Gillings (1972)]) and is of interest also to computer programmers (since computers are especially efficient in multiplying and dividing by 2).

To multiply two numbers a and b set up four columns, labeled "left", "right", "sum" and "summand". In the top row place a in the left column, b in the right column, and 0 in the sum column. If b is odd, place a in the summand column. If b is even, place 0 in the summand column.

Then fill in successive rows of the array. If a, b, s and d are the entries in a given row, then fill in the next row as follows:

If b is even, set the entries in the left, right and sum columns of the next row to be $2a$, $b/2$ and $s+d$.

If b is odd, set the entries in the left, right and sum columns of the next row to be $2a$, $(b-1)/2$ and $s+d$.

Then set the entry in the summand column to be 0 if the entry in the right column (either $b/2$ or $(b-1)/2$) in the same row is even; set the entry in the summand column to be the entry in the left column ($2a$) of the new row if the entry in the right column in the new row (either $b/2$ or $(b-1)/2$) is odd.

left	right	sum	summand
		$\vdots$	
a	b	s	d
$2a$	$b/2$ or $(b-1)/2$	$s+d$	$2a$ or 0
		$\vdots$	

Continue until you reach the row in which the entry in the right column is 0. Then the entry in the sum column is $a \cdot b$.

Here is an example, showing that $116 \cdot 311 = 36076$:

left	right	sum	summand
116	311	0	116
232	155	116	232
464	77	348	464
928	38	812	0
1856	19	812	1856
3712	9	2668	3712
7424	4	6380	0
14848	2	6380	0
29696	1	6380	29696
59392	0	36076	

Given two numbers a and b, prove by induction that for each row,

$$\text{(the left entry)} \cdot \text{(the right entry)} + \text{(the sum entry)} = a \cdot b,$$

and therefore $a \cdot b$ is equal to the last entry in the sum column.

30. The Fibonacci sequence is defined by $a_1 = 1, a_2 = 1$ and for all $n \geq 2$, $a_{n+1} = a_n + a_{n-1}$. Thus the sequence begins

$$1, 1, 2, 3, 5, 8, 13, 21, 34, 55, \ldots.$$

Prove that for all $n \geq 1$, $a_n < \left(\frac{5}{3}\right)^n$.

31. A *composition* of a natural number n is a description of n as an ordered sum of natural numbers. For example, the compositions of 3 are:

$$3, 2 + 1, 1 + 2, 1 + 1 + 1$$

and the compositions of 4 are

$$4, 3 + 1, 2 + 2, 2 + 1 + 1, 1 + 3, 1 + 2 + 1, 1 + 1 + 2, 1 + 1 + 1 + 1.$$

Let $c(n)$ be the number of compositions of n. Guess a formula for $c(n)$ for all $n \geq 1$ and prove your formula by induction.

C. Well-Ordering

The formulations of induction in the two previous sections were developed in the seventeenth century by Pascal and others. However, some results about natural numbers were obtained many centuries earlier, by the ancient Greek mathematicians whose work was collected in Euclid's Elements (300 B.C.) For example, Proposition 4, above, is found in Euclid, Book IX, Proposition 31. Here is how Euclid proved:

Theorem 7. *Every composite number is divisible by some prime number.*

Proof. Suppose A is a composite number. Then A is divisible by some number B. If B is prime, we're done, so assume B is composite. Then B is divisible by some number C, so C is a divisor of A. If C is prime, we're done, so assume C is composite. Then C is divisible by some number.

"Thus, if the investigation be continued in this way, some prime number will be found which will measure [divide] the number before it, which will also measure A. For if it is not found, an infinite series of numbers will measure the number A, each of which is less than the other: which is impossible in numbers."   □

Thus Euclid proves the result by what might be called "infinite descent": there is no infinite descending chain of natural numbers.

The principle of infinite descent can be expressed more affirmatively as the

Theorem 8 (Well-Ordering Principle). *Any nonempty set of natural numbers has a least element.*

We can rephrase Euclid's proof in terms of the well-ordering principle. For any number $A > 2$, let $\mathscr{S}$ be the set of numbers ≥ 2 which divide A. Since A is a positive divisor of itself, $\mathscr{S}$ is nonempty. Euclid's argument using infinite descent is that if we select a strictly decreasing sequence of proper divisors of A, and none is prime, then we get an infinite descending chain of elements of $\mathscr{S}$, impossible. Using well-ordering, we can say: $\mathscr{S}$ has a least element C, that is, A has a least divisor $C \geq 2$. If C is not prime, then C has a smaller divisor $D \geq 2$ which is then a divisor of A, contradicting the assumption that C is least. So C must be prime.

Well-ordering and infinite descent are different forms of induction. We can in fact prove the well-ordering principle using induction. To do so, we prove that if there is a set of natural numbers with no least element, then it must be empty. (This approach uses the standard logical strategy for proving statements of the form "if A then B"-we prove that if B is false, then A must be false. The reason that the strategy works is that the only situation under which the statement "if A then B" is false occurs when A is true and B is false. If we assume B is false and are able to show thereby that A is false, then the situation "A true and B false" cannot occur and so "if A then B" is true.)

Proof of the Well-Ordering Principle. Let $\mathscr{S}$ be a set of natural numbers with no least element. Let $P(n)$ be the statement: "Every number in $\mathscr{S}$ is $>n$." Observe that if $P(m)$ is true, then m is not in $\mathscr{S}$. So by showing that $P(n)$ is true for all n, we will show that $\mathscr{S}$ is empty, which will prove the well-ordering principle.

Evidently $P(1)$ is true, for if not, 1 is in $\mathscr{S}$, and since all natural numbers are ≥ 1, therefore $\mathscr{S}$ would have a least element.

Suppose $P(k)$ is true for some $k > 1$. If $P(k+1)$ is false, then $\mathscr{S}$ contains some number $\leq k+1$. But $P(k)$ is true. So every number in $\mathscr{S}$ is $> k$. But then $k+1$, the only number $\leq k+1$ which is $> k$, would be in $\mathscr{S}$ and would be the least element of $\mathscr{S}$, impossible. Thus if $P(k)$ is true, then $P(k+1)$ is true. By induction, $P(n)$ is true for all $n \geq 1$, and $\mathscr{S}$ is empty. That finishes the proof.   □

One important use of the well-ordering principle is that it permits us to define a number by the property that the number is the smallest number in a certain non-empty set.

For example, consider the set $\mathscr{S}$ of numbers that are multiples of both 24 and 90. That set of common multiples of 24 and 90 is non-empty, for it includes $24 \cdot 90 = 2160$. Thus by well-ordering, the set $\mathscr{S}$ has a smallest number, the *least common multiple* of 24 and 90. Some computation verifies that the least common multiple is 360. But with no computation, well-ordering tells us immediately that

Proposition 9. *Any two numbers a and b have a least common multiple, that is, a number m which is a common multiple of a and b and which is $\leq$ any other common multiple of a and b.*

Proof. Since the set $\mathscr{S}$ of common multiples of a and b contains $a \cdot b$, $\mathscr{S}$ is non-empty. So by well-ordering, $\mathscr{S}$ has a smallest element, which is the least common multiple of a and b. $\qquad\square$

Exercises.

32. Show that there is no rational number b/a whose square is 2, as follows: if $b^2 = 2a^2$, then b is even, so $b = 2c$, so, substituting and canceling 2, $2c^2 = a^2$. Use that argument and well-ordering to show that there can be no natural number $a > 0$ with $b^2 = 2a^2$ for some natural number b.

33. Prove that the well-ordering principle implies induction, as follows: suppose $P(n)$ is a statement which make sense for every $n \geq n_0$. Suppose (a) $P(n_0)$ is true, and (b) for any $n \geq n_0$, if $P(n)$ is true then $P(n+1)$ is true. Let $\mathscr{S}$ be the set of $n \geq n_0$ for which $P(n)$ is false. Using well-ordering, show that $\mathscr{S}$ must be empty.

34. Show that the well-ordering principle is equivalent to "there is no infinite descending chain of natural numbers".

35. Fix N, some integer, and suppose $\mathscr{S}$ is a nonempty set of integers such that every a in $\mathscr{S}$ is $<N$. Show that $\mathscr{S}$ has a maximal element. (Hint: Let $\mathscr{T} = \{n \text{ in } \mathbb{N} | n \geq a \text{ for all } a \text{ in } \mathscr{S}\}$.)

D. The Binomial Theorem

The Binomial Theorem describes the coefficients when the expression $(x+y)^n$ is multiplied out. Recall that $n!$ ("n factorial") is defined by $n! = 1 \cdot 2 \cdot 3 \cdot \ldots \cdot n$ for $n > 0$. We set $0! = 1$.

Theorem 10 (The Binomial Theorem). *For every integer $n \geq 1$,*

$$(x+y)^n = \binom{n}{0}x^n + \ldots + \binom{n}{r}x^{n-r}y^r + \ldots + \binom{n}{n}y^n$$

where

$$\binom{n}{r} = \frac{n!}{r!(n-r)!}$$

for $0 \le r \le n.$

Examples:

$$(x+y)^2 = x^2 + 2xy + y^2; \quad (x+y)^3 = x^3 + 3x^2y + 3xy^2 + y^3.$$

The proof is by induction on n. In order to carry through the argument passing from $n-1$ to n (the induction step) we first set up Pascal's triangle,

$$
\begin{array}{ccccccccc}
 & & & & 1 & & & & \\
 & & & 1 & & 1 & & & \\
 & & 1 & & 2 & & 1 & & \\
 & 1 & & 3 & & 3 & & 1 & \\
1 & & 4 & & 6 & & 4 & & 1
\end{array}
$$

$$\vdots$$

We number the elements $c(n,r)$ of Pascal's triangle by the row n and the position r of the element within the row, both indices starting from 0. Thus Pascal's triangle is labeled

$$
\begin{array}{cccc}
 & c(0,0) & & \\
 c(1,0) & & c(1,1) & \\
 c(2,0) & c(2,1) & & c(2,2) \\
c(3,0) & c(3,1) & c(3,2) & c(3,3)
\end{array}
$$

$$\vdots$$

where

$$c(0,0) = c(n,0) = c(n,n) = 1$$

for all n, and for $1 \le r \le n-1$,

$$c(n,r) = c(n-1,r-1) + c(n-1,r).$$

That is, $c(n,r)$ is the sum of the terms to the upper left and to the upper right:

$$c(n-1,r-1) \quad + \quad c(n-1,r)$$
$$= c(n,r)$$

The entries $c(n,r)$ have a combinatorial interpretation.

Proposition 11. *Let S be a set with n elements. Then $c(n,r)$ is the number of r-element subsets of S.*

Proof. We do this by induction on n, the case $n = 1$ being obvious.

Let S be a set with n elements. The statement is true when $r = 0$ or n, since there is only one subset of S with n elements, namely S, and only one with no elements.

Assume, then, that $n > 1$ and $1 \leq r \leq n - 1$. Let y be an fixed element of S. Let S_0 be the set of all the elements of S except y. S_0 is then a set with $n - 1$ elements. Divide the collection of all r-element subsets of S into two piles, one consisting of those subsets containing y, the other consisting of those subsets not containing y. The first pile consists of exactly those subsets of S obtained by taking an $(r - 1)$-element subset of S_0 and adjoining y. By induction applied to S_0, there are $c(n-1, r-1)$ of these. The second pile consists exactly of the r-element subsets of S_0, of which there are $c(n-1, r)$, again by induction. Thus the number of r-element subsets of S is $c(n-1, r-1) + c(n-1, r) = c(n, r)$, which is what we wished to show. $\qquad\square$

The entries of Pascal's triangle can be computed by the following:

Lemma 12.
$$c(n, r) = \binom{n}{r} = \frac{n!}{r!(n - r!)}$$

Proof. Induction on n. The case $n = 0$ is obvious:

$$\frac{0!}{0!0!} = 1 = c(0, 0),$$

Given $n > 0$, assume that for all r with $0 \leq r \leq n - 1$,

$$c(n - 1, r) = \frac{(n - 1)!}{r!(n - 1 - r)!}.$$

Now
$$c(n, 0) = 1 = \frac{n!}{0!(n - 0)!}, \quad c(n, n) = 1 = \frac{n!}{n!(n - n)!}$$

so the lemma is true for $c(n, r)$ when $r = 0$ or n. For $1 \leq r \leq n - 1$,

$$c(n, r) = c(n - 1, r - 1) + c(n - 1, r)$$
$$= \frac{(n - 1)!}{(r - 1)!(n - r)!} + \frac{(n - 1)!}{(r)!(n - 1 - r)!}$$
$$= \frac{(n - 1)!}{(r - 1)!(n - 1 - r)!} \left[\frac{1}{n - r} + \frac{1}{r} \right]$$
$$= \frac{(n - 1)!}{(r - 1)!(n - 1 - r)!} \cdot \frac{n}{(n - r)r}$$
$$= \frac{n!}{r!(n - r)!}$$

as was to be shown. The lemma is therefore proved by induction. $\qquad\square$

Corollary 13. $\binom{n-1}{r-1} + \binom{n-1}{r} = \binom{n}{r}$.

We therefore know that for each n, $\binom{n}{0} = \binom{n}{n} = 1$, and $\binom{n}{r} = \binom{n-1}{r} + \binom{n-1}{r-1}$ for $1 \leq r \leq n-1$. Using these facts we can prove the Binomial Theorem by induction on n.

Proof of the Binomial Theorem. For $n = 1$, $(x+y) = \binom{1}{0}x + \binom{1}{1}y$ so the binomial theorem is true when $n = 1$. Assume $n > 1$ and the theorem is true for $n-1$, that is,

$$(x+y)^{n-1} = \binom{n-1}{0}x^{n-1} + \binom{n-1}{1}x^{n-2}y^2 + \ldots$$
$$+ \binom{n-1}{r}x^{n-1-r}y^r + \ldots + \binom{n-1}{n-1}y^{n-1}.$$

We compute $(x+y)^n$ as follows:

$$(x+y)^n = (x+y) \cdot (x+y)^{n-1} = x(x+y)^{n-1} + y(x+y)^{n-1}.$$

Multiplying the expansion of $(x+y)^{n-1}$, above, by x and by y, and adding, we get

$$(x+y)^n = \binom{n-1}{0}x^n + \binom{n-1}{1}x^{n-1}y + \ldots + \binom{n-1}{n-1}xy^{n-1}$$
$$+ \binom{n-1}{0}x^{n-1}y + \ldots + \binom{n-1}{n-2}xy^{n-1} + \binom{n-1}{n-1}y^n.$$

Thus the coefficient of $x^{n-r}y^r$ for $r = 1, \ldots, n-1$ is

$$\binom{n-1}{r} + \binom{n-1}{r-1} = \binom{n}{r}$$

by Lemma 3. Since

$$\binom{n-1}{0} = 1 = \binom{n}{0}, \binom{n-1}{n-1} = 1 = \binom{n}{n},$$

we see that

$$(x+y)^n = \binom{n}{0}x^n + \ldots + \binom{n}{r}x^{n-r}y^r + \ldots + \binom{n}{n}y^n,$$

which proves the Binomial Theorem by induction.   □

Exercises.

36. Prove that the sum of the elements of the nth row of Pascal's triangle is 2^n for each n. (How many subsets of a set with n elements are there?)

37. Prove that
$$\binom{s}{s} + \binom{s+1}{s} + \ldots + \binom{n}{s} = \binom{n+1}{s+1}$$
for all s and all $n \geq s$.

38. Prove that for all $n \geq 1$,
$$\binom{n}{0}^2 + \binom{n}{1}^2 + \ldots + \binom{n}{n}^2 = \binom{2n}{n}.$$

Chapter 3
Euclid's Algorithm

This chapter begins with the Division Theorem, a result that describes the result of long division of numbers. Repeated use of the Division Theorem yields the description of any number in base a (e.g. $a = 2$, or $a = 16$ or $a = 60$). Repeated use also yields Euclid's Algorithm for finding the greatest common divisor of two numbers. Euclid's Algorithm dates from the 4th century B. C., but remains one of the fastest and most useful algorithms in modern computational number theory, and has important theoretical consequences for the set $\mathbb{Z}$ of integers.

A. The Division Theorem

Theorem 1 (Division Theorem). *Given nonnegative integers $a > 0$ and b, there exist integers $q > 0$ and r with $0 \leq r < a$ such that $b = aq + r$.*

Based on the roles of a, b, q and r in long division, we call a the *divisor*, b the *dividend*, q the *quotient*, and r the *remainder*.

Proof. We prove the Division Theorem by well-ordering.
Let
$$\mathscr{S} = \{b - ax | x \text{ is a nonnegative integer and } b - ax \geq 0\}.$$

Then $\mathscr{S}$ is a set of nonnegative integers and is non-empty because $b = b - a \cdot 0$ is in $\mathscr{S}$. So by well-ordering, $\mathscr{S}$ has a least element r. Clearly $r = b - aq$ for some integer $q \geq 0$. We must show that $0 \leq r < a$. Since r is in $\mathscr{S}$, $r \geq 0$. Is $r < a$? If not, then $r - a \geq 0$, and

$$r - a = b - qa - a = b - (q+1)a \geq 0.$$

So $r - a$ is in $\mathscr{S}$. This contradicts the assumption that r was the least element of the set $\mathscr{S}$. Thus $r < a$, and so
$$b = qa + r$$

with $0 \leq r < a$, as we wished to show. □

L.N. Childs, *A Concrete Introduction to Higher Algebra*, Undergraduate Texts in Mathematics, © Springer Science+Business Media LLC 2009

To complete the story, we have

Proposition 2 (Uniqueness). *Let $a > 0, b \geq 0$ be integers and suppose $b = aq + r$ for some quotient $q \geq 0$ and some remainder r with $0 \leq r < a$. Then q and r are unique.*

Proof. Suppose $b = aq + r$ and also $b = as + t$, where $q \geq 0, s \geq 0$ and $0 \leq r < a$ and $0 \leq t < a$. Suppose $r \geq t$. Then

$$0 = b - b = (aq + r) - (as + t) = (r - t) - a(s - q),$$

so

$$a(s - q) = r - t.$$

But $0 \leq r - t \leq r < a$. So dividing by a gives $0 \leq s - q = (r - t)/a < 1$. Since $s - q$ is an integer, we must have $s - q = 0$. Hence $s = q$ and $r = t$. $\square$

We can also use well-ordering to prove uniqueness–see Exercise 1.

Given numbers $a > 0$ and b, we say that a *divides* b if $b = aq$ for some integer $q \geq 0$. Thus a divides b if in the equation $b = aq + r$ given by the Division Theorem, the remainder $r = 0$.

Exercises.

1. (i) Prove that if S is a non-empty set of non-negative integers, then the least element of S is unique.

(ii) Use a) to prove Proposition 2, that the quotient and remainder in the Division Theorem are unique.

2. Let a, b be natural numbers. For the fraction $\frac{b}{a}$, let $[\frac{b}{a}]$ be the greatest integer $< \frac{b}{a}$, and let $\{\frac{b}{a}\}$ be the fractional part: $\{\frac{b}{a}\} = \frac{b}{a} - [\frac{b}{a}]$. Thus, for example, $[\frac{22}{7}] = 3, \{\frac{22}{7}\} = \frac{1}{7}$, and $\frac{22}{7} = 3 + \frac{1}{7} = [\frac{22}{7}] + \{\frac{22}{7}\}$. If $b = aq + r$ where $0 \leq r < a$ as in the Division Theorem, how do q and r relate to $[\frac{b}{a}]$ and $\{\frac{b}{a}\}$?

3. Show that if $b = aq + r$ and d is a number that divides both a and b, then d divides r.

4. Let m be the least common multiple of a and b, and let c be a common multiple of a and b. Show that m divides c. Hint: use the division theorem on m and c, and show that the remainder r is a common multiple of a and b, hence $r = 0$.

5. Let

$$J = \{e \text{ in } \mathbb{N} | e = ar + bs \text{ for some integers } r, s\}$$

and let d be the least element of J. Show that d divides a and d divides b.

Bases. As a first application of the Division Theorem, we look at decimal notation.

The normal way we write numbers uses powers of 10. When we take a number b, like $b = $ MCMLXXVI, and write it as $b = 1976$, we mean that

$$b = 1 \times 10^3 + 9 \times 10^2 + 7 \times 10 + 6.$$

We call this way of writing b the *representation of b in base* 10, or *radix* 10. This notation comes from India some 1500 years ago. More generally, for any number $a \geq 2$, if we write a number b using powers of a ($a \geq 2$),

$$b = r_n a^n + r_{n-1} a^{n-1} + \ldots + r_1 a + r_0$$

with each of $r_0, \ldots, r_n$ between 0 and $a - 1$, this is the *representation of b in base* (or *radix*) a.

There is no particular reason except convention, based on physiology, why we have a bias towards a number system based on the number 10. Some cultures have had number systems based on 20, and our culture retains remnants of the ancient (1800 B.C.) Babylonian number system based on 60: for example, the way we measure time in hours, minutes and seconds, or the way we measure angles.

We would write $b = $ MCMLXXVI in base 20 as follows:

$$b = 4 \times 20^2 + 18 \times 20 + 16$$

and in base 60 as

$$b = 32 \times 60 + 56.$$

In the United States, large numbers are often written in base 1000 by inserting commas. For example, the U. S. Public Debt on May 9, 2007 was

$$8,822,507,474,425 = 8 \cdot (10^3)^4 + 822 \cdot (10^3)^3 + 507 \cdot (10^3)^2 + 474 \cdot 10^3 + 425$$

dollars.

Base 2 is commonly used in computing. At the beginning of the evolution of the electronic computer in the mid 1940's, von Neumann recognized that the binary, or base 2 system, was more natural than the decimal system for computers, for several reasons [see Goldstine (1972), p. 260]: the elementary operations of addition, subtraction, and multiplication can be performed more rapidly in base 2, electronic circuitry tends to be binary in character, and the control structure of a computer is logical in nature, not arithmetical, and logic is a binary kind of system (true-false). Our example number $b = $ MCMLXXVI in base 2 is

$$b = 1 \times 2^{10} + 1 \times 2^9 + 1 \times 2^8 + 1 \times 2^7 + 1 \times 2^5 + 1 \times 2^4 + 1 \times 2^3$$

or $b = 11110111000$.

Numbers themselves have no bias in favor of one base or another, as the following theorem shows.

Theorem 3. *Fix a natural number $a \geq 2$. Every integer $b \geq 0$ may be represented in base a: that is, b may be written uniquely as*

$$b = r_n a^n + r_{n-1} a^{n-1} + \ldots + r_1 a + r_0$$

with $0 \leq r_i < a$ for all i.

We denote a number b written in base a as

$$b = (r_n r_{n-1} \ldots r_2 r_1 r_0)_a.$$

Thus $(1976)_{10} = (11110111000)_2$. We omit $(\ \ )_{10}$ in decimal notation when there is no possibility of confusion.

Here is a proof of the theorem, using induction and the division theorem.

Proof. Suppose all numbers $<b$ may be written in base a. To write b in base a, first divide a into b, using the division theorem to get $b = aq + r_0$ for unique numbers q, r_0 with $0 \leq r_0 < a$. By induction, the quotient q may be written in base a:

$$q = r_n a^{n-1} + r_{n-1} a^{n-2} + \ldots + r_2 a + r_1$$

for unique integers $r_1, r_2, \ldots, r_n$ with $0 \leq r_i < a$. Then

$$b = qa + r_0 = a(r_n a^{n-1} + r_{n-1} a^{n-2} + \ldots + r_2 a + r_1)a + r_0$$
$$= r_n a^n + r_{n-1} a^{n-1} + \ldots + r_1 a + r_0$$

with all r_i satisfying $0 \leq r_i < a$.   □

Notice that the proof shows how to get b in base a. First divide b by a, then, successively, divide the quotients by a:

$$b = aq + r_0,$$
$$q = aq_1 + r_1,$$
$$q_1 = aq_2 + r_2,$$
$$\vdots$$
$$q_{n-2} = aq_{n-1} + r_{n-1}$$
$$q_{n-1} = a \cdot 0 + r_n.$$

The process stops when a quotient is reached which is 0. The digits are the remainders: $b = (r_n r_{n-1} \ldots r_2 r_1 r_0)_a$.

Example 1. To get 366 in base 2:

$$366 = 2 \cdot 183 + 0,$$
$$183 = 2 \cdot 91 + 1,$$
$$91 = 2 \cdot 45 + 1,$$

$$45 = 2 \cdot 22 + 1,$$
$$22 = 2 \cdot 11 + 0,$$
$$11 = 2 \cdot 5 + 1,$$
$$5 = 2 \cdot 2 + 1,$$
$$2 = 2 \cdot 1 + 0,$$
$$1 = 1 \cdot 0 + 1,$$

so $366 = (101101110)_2$.

The conversion of a number to base 2 can be set up as a special case of Russian Peasant Arithmetic (RPA) (Chapter 2, Exercise 29). We set up RPA to multiply 1 by 366:

left	right	sum	summand
1	366	0	0
2	183	0	2
4	91	2	4
8	45	6	8
16	22	14	0
32	11	14	32
64	5	46	64
128	2	110	0
256	1	110	256
512	0	366	0

The last entry in the sum column is the sum of the entries in the summand column, namely:

$$366 = 256 + 64 + 32 + 8 + 4 + 2 = (101101110)_2.$$

(In general, to convert m to base 2, the array is constructed as follows. Start with the row

$$1 \ m \ 0 \ 0$$

if m is even, or

$$1 \ m \ 0 \ 1$$

if m is odd. Proceeding down the array, suppose we have a row

$$b \ a \ s \ d.$$

The next row is

$$2b \ \tfrac{a}{2} \ s+d \ 2b$$

if a is even and $\frac{a}{2}$ is odd;

$$2b \ \tfrac{a}{2} \ s+d \ 0$$

if a is even and $\frac{a}{2}$ is even;

$$2b \ \tfrac{a-1}{2} \ s+d \ 2b$$

if a is odd and $\frac{a-1}{2}$ is odd;

$$2b \ \frac{a-1}{2} \ s + d \ 0$$

if a is odd and $\frac{a-1}{2}$ is even. Stop when the entry in the "right" column is 0.)

Operations in base a. We can add, subtract, multiply, and divide in any base, using the same operations learned in grade school. The only change is that to multiply or divide in base a we must know the multiplication table in base a.

It is very easy to remember multiplication tables in base 2. On the other hand, to do multiplication in base 60 is rather more difficult. Babylonian mathematicians kept clay tablets on hand containing base 60 multiplication tables. (See Section 12B for an alternative approach to base 60 multiplication.)

Long division works in any base as well. In fact, in base 2 it is particularly easy, because in determining the correct digits of the quotient, no guesswork is involved.

On the other hand, in base a where a is large it is harder to determine the digits of the quotient. Since computers do multiple precision arithmetic in large bases, some study has been done of long division, and it was observed by D.A. Pope and M.L. Stein in 1960 that the closer the first digit of the divisor is to the base a, the easier guessing the digits of the quotient is. For a more precise description of this phenomenon, see Knuth (1998), Theorem B, p. 272. See also Section 9G.

Exercises.

6. Write 1987 in base 1000.

7. Write 1987 in base 2, and in base 8.

8. In base 2 write the numbers from 29 through 35.

9. One seventh of an hour is 514 seconds, to the nearest second. Write it in minutes and seconds.

10. Write one eleventh of a day in hours, minutes and seconds (to the nearest second).

11. If a runner completes a 50 mile race in 7 hours, 33 minutes and 15 seconds, and he were to run a marathon (26 miles, 385 yards) at the same pace, how long would it take him? (There are 1760 yards to the mile).

12. If an angle θ is one radian, how much is θ in degrees, minutes and seconds (to the nearest second)?

13. Write 176 and 398 in base 2 and multiply them. Check the multiplication by multiplying them in base 10 and converting the answer to base 2.

14. Multiply $(253)_8$ and $(601)_8$. Check the answer.

15. If your calculator carries 8 decimal digits, and you wish to multiply two 20 digit numbers, you have to write the numbers in base 10^n for some n and write a program to do the multiplication in base 10^n. Which n would be appropriate? Can you use $n = 8$?

16. Divide $(110110011)_2$ into $(1100000100101)_2$ using long division in base 2.

17. Divide $(1,4,25,46)_{60}$ by $(1,38)_{60}$, using long division in base 60. Then multiply both numbers by 32 and do the division. Is it any easier?

B. Greatest Common Divisors

What do we mean by the greatest common divisor of two numbers? We deal with the three words, "greatest", "common", "divisor", in reverse order:

Let a, b be integers, with a not equal to zero. Say that a *divides* b, or a *is a divisor of* b, if $b = aq$ for some integer q, that is, b is equal to some integer multiple of a.

Thus, 15 divides 45, because $45 = 15 \cdot 3$, whereas 15 does not divide 42, because there is no integer q so that $42 = 15q$. (It is true that $42 = 15 \cdot (14/5)$, but 14/5 is not an integer.)

The words "divisor" and "factor" mean the same thing.

If we are looking for divisors of a natural number, once we find the positive divisors, then the negative divisors will just be the negatives of the positive divisors. For example, 6 has divisors 1, 2, 3, 6 and also $-1, -2, -3$ and -6. So when looking for divisors of a number in this section, we'll just write down the positive divisors.

We can find the divisors of a small number easily. For example,

the divisors of 15 are 1, 3, 5, and 15;

the divisors of 28 are 1, 2, 4, 7, 14, and 28;

the divisors of 42 are 1, 2, 3, 6, 7, 14, 21, and 42.

Notice that 1 divides every integer (as does -1), and every integer divides 0.

We will often use the notation $a \mid b$ to mean, a divides b. Thus $15 \mid 45$.

(The notation $\mid$ can be confusing. If we write $4 \mid 28$, it is a statement, "4 divides 28." If we write 4/28, with / rather than $\mid$, we are writing a fraction, a number, "4 over 28." So $\mid$ is a verb, "divides", while / is a preposition, "over".)

If a and b are integers, a *common divisor* of a and b is an integer e such that e divides a and e divides b.

For example, the common divisors of 28 and 42 are 1, 2, 7 and 14 (and their negatives), as we can see by comparing the lists of divisors of 28 and 42.

The common divisors of 15 and 42 are 1 and 3.

Note that 1 is a common divisor of any two integers.

A number d is the *greatest common divisor (g.c.d.)* of a and b if:

(i) d is a common divisor of a and b and

(ii) no common divisor of a and b is larger than d.

We denote the greatest common divisor of a and b by (a,b).

To continue the examples above, the greatest common divisor of 28 and 42 is 14, that is, $(28,42) = 14$. Also, $(15,42) = 3$, while $(28,15) = 1$.

Since 1 is a common divisor of every number, every two numbers a and b always have a common divisor. On the other hand, there can be only a finite number of common divisors of a and b since every common divisor of a must divide a, hence is $<a$. So there is a *greatest* common divisor of a and b.

One final bit of terminology. Two numbers a and b are *coprime* or *relatively prime* if their greatest common divisor is 1. Thus 15 and 28 are coprime, but 15 and 42 are not coprime, and 28 and 42 are not coprime.

Exercises.

18. Find all positive common divisors of:
(i) 16 and 48,
(ii) 30 and 45,
(iii) 18 and 65.

19. Find the greatest common divisor of:
(i) 35 and 65,
(ii) 135 and 156,
(iii) 49 and 99.

20. Find the greatest common divisor of 17017 and 19210.

21. Find the greatest common divisor of 21331 and 43947. (You may wish to use a calculator.)

22. Find the greatest common divisor of 210632 and 423137. (You may wish to use a computer.)

23. Show that for any number n, n and $n+1$ are coprime.

24. Show that if $a \mid b$, then $(a,b) = a$.

25. Given numbers a and b, suppose there are integers r,s so that $ar + bs = 1$. Show that a and b are coprime.

26. Show that the greatest common divisor of a and b is equal to the greatest common divisor of a and $-b$.

27. Show that $(a,m) \leq (a,mn)$ for any integers a, m and n.

28. Show that if $(a,b) = 1$ and c divides a, then $(c,b) = 1$.

29. Show that of any three consecutive integers, exactly one is divisible by 3.

30. Show that of any m consecutive integers, exactly one is divisible by m.

31. Show that for all numbers $a > 0$, b, b', c, c', if $a \mid b - c$ and $a \mid b' - c'$, then $a \mid bb' - cc'$.

32. Let $m > n > 1$ be natural numbers. Show that there is some t with $n \leq t < m$, such that $m - n$ divides t.

C. Euclid's Algorithm

If you tried to do some of the exercises above, such as finding the greatest common divisor of 21331 and 43947, you will appreciate a method of Euclid to find the greatest common divisor of two numbers, now known as Euclid's Algorithm. It works as follows. Suppose the two numbers are a and b, with $a \leq b$. The next two paragraphs are paraphrased from Euclid's Elements, Book VII, Proposition 2.

If a divides b, a is a common divisor of b and it is manifestly also the greatest, for no number greater than a will divide a. But if a does not divide b then, the lesser of the numbers a, b being continually subtracted from the greater, some number will be left which will divide the one before it.

This number which is left is the greatest common divisor of b and a, and any common divisor of b and a divides the greatest common divisor of b and a.

We illustrate Euclid's method with 18 and 7.

Subtract 7 from 18 to get 11, leaving 11 and 7.
Subtract 7 from 11 to get 4, leaving 4 and 7.
Subtract 4 from 7 to get 3, leaving 3 and 4.
Subtract 3 from 4 to get 1, leaving 3 and 1.
Now 1 divides 3, so 1 is the greatest common divisor of 18 and 7.

Or consider 84 and 217.

$$217 - 84 = 133,$$
$$133 - 84 = 49,$$
$$84 - 49 = 35,$$
$$49 - 35 = 14,$$
$$35 - 14 = 21,$$
$$21 - 14 = 7,$$

and 7 divides 14. So 7 is the greatest common divisor of 84 and 217.

Use of Euclid's Algorithm is aided by division. When we divide 84 into 217 (for example, using long division) the quotient is the number of times we subtract 84 from 217 before we end up with a number less than 84. Thus, $217 = 2 \cdot 84 + 49$, so $217 - 2 \cdot 84 = 49$: that is, after we subtract 84 two times from 217 we obtain a number (49) less than 84.

Thus we can describe the algorithm of Euclid more compactly by replacing the repeated subtraction in Euclid's original formulation by repeated uses of the Division Theorem. For the example of 217 and 84 we have:

$$217 = 84 \cdot 2 + 49$$
$$84 = 49 \cdot 1 + 35,$$
$$49 = 35 \cdot 1 + 14$$
$$35 = 14 \cdot 2 + 7,$$
$$14 = 7 \cdot 2 + 0.$$

and so 7 is the greatest common divisor of 217 and 84.

In words, here is Euclid's Algorithm with division:

If a divides b, a is a common divisor of b and it is manifestly also the greatest, for no number greater than a will divide a. But if a does not divide b then divide a into b and get a remainder. then divide that remainder into a and get a new remainder. Continually divide the new remainder into the previous remainder until the new remainder is zero. The last non-zero remainder is the greatest common divisor of a and b, and any common divisor of b and a divides the greatest common divisor of b and a.

Here it is, in mathematical symbolism:

Euclid's Algorithm. Given natural numbers a and b, apply the Division Theorem successively as follows:

$$b = aq_1 + r_1,$$
$$a = r_1 q_2 + r_2,$$
$$r_1 = r_2 q_3 + r_3,$$
$$r_2 = r_3 q_4 + r_4,$$
$$\vdots$$
$$r_{n-2} = r_{n-1} q_n + r_n$$
$$r_{n-1} = r_n q_{n+1} + 0.$$

When we reach the point where r_n divides r_{n-1}, then r_n is the greatest common divisor of a and b.

We shall prove this last statement carefully in the next section.

Exercises.

33. Try Euclid's Algorithm on
(i) 135 and 156.
(ii) 17017 and 19210.
(iii) 21331 and 43947.

34. Using Euclid's Algorithm (with division), find the greatest common divisor of:
  (i) 121 and 365,
  (ii) 89 and 144,
  (iii) 295 and 595,
  (iv) 1001 and 1309.

35. Using Euclid's Algorithm (with division), find the greatest common divisor of:
  (i) 17017 and 18900,
  (ii) 21063 and 43137,
  (iii) 210632 and 423137,
  (iv) 92263 and 159037,
  (v) 112345 and 112354.

36. Show that if e divides a and e divides b, then e divides $ar + bs$ for any integers r, s.

37. Show that if $b = aq + r$, then $(b, a) = (a, r)$.

38. Using this last exercise, prove by induction the statement that r_n is the greatest common divisor of a and b.

D. Bezout's Identity

Euclid's Algorithm is more useful than simply giving an efficient way to determine the greatest common divisor of two numbers. It also yields a relationship between two numbers and their greatest common divisor that is of great importance, both practically and theoretically, as we shall see. The relationship is called:

Theorem 4 (Bezout's Identity). *If the greatest common divisor of a and b is d, then $d = ar + bs$ for some integers r and s.*

Example 2. Before we show how to find r and s, hide the rest of this page and try the numbers 365 and 1876. It is easy to see they are relatively prime (the divisors of 365 are 1, 5, 73 and 365 and none of 5, 73 and 365 divides 1876). Try to write $1 = 365x + 1876y$ for some integers x and y.

It is not obvious how to do this!

Here is how. Do Euclid's Algorithm:

$$1876 = 365 \cdot 5 + 51,$$
$$365 = 51 \cdot 7 + 8,$$
$$51 = 8 \cdot 6 + 3,$$
$$8 = 3 \cdot 2 + 2,$$
$$3 = 2 \cdot 1 + 1;$$

then 1 is the greatest common divisor (as we already knew).

Now solve for the remainders,

$$1 = 3 - 2 \cdot 1,$$
$$2 = 8 - 3 \cdot 2,$$
$$3 = 51 - 8 \cdot 6,$$
$$8 = 365 - 51 \cdot 7,$$
$$51 = 1876 - 365 \cdot 5,$$

and successively substitute the remainders into the equation $1 = 3 - 2 \cdot 1$, starting with 2:

$$1 = 3 - 2$$
$$= 3 - (8 - 3 \cdot 2) = 3 \cdot 3 - 8$$
$$= 3(51 - 8 \cdot 6) - 8 = 3 \cdot 51 - 8 \cdot 19$$
$$= 3 \cdot 51 - 19(365 - 51 \cdot 7) = 136 \cdot 51 - 19 \cdot 365$$
$$= 136(1876 - 5 \cdot 365) - 19 \cdot 365 = 136 \cdot 1876 - 699 \cdot 365.$$

So $x = -699$, $y = 136$.

Solving Bezout's Identity by Euclid's Algorithm is often called the *Extended Euclidean Algorithm*. But as we have just described it, the procedure is confusing, since it is hard to keep track of the numbers that are remainders, to be substituted for, and the numbers that are quotients, to be kept. (In the equation $1 = 3 \cdot 3 - 8$ in the example above, one 3 is kept, the other is replaced by $51 - 8 \cdot 6$.)

It is easier to do the computations by starting at the top of Euclid's Algorithm, rather than the bottom, and successively write the original two numbers and all the remainders as linear combinations of the two original numbers. Let's look at the same example.

Example 3. We first have the obvious equations:

$$1876 = 0 \cdot 365 + 1 \cdot 1876,$$
$$365 = 1 \cdot 365 + 0 \cdot 1876.$$

Then the equation that gives the first remainder in Euclid's Algorithm for 1876 and 365, namely,

$$51 = 1876 - 5 \cdot 365,$$

can be obtained by multiplying the 365 equation by 5:

$$5 \cdot 365 = 5(1 \cdot 365 + 0 \cdot 1876) = 5 \cdot 365 + 0 \cdot 1876;$$

then subtracting the equation for $5 \cdot 365$ from the equation for 1876:

$$51 = 1876 - 5 \cdot 365$$
$$= (0 \cdot 365 + 1 \cdot 1876) - 5(1 \cdot 365 + 0 \cdot 1876)$$
$$= -5 \cdot 365 + 1 \cdot 1876.$$

The next remainder, 8, in Euclid's Algorithm satisfies

$$8 = 365 - 7 \cdot 51$$

so we substitute in the equations for 365 and 51 and collect coefficients:

$$8 = (1 \cdot 365 + 0 \cdot 1876) - 7(-5 \cdot 365 + 1 \cdot 1876)$$
$$= (1 + 7 \cdot 5)365 + (0 - 7 \cdot 1)1876$$
$$= 36 \cdot 365 - 7 \cdot 1876.$$

The next remainder, 3, satisfies $3 = 51 - 6 \cdot 8$, so we substitute for 51 and 8 and collect coefficients of 365 and 1876:

$$3 = 51 - 6 \cdot 8$$
$$= (-5 \cdot 365 + 1 \cdot 1876) - 6(36 \cdot 365 - 7 \cdot 1876)$$
$$= (-5 - 6 \cdot 36)365 + (1 + 6 \cdot 7)1876$$
$$= -221 \cdot 365 + 43 \cdot 1876.$$

Continuing,

$$2 = 8 - 2 \cdot 3$$
$$= (36 \cdot 365 - 7 \cdot 1876) - 2(-221 \cdot 365 + 43 \cdot 1876)$$
$$= (36 + 442)365 + (-7 - 86)1876$$
$$= 478 \cdot 365 - 93 \cdot 1876$$

and finally

$$1 = 3 - 2$$
$$= (-221 \cdot 365 + 43 \cdot 1876) - (478 \cdot 365 - 93 \cdot 1876)$$
$$= (-221 - 478)365 + (43 + 93)1876$$
$$= -699 \cdot 365 + 136 \cdot 1876.$$

All of this can be done efficiently by setting up a matrix of three columns, one for the remainders, one for the coefficients of 365 and one for the coefficients of 1876, and just keeping track of the coefficients as we proceed down Euclid's Algorithm as above. In that format, to solve $e = 365x + 1876y$ for $e = 1$, the computations look as follows:

e	$x = $ coeff. of 365	$y = $ coeff. of 1876
1876	0	1
365	1	0
$365 \cdot 5$	5	0
$51 = 1876 - 365 \cdot 5$	-5	1
$51 \cdot 7$	-35	7
$8 = 365 - 51 \cdot 7$	36	-7
$8 \cdot 6$	216	-42

$$3 = 51 - 8 \cdot 6 \quad -221 \quad 43$$
$$3 \cdot 2 \quad\quad\quad -442 \quad 86$$
$$2 = 8 - 3 \cdot 2 \quad 478 \quad -93$$
$$1 = 3 - 2 \quad\quad -699 \quad 136$$

We call this matrix the *EEA matrix*. Each row (e, x, y) of the EEA matrix represents the equation

$$e = 365x + 1876y.$$

Thus the row headed by 8 represents the equation

$$8 = 365 \cdot 36 + 1876 \cdot (-7)$$

and the last row represents

$$1 = 365 \cdot (-699) + 1876 \cdot 136.$$

Once we write down the first two rows of the EEA matrix, the other rows are obtained by multiplying a previous row by a constant, or subtracting one row from another row, just as in Gaussian elimination with matrices. The EEA matrix collects only the relevant data from the computation we did just above, without having to write down 1876 and 365 many times. (In that computation, the original numbers 1876 and 365 act as placeholders for the coefficients. The EEA matrix eliminates the need for the placeholders.)

To show that this procedure always yields the greatest common divisor as a linear combination of the two original numbers, here is a proof by induction of the last assertion of Euclid's Algorithm and of Bezout's Identity:

Theorem 5. *Let a and b be natural numbers. If a divides b then a is the greatest common divisor of a and b. If a does not divide b, and r_n is the last nonzero remainder in Euclid's Algorithm for a and b, then r_n is the greatest common divisor of a and b. If d is the greatest common divisor of a and b, then $d = ax + by$ for some integers x and y.*

Proof. If a divides b then a is clearly the greatest common divisor of a and b, and $a = a \cdot 1 + b \cdot 0$, so the theorem is true in that case. So assume that a does not divide b. Then Euclid's Algorithm for a and b involves at least two divisions. Suppose Euclid's Algorithm contains $n + 1$ divisions ($n \geq 1$) so that r_n is the last nonzero remainder in Euclid's Algorithm for a and b. We prove the theorem by induction on n.

If $n = 1$, then Euclid's Algorithm for a and b has the form:

$$b = aq_1 + r_1,$$
$$a = r_1 q_2 + 0.$$

Then r_1 divides a, so $(r_1, a) = r_1$. It is easy to verify from the first equation that $(r_1, a) = (a, b)$ (Exercise 37), so r_1 is the greatest common divisor of a and b. Also $r_1 = b \cdot 1 + a \cdot (-q_1)$, so Bezout's Identity holds.

Assume the theorem is true for $n = k - 1$, so that the theorem is true for any two numbers whose Euclid's Algorithm involves k divisions. Suppose Euclid's Algorithm for a and b involves $k + 1$ divisions:

$$b = aq_1 + r_1,$$
$$a = r_1 q_2 + r_2,$$
$$r_1 = r_2 q_3 + r_3,$$
$$\vdots$$
$$r_{k-2} = r_{k-1} q_k + r_k$$
$$r_{k-1} = r_k q_{k+1} + 0.$$

Notice that if we omit the first line of Euclid's Algorithm for a and b, what is left is Euclid's Algorithm for r_1 and a, and that algorithm involves k divisions. So by the induction assumption, r_k is the greatest common divisor of r_1 and a, and $r_k = au + r_1 v$ for some integers u and v.

Now $b = aq_1 + r_1$, so $(b,a) = (a, r_1) = r_k$ (again by Exercise 37, above). Moreover, substituting $r_1 = b - aq_1$ into the equation $r_k = au + r_1 v$ gives

$$r_k = au + (b - aq_1)v$$
$$= bv + a(u - q_1 v).$$

Hence Bezout's Identity holds for a and b. The theorem is true by induction.   □

Corollary 6. *Two numbers a and b are coprime iff there are integers r and s so that $ar + bs = 1$.*

(Notation: "iff" means "if and only if". An "iff" assertion, as in the corollary, requires two proofs.)

Proof. First assume that $(a,b) = 1$. Then, from Theorem 5, there are integers r and s so that $1 = ar + bs$. Conversely, if $1 = ar + bs$ for some integers r and s, and d divides a and b, then d divides $ar + bs = 1$, so $d = 1$ or -1. So the greatest common divisor of a and b is 1.   □

Here is a useful consequence of Bezout's Identity.

Corollary 7. *If e divides a and e divides b, then e divides (a,b).*

Proof. Write $d = (a,b)$ as $d = ar + bs$. If e divides a and b, then $a = ef, b = eg$ for some integers f, g. Then

$$d = ar + bs = efr + egs = e(fr + gs),$$

so e divides d.   □

A consequence of great importance, as we shall see below and in the next chapter, is:

Corollary 8. *If a divides bc, and a and b are coprime, then a divides c.*

Proof. Bezout's Identity is

$$ar + bs = 1$$

for some integers r, s. Multiply that equation by c:

$$acr + bcs = c.$$

Now a obviously divides acr. If a divides bc, then a divides bcs, so a divides $acr + bcs = c$. □

Another application of Bezout's Identity is the following result, which is useful for factoring large numbers.

Proposition 9. *For every integers a, b, m, (ab, m) divides $(a, m)(b, m)$. If a and b are coprime, then*

$$(a, m)(b, m) = (ab, m).$$

Proof. Let $(a, m) = ra + sm$, $(b, m) = tb + vm$. Then

$$(a, m)(b, m) = (ra + sm)(tb + vm) = rtab + zm$$

for $z = rav + stb + svm$. Since (ab, m) divides ab and m, therefore (ab, m) divides $(a, m)(b, m)$.

For the second part, notice that (a, m) divides (ab, m). Write

$$(ab, m) = (a, m)e$$

for some integer e. Also, (b, m) divides (ab, m). Since a and b are coprime, so are (a, m) and (b, m), and so by Corollary 8, (b, m) divides e. Thus

$$(ab, m) = (a, m)(b, m)f$$

for some integer f. Since (ab, m) divides $(a, m)(b, m)$, we must have $f = 1$. □

Exercises.

39. Using the EEA matrix, find d, the greatest common divisor, and find r, s so that $ar + bs = d$, where a and b are:
   (i) 270 and 114,
   (ii) 242 and 1870,
   (iii) 600 and 11312,
   (iv) 11213 and 1001,
   (v) 500 and 3000.

40. Give six counterexamples to the assertion: if a divides bc and a does not divide b, then a divides c.

41. Suppose a is a number >1 with the following property: for all b,c, if a divides bc and a does not divide b, then a divides c. Show that a must be prime.

42. Observe that $98^2 - 31^2 = 9604 - 961 = 8643 = 2881 \cdot 3$, so $(98^2 - 31^2, 2881) = 2881$. Use Proposition 9 to factor 2881.

43. Prove that for all numbers a and b, if $(a,b) = d$ and $a = df, b = dg$, then $(f,g) = 1$.

44. Prove that for all numbers a,b,m, if $(a,m) = 1$ and $(b,m) = 1$, then $(ab,m) = 1$.

45. Prove that for all numbers a,b,m,n, if $am + bn = e$ for some e, then (a,b) divides e.

46. Prove that for all numbers a,b, if $d = (a,b)$ and $ra + sb = d$, then $(r,s) = 1$.

47. Prove the converse of Corollary 8: Suppose a and b are any numbers with $(a,b) > 1$. Then there is a number c so that a divides bc and a does not divide c.

48. Prove that for every numbers $m, a, b > 0$, $m(a,b) = (ma, mb)$. Do it in two parts: $m(a,b) \leq (ma, mb)$ and $(ma, mb) \leq m(a,b)$. In addition to the definition of greatest common divisor, you may find it convenient to use Bezout's Identity.

49. Prove: for all numbers a,b,m, if $(a,m) = d$ and $(b,m) = 1$, then $(ab,m) = d$.

50. Prove that for all numbers a,b,c, if a divides bc, then $a/(a,b)$ divides c.

51. Prove that for all numbers a,b, there are integers r,s so that

$$\frac{1}{ab} = \frac{s}{a} + \frac{r}{b}$$

if and only if $(a,b) = 1$.

52. Prove that if m is an integer and there is a rational number r/s so that $(r/s)^2 = m$, then there is an integer n so that $n^2 = m$.

53. Define the greatest common divisor of three numbers a,b and c. Call it (a,b,c). Show that $(a,b,c) = (a,(b,c))$.

54. Show that $(a,b,c) = ax + by + cz$ for some integers x,y,z.

55. For a,b natural numbers, consider the set J of all positive integers of the form $ar + bs$ for integers r,s. Since J is a nonempty set of natural numbers, by well-ordering J has a least element c. Show that c is the greatest common divisor of a and b.

E. Linear Diophantine Equations

For numbers a, b, e, Bezout's Identity can be used to decide whether or not there are integer solutions of equations of the form $ax + by = e$, and to find a solution if there is a solution.

Proposition 10. *Given integers a, b, e, there are integers m and n with $am + bn = e$ if and only if (a, b) divides e.*

Proof. If $am + bn = e$ for some integers m, n, then the greatest common divisor of a and b divides e.

Conversely, if $d = (a, b)$ divides e, then by Bezout's Identity we can find integers r, s so that $ar + bs = d$. If $e = dm$ for some integer m, then $x = rm, y = sm$ solves the equation $ax + by = e$.   □

The proof shows how to find a solution of $ax + by = e$ if (a, b) divides e.

Example 4. Suppose we want to find a solution to

$$365x + 1876y = 24.$$

We know that $(365, 1876) = 1$, and using the EEA matrix, we found in Example 3 that

$$1 = -699 \cdot 365 + 136 \cdot 1876.$$

Hence

$$24 = -(24 \cdot 699) \cdot 365 + (24 \cdot 136) \cdot 1876.$$

We can sometimes solve an equation like $365x + 1876y = 24$ more efficiently. If we apply the EEA matrix as we did in Example 3 to find r, s so that $ar + bs = d$, then to solve $ax + by = e$ with e a multiple of d, we can stop as soon as we find a remainder c that divides e. Multiplying the row in the EEA matrix headed by c by the integer e/c will then give a solution to $ax + by = e$.

Example 5. To solve

$$24 = 365x + 1876y$$

notice that we found that

$$8 = 36 \cdot 365 + (-7) \cdot 1876.$$

Multiplying that equation (or the corresponding row of the EEA matrix) by 3 gives the equation

$$24 = 108 \cdot 365 + (-21) \cdot 1876:$$

e	$x =$ coeff. of 365	$y =$ coeff. of 1876
1876	0	1
365	1	0
$365 \cdot 5$	5	0

$$51 = 1876 - 365 \cdot 5 \qquad -5 \qquad 1$$
$$51 \cdot 7 \qquad -35 \qquad 7$$
$$8 = 365 - 51 \cdot 7 \qquad 36 \qquad -7 \; \cdot$$
$$24 = 3 \cdot 8 \qquad 108 \qquad -21$$

Thus $x = 108, y = -21$ solves $24 = 365x + 1876y$.

Or if we want to solve $35 = 365 + 1876$, we may notice that $35 = 51 - 2 \cdot 8$, so we can add to the EEA matrix two more rows:

$$\vdots$$
$$51 = 1876 - 365 \cdot 5 \qquad -5 \qquad 1$$
$$51 \cdot 7 \qquad -35 \qquad 7$$
$$8 = 365 - 51 \cdot 7 \qquad 36 \qquad -7 \; ,$$
$$16 = 2 \cdot 8 \qquad 72 \qquad -14$$
$$35 = 51 - 16 \qquad -77 \qquad 15$$

so that $35 = 365 \cdot (-77) + 1876 \cdot 15$.

Once we know that there is a solution of $ax + by = c$, then we can ask, can we describe all the solutions of $ax + by = c$?

Suppose

$$ax_0 + by_0 = c$$
$$ax_1 + by_1 = c.$$

Then subtracting the first equation from the second yields

$$a(x_1 - x_0) + b(y_1 - y_0) = 0.$$

Thus any two solutions of $ax + by = c$ differ by a solution of $ax + by = 0$. Conversely, if

$$az + bw = 0$$
$$ax + by = c$$

then

$$a(x + z) + b(y + w) = c.$$

In short,

Proposition 11. *Let x_0, y_0 be a solution of $ax + by = c$. Then the general solution of $ax + by = c$ is of the form $x = x_0 + z, y = y_0 + w$, where z, w is any solution of $ax + by = 0$.*

This proposition is similar to comparable results in linear algebra and in differential equations: to find the general solution of a non-homogeneous equation (like $ax + by = c$), find some particular solution to the non-homogeneous equation and add to it the general solution of the corresponding homogeneous equation (like $ax + by = 0$).

For the homogeneous equation we have:

Proposition 12. *Let* $d = (a,b)$. *Then the general solution of* $ax + by = 0$ *is*

$$x = \frac{b}{d}k$$
$$y = -\frac{a}{d}k$$

for k any integer.

Proof. Suppose

$$ax + by = 0$$

Divide both sides by d to get

$$\frac{a}{d}x = -\frac{b}{d}y.$$

Since the integers a/d and b/d are coprime (by Exercise 43), a/d divides y by Corollary 8. Hence $y = \frac{a}{d}k$ for some integer k. Then

$$\frac{a}{d}x = -\frac{b}{d}\left(\frac{a}{d}k\right),$$

hence

$$x = -\frac{b}{d}k.$$

$\square$

Corollary 13. *If* x_0, y_0 *is a solution of* $ax + by = c$, *then the solutions of* $ax + by = c$ *are*

$$x = x_0 + \frac{b}{d}k$$
$$y = y_0 - \frac{a}{d}k$$

for every k in $\mathbb{Z}$.

A historical note: The French mathematician Etienne Bezout obtained "Bezout's Identity" for polynomials in 1779, but "Bezout's Identity" for relatively prime integers goes back to Bachet (1621). See: Mehl: serge.mehl.free.fr/chrono/Bachet.html

Exercises.

56. Using the EEA matrix, find a solution of
  (i) $83x + 35y = 24$,
  (ii) $100x + 167y = 33$,
  (iii) $49x + 117y = 36$,

57. Find all solutions of
  (i) $114x + 270y = 0$,
  (ii) $114x + 270y = 24$.

58. Find the solution with the smallest $x > 0$ of
  (i) $66x + 45y = 0$,
  (ii) $1001x + 143y = 0$.

59. (i) Find all solutions of $34x - 62y = 8$ with $x, y \geq 0$.
  (ii) Find all solutions of $62y - 34x = 8$ with $x, y \geq 0$.

60. Find all solutions:
  (i) $242x + 1870y = 66$,
  (ii) $327x + 870y = 66$
  (iii) $327x + 870y = 56$.

61. Find $d = (3731, 1894)$ and write $d = 3731r + 1894s$ where
  (i) $r > 0$ and $s < 0$;
  (ii) $r < 0$ and $s > 0$.

62. Decide if each of the following has a solution or not. If so, find the solution with the smallest possible $x \geq 0$:
  (i) $133x + 203y = 38$,
  (ii) $133x + 203y = 40$,
  (iii) $133x + 203y = 42$,
  (iv) $133x + 203y = 44$.

63. Fahrenheit and centigrade temperatures are related by the formula

$$f = \frac{9}{5}c + 32.$$

Thus $c = 0°$ is the same as $f = 32°$, and $f = 40°$ is the same as $c = (40/9)°$. Find all solutions of $f = \frac{9}{5}c + 32$ where both f and c are integers.

64. You are given two "hour" glasses: a 6-minute hourglass and an 11-minute hourglass, and you wish to measure 13 minutes. How do you do it?

65. You take a 13 quart jug and a 16 quart jug to a stream and want to bring back 5 quarts of water. How do you do it?

66. You take an a quart jug and a b quart jug to the stream and want to bring back c quarts of water. For which c can it be done? How?

67. Suppose $2 < a < b$ are natural numbers, $(a, b) = d$, and $d = ar + bs$, where r and s are obtained by Euclid's Algorithm. Show that $-b/2 < r < b/2$ and $-a/2 < s < a/2$.

F. The Efficiency of Euclid's Algorithm

Consider how we might determine the greatest common divisor of 92263 and 159037 if we did not know Euclid's Algorithm. One way would be to search for divisors of 92263, and each time we found one, see if it is also a divisor of 159037. But if we began searching for divisors of 92263 we would find that divisors, or factors, of 92263 are not easily found. Perhaps we would write a program to divide 92263 by each odd number starting with 3 until a divisor of 92263 were found. If we do this, we would, after 128 divisions, find that 257 divides 92263: $92263 = 257 \cdot 359$. Then checking each factor, we would find that 359 divides 159037, and so 359 is the greatest common divisor of 92263 and 159037.

Seeking the greatest common divisor of 92263 and 159037 in this way takes 129 divisions to find the factor 257 of 92263.

How much more efficient is Euclid's Algorithm! If we try it with 92263 and 159037, we find that 359 is the last non-zero remainder, and hence is the greatest common divisor of 92263 and 159037, in a total of ten divisions.

In this section, we explore how efficient Euclid's Algorithm is on any two numbers a, b.

Let $N(a, b)$ denote the number of steps needed to obtain the last non-zero remainder of a and b ($a < b$) in Euclid's Algorithm using division and not just subtraction. Thus, as the algorithm is laid out in section C, $N(a, b) = n$.

The size of $N(a, b)$ relates to how quickly the sequence $r_1, r_2, \ldots, r_n$ of remainders decreases, and in turn to the size of the quotients. A large decrease in a remainder means that the next quotient will be large. So large quotients correspond to a rapid decrease in the remainders, and that implies that $N(a, b)$ will be small. For example, if $a = 63725, b = 125731$, then Euclid's Algorithm includes the quotients 36 and 14, and $N(a, b) = 5$:

$$125731 = 63725 \cdot 1 + 62006,$$
$$63725 = 62006 \cdot 1 + 1719,$$
$$62006 = 1719 \cdot 36 + 122,$$
$$1719 = 122 \cdot 14 + 11,$$
$$122 = 11 \cdot 11 + 1.$$

On the other hand, for $a = 55$, $b = 89$, then even though $r_1 = 34$ is a much smaller remainder than the first remainder (62006) in the previous example, we have $N(a, b) = 8$:

$$89 = 55 \cdot 1 + 34,$$
$$55 = 34 \cdot 1 + 21,$$
$$34 = 21 \cdot 1 + 13,$$
$$21 = 13 \cdot 1 + 8,$$
$$13 = 8 \cdot 1 + 5,$$

$$8 = 5 \cdot 1 + 3,$$
$$5 = 3 \cdot 1 + 2,$$
$$3 = 2 \cdot 1 + 1.$$

The second example is one where in Euclid's Algorithm, none of the quotients is greater than one.

Perhaps you recognized the remainders in Euclid's Algorithm for 55 and 89. They form part of the Fibonacci sequence.

The Fibonacci sequence is so named because it arose in the Liber Abaci [Sigler (2003)] of Leonardo of Pisa, also known as Fibonacci, in connection with the following problem:

Suppose a man has one pair of rabbits. How many pairs of rabbits can be bred from the initial pair in one year if each pair begins to breed in the second month after their birth, each month producing a new pair, and no deaths occur?

It is not hard to see that in the first, second, third, etc. months there are the following numbers of pairs of rabbits:

$$1, 1, 2, 3, 5, 8, 13, 21, 34, 55, 89, 144, 233, \ldots.$$

This sequence is the Fibonacci sequence, $F_1 = 1, F_2 = 1, F_3 = 2$, etc. We can start the sequence also with $F_0 = 0$. Then the sequence of Fibonacci numbers is defined by

$$F_0 = 0, F_1 = 1$$
and
for any $n > 0$, $F_{n+1} = F_n + F_{n-1}$.

The next number in the Fibonacci sequence is the sum of the previous two.

If we apply Euclid's Algorithm to F_{n+1} and F_n, the first remainder $r_1 = F_{n-1}$, the next remainder is $r_2 = F_{n-2}$, etc. So in Euclid's Algorithm for any two consecutive Fibonacci numbers, the sequence of remainders consists of all the previous numbers in the Fibonacci sequence, until we get a remainder of $1 = F_2$. Thus Euclid's Algorithm for two consecutive numbers F_n and F_{n+1} in the Fibonacci sequence requires $n - 2$ steps to find the last nonzero remainder. In notation,

$$N(F_n, F_{n+1}) = n - 2.$$

All of the quotients in Euclid's Algorithm for two consecutive Fibonacci numbers are 1 until the last non-zero remainder is reached.

The examples above suggest that Euclid's Algorithm takes fewer steps when the quotients are large, than when the quotients are small. Thus Euclid's Algorithm would appear to be less efficient on Fibonacci numbers than on other numbers of similar size.

This is in fact the case, as Lamé proved in the nineteenth century.

Theorem 14 (Lamé's Theorem). *Let a and $b > a$ be two natural numbers. Suppose $a < F_n$, where F_n is the nth term in the Fibonacci sequence. Then $N(a,b) \leq n - 3 < n - 2 = N(F_n, F_{n+1})$.*

For example, if $a < 8 = F_6$ and b is any number larger than a, then according to Lamé's Theorem, Euclid's Algorithm for a and b takes at most 3 steps to find the last nonzero remainder.

The proof of Lamé's theorem is by induction, and we leave the proof as an exercise, below. We note that the induction argument has an interesting wrinkle to it that you may not find unless you do some examples.

In order to translate Lamé's theorem into usable form, we need to know how many digits the Fibonacci number F_n has. We can get some idea, by finding the smallest Fibonacci number of a given number of digits. For example:

$F_1 = 1$ is the smallest with 1 digit;

$F_7 = 13$ is the smallest with 2 digits;

$F_{12} = 144$ is the smallest with 3 digits;

$F_{17} = 1597$ is the smallest with 4 digits; etc.

You might guess that every fifth Fibonacci number thereafter gains another digit, and that is the case. We leave the verification as an exercise, below.

Now if F_{5d+2} has at least $d + 1$ decimal digits, then any number a with d digits satisfies $a < F_{5d+2}$. So from Lamé's theorem we get:

Corollary 15. *If $a < b$ and a has d digits, then*

$$N(a,b) \leq (5d + 2) - 3 < 5d.$$

The corollary shows how efficient Euclid's Algorithm is. Even on the worst possible examples, Euclid's Algorithm takes less than $5d$ steps, where d is the number of decimal digits of the smaller of the two numbers being computed. Thus for example, if we want to find the greatest common divisor of a and $b > a$, where a has 200 digits, Euclid's Algorithm will take at most 1000 steps. A fast computer can do this in less than a thousandth of a second. By contrast, to factor the number a could take weeks.

For a study of the average behavior of Euclid's Algorithm (as opposed to the worst-case behavior), see Knuth (1998), Section 4.5.3, where it is shown that if $a < b$ are randomly chosen and a has d digits, then the number of steps in of Euclid's Algorithm on a and b is approximately $2d$.

The Fibonacci numbers are such a interesting set of numbers that a mathematics journal, the Fibonacci Quarterly, was founded in 1963 to publish results related to the Fibonacci series. We will simply hint at the richness of this set of numbers in the exercises.

Exercises.

68. (i) Show that if $b > a$ and $b' = b +$ (positive multiple of) a, then $N(a,b') = N(a,b)$.

(ii) Verify that if $a < 8$ and $b > a$, then $N(a,b)$ takes at most three steps.

69. Prove Lamé's theorem. (See the remark following Lamé's theorem.)

70. (i) Prove that

$$F_n = \frac{[(1+\sqrt{5})/2]^n - [(1-\sqrt{5})/2]^n}{\sqrt{5}}$$

by induction on n.

(ii) Prove that $F_{n+5}/F_n \geq 10$ for all $n \geq 5$.

(iii) Prove that F_{5d+2} has at least $d+1$ decimal digits.

71. Verify that the Fibonacci sequence gives the size of the rabbit population each month.

72. Prove by induction that

$$F_n^2 - F_{n+1}F_{n-1} = (-1)^{n+1},$$

so that consecutive Fibonacci numbers are coprime.

73. (Askey). Show that

$$\begin{pmatrix} 1 & 1 \\ 1 & 0 \end{pmatrix}^n = \begin{pmatrix} F_{n+1} & F_n \\ F_n & F_{n-1} \end{pmatrix}.$$

Take the determinant of both sides and reprove the last exercise.

74. The Fibonacci numbers and the golden ratio:

The golden ratio is the ratio $b : a$ (with $b > a$) so that $b : a = (a+b) : b$, or $\frac{b}{a} = \frac{a+b}{b}$. The golden ratio was considered by the ancient Greeks to be the most perfect proportion for the lengths of the sides of rectangles, such as portraits. Show that if $b : a$ is the golden ratio, then $\phi = b/a = (1+\sqrt{5})/2$.

75. Show that

$$\frac{F_2}{F_1} < \frac{F_4}{F_3} < \cdots < \frac{F_{2n}}{F_{2n-1}} < \cdots < \phi < \cdots < \frac{F_{2n+1}}{F_{2n}} < \cdots < \frac{F_3}{F_2}.$$

76. Using Exercises 72 and 75, prove that

$$\left| \frac{F_{n+1}}{F_n} - \phi \right| \leq \frac{1}{F_n F_{n-1}},$$

and hence, for all d,

$$\left| \frac{F_{5d+4}}{F_{5d+3}} - \phi \right| < \frac{1}{10^{2d}}.$$

Thus the golden ratio can be approximated as closely as desired by ratios of consecutive Fibonacci numbers.

77. Let F_k be the kth Fibonacci number. Prove that if $d = (r, s)$, then $F_d = (F_r, F_s)$, as follows.

(i) For any $m, k, m > k$, $F_m = F_k F_{m-k+1} + F_{k-1} F_{m-k}$ (induction on k).

(ii) F_{nd} is divisible by F_d for all n and d (try induction on n: set $m = nd$, $k = d$ in (i)).

(iii) F_d is a common divisor of F_{kd} and F_{ld}, for any k, l (use (ii)).

(iv) $(F_m, F_{m+1}) = 1$ for all m (use Exercise 72).

(v) If e divides F_r and e divides F_s and $d = (r, s)$, then e divides F_d. (Write $d = ar - bs$, $r, s > 0$, use (i) with $m = ar$, $k = bs$, then use (ii) and (iv)).

Chapter 4
Unique Factorization

This chapter uses Bezout's identity and induction to prove the Fundamental Theorem of Arithmetic, that every natural number factors uniquely into a product of prime numbers. After exploring some initial consequences of the Fundamental Theorem, we introduce the study of prime numbers, a deep and fascinating area of number theory.

A. The Fundamental Theorem of Arithmetic

A natural number $p > 1$ is *prime* if the only divisor of p greater than 1 is p itself: Note: 1 is not prime, by convention.

Primes are the building blocks of natural numbers, for

Theorem 1. *Every natural number >1 factors into a product of primes.*

In this theorem and the remainder of the chapter, we use the convention that a product of primes may consist of only one factor. Thus the following are factorizations of the numbers from 4 through 11 into products of primes:

$$4 = 2 \cdot 2, \quad 5 = 5, \quad 6 = 2 \cdot 3, \quad 7 = 7,$$
$$8 = 2 \cdot 2 \cdot 2, 9 = 3 \cdot 3, 10 = 2 \cdot 5, 11 = 11.$$

The proof of Theorem 1 is an application of complete induction that you may have done in Chapter 2:

Proof. If $n > 1$ is prime, then n is a product of primes. Otherwise, $n = ab$ with $1 < a < n$ and $1 < b < n$. By complete induction, $a = p_1 \ldots p_r$, a product of primes, and also $b = q_1 \ldots q_s$, a product of primes. So

$$n = ab = p_1 \ldots p_r \cdot q_1 \ldots q_s,$$

a product of primes. □

L.N. Childs, *A Concrete Introduction to Higher Algebra*, Undergraduate Texts in Mathematics, © Springer Science+Business Media LLC 2009

The Fundamental Theorem of Arithmetic says that factorization of natural number >1 into a product of primes is unique. What does "unique" mean?

Suppose a is a natural number. If $a = p_1 \ldots p_n$ and also $a = q_1 \ldots q_m$ are factorizations of a into products of primes, we shall say that the factorizations are the same if the set of p_i's is the same as the set of q_j's (including repetitions). That is, $m = n$ and each prime occurs exactly as many times among the p_i's as it occurs among the q_j's. Thus we consider the factorizations $2 \cdot 2 \cdot 3 \cdot 31$ and $3 \cdot 2 \cdot 31 \cdot 2$ to be the same, because each prime occurs an equal number of times in each factorization. On the other hand, the factorizations $2 \cdot 3 \cdot 3 \cdot 31$ and $2 \cdot 3 \cdot 2 \cdot 31$ are different. The factorization of a is *unique* if any two factorizations of a are the same.

Theorem 2 (Fundamental Theorem of Arithmetic). *Any natural number $n > 2$ factors uniquely into a product of primes.*

We've proved that there is a factorization. We need only prove uniqueness.

Proof. We use complete induction. Suppose that the result is true for all numbers $<a$. Suppose $a = p_1 \ldots p_n$ and also $a = q_1 \ldots q_m$ are factorizations of a into products of primes. We want to show that the two factorizations are the same.

If $a = p_1$ is prime, then both m and $n = 1$ and $p_1 = q_1$, since a prime cannot factor into a product of two or more primes, by definition. So the theorem is true if a is prime.

Now assume that a is not prime. Suppose that p_1, the leftmost prime in the first factorization of a, is equal to some prime q_i in the second factorization. (We'll show shortly that this must be so.) Then a/p_1 is a natural number ≥ 2 (since a is not prime) and, of course, $a/p_1 < a$.

Since $p_1 = q_j$, we get two factorizations of a/p_1, namely:

$$\frac{a}{p_1} = p_2 \cdot \ldots \cdot p_n$$

and

$$\frac{a}{p_1} = q_1 \cdot \ldots \cdot q_{j-1} \cdot q_{j+1} \cdot \ldots \cdot q_m$$

By the induction assumption, the two factorizations of a/p_1 are the same. That is, the set of primes $\{p_2, \ldots, p_n\}$ is the same as the set of primes $\{q_1, \ldots, q_{j-1}, q_{j+1}, \ldots, q_m\}$. But since $p_1 = q_j$, the set of primes

$$\{p_1, p_2, \ldots, p_n\}$$

is then the same as the set of primes

$$\{q_1, \ldots, q_{j-1}, q_j, q_{j+1}, \ldots, q_m\}.$$

But then the two factorizations of a are the same, and the result is true for the number a. That would prove the theorem by complete induction.

We prove that if $p_1 \cdot \ldots \cdot p_n = q_1 \cdot \ldots \cdot q_m$, then $p_1 = q_j$ for some j by means of the following important lemma:

Lemma 3. *If p is prime and p divides bc, then p divides b or p divides c.*

Proof. We use Corollary 8 of Chapter 3, an application of Bezout's Identity, namely: if a divides bc and $(a,b) = 1$, then a divides c.

Suppose p is a prime and p divides bc. Since p is prime then either p divides b, or $(p,b) = 1$. If $(p,b) = 1$, then by the corollary, p divides c. □

From Lemma 3 it follows by induction (see Exercise 2, below) that if a prime divides a product of m numbers it must divide one of the factors.

To complete the proof of uniqueness of factorization, suppose we have $p_1 \cdot p_2 \cdot \ldots \cdot p_n = q_1 \cdot \ldots \cdot q_m$. Then p_1 divides $q_1 q_2 \cdot \ldots \cdot q_m$. Since p_1 is prime, p_1 must divide one of the q's, say q_j. Since q_j is prime, q_j is divisible by only itself and 1. Since $p_1 \neq 1, p_1 = q_j$.

Thus the induction argument described above for proving uniqueness of factorization can always be used, and the proof of uniqueness of factorization is complete. □

The proof of the Fundamental Theorem of Arithmetic depends on the property that if a prime number p divides a product bc of two numbers b, c then p divides b or p divides c.

In turn, assuming that the Fundamental Theorem of Arithmetic is true, the statement about primes follows. For if p divides bc then $bc = pf$ for some number f. If we factor f into a product of primes, then bc has a factorization into a product of primes, one of which is p. On the other hand, if p does not divide b and p does not divide c, then the prime factorizations of b and of c would not include p, so bc has a factorization (namely, the product of the factorizations of b and of c) that does not include p. Thus bc would have two factorizations, one including p, one excluding p. This would violate the Fundamental Theorem.

The problem of factoring a natural number n into a product of primes is much harder in practice than the problem of finding the greatest common divisor of two numbers. One needs to find divisors of n, and that is often difficult. The naive approach is simply to try to find prime divisors of n by trial division.

For example, to factor 3372, we first see that 2 is a divisor of 3372: $3372 = 2 \cdot 1686$. Then we see that 2 is a divisor of 1686: $1686 = 2 \cdot 843$, so $3372 = 2 \cdot 2 \cdot 843$. Then we see that 3 divides 843: $843 = 3 \cdot 281$, so $3372 = 2 \cdot 2 \cdot 3 \cdot 281$. Finally, we check that 281 is not divisible by 2, 3, 5, 7, 11 or 13, and observe that $17 > \sqrt{281}$, so 281 must be prime by Exercise 5, below, and we have the factorization of 3372 into a product of primes,

$$3372 = 2 \cdot 2 \cdot 3 \cdot 281.$$

There are obvious tricks for testing a number n for divisibility by 2, 3, or 5. Later we shall see tests for 7, 11, and 13. In general, however, unless n happens to be prime it is a slow process looking for divisors. Consider, for example, trying $92263 = 257 \cdot 359$ by hand. Even with methods that are much more efficient than trial division, it was claimed in 1977 that to factor a certain 129 digit number N (known as RSA-129) that was the secret product of a 64-digit prime number and a 65-digit prime number, using the best methods and computers then available, would take about 40×10^{15} years. By comparison, finding the greatest common divisor of two 129-digit numbers would take under a second, as we showed in Section 3F.

The slowness of all known methods for factoring large numbers is the basis for the effectiveness of a remarkable application of number theory to secret codes, RSA cryptography. See Chapter 10A, below. Because of the relationship between factoring large numbers and cryptography, intensive research has taken place since 1977 on the problem of factoring large numbers, involving both theory and computer hardware. As a result of this effort, by 1994 there was enough progress in factoring algorithms and computing power that A. Lenstra of Bellcore and a team of 600 volunteers were able to factor RSA-129 using hundreds of computers on six continents over a period of eight months. (For their efforts they won a $100 prize and newspaper stories worldwide.)

We will examine some factoring methods in later chapters.

Exercises.

1. Prove (without going back to Chapter 3) that if a divides bc and $(a,b) = 1$, then a divides c.

2. Prove by induction that if a prime number p divides $a_1 \cdot a_2 \cdot \ldots \cdot a_n$, then p must divide one of the factors a_j.

3. Prove that a and b are coprime if and only if no prime number divides both a and b.

4. Let n, q be numbers ≥ 2. Show that for every number r, $(n, q^r) = 1$ if and only if $(n,q) = 1$.

5. Show that if n is not prime, n has a prime divisor $< \sqrt{n}$.

6. Is 2021 prime?

7. Let $2\mathbb{N}$ denote the even integers > 0. Say that a number a in $2\mathbb{N}$ is *irreducible* if there are no numbers b, c in $2\mathbb{N}$ so that $a = bc$.

(i) Show that if n is an odd number, then $2n$ is in $2\mathbb{N}$ and is irreducible. Conversely, show that every irreducible number in $2\mathbb{N}$ is twice an odd number.

(ii) Show that every number a in $2\mathbb{N}$ factors into a product of irreducible numbers in $2\mathbb{N}$.

(iii) Show that factorization of numbers in $2\mathbb{N}$ into products of irreducibles in $2\mathbb{N}$ is not unique.

(iv) Show that the analogue of Lemma 3 fails in $2\mathbb{N}$.

8. Let $3\mathbb{N}$ denote the numbers > 0 that are multiples of 3. Say that a number a in $3\mathbb{N}$ is irreducible if there are no numbers b, c in $3\mathbb{N}$ so that $a = bc$.

(i) Characterize the irreducible numbers in $3\mathbb{N}$.

(ii) Show that every number a in $3\mathbb{N}$ factors into a product of irreducible numbers in $3\mathbb{N}$.

(iii) Show that factorization of numbers in $3\mathbb{N}$ into products of irreducible numbers in $3\mathbb{N}$ is not unique.

9. Let $\mathbb{Z}[\sqrt{-23}]$ denote the set of complex numbers of the form $a + b\sqrt{-23}$, where a and b are integers.

(i) Show that every element of $\mathbb{Z}[\sqrt{-23}]$ may be written uniquely in the form $a + b\sqrt{-23}$ for integers a and b.

(ii) Verify that $3 \cdot 3 \cdot 3 = (2 + \sqrt{-23})(2 - \sqrt{-23})$.

(iii) Verify that $2 + \sqrt{-23}$ and $2 - \sqrt{-23}$ are not multiples of 3 in $\mathbb{Z}[\sqrt{-23}]$.

(iv) Show that 3 is a "prime" in $\mathbb{Z}[\sqrt{-23}]$ in the sense that the only elements of $\mathbb{Z}[\sqrt{-23}]$ that divide 3 are 3, -3, 1 and -1. Thus Lemma 3 fails in $\mathbb{Z}[\sqrt{-23}]$.

B. Exponential Notation

In writing the prime factorization of a number a it is convenient to collect together the various prime factors in increasing order and use exponential notation. Thus instead of writing 144 as $2 \cdot 2 \cdot 2 \cdot 2 \cdot 3 \cdot 3$, we can write it as $2^4 3^2$. Other examples:

$$975 = 3 \cdot 5^2 \cdot 13 = 2^0 \cdot 3^1 \cdot 5^2 \cdot 7^0 \cdot 11^0 \cdot 13^1,$$

$$1000 = 2^3 \cdot 5^3 = 2^3 \cdot 3^0 \cdot 5^3$$

$$3372 = 2^2 \cdot 3 \cdot 281.$$

The factorization of 975 illustrates that we can include in the factorization primes that do not actually divide the number a, as long as we give them the exponent zero.

In general, we write the number a as

$$a = p_1^{e_1} p_2^{e_2} \cdots p_r^{e_r}.$$

Uniqueness of factorization says that there is only one way to write a number a in this way, except for the inclusion of extra primes with exponent zero.

Here are three applications of exponential notation and the uniqueness of the exponents given by the Fundamental Theorem of Arithmetic: to irrationality, to divisibility, and to least common multiples.

I. Irrationality. In Chapter 2, Exercise 32, we proved that $\sqrt{2}$ is irrational, that is, there is no rational number $\frac{a}{b}$ so that $\sqrt{2} = \frac{a}{b}$, using infinite descent (well-ordering). Using the Fundamental Theorem of Arithmetic we get such results easily. For example:

Theorem 4. $\sqrt{3}$ *is irrational.*

Proof. If not, $\sqrt{3} = a/b$, a, b natural numbers. Multiplying both sides by b and squaring, we get $3b^2 = a^2$. Let 3^e be the power of 3 appearing in the factorization of a, and 3^f the power of 3 appearing in the factorization of b. Then since $3b^2 = a^2$, we have $2f + 1 = 2e$. But the left side of this equation is odd, and the right side even, impossible. $\square$

Exercises.

10. Prove that if p is a prime number, then $\sqrt{p}$ is irrational, using an argument like that for $\sqrt{3}$.

11. Prove that if p is a prime number, then $\sqrt{p}$ is irrational, by writing $\sqrt{p} = a/b$ for a and b integers, then writing $pb^2 = a^2$ and counting the number of primes (including repetitions) on the left and right sides. (Would that argument work for $\sqrt{22}$?)

12. (i) Prove that the natural number a is a cube iff the exponent of each prime factor of a is a multiple of 3.

(ii) Prove that if the natural number a is not a cube, then $a^{1/3}$ is irrational.

13. Show that $(100)^{1/5}$ is irrational.

14. (i) If a, b are natural numbers, $(a, b) = 1$, and ab is a square, show that a and b are squares.

(ii) If a and b are integers with $(a, b) = 1$ and ab is a square, is a necessarily a square?

15. If a, b are integers with $(a, b) = 1$ and $ab = c^r$ where r is an odd integer ≥ 1, show that both a and b are rth powers.

II. Divisibility. We can interpret notions of divisibility in terms of exponential notation. Suppose

$$a = p_1^{e_1} p_2^{e_2} \cdots p_n^{e_n}$$

and

$$b = p_1^{f_1} p_2^{f_2} \cdots p_n^{f_n}$$

where $p_1, \ldots, p_n$ include all primes that divide either a or b, and some of the exponents e_i or f_i may be zero.

Proposition 5. *With a, b as above, a divides b iff $e_i \leq f_i$ for all $i = 1, \ldots, r$.*

Proof. If for all $i = 1, \ldots, r$ we have $e_i \leq f_i$, then $c_i = f_i - e_i \geq 0$. Hence

$$q = p_1^{c_1} p_2^{c_2} \cdots p_n^{c_n}$$

is an integer and $b = aq$.

Conversely, if a divides b, then $b = aq$ for some natural number q. Then every prime that divides q also divides b. Write q as a product of primes, as above. Then $c_i \geq 0$ for $i = 1, \ldots, r$, and $aq = b$ means that $e_i + c_i = f_i$ for each i, hence $e_i \leq f_i$. $\qquad \square$

Using Proposition 5, we can see easily that the greatest common divisor of two numbers a and b has a prime factorization in which the exponent of each prime p is the smaller of the exponents of p in a and in b. For if e is a common divisor of a and b, then for each prime p dividing e, the exponent of p in e must be $\leq$ the exponent of p in a, and $\leq$ the exponent of p in b. If d is the greatest common divisor, then the exponent of p in d must be as large as possible, hence must equal the smaller of the exponents of p in a and in b.

It is convenient to use the notation $p^e \| a$ if p^e is the power of p in the prime factorization of a. Thus, $p^e \| a$ if p^e divides a but p^{e+1} does not. Using this notation, a divides b if and only if for all primes p, if $p^e \| a$ and $p^f \| b$, then $e \leq f$. The greatest common divisor (a, b) of two numbers has the property that if $p^e \| a$ and $p^f \| b$, then $p^{\min\{e,f\}} \| (a, b)$.

Exercises.

16. Find the greatest common divisor of $2^7 3^2 5^6$ and $2^4 3^5 5^6 7$.

17. Find the greatest common divisor of $2^2 3^2 4^5 5^6 6^5$ and $2^4 3^5 4^3 5^3 6^7$.

18. If $(a, b) = p^3$, p a prime, what is (a^2, b^2)?

19. If $(a, b) = 8$, what are the possible values of (a^4, b^5)?

20. Prove that $(a, b) = 1$ if and only if no prime divisor of a divides b.

21. Prove that if $(a, c) = 1$ and $(b, c) = 1$ then $(ab, c) = 1$.

The next five exercises are exercises from earlier sections that you can now try using the exponential description of the greatest common divisor:

22. Show that (a, m) divides (a, mn) for every a, m, n, by showing that for each prime p, if $p^e \| a$, $p^s \| m$ and $p^t \| n$, then $\min\{e, s\} \leq \min\{e, s + t\}$.

23. Show that if $(a, b) = 1$ and $c | a$, then $(c, b) = 1$.

24. Show if $(a, m) = d$ and $(b, m) = 1$, then $(ab, m) = d$.

25. Show that $(ma, mb) = m(a, b)$ by showing that for each prime p, if $p^e \| a$, $p^f \| b$ and $p^t \| m$, then $t + \min\{e, f\} = \min\{t + e, t + f\}$.

26. Show that if $(a, b) = 1$, then $(n, ab) = (n, a)(n, b)$.

27. Prove that if $a | bc$ and $(a, b) | c$, then $a | c^2$.

28. Prove that if $(a, b) = 1$ and c is any integer, then there is some integer m so that $(a + bm, c) = 1$.

29. Each of the following three statements is claimed to be true for all natural numbers a, b, c, m. In each case, prove the statement or give an example to show it is false:

(i) If d is the greatest common divisor of a and b, then the greatest common divisor of a and mb is md.

(ii) If a divides bc and a doesn't divide b, then a divides c.

(iii) If d is the greatest common divisor of a and b, then d^3 is the greatest common divisor of a^3 and b^3.

III. Least Common Multiples. Given natural numbers a, b, a number $m > 0$ is a *common multiple* of a and b if $m = ar$ for some natural number r, and also $m = sb$ for some natural number s. In terms of divisibility, m is a common multiple of a and b if a divides m and b divides m. One example of a common multiple of a and b is their product, ab. If we consider the two numbers 12 and 30, then $12 \cdot 30 = 360$ is a common multiple, but so are 180, 240, 720, 3600, 60, 120, 2400, etc. In fact, there are infinitely many common multiples of 12 and 30.

Any two integers a and b have infinitely many common multiples. The least common multiple of a and b is the smallest positive number in the set of common multiples of a and b.

It's easy to check that the least common multiple of 12 and 30 is 60.

The least common multiple of two natural numbers a and b exists as a consequence of the well-ordering principle (see Section 2C). The set S consisting of all positive common multiples of a and b is a subset of the natural numbers, and is non-empty because the number ab is in S. So by well-ordering, S has a least element, namely, the least common multiple of a and b.

We denote the least common multiple of a and b by $[a, b]$.

Some other examples: $[4, 6] = 12, [4, 7] = 28, [15, 20] = 60, [7, 14] - 14$.

The least common multiple of two numbers a and b arises in connection with adding fractions. Suppose you wish to add $1/6$ and $1/10$. To do so, you need to find a common denominator, for example, 60:

$$\frac{1}{6} + \frac{1}{10} = \frac{10}{60} + \frac{6}{60} = \frac{10 + 6}{60} = \frac{16}{60} = \frac{4}{15}$$

Any common multiple of the denominators of two fractions will be a common denominator. The least common multiple of the denominators is the least common denominator.

In this example, if we choose the smaller common denominator $30 = [6,10]$, we get:

$$\frac{1}{6} + \frac{1}{10} = \frac{5}{30} + \frac{3}{30} = \frac{5+3}{30} = \frac{8}{30} = \frac{4}{15}.$$

Both denominators 30 and 60 can be used, but 30 keeps the numbers in the computation smaller.

Here is how to find the least common multiple.

Proposition 6. *a) The least common multiple of a and b is the product divided by the greatest common divisor. In symbols:*

$$[a,b] = \frac{ab}{(a,b)}.$$

b) The least common multiple of a and b divides every common multiple of a and b.

Proof. We first show both parts if a and b are coprime. Clearly ab is a common multiple of a and b. We now show that any common multiple of a and b is a multiple of ab. Suppose $m > 0$ is a common multiple of a and b. Then $m = as$ for some number $s > 0$, and b divides $m = as$. But then, since a and b are coprime, b divides s, so $s = bt$ for some integer $t > 0$. Thus $m = as = abt$. Since t is a positive integer, $m \geq ab$ and m is a multiple of ab. Thus ab is the least common multiple of a and b, and ab divides any other common multiple. The proposition is true if $(a,b) = 1$.

Now suppose the greatest common divisor of a and b is d. Let $a = da'$, $b = db'$. Then a' and b' are coprime (since if $ar + bs = d$ from Bezout's identity, then $a'r + b's = 1$). So the least common multiple of a' and b' is $a'b'$, and any common multiple of a' and b' is $a'b't$ for some positive integer t.

Now $\frac{ab}{d} = a'b'd = ab' = a'b$, so is a common multiple of a and b. We show that any common multiple of a and b is a multiple of $a'b'd$. Suppose $m > 0$ is a common multiple of a and b. Then m is a multiple of d, so write $m = dm'$ for some number m'. Then a divides m, so a' divides m'; also b divides m, so b' divides m'. Now $(a',b') = 1$, so by the first part of the proof, $a'b'$ divides m'. But then

$$\frac{ab}{d} = da'b' \text{ divides } dm' = m.$$

Thus $\frac{ab}{d}$ divides every common multiple of a and b, so is the least common multiple of a and b. □

Here is an alternate proof, only using the Division Theorem, of part b) of this proposition.

Proposition 7. *The least common multiple of a and b divides every common multiple of a and b.*

Proof. Let m be the least common multiple of a and b, and suppose $n > 0$ is any common multiple of a and b. Write $n = mq + r$ with $0 \leq r < m$, using the division theorem.

Since a divides m and a divides n, then a divides $n - mq = r$.

Since b divides m and b divides n, then b divides $n - mq = r$.

Thus r is a common multiple of a and b. But m was the least positive integer that is a common multiple of a and b, and $r < m$ Thus r must be 0. That means, m divides n.   □

Here is a proof of part a) that uses exponential notation.

Proposition 8. $[a,b] = \frac{ab}{(a,b)}$.

Proof. Let p be a prime number that divides a or b. Let $p^e \| a, p^f \| b$. Then

$$p^{\max\{e,f\}} \| [a,b] \text{ and } p^{\min\{e,f\}} \| (a,b).$$

The formula $[a,b](a,b) = ab$ then follows from the easily verified relation

$$e + f = \max\{e,f\} + \min\{e,f\}.$$

□

This last argument shows how to find the least common multiple of two numbers a and b given their factorizations into products of primes. However, the formula $[a,b] = ab/(a,b)$ has the virtue that we don't need to be able to factor a and b to find $[a,b]$–we only need to find the greatest common divisor, using Euclid's Algorithm.

Exercises.

30. Find the least common multiple of
   (i) 96 and 240
   (ii) 210 and 126
   (iii) 72 and 105

31. (i) Show that if a and b are coprime and you add $1/a$ and $1/b$ by using the common denominator ab, the resulting fraction $\frac{a+b}{ab}$ is reduced.

(ii) Show that if a and b are not coprime, and you add $1/a$ and $1/b$ by using the common denominator ab, the resulting fraction $\frac{a+b}{ab}$ is not reduced. (Can you say anything about the sum of $1/a$ and $1/b$ if $(a,b) > 1$ and you use the common denominator $[a,b]$?)

32. Prove that $r[a,b] = [ra,rb]$ for all positive integers r, a, b.

33. Show that $[a,m] = m$ if and only if a divides m.

34. Show that if e divides g and f divides h, then
   (i) $[e,f]$ divides $[g,h]$, and
   (ii) (e,f) divides (g,h).

35. Suppose that $ar + bs = c$ for some integers r, s. Show that the general solution of the equation $ax + by = c$ is $x = r + n[a,b]/a$, $y = s - n[a,b]/b$ for all integers n.

36. Find all solutions of $117x + 1001y = 65$.

37. Find the least common multiple of

$$6^2 7^3 8^4 9^5 10^6 \text{ and } 6^5 7^6 8^2 9^3 10^4.$$

38. Find the smallest $k > 0$ so that
(i) 15 divides $10k$;
(ii) 15 divides $11k$;
(iii) 15 divides $12k$.

39. Show that the smallest $k > 0$ so that a divides bk is $k = [a,b]/b$.

40. If a and b are two natural numbers so that $(a,b) = 10, [a,b] = 100$, what can a and b be?

41. Given natural numbers d and m, show that there are natural numbers a and b so that $(a,b) = d$ and $[a,b] = m$, if and only if d divides m.

42. How should we define $[a,b,c]$? Is $[a,b,c] = abc/(a,b,c)$? Explain.

C. Primes

Prime numbers have been of continuing interest in mathematics since the time of Pythagoras, 500 B.C. In the remainder of this chapter we survey some of the most famous results about prime numbers.

I. Euclid's Theorem. We showed in Section 4A, Theorem 1, that every natural number (except 1) is a product of primes. Thus an obvious question to ask is, how many primes are there?

The ancient Greeks found the answer–it is in Euclid (Book IX, Proposition 20 of the Elements (300 B.C.)):

Theorem 9. *There are infinitely many primes.*

Proof. (Euclid) Suppose the set of primes is finite in number: suppose $p_1, p_2, \ldots, p_r$ are all the primes. Consider the number $m = p_1 p_2 \cdots p_r + 1$. It must have a prime divisor q. If q were one of the primes $p_1, p_2, \ldots, p_r$, then q would divide the product $p_1 p_2 \cdots p_r = m - 1$, and so q would divide $m - (m - 1) = 1$, impossible. So q cannot be one of the primes $p_1, p_2, \ldots, p_r$, and so must be a new prime. This contradicts the assumption that $p_1, \ldots, p_r$ were all the primes. So the number of primes cannot be finite. $\qquad\square$

One way to show that there are infinitely many primes is to define an infinite sequence of numbers that must be prime. It is common to define an infinite sequence of numbers by some kind of inductive process: once you have defined the sequence up to the nth element of the sequence, you define the $(n+1)$st element of the sequence in terms of the elements of the sequence you have previously defined. The Fibonacci sequence (Section 3F) is an example of an inductively defined sequence of natural numbers, as is the sequence $f(n) = n!$, defined inductively by $f(0) = 0! = 1$, $f(n+1) = (n+1)! = n!(n+1) = f(n)(n+1)$.

Here is a general procedure to get an infinite sequence of primes:

Proposition 10. *Suppose given an infinite sequence of numbers*

$$a_1, a_2, \ldots, a_n, \ldots$$

with the property that for each $m \neq n$, a_n and a_m are coprime. For each $n \geq 1$, let p_n be the smallest prime factor of a_n. Then the sequence

$$p_1, p_2, \ldots$$

is an infinite sequence of distinct prime numbers.

Proof. If $m \neq n$, $p_n \neq p_m$, for otherwise a_m and a_n would not be coprime.   □

Example 1. Related to the idea of Euclid's proof, let $a_1 = 2$ and for all $n > 1$, let $a_n = a_1 a_2 \cdot \ldots \cdot a_{n-1} + 1$. Then for all $m \neq n$, a_m and a_n are coprime.

Fermat numbers. Some mathematicians have tried to improve on the strategy of the last proposition: they wished to find a simple function on the natural numbers whose values were distinct primes. This idea goes back at least to Fermat (1630's), who proposed

$$F(n) = 2^{2^n} + 1$$

as such a function. The numbers $F(n)$ are called Fermat numbers. Here are the first five. They are all prime:

$$F(0) = 2 + 1 = 3,$$
$$F(1) = 2^2 + 1 = 5,$$
$$F(2) = 2^4 + 1 = 17,$$
$$F(3) = 2^8 + 1 = 257,$$
$$F(4) = 2^{16} + 1 = 65537.$$

Based on the evidence that the first five Fermat numbers are prime, and other evidence, Fermat conjectured that $F(n)$ was prime for all n. This conjecture turned out to be one of the least accurate famous conjectures in the history of mathematics. It was first disproved by Euler in 1732, who showed that

$$F(5) = 2^{32} + 1 = 4294967297$$

factors as $641 \cdot 6700417$. (See Section 10B.) Then in 1880 the otherwise obscure mathematician Landry factored

$$F(6) = 2^{64} + 1 = 274177 \cdot 67280421310721,$$

a remarkable feat without a computer.

It was known as of April, 2007, that $F(n)$ is composite at least for $5 \leq n \leq 32$. No Fermat number $> F(4)$ has been found to be prime. (See Wilfrid Keller's web page, www.prothsearch.net/fermat.html, for recent information on the factoring of Fermat numbers.)

Beyond Fermat's conjecture, Fermat numbers are interesting for at least three reasons.

(a) Fermat numbers relate to geometric constructions by straightedge and compass. Gauss, around 1800, proved that a regular polygon of n sides can be constructed with straightedge and compass if and only if $n = 2^r . p_1 p_2 \cdots p_n$ where $p_1, p_2, \ldots, p_n$ are distinct prime Fermat numbers. Thus knowing if $F(n)$ is prime is of (at least theoretical) geometric interest.

(b) Fermat numbers grow large very quickly with n. The Fermat number $F(n)$ has approximately $3 \cdot 2^n / 10$ decimal digits. Thus $F(7)$ has 39 digits, $F(8)$ 78 digits, $F(9)$ 155 digits. They have no obvious small prime factors, so showing they are not prime, and finding factors of them, is a substantial challenge. This challenge has led to the discovery and application of new factoring methods during the past 30 years.

Here are some references:

Morrison and Brillhart (1975) developed a new factoring algorithm and used it to factor $F(7)$.

Brent and Pollard (1981) developed an improved factoring algorithm that they used to find the smallest prime factor of $F(7)$ in just under 7 hours on a UNIVAC, and the smallest prime factor of $F(8)$ in 2 hours.

Lenstra, Lenstra, Manasse, and Pollard (1993) applied a new algorithm, the number field sieve, to find the three factors of $F(9)$.

Young and Buell (1988) proved that the Fermat number $F(20)$ is composite. $F(20)$ has just over one million binary digits, or 315,653 decimal digits. The computation was done on a Cray-2 supercomputer and took about 10 days. It was verified on another Cray in 82 hours. Both computations were done via single-processor computer programs. (In fact, as the authors state, "The program itself was very simple and only about 200 lines long, much of which was used for checkpointing and restarting the program.") One objective of the computation was to verify the hardware reliability of the computer used. The authors conclude their report, "We remark that this 10-day computation on a supercomputer may well be the largest computation ever performed whose result is a single bit answer. Never have so many circuits labored for so many cycles to produce so few output bits."

The Young-Buell computation, which was listed in the 1993 Guinness Book of Records, may remain one of the longest single-processor computations in the history of number theory. Recent approaches to factoring large numbers, and in particular, Fermat numbers, use parallel processing, so doing large scale computations by single processor is fast becoming obsolete.

As of 2008, the largest known prime factor of a Fermat number was

$$3 \cdot 2^{2478785} + 1,$$

a 746190-digit factor of F(2478782).

(c) The Fermat numbers, although not always prime, are nonetheless pairwise coprime. This property was observed by G. Polya, and yields another proof that there are infinitely many primes, by Proposition 10, above. We simply let p_n, be the smallest prime factor of $F(n)$, for each n. Then the sequence $p_1, p_2, \ldots$ is an infinite sequence of primes, and so there are infinitely many primes.

Here is a proof of Polya's observation:

Proposition 11. *If $m \neq n$, then $F(m)$ and $F(n)$ are coprime.*

Proof. Let $m < n$. We'll start Euclid's algorithm with $F(m)$ and $F(n)$. Write $r = n - m$, then $2^n = 2^m \cdot 2^r$. Let $a = 2^{2^m}$, so that $a + 1 = F(m)$. Now

$$F(n) - 2 = 2^{2^n} - 1 = (2^{2^m})^{2^r} - 1 = a^{2^r} - 1$$
$$= (a-1)(1 + a + a^2 + a^3 + \ldots + a^{2^r - 2} + a^{2^r - 1}).$$

Pair off the terms in the right factor as follows:

$$= (a-1)[(1+a) + (a^2 + a^3) + \ldots + (a^{2^r - 2} + a^{2^r - 1})]$$
$$= (a-1)[(1+a) + a^2(1+a) + \ldots + a^{2^r - 2}(1+a)],$$

which is a multiple of $1 + a = F(m)$. Thus $F(n) - 2 = F(m)q$ for some number q, so dividing $F(n)$ by $F(m)$ leaves a remainder of 2. Since $F(m)$ and $F(n)$ are odd, therefore the greatest common divisor of $F(m)$ and $F(n)$ is 1. $\square$

Taking p_n to be the smallest prime factor of $F(n)$ for each n, the first 14 terms of the sequence $\{p_n\}$ (taken from Brent and Pollard (1981)) are:

$$p_0 = 3 \qquad p_7 = 59649589127497217$$
$$p_1 = 5 \qquad p_8 = 1238926361552897$$
$$p_2 = 17 \qquad p_9 = 2424833,$$
$$p_3 = 257 \qquad p_{10} = 45592577$$
$$p_4 = 65537 \qquad p_{11} = 319489$$
$$p_5 = 641 \qquad p_{12} = 114689$$
$$p_6 = 274177 \qquad p_{13} = 2710954639361$$

As of 2008, F_{14} is known to be composite, but no factors are known. Thus p_{14} is unknown.

We will consider other large primes later in the book.

Exercises.

43. Use the ideas of Euclid's proof to prove that there are infinitely many primes of the form $4n - 1$. (Hint: Consider $4p_1 \cdots p_r - 1$.)

44. Try the same for numbers of the form $6n - 1$.

45. Define a sequence of numbers inductively, as follows: Let $a_1 = 2$, and for each $n > 1$, define $a_{n+1} = a_n(a_n - 1) + 1$. Prove that

$$a_n = a_1 \cdot a_2 \cdots a_{n-1} + 1.$$

Prove that for all $m \neq n$, a_m and a_n are coprime.

46. Here is another proof of Polya's theorem.
  (i) Show that $F(m+1) = F(m)(F(m) - 2) + 2$ for all $m \geq 0$.
  (ii) Show that
$$F(m+1) = F(0)F(1) \cdots F(m) + 2.$$

(iii) Use (ii) to show that $F(m)$ and $F(n)$ are coprime if $m \neq n$.

47. Let $a_1 \geq 2$, and for every $n > 1$, define a_n by $a_n = a_{n-1}(a_{n-1} + 1)$. Show that for all n, a_n is divisible by at least n distinct primes. [Saidak (2006)]

48. Prove by induction that $F(n) - 2 = 2^{2^n} - 1$ is divisible by at least n distinct primes, thereby giving another proof that there are infinitely many primes.

II. Primes in an interval. How many primes are less than some given number n? This question was studied throughout the nineteenth century by some of the greatest mathematicians of the century, including Gauss and Riemann.

Definition. Let $\pi(x)$ be the function defined for all real numbers $x > 0$ by $\pi(x) = $ the number of prime numbers $\leq x$.

The sequence of primes begins

$$2, 3, 5, 7, 11, 13, 17, 19, 23, 29, 31, 37, 41, 43, 47, \ldots$$

so $\pi(3) = 2, \pi(10) = 4, \pi(\sqrt{200}) = 6, \pi(19) = 8, \pi(50) = 15$.

The Prime Number Theorem, proved in 1896 by Hadamard and de la Vallee Poussin, is

$$\lim_{x \to \infty} \frac{\pi(x)}{x / \ln(x)} = 1.$$

Here $\ln(x)$ is the natural logarithm of x.

For estimating how many primes there are of a given size, fairly precise numerical results have been found: Chebyshev in 1850 proved that $\pi(x) < 1.10555(x/\ln x)$ for all x, and in 1962 Rosser and Schoenfeld proved that $x/\ln(x) \le \pi(x)$ for all $x > 17$. Thus the number of primes $< x$ is squeezed between two computable quantities: for all $x > 17$,

$$\frac{x}{\ln(x)} \le \pi(x) \le (1+\varepsilon)\frac{x}{\ln(x)}$$

where $1 + \varepsilon < 1.10555$ for all x and ε approaches 0 as x goes to infinity.

Dividing these inequalities by x yields

$$\frac{1}{\ln(x)} \le \frac{\pi(x)}{x} \le (1+\varepsilon)\frac{1}{\ln(x)}$$

This says that on average, one of every $\ln x$ numbers less than x is prime. If $x = 10^r$, then on average, one of every $\ln 10^r = r\ln 10$ numbers is prime.

For example, if we let $x = 10^{10}$, then since $\ln 10 = 2.3026$, among all the numbers less than 10^{10}, one of every $10\ln(10) = 23$ numbers is prime. Setting $x = 10^{80}$, among numbers of 80 digits or less, one of every $80\ln 10 = 184$ numbers is prime. Setting $x = 10^{100}$, among numbers of 100 digits or less, one of every $100\ln 10 = 230$ numbers is prime.

These results (whose proofs are well beyond the scope of this book) are much stronger than simply indicating that there are infinitely many primes. They show that large primes are not at all scarce. Thus if we need to have some prime numbers of around 80 digits (we'll show later that such primes are of practical value) and have a method for quickly checking whether a given large number is prime or not (we'll show later that such a method exists), then if we randomly select numbers of 80 digits or less, we should expect that about 1 of every 184 numbers we select will in fact be prime.

Suppose however that instead of looking for primes of 80 digits or less, we want to find primes of exactly 80 digits. Primes are not uniformly distributed among numbers–primes are more dense among small numbers than among large numbers. For example, there are more primes between 0 and 100 (25 primes) than there are between 5000 and 5100 (12 primes). (A list of the first 1000 primes may be found at primes.utm.edu) So we need to ask if the ratio $\pi(10^r)/10^r$ is reasonably close to the ratio of primes of exactly r digits among numbers of exactly r digits.

As an example, we look at $r = 10$. How much more dense are primes among the 10^{10} numbers of 10 or fewer digits than among the $9 \cdot 10^9$ numbers of exactly 10 digits?

We can answer this precisely, because $\pi(10^9)$ and $\pi(10^{10})$ are known:

$$\pi(10^9) = 50,847,534,$$

while

$$\pi(10^{10}) = 455,052,511,$$

and so

$$\pi(10^{10})/10^{10} = .0455$$

while
$$\pi(10^9)/10^9 = .0508.$$

The number of 10 digit primes, divided by the number of 10 digit numbers, is

$$\frac{\pi(10^{10}) - \pi(10^9)}{9 \cdot 10^9} = .0449,$$

Thus at least among 10 digit numbers, the density of primes for numbers of 10 digits or less (namely, .0455) is very close to the density of primes of exactly 10 digits (namely .0449).

We can use the Chebyshev and Rosser-Schonfeld results to get a lower bound on the proportion of primes among all numbers of exactly r digits. There are $9 \cdot 10^{r-1}$ numbers of exactly r digits. We have

$$\pi(10^r) > \frac{10^r}{\ln 10^r} \text{ for } r > 2$$

(Rosser-Schonfeld), and

$$\pi(10^{r-1}) < (1+\varepsilon)\frac{10^{r-1}}{\ln 10^{r-1}} \text{ for all } r$$

(Chebyshev). So

$$\frac{\pi(10^r) - \pi(10^{r-1})}{9 \cdot 10^{r-1}} \geq \frac{1}{9 \cdot 10^{r-1}}\left(\frac{10^r}{r\ln 10} - \frac{(1+\varepsilon)10^{r-1}}{(r-1)\ln 10}\right)$$

$$= \frac{1}{r\ln 10}\left(\frac{10}{9} - \frac{(1+\varepsilon)r}{9(r-1)}\right)$$

$$= \frac{1}{r\ln 10}C$$

where

$$C = \frac{10}{9} - \frac{(1+\varepsilon)(r-1)}{9(r-1)}$$

$$= \frac{10}{9} - \frac{(1+\varepsilon)(r-1)}{9(r-1)} - \frac{(1+\varepsilon)}{9(r-1)}$$

$$= \left(\frac{9-\varepsilon}{9} - \frac{1+\varepsilon}{9(r-1)}\right).$$

Using Chebyshev's upper bound $\varepsilon \leq .10555$, we have

$$1 > C > \frac{9 - .10555}{9} - \frac{1.10555}{9}\frac{1}{r-1}$$

$$= .9883 - .1228\frac{1}{r-1}.$$

From this we see that we should be able to find primes with a given large number of digits without too much work. For example, we find that $C \geq .987$ for $r \geq 100$, and so the proportion of primes among 100 digit numbers is at least $\frac{.987}{100 \ln 10} = .00429 > \frac{1}{234}$. On average, at least 1 of every 234 numbers of exactly 100 digits is prime.

There are accessible proofs that there exist constants A and B with $Ax/\ln x < \pi(x) < Bx/\ln x$. See Zagier (1977).

Exercises.

49. Prove that for every n there exist n consecutive natural numbers none of which are prime. (Hint: Start with $(n+1)!+2$.)

50. Prove that for every n there exists a prime p with $n < p \leq n!+1$.

51. Use the Chebyshev and Rosser-Schonfeld estimates to prove that for all $n > 17$,

$$\pi(2n) - \pi(n) > 1.$$

Then prove Bertrand's Postulate: for every $n > 1$, there is a prime p with $n < p < 2n$.

Chapter 5
Congruence

This chapter is devoted to defining and studying Gauss's useful notion of congruence for integers.

A. Congruence Modulo m

Congruence is related to the notion of divisibility.

Definition. Two integers a and b are *congruent modulo m*, written

$$a \equiv b \pmod{m}$$

if m divides $a - b$, or equivalently, if $b = a +$ some multiple of m.

A special case is that a number a is congruent to $0 \pmod{m}$, written $a \equiv 0 \pmod{m}$, if and only if m divides a. But the value of the congruence notation is not in providing an alternate for the notation $m|a$, but in providing a highly suggestive notation to use in place of $m|(a - b)$.

Here are some examples of congruences that you may verify:

$$143 \equiv 0 \pmod{13}$$
$$143 \equiv 13 \pmod{10}$$
$$-114 \equiv 7 \pmod{11}$$
$$4726 \equiv 1 \pmod{9}$$
$$-35 \equiv -335 \pmod{6}.$$

In a congruence mod m, the number m is called the *modulus* (plural: "moduli"). Any two integers are congruent modulo 1, so the modulus 1 is not of much interest. For this reason the modulus will normally be ≥ 2.

L.N. Childs, *A Concrete Introduction to Higher Algebra*, Undergraduate Texts in Mathematics, © Springer Science+Business Media LLC 2009

The set of integers to which the integer a is congruent modulo m is the set $\{a+mk \mid k \text{ any integer}\}$. For example, the integers congruent to 5 modulo 11 are the integers of the form

$$5+11k, k = \ldots, -3, -2, -1, 0, 1, 2, 3, \ldots;$$

that is, the integers

$$\ldots, -28, -17, -6, 5, 16, 27, 38, 49, 60, \ldots.$$

We can visualize the numbers congruent to a number a modulo m by taking a circle of circumference m and wrapping the real line around it. The figure on the front cover shows this for $m = 6$. Two integers are congruent modulo m if they lie on the same spoke. Thus the integers $\ldots, -7, -1, 5, 11, \ldots$ are all congruent to each other modulo 6, and none are congruent to an integer on any other spoke, such as -5 or 2 or 10.

The Division Theorem for two numbers a and m asserts that $a = mq + r$, with $0 \le r < m$. In terms of congruences, this last equation says that $a \equiv r \pmod{m}$: modulo the divisor, the dividend is congruent to the remainder. Since the remainder in the Division Theorem is unique, we have:

Proposition 1. *Let m be a natural number >1. Every natural number is congruent modulo m to exactly one number in the set $\{0, 1, \ldots, m-1\}$.*

The smallest number ≥ 0 that is congruent to a given integer a modulo m is called the *least non-negative residue* of $a \pmod{m}$. Thus the least non-negative residue of 234047 modulo 10 is 7. The least non-negative residue of -27 modulo 8 is $-27 + 32 = 5$.

The Division Theorem also gives a criterion for two numbers to be congruent modulo m:

Proposition 2. *Let a and b be two natural numbers, and suppose the remainder on dividing a by m is r, and the remainder on dividing b by m is s. Then $a \equiv b \pmod{m}$ if and only if $r = s$.*

Proof. We have $a = mq + r$ and $b = mt + s$ for some natural numbers q, t.

If $r = s$, then $a - mq = b - mt$, so $a - b = m(q - t)$, and so $a \equiv b \pmod{m}$.

Conversely, if $a \equiv b \pmod{m}$, then $b = a + mk$ for some k, so if $a = mq + r$ is the result of dividing a by m, then $b = a + mk = mk + mq + r = m(k + q) + r$. Since $0 < r < m$, this expression for b is what is obtained from the Division Theorem when b is divided by m. By the uniqueness of the quotient and remainder in the Division Theorem, $s = r$. □

To express Proposition 2 another way, two numbers are congruent modulo m iff their least non-negative residues are equal.

Often the notation "a mod m" is used to denote the least non-negative residue of the integer a. For example, the command

$$> -38284 \text{ mod } 37;$$

in MAPLE gives the least non-negative residue of -38284 modulo 37, namely 11. Using that notation, the Division Theorem reads,

there is some number q so that

$$a = mq + (a \text{ mod } m),$$

and Proposition 2 reads:

$a \equiv b \pmod{m}$ *if and only if* $(a \text{ mod } m) = (b \text{ mod } m)$.

There should be no problem using the notation "a mod m" as long as one remembers:

$$\text{"}a \equiv b \pmod{m}\text{"}$$

is a sentence, while

$$\text{"}a \text{ mod } m\text{"}$$

is a number.

Exercises.

1. Show that every integer is congruent modulo m to exactly one of the numbers in the set $\{0, 1, \ldots, m-1\}$.

2. Find
  (i) 3412 mod 5;
  (ii) 177 mod 11;
  (iii) 31 mod 9;
  (iv) 31 mod 35.

3. Find:
  (i) 365 mod 5;
  (ii) -4124 mod 3;
  (iii) 3122182546 mod 10;
  (iv) -2345678 mod 10.

4. New Years Day fell on a Sunday in the year 2006. On what day of the week did New Years Day fall on in the year 2007? (Think modulo 7.)

5. Find all numbers b with $1900 < b < 2000$ that are congruent to a modulo m where
  (i) $a = 1, m = 13$;
  (ii) $a = 1776, m = 25$;
  (iii) $a = 1914, m = 27$.

6. Find a number a which satisfies $a \equiv 5 \pmod 8$ and $a \equiv 3 \pmod 7$ simultaneously.

7. Show that if $m > 4$ is not prime, then $(m-1)! \equiv 0 \pmod m$.

8. The first theorem in Gauss's Disquisitiones Arithmeticae (1801) is the following: "Let m successive integers $a, a+1, a+2, \ldots, a+m-1$ and another integer A be given. Then one and only one of these integers will be congruent to A modulo m." Prove this theorem.

B. Basic Properties

The congruence symbol looks like an "equals" symbol. This is not an accident. We can view congruence as a kind of equality. Most of the manipulations we can do with equality we can do with congruence modulo m. In particular, congruence is an equivalence relation:

Proposition 3. *Congruence modulo m is:*
Reflexive: $a \equiv a \pmod m$ for all integers a;
Symmetric: for all integers a, b, if $a \equiv b \pmod m$, then $b \equiv a \pmod m$;
Transitive: for all integers a, b, c, if $a \equiv b \pmod m$ and $b \equiv c \pmod m$, then $a \equiv c \pmod m$.

All of these are familiar properties of equality. We prove transitivity:

Proof. If $a \equiv b \pmod m$, then $a = b + sm$ for some integer s. If $b \equiv c \pmod m$, then $b = c + tm$ for some integer t. Substituting, we get $a = (c+tm) + sm = c + (t+s)m$, so $a \equiv c \pmod m$. $\qquad\square$

Also, congruence respects addition and multiplication:

Proposition 4. *For all integers a, b, c, a', b', k and m:*
  (i) if $a \equiv b \pmod m$ then $ka \equiv kb \pmod m$;
if $a \equiv b \pmod m$ and $a' \equiv b' \pmod m$ then:
  (ii-a) $a + a' \equiv b + b' \pmod m$, and
  (ii-m) $aa' \equiv bb' \pmod m$.

These follow easily from the condition that $a \equiv b \pmod m$ iff $a = b + mq$ for some integer q. We prove (ii-m).

Proof of (ii-m). If $a \equiv b \pmod m$ and $a' \equiv b' \pmod m$, then $a = b + sm$ and $a' = b' + tm$ for some integers s and t. Then

$$aa' = bb' + msb' + tbm + stmz = bb' + (\text{multiple of } m).$$

So $aa' \equiv bb' \pmod m$. $\qquad\square$

The proof of (ii-m) can also be done using (i) and transitivity. If $a \equiv b$ (mod m) then $aa' \equiv ba'$ (mod m); if $a' \equiv b'$ (mod m) then $ba' \equiv bb'$ (mod m). Hence, by transitivity, $aa' \equiv bb'$ (mod m).

Property (ii-m) shows an advantage of the congruence notation. The statement (ii-m) seems natural in congruence notation because the analogous result for equal quantities:

"if $a = b$ and $a' = b'$ then $aa' = bb'$",

is so familiar. By comparison, the divisibility version of (ii-m): "if $m \mid a - b$ and $m \mid a' - b'$ then $m \mid aa' - bb'$", is not familiar.

The main property of ordinary equality which is lacking in general for congruences mod m is cancellation. If $ab \equiv ac$ (mod m), the congruence $b \equiv c$ (mod m) does not necessarily follow. For example, $2 \cdot 1 \equiv 2 \cdot 3$ (mod 4), but $1 \not\equiv 3$ (mod 4). Similarly, $6 \cdot 1 \equiv 6 \cdot 4$ (mod 9), but $1 \not\equiv 4$ (mod 9).

We shall postpone for now the rules which replace the usual rule of cancellation. Without looking ahead, see if you can find a useful rule of your own. Look at some examples.

We also have

Proposition 5. *If* $a \equiv b$ (mod m) *and* d *divides* m, *then* $a \equiv b$ (mod d)

Proof. If m divides $a - b$ and d divides m, then d divides $a - b$. □

A consequence of (ii-m) is

Proposition 6. *For all natural numbers* e *and all integers* a, b:
if $a \equiv b$ (mod m), *then* $a^e \equiv b^e$ (mod m)

The proof is an easy induction argument. See Exercise 11.

We can use Proposition 6 to help us find the least non-negative residue of some very large numbers. When finding least non-negative residues of high powers of numbers, the main idea is to keep the intermediate computations as close to zero as possible by reducing modulo the modulus after each addition or multiplication.

Example 1. 12^{39} mod 13: 12^{39} is a 42 digit number. But since $12 \equiv -1$ (mod 13),

$$12^{39} \equiv (-1)^{39} \equiv -1 \equiv 12 \quad (\text{mod } 13).$$

Example 2. 6^{37} mod 13: 6^{37} is a 29 digit number. We find

$$6^2 \equiv 36 \equiv -3 \quad (\text{mod } 13)$$
$$6^6 = (6^2)^3 \equiv (-3)^3 \equiv -27 \equiv -1 \quad (\text{mod } 13)$$
$$6^{12} = (6^6)^2 \equiv (-1)^2 \equiv 1 \quad (\text{mod } 13)$$
$$6^{36} = (6^{12})^3 \equiv (1)^3 \equiv 1 \quad (\text{mod } 13)$$
$$6^{37} = 6 \cdot 6^{36} \equiv 6 \cdot 1 \equiv 6 \quad (\text{mod } 13)$$

Example 3. 85^{85} mod 19: 85^{85} is a 164 digit number. We find

$$85 \equiv -10 \pmod{19}$$

$$85^2 \equiv (-10)^2 = 100 \equiv 5 \pmod{19}$$

$$85^4 = (85^2)^2 \equiv (5)^2 \equiv 25 \equiv 6 \pmod{19}$$

$$85^8 = (85^4)^2 \equiv 6^2 = 36 \equiv -2 \pmod{19}$$

$$85^{16} = (85^8)^2 \equiv (-2)^2 = 4 \pmod{19}$$

$$85^{18} = 85^{16}85^2 \equiv 4 \cdot 5 \equiv 1 \pmod{19}$$

$$85^{72} \equiv (85^{18})^4 \equiv 1^4 = 1 \pmod{19}$$

$$85^{85} = 85^{72} \cdot 85^8 \cdot 85^4 \cdot 85$$

$$\equiv 1 \cdot (-2) \cdot 6 \cdot (-10) = 120 = 19 \cdot 6 + 6 \equiv 6 \pmod{19}$$

Notice in this last example, we usually chose not the least non-negative residue at each stage, but the integer that is smallest in absolute value: thus when we reached 36 in the fourth line, we used $36 \equiv -2 \pmod{19}$ instead of $36 \equiv 17 \pmod{19}$, because -2 is closer to 0.

Also, after finding $85^{16} \equiv 4 \pmod{17}$, we noticed that $4 \cdot 5 = 20 \equiv 1 \pmod{19}$ and $85^2 \equiv 5 \pmod{19}$, so we determined $85^{18} \equiv 1$ at that point. Once we found that $85^{18} \equiv 1$, we can then divide the exponent 85 by 18 to get a remainder of 13; then $85^{85} \equiv 85^{13} \pmod{19}$. The last line computed 85^{13} by writing $13 = 8 + 4 + 1$ and multiplying the residues of the corresponding powers of 85.

Example 4. 8^{35} mod 20: we find

$$8^2 \equiv 64 \equiv 4 \pmod{20}$$

$$8^4 \equiv 4^2 = 16 \equiv -4 \pmod{20}$$

$$8^8 \equiv (-4)^2 = 16 \equiv -4 \pmod{20}$$

$$8^{16} \equiv -4 \pmod{20}$$

$$8^{32} \equiv -4 \pmod{20}$$

and so

$$8^{35} = 8^{32} \cdot 8^2 \cdot 8 \equiv (-4) \cdot 4 \cdot 8 \equiv -16 \cdot 8 \equiv 4 \cdot 8 \equiv 12 \pmod{20}.$$

In the last step, we continued the idea of keeping the numbers as small as possible. Rather than multiplying $-4, 4$ and 8 together (to get -128) and then finding the least non-negative residue of -128 modulo 20, we multiplied -4 and 4 together first to get -16, found the least non-negative residue of -16, namely 4, and then multiplied 4 by 8 to get 32, whose least non-negative residue is 12.

Here is an unusual example:

Example 5. 9^{100} mod 24: we find

$$9^2 = 81 \equiv 9 \pmod{24}$$
$$9^3 \equiv 9^2 \equiv 9 \equiv 9 \pmod{24};$$

at which point one can see by a very easy induction argument using transitivity of congruence that

$$9^e \equiv 9 \pmod{24}$$

for every positive exponent e.

Exercises.

9. Prove (i) and (ii-a).

10. Prove that for all integers a

$$a \equiv a \pmod{m} \text{ for all } a,$$

and for all a and b,

$$\text{if } a \equiv b \pmod{m}, \text{ then } b \equiv a \pmod{m}.$$

11. Use (ii-m) of Proposition 4 to prove that if $a \equiv b \pmod{m}$, then $a^e \equiv b^e$ $\pmod{m}$ for every number $e > 0$.

12. Suppose a is an odd number. Is it true that if $ab \equiv ac \pmod{12}$ then $b \equiv c$ $\pmod{12}$?

13. For which numbers a is it true that if $15a \equiv ca \pmod{25}$, then $15 \equiv c$ $\pmod{25}$?

14. Compute the least non-negative residue of $4^n \pmod 9$ for $n = 1, 2, 3, 4, 5, \ldots$. Prove that $6 \cdot 4^n \equiv 6 \pmod 9$ for every $n > 0$.

15. Show that $7^{16} \equiv 1 \pmod{17}$ and use that congruence to find the least non-negative residue of 7^{546} modulo 17.

16. Find (the least nonnegative residue of):
   (i) 5^{18} mod 11;
   (ii) 68^{105} mod 7;
   (iii) 4^{47} mod 12;
   (iv) 66^{75} mod 19.

17. Show that 1 is the least nonnegative residue of $a^6 \pmod 7$ for each number a, $1 \le a \le 6$.

18. Show that $5^e + 6^e \equiv 0 \pmod{11}$ for all odd numbers e, but not for any even number e.

19. (i) Show that $(a+b)^2 \equiv a^2 + b^2 \pmod 2$ for all two integers a, b.
(ii) Show that for all integers $a_1, \ldots, a_n$,

$$(a_1 + a_2 + \ldots + a_n)^2 \equiv a_1^2 + a_2^2 + \ldots + a_n^2 \pmod 2.$$

20. Show that $a^2 \equiv a \pmod 2$ for every integer a.

21. Find all numbers b, $1 < b \le 15$ for which $b^2 \equiv b \pmod{15}$.

22. Show that for $0 \le a \le 6$, the least non-negative residue of a^{67} modulo 7 is a.

C. Divisibility Tricks

Congruence can illuminate an old trick, called "casting out nines," used to detect errors in addition and multiplication: sum the digits and do the operation on the sum of the digits.

Example 6. Suppose we multiplied 3589 and 4363 and obtained 15256397. We check the multiplication as follows:

We sum the digits of 3589 to obtain $3 + 4 + 8 + 9 = 24$, then sum the digits of 24: $2 + 4$, to obtain 6.

Then we sum the digits of 4373 to obtain $4 + 3 + 7 + 3 = 17$, then find $1 + 7$ to obtain 8.

We multiply 6 and 8 to obtain 48, then sum the digits of 48, $4 + 8 = 12$, then sum the digits of 12, $1 + 2 = 3$ to obtain 3.

Then we sum the digits of our result of multiplying 3589 and 4373, namely 15256397: $1 + 5 + 2 + 5 + 6 + 3 + 9 + 7 = 38$, then sum the digits of 38: $3 + 8 = 11$, then sum the digits of 11 to obtain 2.

We see if $3 = 2$. Since that is false, we conclude that the correct result of multiplying 3589 and 4373 is not 15256397 .

Casting out nines as a device for checking errors in computations was known to the medieval Arabs, and brought to Europe along with Hindu-Arabic numerals by Leonardo of Pisa (also known as Fibonacci) in his book, Liber Abaci [Sigler (2003)].

Here is why casting out nines works.

$$10 \equiv 1 \pmod 9,$$

so by Proposition 6, for every $n > 0$,

$$10^n \equiv 1 \pmod 9.$$

So for every number a, we have by Proposition 4 that

$$a \cdot 10^n \equiv a \pmod 9.$$

If a is a number that in ordinary base 10 notation is written as

$$r_n r_{n-1} \ldots r_2 r_1 r_0,$$

so that

$$a = r_n 10^n + r_{n-1} 10^{n-1} + \ldots + r_2 10^2 + r_1 10 + r_0,$$

then by Proposition 4,

$$a \equiv r_n + r_{n-1} + \ldots + r_2 + r_1 + r_0 \pmod 9.$$

Thus every number is congruent to the sum of its digits modulo 9.

Thus, for example,

$$3589 = 3 \cdot 10^3 + 5 \cdot 10^2 + 8 \cdot 10 + 9$$
$$\equiv 3 + 5 + 8 + 9 = 24 \pmod 9$$

and

$$24 = 2 \cdot 10 + 4 \equiv 2 + 4 = 6 \pmod 9.$$

and so by transitivity of congruence, $3589 \equiv 6 \pmod 9$. Replacing 3589 by 6 involves subtracting, or "casting out", many nines. In fact, 6 is the remainder when you divide 3589 by 9. Similarly,

$$4373 \equiv 8 \pmod 9$$

and

$$15256397 \equiv 3 \pmod 9$$

by the same "sum the digits modulo 9" computation. Now if $3589 \equiv 6 \pmod 9$, and $4373 \equiv 8 \pmod 9$, then by Proposition 4,

$$3589 \cdot 4373 \equiv 6 \cdot 8 = 48 \equiv 3 \pmod 9.$$

So if we compute $3589 \cdot 43732$ and get a number that is congruent to 2 modulo 9, then we must have made a mistake in the computation. (Note that casting out nines does not detect all erroneous computations. For example, it won't uncover errors caused by transposing digits, such as $21 \cdot 38 = 978$.)

To sum up the results we showed about numbers modulo 9:

Proposition 7. *Any number a is congruent to the sum of its digits modulo 9.*

In particular,

Corollary 8. *A number a is divisible by 9 if and only if 9 divides the sum of the digits of a.*

Proof. (Recall that "iff" means "if and only if"–see Section 3D, Corollary 6.) The number a is divisible by m iff a is congruent to 0 modulo m. Thus 9 divides a iff $a \equiv 0 \pmod 9$, iff the sum of the digits of a is congruent to 0 (mod 9), iff 9 divides the sum of the digits of a.   □

Here are some other divisibility criteria:

Proposition 9. *3 divides a iff 3 divides the sum of the digits of a.*

Proof. Since 9 divides $10^e - 1$ for all e, so does 3. The proof is the same as for 9.   □

Proposition 10. *2 divides a iff 2 divides the units digit r_0 of a.*

Proof. Since 2 divides 10, $10^e \equiv 0 \pmod 2$ for all $e \geq 1$. So $a \equiv r_0 \pmod 2$.   □

Proposition 11. *5 divides a iff 5 divides the units digit of a.*

Proof. The same as for 2.   □

Proposition 12. *11 divides a iff 11 divides the alternating sum of the digits of a.*

Proof. Since $10 \equiv -1 \pmod{11}$, $10^e \equiv (-1)^n \pmod{11}$ for all e. Then

$$a = r_n 10^n + r_{n-1} 10^{n-1} + \ldots + a_2 10^2 + a_1 10 + a_0$$
$$\equiv a_n(-1)^n + a_{n-1}(-1)^{n-1} + \ldots + a_2(-1)^2 + a_1(-1) + a_0 \pmod{11}.$$

□

Proposition 13. *7 (respectively, 11, 13) divides a iff 7 (respectively, 11, 13) divides the alternating sum of the "digits" of a in base 1000.*

Proof. Suppose

$$a = b_m 1000^m + b_{m-1} 1000^{m-1} + \ldots + b_1 1000 + b_0$$

(where $0 \leq b_k < 1000$ for all $0 \leq k \leq m$). Now $1000 = 1001 - 1$, and $1001 = 7 \cdot 11 \cdot 13$. So

$$1000 \equiv -1 \pmod 7 \text{ and also mod 11 and mod 13}.$$

Thus $1000^e \equiv (-1)^e \pmod{1001}$ for all $e \geq 1$, and we have

$$a \equiv b_m(-1)^m + b_{m-1}(-1)^{m-1} + \ldots + b_1(-1) + b_0 \pmod{1001}.$$

Since 7 divides 1001, it follows by Proposition 5 that

$$a \equiv b_m(-1)^m + b_{m-1}(-1)^{m-1} + \ldots + b_1(-1) + b_0 \pmod 7.$$

(and also mod 11 and mod 13). Then 7 divides a iff 7 divides $b_0 - b_1 + b_2 + \ldots + (-1)^m b_m$.   □

Reviewing the results in this section, we have divisibility tests of large numbers by 2, 3, 5, 7, 9, 10, 11 and 13. What about divisibility by other small numbers? See the exercises.

Exercises.

23. Find a test for divisibility by 4.

24. Find a test for divisibility by 6.

25. Show that every number a is congruent modulo 8 to its units digit in base 1000.

26. (i) Show that 7 divides $10a + b$ if and only if 7 divides $a - 2b$.

(ii) Use (i) to show that 7 divides $821528 = 82152 \cdot 10 + 8$ if and only if 7 divides $82152 - 16 = 82136$. Use (i) three more times to decide whether or not 7 divides 821528.

(iii) Use (i) to decide whether or not 7 divides
(a) 904589
(b) 1036673 .

27. (i) Show that 19 divides $10a + b$ iff 19 divides $a + 2b$.

(ii) Use (i) as in the last exercise to decide whether or not 19 divides
(a) 821534
(b) 1165726.

28. Show that 11 divides $10a + b$ iff 11 divides $a - b$. Use this to show that 11 divides 232595.

29. Come up with a strategy like that in the last three exercises to decide whether a number is divisible by 13.

30. Find the least non-negative residues mod 7, 11 and 13 of:
(i) 11233456;
(ii) 58473625; and
(iii) $100,000,000,000,000,001$.

The next exercises assume you are comfortable with numbers in bases other than 10 (see Section 3A).

31. If we are doing base 12 arithmetic can we check it by casting out 11's? Explain.

32. Find nice tests for divisibility of numbers in base 34 by each of 2, 3, 5, 7, 11, and 17.

33. Prove: if $x \equiv y \pmod{m}$, then $(m,x) = (m,y)$.

34. (i) Use the last exercise to prove the following result, due to Graham (1984):
Suppose a, b and $c_i, i = 0, 1, \ldots, n$, are integers and $(a,b) = 1$. Then

$$(a - b, c_0 a^n + c_1 a^{n-1} b + \ldots + c_{n-1} a b^{n-1} + c_n b^n)$$
$$= (a - b, c_0 + c_1 + \ldots + c_n).$$

(ii) Set $a = 10, b = 1$, and derive the test for divisibility by 9 from this formula.

35. Explain this "pick a number" puzzle:

Pick a number from 1 to 9;

Add it to 4;

Multiply the result by 9;

Add 6 to the result;

Sum the digits of the result;

Multiply the result by 2;

Divide the result by 3;

Add 4 to the result to get a number, call it α;

Pick the largest (in area) European country whose name starts with the α-th letter of the alphabet (for example, France starts with F, the sixth letter of the alphabet);

Pick a popular website with first letter equal to the last letter of the country;

Pick a fruit with first letter equal to the last letter of the website (omit the ".com").

Then the last letter of the fruit is E.

D. Luhn's Formula

The detecting or correcting of errors is an important application of algebra and number theory in the present "information age". Casting out nines may be the oldest known example of an error detection scheme. In this section we look at a much newer example, a method of checking the validity of credit card numbers and other identification numbers, proposed in the 1960's by H. P. Luhn of IBM and known as Luhn's formula.

To begin with an example, suppose we consider the following number:

$$M = 5678\ 9012\ 4567\ 8901.$$

This looks like it might be a credit card number. But not every sixteen digit number beginning with 56 can be a valid credit card number. The digits of the number must satisfy Luhn's formula. Here is how it works.

First, define a function $p(x)$ on the digits $0, 1, 2, \ldots, 8, 9$ by

$$p(0) = 0$$
$$p(9) = 9$$
$$p(n) = 2n \bmod 9 \text{ for } 1 \leq n \leq 8.$$

Thus for $1 \leq n \leq 8$, $p(n)$ is the least non-negative residue of $2n$ modulo 9.

Here is the table of values of $p(x)$:

n	0	1	2	3	4	5	6	7	8	9
$p(n)$	0	2	4	6	8	1	3	5	7	9

Given a credit card number of n digits, number the digits starting from the right:

$$a_n a_{n-1} a_{n-2} \ldots a_3 a_2 a_1.$$

Beginning with a_2, the second digit from the right, apply the function $p(x)$ to every second digit, that is, to $a_2, a_4, a_6, \ldots$. Then sum all of the modified and unmodified digits:

$$S = a_1 + p(a_2) + a_3 + p(a_4) + a_5 + \ldots.$$

Then M is an invalid credit card number if $S \not\equiv 0 \pmod{10}$.

For our 16 digit example $M = 5678\ 9012\ 4567\ 8901$, we apply $p(x)$ to every other digit, then sum the resulting digits:

$$
\begin{aligned}
S &= p(5) + 6 + p(7) + 8 + p(9) + 0 + p(1) + 2 \\
&\quad + p(4) + 5 + p(6) + 7 + p(8) + 9 + p(0) + 1 \\
&= 1 + 6 + 5 + 8 + 9 + 0 + 2 + 2 + 8 + 5 + 3 + 7 + 7 + 9 + 0 + 1
\end{aligned}
$$

Then $S \equiv 3 \pmod{10}$, so M is not a valid credit card number.

Luhn's formula is a quick computation that retailers can perform before submitting a credit card number to a central agency for validation. A Luhn formula check is particularly useful for internet purchases, where customers type in their own credit card numbers. The retailer can instantly detect if the customer made a simple error when keying in the card number, and ask the customer to reenter the number immediately, rather than waiting for validation and perhaps losing the sale because of the delay.

Luhn's formula detects a single instance of two of the most common errors in typing numbers, mistyping a digit, or (with one exception) transposing two adjacent digits. See Exercises 37 and 38.

Exercises.

36. Check the validity of the following numbers by Luhn's formula. If any of them is invalid, change the rightmost digit so that the resulting number satisfies Luhn's formula.
   (i) 4356 2678 9889 6473
   (ii) 346 8965 1938 7647
   (iii) 6011 8665 5575 1270

37. Show that if you try to type in a valid 16 digit credit card number, but mistype one of the 16 digits, the resulting number will be shown invalid by Luhn's formula.

38. Show that if you try to type in a valid 16 digit credit card number, but transpose two adjacent digits, then the resulting number will be shown invalid by Luhn's formula unless the two transposed adjacent digits are 0 and 9.

E. More Properties of Congruence

The equivalence properties of Proposition 3, Section B, show that congruence to a fixed modulus m is much like equality, except for canceling. In this section we list properties which relate congruences to different moduli, and describe how to cancel.

First we look at different moduli.

We already know Proposition 5:

If $a \equiv b \pmod{m}$ and d divides m, then $a \equiv b \pmod{d}$.

We used this property in discussing the divisibility test for 7, 11, and 13, for we observed that if $a \equiv b \pmod{1001}$, then $a \equiv b \pmod{7}$ since 7 divides 1001.

Proposition 14. *If $a \equiv b \pmod{r}$ and $a \equiv b \pmod{s}$ then $a \equiv b \pmod{[r,s]}$.*

Proof. We know that the least common multiple $[r,s]$ of two numbers r and s divides every common multiple of r and s. Thus if r divides $a - b$, and s divides $a - b$, then $[r,s]$ divides $a - b$. □

Example 7. if $a \equiv b \pmod{7}$ and $a \equiv b \pmod{11}$ then $a \equiv b \pmod{77}$. If also $a \equiv b \pmod{13}$ then $a \equiv b \pmod{[77,13]}$, that is, $a \equiv b \pmod{1001}$.

Example 8. We show:
$$2^{340} \equiv 1 \pmod{341}. \tag{1}$$

To see this, we observe that $341 = 11 \cdot 31 = [11,31]$, so by Proposition 14, to show (1) we only need to show:

$$2^{340} \equiv 1 \pmod{11} \text{ and } 2^{340} \equiv 1 \pmod{31}.$$

Now
$$2^5 = 32 \equiv -1 \pmod{11}$$

so
$$2^{340} = (2^5)^{68} \equiv (-1)^{68} \equiv 1 \pmod{11};$$

also
$$2^5 = 32 \equiv 1 \pmod{31}$$

so
$$2^{340} = (2^5)^{68} \equiv 1^{68} \equiv 1 \pmod{31}.$$

Thus (1) follows.

The cancellation properties of congruences are summed up by the following. Here $r \neq 0$.

Proposition 15. *If $ra \equiv rb \pmod{m}$, then $a \equiv b \pmod{\frac{m}{(r,m)}}$.*

For example, $12 = 4 \cdot 3 \equiv 4 \cdot 8 = 32 \pmod{10}$, so since $10/(10,4) = 5$, we may conclude that $3 \equiv 8 \pmod{5}$.

Two special cases of Proposition 15 are when r divides the modulus m or when $(r,m) = 1$:

Proposition 16. *If $ra \equiv rb$ (mod rm) then $a \equiv b$ (mod m).*

Proof. If $ra \equiv rb$ (mod rm) then $ra - rb = rmc$ for some c; canceling r gives $a - b = mc$, so $a \equiv b$ (mod m). $\qquad\qquad\square$

Proposition 17. *If $ra \equiv rb$ (mod m) and $(r,m) = 1$, then $a \equiv b$ (mod m).*

Proof. If m divides $ra - rb = r(a - b)$, then, since m and r are coprime, we can conclude that m divides $a - b$. $\qquad\qquad\square$

For example, since $12 = 4 \cdot 3 \equiv 4 \cdot 8 = 32$ (mod 5) and $(4,5) = 1$, it follows that $3 \equiv 8$ (mod 5).

Exercises.

39. Show that $2^{560} \equiv 1$ (mod 561).

40. Show that $3^{1728} \equiv 1$ (mod 1729).

41. Find all numbers $a \le 20$ so that $6a \equiv 16$ (mod 20).

42. Find all numbers $a \le 36$ so that $16a \equiv 0$ (mod 36).

43. Prove Proposition 15.

F. Linear Congruences and Bezout's Identity

In secondary school algebra you learn to solve equations involving an unknown quantity. Here we begin considering the problem of solving congruences containing unknowns.

The simplest such congruence is

(i) $x + c \equiv d$ (mod m),

which is easy to solve: simply add $-c$ to both sides to get $x \equiv d - c$ (mod m).

The next simplest is

(ii) $ax \equiv b$ (mod m).

If this were an equality, $ax = b$, and x, a and b were required to be integers, we would be able to solve this if and only if a divides b. But (ii) is a congruence, and so we need to find an integer x so that $b = ax +$ (multiple of m), or equivalently, find integers x and y so that $b = ax + my$.

We know how to handle such equations, from Section 3E:

If the greatest common divisor (a, m) of a and m does not divide b, then there are no integers x and y with $ax + my = b$, and so $ax \equiv b \pmod{m}$ has no solution.

If (a, m) divides b, then we can solve $b = ax + my$ for x and y using Bezout's Identity. Thus:

Proposition 18. *The congruence $ax \equiv b \pmod{m}$ is solvable iff (a, m) divides b.*

Example 9. We solve

$$10x \equiv 14 \pmod{15}.$$

Here $(10, 15) = 5$, which does not divide 14. So there is no solution. (For if there were integers x, y with $10x + 15y = 14$, then 5 would divide the left side, so 5 would divide 14.)

$$10x \equiv 14 \pmod{18}.$$

Here $(10, 18) = 2$ divides 14. So there is a solution of $10x + 18y = 14$. We can find a solution by the extended Euclidean algorithm. We set up the EEA matrix as in Chapter 3:

e	$x = $ coeff. of 10	$y = $ coeff. of 18
18	0	1
10	1	0

$$\vdots$$

except that since we just want x, the coefficient of 10, we may omit the y-column:

e	$x = $ coeff. of 10
18	0
10	1
20	2
$2 = 20 - 18$	2
4	4
$14 = 10 + 4$	5

Hence $10 \cdot 5 \equiv 14 \pmod{18}$.

A particularly interesting case is the congruence $ax \equiv 1 \pmod{m}$.

Proposition 19. *If $(a, m) = 1$, then $ax \equiv 1 \pmod{m}$ has a unique solution modulo m.*

Proof. The congruence $ax \equiv 1 \pmod{m}$ is equivalent to the equation $ax + my = 1$. If $(a, m) = 1$, then there are integers r, s so that $ar + ms = 1$, and so $x = r$ is a solution to the congruence $ax \equiv 1 \pmod{m}$.

If also $ar' \equiv 1 \pmod{m}$, then

$$a(r - r') \equiv 0 \pmod{m}$$

and so $m \mid a(r-r')$. Since m and a are coprime, it follows by Corollary 8 of Chapter 3 that m divides $r - r'$, and so

$$r \equiv r' \pmod{m}.$$

Thus the solution of $ax = 1 \pmod{m}$ is unique modulo m. □

The solution r of $ar \equiv 1 \pmod{m}$ is the *inverse* of a modulo m. For example, $27x \equiv 1 \pmod{31}$ has a unique solution since $(27, 31) = 1$. By Bezout's identity, $-8 \cdot 27 + 7 \cdot 31 = 1$, so $x \equiv -8 \pmod{31}$ (or $x \equiv 23 \pmod{31}$) is the inverse of 27 modulo 31.

Corollary 20. *If $(a, m) = 1$, then $ax \equiv b \pmod{m}$ has a solution for all b.*

Proof. Find the inverse r of $a \bmod m$ and set $x = rb$. □

Just as with linear diophantine equations (Section 3E), to find the general solution of $ax \equiv b \pmod{m}$, we need to find any solution, then add to it the general solution of the homogeneous congruence

$$ax \equiv 0 \pmod{m}.$$

Then x is a solution of the homogeneous congruence iff m divides ax, iff

$$\frac{m}{d} \text{ divides } \frac{a}{d}x,$$

where $d = (a, m)$. Since $\frac{m}{d}$ and $\frac{a}{d}$ are coprime, that last statement is equivalent to

$$\frac{m}{d} \text{ divides } x,$$

hence

$$x \equiv \frac{m}{d}k \text{ for some integer } k.$$

Modulo m, we then have $d = (a, m)$ solutions to the homogeneous congruence $ax \equiv 0 \pmod{m}$, namely,

$$x = \frac{m}{d}k \text{ for } k = 0, \ldots, d - 1.$$

Example 10. Consider

$$18x \equiv 12 \pmod{20}.$$

We cancel 2 from everything to get

$$9x \equiv 6 \pmod{10}.$$

Then the inverse of 9 modulo 10 is 9, so

$$x \equiv 9 \cdot 6 \equiv 4 \pmod{10}.$$

So $x \equiv 4$ or $14 \pmod{20}$.

Or, we can observe that $12 \equiv 72$ (mod 20), so one solution of $18x \equiv 12$ (mod 20) is $x = 4$. Then the general solution is found by finding all solutions of $18z \equiv 0$ (mod 20):

$$18z \equiv 0 \quad (\text{mod } 20)$$
$$9z \equiv 0 \quad (\text{mod } 10)$$
$$z \equiv 0 \quad (\text{mod } 10),$$

so $z \equiv 0$ or 10 (mod 20), and then $x \equiv 4 + 0$ or $4 + 10$ (mod 20).

Exercises.

44. Decide whether each of the following congruences has a solution. If so, find the least nonnegative solution:

(i) $12x \equiv 5$ (mod 29);
(ii) $12x \equiv 5$ (mod 38);
(iii) $12x \equiv 5$ (mod 47);
(iv) $12x \equiv 5$ (mod 56);
(v) $12x \equiv 5$ (mod 65).

45. Same question with $12x \equiv 42$ mod
(i) 21; (ii) 22; (iii) 23; (iv) 24; (v) 25.

46. Same question with $9x \equiv 1$ mod
(i) 20; (ii) 21; (iii) 22.

47. Find a solution of $9x \equiv 24$ (mod 21).

48. Solve $313x \equiv 1$ (mod 463).

49. Solve $7x \equiv 1$ (mod 218).

50. Solve $7x \equiv 13$ (mod 218).

When trying to solve quadratic congruences, that is, congruences of the form

$$ax^2 + bx + c \equiv 0 \quad (\text{mod } m) \tag{5.2}$$

the theory becomes very subtle. Even the simplest case:

$$x^2 \equiv a \quad (\text{mod } m) \tag{5.3}$$

is very interesting: how do you decide whether a is a square modulo m? Gauss (1801) was the first to give a complete treatment of the solution of (2) by means of the famous law of quadratic reciprocity. See Chapter 21.

51. Suppose m is odd and $(a,m) = 1$. Show that

$$ax^2 + bx + c \equiv 0 \quad (\text{mod } m)$$

has an integer solution $x \equiv r$ (mod m) if and only if $b^2 - 4ac$ is a square modulo m.

52. Find all solutions $a, b, c \geq 1$ of the following set of congruences:

$$a \equiv b \pmod{c},$$
$$b \equiv c \pmod{a},$$
$$c \equiv a \pmod{b}.$$

53. Show that a number n is the difference of two squares, $n = a^2 - b^2$ with a, b in $\mathbb{Z}$, if and only if n is not congruent to 2 modulo 4.

54. Suppose a, b, c are positive integers and $a^2 + b^2 = c^2$. If a is not a multiple of 4, show that b is a multiple of 4 and c is not a multiple of 4.

Part II
Congruence classes and rings

Chapter 6
Congruence Classes

The idea in this chapter is to use congruence to split up the set $\mathbb{Z}$ of integers into a finite collection of disjoint subsets, think of the subsets as objects, and then see if the arithmetic operations on $\mathbb{Z}$ can induce arithmetic operations on the new objects in a way that makes sense. To see how this might work, we first look at two examples.

A. Two Examples

$\mathbb{Z}/2\mathbb{Z}$. The notion of even and odd integers goes back at least to the Pythagoreans (500 B.C.). A number is even if it is a multiple of 2. A number is odd if when divided by 2, it leaves a remainder of 1. Every number is either even or odd, and no number is both even and odd. So the set $\mathbb{Z}$ splits into two disjoint sets, *even*, the set of even numbers, and *odd*, the set of odd numbers. Let us call the set $\{even, odd\}$ by $\mathbb{Z}/2\mathbb{Z}$. The reason for this notation will become clear in the next section.

We want to think of the two sets *even* and *odd* as objects to add or multiply. In fact, Euclid (300 B.C.) showed us how to do this. In the Elements, Book IX, Euclid proves such facts as: an even number plus an odd number is odd, an odd number plus an odd number is even, an odd number times an even number is even, and so on. Book IX of Euclid can be viewed as a treatise on doing arithmetic with the objects of the set $\mathbb{Z}/2\mathbb{Z}$.

We may collect Euclid's rules into two tables, one for addition, one for multiplication:

+	even	odd
even	even	odd
odd	odd	even

·	even	odd
even	even	even
odd	even	odd

Once we reinterpret the sets *even* and *odd* in terms of congruence, we can generalize this example, and will do so in the next section.

S. Another way to classify integers is by where they lie on the real line: positive integers, negative integers and $\{0\}$. So let us define the three classes:

$$pos = \{1, 2, 3, \ldots\}$$
$$zero = \{0\}$$
$$neg = \{-1, -2, -3, \ldots\}$$

and let $\mathbb{S} = \{pos, zero, neg\}$. Again we want to think of *pos*, *zero* and *neg* as objects and define addition and multiplication on these objects. Let us start with multiplication.

Multiplication. To define multiplication, we recall the well-known facts about multiplying integers:

$$positive \times positive = positive,$$
$$positive \times negative = negative,$$
$$negative \times negative = positive,$$
$$zero \times \text{ any integer } = zero.$$

So multiplication in $\mathbb{S}$ makes sense. We can collect these facts into a multiplication table for $\mathbb{S}$:

·	neg	zero	pos
neg	pos	zero	neg
zero	zero	zero	zero
pos	neg	zero	pos

Addition. Now let us set up an addition table for $\mathbb{S}$. Most of it is easy:

+	neg	zero	pos
neg	neg	neg	?
zero	neg	zero	pos
pos	?	pos	pos

The entries we have filled in correspond to

$$positive + positive = positive$$
$$positive + zero = zero + positive = positive$$
$$negative + negative = negative$$
$$negative + zero = zero + negative = negative$$
$$zero + zero = zero.$$

But what about the entries marked ?: positive + negative = ?

A moment's thought tells us that every integer can be expressed as a positive integer plus a negative integer. So in contrast to *even + odd*, which equals *odd* because any even integer + any odd integer is an odd integer, *pos + neg* does not give us one of the sets *pos*, *zero* or *neg*. If we choose different elements in the sets *pos* and *neg*, we can end up in any of the three sets. (For example, $2 + (-1)$ is in *pos*, $3 + (-3)$ is in *zero* and $4 + (-6)$ is in *neg*.) So we are stuck. There is no way to define *pos* + *neg* as a single object in $\mathbb{S}$ so that

(*) the sum of any integer in *pos* and any integer in *neg* is in the object we define *pos* + *neg* to be.

Because this ambiguity occurs in trying to define addition, we say that addition in $\mathbb{S}$ is *not well-defined*.

This ambiguity did not occur with $\mathbb{Z}/2\mathbb{Z}$ above. One of the things we will need to show below is that when we define $\mathbb{Z}/m\mathbb{Z}$, as a generalization of $\mathbb{Z}/2\mathbb{Z}$, addition and multiplication will be well-defined: that is, will satisfy the analogues of property (*).

B. Congruence Classes and $\mathbb{Z}/m\mathbb{Z}$

Now we use congruence modulo m to split up the integers into sets, analogous to *even* and *odd*, on which we can do arithmetic.

We proved in Section 5B, Proposition 3, that congruence modulo m is an equivalence relation: for any integers a, b and c, we showed that $\equiv$ (mod m) is

$$\text{Reflexive: } a \equiv a \quad (\text{mod } m)$$
$$\text{Symmetric: if } a \equiv b \quad (\text{mod } m), \text{ then } b \equiv a \quad (\text{mod } m)$$
$$\text{Transitive: if } a \equiv b \quad (\text{mod } m), \text{ and } b \equiv c \quad (\text{mod } m),$$
$$\text{then } a \equiv c \quad (\text{mod } m).$$

When a set S has an equivalence relation on it, then the equivalence relation splits up, or *partitions*, the set S into subsets, called equivalence classes, defined by the property that two elements are in the same equivalence class if they are equivalent.

In particular, if S is the set of integers $\mathbb{Z}$ and m is a positive integer, then congruence modulo m partitions $\mathbb{Z}$ into equivalence classes. The equivalence class of the integer a is called the *congruence class of a* (mod m), written $[a]_m$. Thus $[a]_m$ is the set of all integers that are congruent to a modulo m, that is, all integers of the form

$$a + (\text{multiple of } m).$$

The set of congruence classes modulo m is denoted by $\mathbb{Z}/m\mathbb{Z}$.

A congruence class may described by any element in the class:

Proposition 1. *For a, a' in $\mathbb{Z}$, $a' \equiv a$ (mod m) if and only if $[a]_m = [a']_m$.*

Proof. If $[a']_m = [a]_m$, then since a' is in $[a']_m$ (by reflexivity), a' is in $[a]_m$, so $a' \equiv a$ (mod m). Conversely, if $a' \equiv a$ (mod m), then $a \equiv a'$ (mod m) by symmetry. Then $[a']_m \subseteq [a]_m$: for if d is in $[a']_m$, then $d \equiv a'$ (mod m), hence by transitivity, $d \equiv a$ (mod m), so d is in $[a]_m$. The same argument shows that $[a]_m \subseteq [a']_m$. Hence $[a]_m = [a']_m$. □

To restate Proposition 1, congruence (mod m) is the same as equality of congruence classes: that is why the notation

$$\equiv \quad (\text{mod } m)$$

shares so many properties of equality.

As described in Chapter 1, the three properties of an equivalence relation imply that if two equivalence classes have any elements at all in common, then they coincide.

Proposition 2. *Suppose $[a]_m$ and $[b]_m$ are two congruence classes and c in $\mathbb{Z}$ is in both $[a]_m$ and $[b]_m$. Then $[a]_m = [b]_m$.*

Proof. If c is in $[a]_m$, then $c \equiv a$ (mod m), so by Proposition 1, $[c]_m = [a]_m$. If c is in $[b]$, then $c \equiv b$ (mod m), so $[c]_m = [b]_m$. Hence $[a]_m = [b]_m$.   □

There are exactly m congruence classes in $\mathbb{Z}/m\mathbb{Z}$. To see this, recall that every integer a is congruent modulo m to exactly one of the numbers $0, 1, 2, \ldots, m-1$. If $a \equiv r$ (mod m) with $0 \leq r \leq m-1$, then $[a]_m = [r]_m$, and so every congruence class mod m is equal to one of $[0]_m, [1]_m, \ldots, [m-1]_m$. These classes are all different, so there are exactly m congruence classes modulo m.

Thus

$$\mathbb{Z}/m\mathbb{Z} = \{[0]_m, [1]_m, \ldots, [m-1]_m\}.$$

In particular, when $m = 2$, $\mathbb{Z}/2\mathbb{Z} = \{[0]_2, [1]_2\}$. The congruence class $[1]_2$ is the set of all integers congruent to 1 modulo 2. Thus $[1]_2$ is the set *odd*. Similarly, the congruence class $[0]_2$ is the set of all even integers: $[0]_2 = even$.

Any element b of a congruence class mod m is called a *representative* of that class. By Proposition 1, we may label a congruence class by any representative of the class. Thus in $\mathbb{Z}/6\mathbb{Z}$, $[-7]_6 = [-1]_6 = [5]_6 = [11]_6$, etc.

It is often convenient to label a congruence class by the least nonnegative, or least positive element of the class, but on occasion other sets of labels are more convenient.

Now that we understand $\mathbb{Z}/m\mathbb{Z}$ as a set, we want to define arithmetic on $\mathbb{Z}/m\mathbb{Z}$. We do so by:

$$[a]_m + [b]_m = [a+b]_m,$$
$$-[a]_m = [-a]_m ,$$
$$[a]_m \cdot [b]_m = [ab]_m .$$

Thus, with $m = 12$,

$$[7]_{12} + [9]_{12} = [7+9]_{12} = [16]_{12},$$
$$-[3]_{12} = [-3]_{12},$$
$$[7]_{12} \cdot [9]_{12} = [7 \cdot 9]_{12} = [63]_{12}.$$

In defining $[a]_m + [b]_m = [a+b]_m$, we must make sure that the addition is well-defined. This means, if we take any integer in the set $[a]_m$, and add to it any integer in the set $[b]_m$, the sum should be in the set $[a+b]_m$. For example, -3 is in $[9]_{12}$ and 31 is in $[7]_{12}$: is $-3+31 = 28$ in $[9+7]_{12}$? Yes: because $16 \equiv 28$ (mod 12).

In general, we must show:
if $[a]_m = [a']_m$ and $[b]_m = [b']_m$, then

$$[a+b]_m = [a'+b']_m,$$
$$[-a]_m = [-a']_m,$$

and

$$[ab]_m = [a'b']_m.$$

To prove these facts, we translate into congruence notation, using Proposition 1. The first becomes

if $a \equiv a' \pmod{m}$ and $b \equiv b' \pmod{m}$, then $a+b \equiv a'+b' \pmod{m}$.

Similarly, the fact about negatives translates into

if $a \equiv a' \pmod{m}$, then $-a \equiv -a' \pmod{m}$,

The fact about multiplication becomes:

if $a \equiv a' \pmod{m}$ and $b \equiv b' \pmod{m}$, then $ab \equiv a'b' \pmod{m}$.

All of these statements about congruence are true–see Section 5B, Proposition 4. The congruence classes $[0]_m$ and $[1]_m$ are special, in that

$$[0]_m + [b]_m = [b]_m \text{ for all } b,$$
$$[0]_m[b]_m = [0]_m, \text{ for all } b,$$
$$[1]_m[b]_m = [b]_m \text{ for all } b.$$

Thus $[0]_m$ and $[1]_m$ act just like 0 and 1 do in $\mathbb{Z}$, which is hardly surprising since $[0]_m$ contains 0, $[1]_m$ contains 1 and addition and multiplication in $\mathbb{Z}/m\mathbb{Z}$ are defined by means of representatives of the congruence classes in $\mathbb{Z}$.

We look at three examples.

$\mathbb{Z}/2\mathbb{Z}$–**even and odd.** The set $\mathbb{Z}/2\mathbb{Z}$ consists of two congruence classes, $[0]_2$ and $[1]_2$, where

$$[0]_2 = \{\ldots, -4, -2, 0, 2, 4, 6, \ldots\} = even$$

and

$$[1]_2 = \{\ldots, 5, -3, -1, 1, 3, 5, 7, \ldots\} = odd.$$

The tables for addition and multiplication that we found for $\mathbb{Z}/2\mathbb{Z}$ in the previous section become, in the new notation,

+	$[0]_2$	$[1]_2$
$[0]_2$	$[0]_2$	$[1]_2$
$[1]_2$	$[1]_2$	$[0]_2$

$\cdot$	$[0]_2$	$[1]_2$
$[0]_2$	$[0]_2$	$[0]_2$
$[1]_2$	$[0]_2$	$[1]_2$

All we did was substitute $[0]_2 = even$ and $[1]_2 = odd$ in the earlier tables.

$\mathbb{Z}/12\mathbb{Z}$–"clock arithmetic". The set $\mathbb{Z}/12\mathbb{Z}$ of congruence classes mod 12 contains twelve congruence classes. The congruence class $[a]_{12}$ consists of all integers which are congruent to a (mod 12). Thus, for example, the congruence class modulo 12 containing 5 is

$$[5]_{12} = \{\ldots, -31, -19, -7, 5, 17, 29, 41, \ldots\},$$

the congruence class containing 2 is

$$[2]_{12} = \{\ldots, -34, -22, -10, 2, 14, 26, 38, \ldots\},$$

the congruence class containing 12 is

$$[12]_{12} = \{\ldots, -36, -24, -12, 0, 12, 24, 36, 48, \ldots\},$$

the set of all multiples of 12.

Each integer is congruent modulo 12 to exactly one of the numbers $1, \ldots, 11, 12$, and so $\mathbb{Z}/12\mathbb{Z}$ consists of the 12 distinct congruence classes:

$$\mathbb{Z}/12\mathbb{Z} = \{[1]_{12}, [2]_{12}, \ldots, [11]_{12}, [12]_{12}\}.$$

We can use a clock to describe $\mathbb{Z}/12\mathbb{Z}$ visually. Take the real number line and wrap it around a circle of circumference 12, so that the numbers 1 through 12 are located as usual on a clock. Since the real line is infinitely long, it will wrap infinitely often around the circle, and so each "hour" point on the clock will coincide with infinitely many integers:

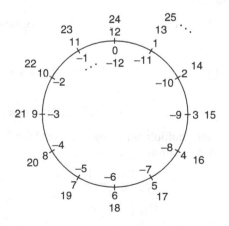

The collection of integers corresponding to a given hour on the clock is the congruence class (mod 12) of the given hour. Thus for example, the integers located at the hour 5 consist of the numbers

$$\{..-31,-19,-7,5,17,...\} = [5]_{12}$$

and the integers located at the hour 12 are all those congruent to 12 (mod 12), that is, $[12]_{12}$, all multiples of 12.

We add congruence classes in $\mathbb{Z}/12\mathbb{Z}$ by

$$[a]_{12} + [b]_{12} = [a+b]_{12}.$$

For example,

$$[9]_{12} + [8]_{12} = [9+8]_{12} = [17]_{12} = [5]_{12}.$$

Multiplication of congruence classes is similar:

$$[a]_{12}.[b]_{12} = [ab]_{12}.$$

For example,

$$[7]_{12} \cdot [5]_{12} = [7 \cdot 5]_{12} = [35]_{12} = [11]_{12}.$$

Addition and multiplication of congruence classes modulo 12 is sometimes called "clock arithmetic", because addition modulo 12 relates to adding time in hours : 6 hours after 11 o'clock is 5 o'clock, and $[11]_{12} + [6]_{12} = [11+6]_{12} = [17]_{12} = [5]_{12}$. Those accustomed to U. S. military time or European or Canadian train times should work in $\mathbb{Z}/24\mathbb{Z}$. Then 14 hours past 17:00 is 7:00, which corresponds to $[17]_{24} + [14]_{24} = [17+14]_{24} = [31]_{24} = [7]_{24}$.

$\mathbb{Z}/9\mathbb{Z}$–**"casting out nines".** We can reinterpret casting out nines in terms of operations in $\mathbb{Z}/9\mathbb{Z}$.

Let $a = r_n 10^n + r_{n-1} 10^{n-1} + \ldots r_1 10 + r_0$. Then, using that $[a+b]_9 = [a]_9 + [b]_9$ and $[ab]_9 = [a]_9 \cdot [b]_9$, we have

$$[a]_9 = [r_n]_9 [10^n]_9 + [r_{n-1}]_9 [10^{n-1}]_9 + \ldots + [r_1]_9 [10]_9 + [r_0]_9.$$

Now since $10^k = 1$ (mod 9) for all $k \geq 0$, we have $[10^k]_9 = [1]_9$ for all k, and so

$$\begin{aligned}
[a]_9 &= [r_n]_9 [1]_9 + [r_{n-1}]_9 [1]_9 + \ldots + [r_1]_9 [1]_9 + [r_0]_9. \\
&= [r_n]_9 + [r_{n-1}]_9 + \ldots + [r_1]_9 + [r_0]_9. \\
&= [r_n + r_{n-1} + \ldots + r_1 + r_0]_9.
\end{aligned}$$

That says that the congruence class of a (mod 9) is equal to the congruence class of the sum of the digits of a (mod 9). In particular, 9 divides a if and only if $[a]_9 = [0]_9$, if and only if the congruence class mod 9 of the sum of the digits of a is equal to $[0]_9$, iff 9 divides the sum of the digits of a.

Coda. Given integers a, b and a natural number $m > 1$, all of these notations describe the same relationship:

$$a = b + (\text{multiple of } m),$$

$$m \text{ divides } b - a: \text{ notation: } m \mid (b - a),$$

$$a \equiv b \pmod{m}.$$

The notion of congruence classes gives us another way to express the same relationship:

$$[a]_m = [b]_m.$$

The main point of congruence classes is not to just come up with yet another way to express that relationship, but rather that the set of congruence classes modulo m, $\mathbb{Z}/m\mathbb{Z}$ is a set on which we can do arithmetic operations in a natural manner. The usefulness of $\mathbb{Z}/m\mathbb{Z}$ will, we hope, be made clear by the applications to be presented later.

Exercises.

1. What is the hour:
   (i) 8 hours after 11 A.M.?
   (ii) 15 hours after 11 P.M.?
   (iii) 21 hours after 6 A.M.?

2. If you leave San Antonio at 7 A.M. CST by rail on the "Texas Eagle", at what hour will you arrive at your destination if it is:
   (i) Chicago and the trip takes 31 hours?
   (ii) St. Paul and the trip takes 63 hours?
   (iii) Winnipeg and the trip takes 106 hours?
   (iv) Churchill and the trip takes 146 hours?
   (All destinations are in the same time zone as San Antonio.)

C. Arithmetic Modulo m

Recall from Section 5A that the least non-negative residue of an integer a modulo m is the unique number r with $0 \leq r < m$ so that $a \equiv r \pmod{m}$. If $a \geq 0$, then r is the remainder when a is divided by m. The least non-negative residue of a modulo m is so useful that most computer languages have a special command for it.

In Maple, it is a mod m.

In Mathematica and Excel, it is $mod(a, m)$.

In C, C++, Java and related programming languages, it is $a\%m$.

Thus
$$35 \bmod 13 = 9; \quad \bmod(35,13) = 9; \quad 35\%13 = 9.$$

We'll use the Maple notation $a \bmod m$.

Then by Section 5A, $a \bmod m = b \bmod m$ iff $a \equiv b \pmod{m}$ iff $[a]_m = [b]_m$. Thus the function ( ) mod m yields a one-to-one correspondence between $\mathbb{Z}/m\mathbb{Z}$ and the set $\{0,1,2,\ldots,m-1\}$.

The operations of addition, negation, and multiplication on $\mathbb{Z}/m\mathbb{Z}$ induce operations on the set $\{0,1,2,\ldots,m-1\}$, as follows:

If a,b are integers, then

$$(a \bmod m) + (b \bmod m) = (a+b) \bmod m;$$
$$-(a \bmod m) = (-a) \bmod m = m - a;$$
$$(a \bmod m) \cdot (b \bmod m) = a \cdot b \bmod m.$$

Thus the operations on $\{0,1,2,\ldots,m-1\}$ work by doing the operation in $\mathbb{Z}$ and then, if the result of the operation is outside the set $\{0,1,\ldots,m-1\}$, we find the least non-negative residue modulo m of the result.

These operations define what we may call *arithmetic mod m*.

To illustrate it, here are tables of addition and multiplication for $m = 3$:

+ mod 3	0	1	2
0	0	1	2
1	1	2	0
2	2	0	1

· mod 3	0	1	2
0	0	0	0
1	0	1	2
2	0	2	1

Only a few entries are different from ordinary addition and multiplication: $(1+2) \bmod 3 = 0$ because the remainder on dividing $1 + 2 = 3$ by 3 is 0. Similarly, $(2 \cdot 2) \bmod 3 = 1$.

Here are the addition and multiplication tables for arithmetic mod 6:

+ mod 6	0	1	2	3	4	5
0	0	1	2	3	4	5
1	1	2	3	4	5	0
2	2	3	4	5	0	1
3	3	4	5	0	1	2
4	4	5	0	1	2	3
5	5	0	1	2	3	4

· mod 6	0	1	2	3	4	5
0	0	0	0	0	0	0
1	0	1	2	3	4	5
2	0	2	4	0	2	$\bar{4}$
3	0	3	0	3	0	3
4	0	4	2	0	4	2
5	0	5	4	3	2	1

For example, the barred entry 1 in the addition table means that $(3+4) \bmod 6 = 1$. In the multiplication table the barred entry 4 means that $(2 \cdot 5) \bmod 6 = 4$. These correspond to the results in $\mathbb{Z}/6\mathbb{Z}$ because $[3]_6 + [4]_6 = [7]_6 = [1]_6$, and $[2]_6 \cdot [5]_6 = [10]_6 = [4]_6$. From the addition table one also reads that $(-2) \bmod 6 = 4$, because $(2+4) \bmod 6 = 0$.

Or consider arithmetic mod 13: here are some examples of sums and products mod 13:

$(12 + 8) \bmod 13 = 7$ because $[20]_{13} = [7]_{13}$,

$(6 + 5) \bmod 13 = 11$ (when a sum is < 13, addition mod 13 is ordinary addition),

$(10 + 5) \bmod 13 = 2$ because $[15]_{13} = [2]_{13}$,

$(12 \cdot 8) \bmod 13 = 5$ because $[96]_{13} = [5]_{13}$,

$(6 \cdot 5) \bmod 13 = 4$ because $[30]_{13} = [4]_{13}$,

$(10 \cdot 5) \bmod 13 = 11$ because $[50]_{13} = [11]_{13}$.

If $\mathbb{Z}/m\mathbb{Z}$ looks just like arithmetic on $\{0, 1, \ldots, m-1\}$ mod m, then why should we define $\mathbb{Z}/m\mathbb{Z}$ at all? An analogy with fractions may help to explain why.

The relationship between the set $\{0, 1, \ldots, m-1\}$ with operations mod m and

$$\mathbb{Z}/m\mathbb{Z} = \{[0]_m, [1]_m, \ldots, [m-1]_m\}$$

with operations $+, \cdot, -$, is similar to the relationship between fractions that are reduced, that is, have relatively prime numerator and denominator, and arbitrary fractions.

To multiply two reduced fractions, such as 5/12 and 3/10, we multiply numerators and denominators together, to get

$$\frac{5}{12} \cdot \frac{3}{10} = \frac{5 \cdot 3}{12 \cdot 10} = \frac{15}{120}.$$

Often, as in this case, the fraction is not reduced, so we reduce it: $15/120 = 1/8$. Reducing is analogous to taking the least nonnegative residue modulo m, as for example with $m = 12$: we set $9 \cdot 7 \bmod 12$ to be the remainder upon dividing $9 \cdot 7 = 63$ by 12, namely 3.

However, adding fractions is usually impossible without using nonreduced fractions. For example, to add 5/12 and 3/10, we find a common denominator, e.g. 60, then add as follows:

$$\frac{5}{12} + \frac{3}{10} = \frac{25}{60} + \frac{18}{60} = \frac{25 + 18}{60} = \frac{43}{60}:$$

we replace 5/12 by 25/60 and 3/10 by 18/60, then add the numerators of the nonreduced fractions with equal denominators: 25/60 + 18/60 = 43/60.

Similarly, in working with $\mathbb{Z}/m\mathbb{Z}$, it is often desirable to use numbers other than $\{0, 1, \ldots, m - 1\}$ to represent the elements of $\mathbb{Z}/m\mathbb{Z}$. For example, if we want to solve the equation

$$[x]_{31}^2 = [2]_{31},$$

a solution becomes obvious if we observe that $[2]_{31} = [64]_{31}$. To solve

$$[3]_5 [x]_5 = [2]_5,$$

the problem becomes easy once we note that $[2]_5 = [12]_5$. To solve

$$[8]_{13} [x]_{13} = [6]_{13},$$

observe that $[6]_{13} = [32]_{13}$.

Or if we want to describe multiplication in $\mathbb{Z}/m\mathbb{Z}$, it becomes easy if we discover that

$$\mathbb{Z}/13\mathbb{Z} = \{[0], [2], [2^2], [2^3], \ldots, [2^{11}], [2^{12}] = [1]\},$$

for then $[2^r] \cdot [2^s] = [2^{r+s}]$–multiplication is transformed into addition of exponents modulo 12.

Thus viewing $\mathbb{Z}/m\mathbb{Z}$ as congruence classes allows us the freedom, as with fractions, to pick any element of an equivalence class to represent that class. Often the least nonnegative representative may not be the representative of choice.

Round robin tournaments. Addition modulo m can be used to design round robin tournaments.

A round robin tournament is a competition involving m players (or teams) in which each player plays every other player exactly once. For example, with four players A, B, C and D, the tournament would consist of six matches–A vs. B, A vs. C, A vs. D, B vs. C, B vs. D, and C vs. D. In a round robin tournament matches are scheduled into "rounds", time periods in which several matches occur simultaneously. An objective of scheduling is to minimize the number of rounds.

For example, a four player tournament can be run in three rounds, where every player competes in each round, as follows:

Round 1: A vs. B, C vs. D
Round 2: A vs. C, B vs. D
Round 3: A vs. D, B vs. C.

There cannot be fewer than three rounds, because each player must play all three opponents, and (excepting chess?) each player cannot play more than one opponent at a time.

For m odd, the addition table mod m can be used to design a round robin tournament for m or $m + 1$ players. To illustrate, consider a tournament with five players, numbered 1, 2, 3, 4 and 5. Write down the addition table for addition modulo 5 (except we use 5 instead of 0):

+	1	2	3	4	5
1	2	3	4	5	1
2	3	4	5	1	2
3	4	5	1	2	3
4	5	1	2	3	4
5	1	2	3	4	5

Interpret the entries of this addition table as follows: if a is a player listed in the leftmost column, and b is a player listed in the top row, then Player a plays Player b in round $a + b$.

Each column of the table describes the round in which the player at the top of the column plays each of the players in the leftmost column. For example, the column for player 3 is:

+	3
1	4
2	5
3	1
4	2
5	3

Thus player 3 plays player 1 in round 4, player 2 in round 5, player 3 in round 1, player 4 in round 2 and player 5 in round 3.

But, you say, how can player 3 play player 3 in round 1? Obviously, she can't. So player 3 sits out in round 1. With an odd number of players, one player must sit out in each round, or, as they say, receives a bye. So in round 1 the matches are 1 vs. 5, 2 vs. 4, and 3 gets a bye. In round 4, the matches are 1 vs. 3, 4 vs. 5 and 2 vs. 2–that is, 2 gets a bye.

In this tournament design, there are m rounds. Each column of the table for addition modulo m contains all the numbers between 1 and m exactly once, so each player plays each other player in a different round, and each round has $\frac{m-1}{2}$ matches.

Now suppose there is an even number $2n$ of players. Then the design is even more efficient. Pick one player and call him "Bye". Number the remaining $m = 2n - 1$ players by $1, 2, \ldots, m$. Write down the addition table for addition modulo m and use it to assign matches to rounds for the players $1, 2, \ldots, m$. In any round, if a player b is assigned to play himself, then instead of sitting out that round, b plays Bye. It's easy to see from the table when Bye plays each opponent–just read the round from the diagonal of the table. For example, with $2n = 6$:

+	1	2	3	4	5
1	2				
2		4			
3			1		
4				3	
5					5

Bye plays player 1 in round 2, player 2 in round 4, player 3 in round 1, player 4 in round 3, and player 5 in round 5.

With an even number $2n$ of players, there are $2n - 1$ rounds, and each player plays in each round.

Exercises.

3. Compute $13 + 19 \bmod 23$.

4. Compute $13 \cdot 19 \bmod 23$.

5. Write down the addition and multiplication tables for arithmetic modulo 4.

6. Write down the addition and multiplication tables for arithmetic modulo 5.

7. Write down the addition and multiplication tables for arithmetic modulo 8.

8. Solve the equation $[a]_m[x]_m = [b]_m$ by finding convenient representatives for $[a]$ and $[b]$:
  (i) $[6]_{10}[x]_{10} = [4]_{10}$,
  (ii) $[5]_7[x]_7 = [3]_7$,
  (iii) $[4]_7[x]_7 = [2]_7$,
  (iv) $[4]_{17}[x]_{17} = [2]_{17}$,
  (v) $[13]_{19}[x]_{19} = [16]_{19}$.

9. Solve
$$[3]_{11}[x]_{11}^2 = [4]_{11}$$

10. Solve
$$[11]_{13}[x]_{13}^2 = [7]_{13}$$

11. Solve
$$[x]_{11}^2 = [3]_{11}$$

12. Describe the fourth round of a round robin tournament with 10 players.

13. Describe the fifth round of a round robin tournament with 12 players.

14. Describe Bye's opponent in each round of a 10 player tournament.

15. Show that in a round robin tournament with an even number $2n$ of players, player a plays Bye in round $2a \bmod 2n - 1$.

16. Let m be even. Try to use the addition table mod m to design a round robin tournament. What (if anything) goes wrong?

D. Complete Sets of Representatives

Definition. A *complete set of representatives* for $\mathbb{Z}/m\mathbb{Z}$ (or "modulo m") is a set of integers $\{r_1, \ldots, r_m\}$ so that every integer is congruent modulo m to exactly one of the numbers in the set.

If $\{r_1, \ldots, r_m\}$ is a complete set of representatives for $\mathbb{Z}/m\mathbb{Z}$, then

$$\mathbb{Z}/m\mathbb{Z} = \{[r_1]_m, [r_2]_m, \ldots, [r_m]_m\}.$$

Thus $\{0, 1, 2, \ldots, m-1\}$ is a complete set of representatives for $\mathbb{Z}/m\mathbb{Z}$. So is $\{1, 2, \ldots, m-1, m\}$, or any set of m consecutive integers (see Chapter 5, Exercise 8). But there are many others.

For example, we will prove later in the book the following:

Theorem 3 (Primitive Root Theorem). *Let p be a prime number. There exists some integer b so that*

$$\{0, b, b^2, b^3, \ldots, b^{p-1}\}$$

is a complete set of representatives for $\mathbb{Z}/p\mathbb{Z}$.

An integer b satisfying the Primitive Root Theorem is called a *primitive root modulo p*. Here are some examples:

Modulo 5, the numbers 2 and 3 are primitive roots. In particular, modulo 5,

$$0 = 0, 1 \equiv 3^4, 2 \equiv 3^3, 3 \equiv 3^1, 4 \equiv 3^2,$$

so $\{0, 3, 3^2, 3^3, 3^4\}$ is a complete set of representatives modulo 5.

Modulo 7, the numbers 3 and 5 are primitive roots, but 1, 2, 4 and 6 are not.

Modulo 17, the numbers 3, 5, 6, and 7 are some of the primitive roots.

Other examples are in the exercises.

When we represent the non-zero congruence classes modulo p by the powers of a primitive root, then multiplication of congruence classes turns into addition of exponents modulo $p-1$. For example, 3 is a primitive root modulo 17, and we can show that

$$[12]_{17} = [3^{13}]_{17}$$

and

$$[11]_{17} = [3^7]_{17}.$$

So

$$[12]_{17} \cdot [11]_{17} = [3^{13}]_{17} \cdot [3^7]_{17} = [3^{20}]_{17}.$$

Now it turns out that $[3^{16}]_{17} = [1]_{17}$ and $[3^4]_{17} = [13]_{17}$. So

$$[3^{20}]_{17} = [3^{16}]_{17}[3^4]_{17} = [1]_{17}[3^4]_{17} = [13]_{17}.$$

In doing multiplication in this way, a "logarithm to the base 3" table is very convenient, where $\log_3 n = r$ means $3^r \equiv n \pmod{17}$:

$3^r = n$	1	2	3	4	5	6	7	8	9	10	11	12	13	14	15	16
$r = log_3 n$	16	14	1	12	5	15	11	10	2	3	7	13	4	9	6	8

Such a log table is especially convenient for evaluating polynomials.

Example 1. Suppose $f(x) = x^4 + 5x^3 + 8x^2 + x + 15$ and we wish to compute $f(12)$ modulo 17. So we want

$$12^4 + 5 \cdot 12^3 + 8 \cdot 12^2 + 12 + 15 \pmod{17}.$$

Since $12 \equiv 3^{13}, 5 \equiv 3^5$ and $8 \equiv 3^{10}$, this is

$$3^{52} + 3^5 3^{39} + 3^{10} 3^{26} + 12 + 15.$$

Since $3^{16} \equiv 1 \pmod{17}$, we find (again from the log table),

$$3^{52} \equiv 3^{48} 3^4 \equiv 3^4 \equiv 13,$$
$$3^5 3^{39} = 3^{44} \equiv 3^{12} \equiv 4,$$
$$\text{and}$$
$$3^{10} 3^{26} = 3^{36} \equiv 3^4 \equiv 13 \pmod{17}.$$

Hence

$$f(12) \equiv 13 + 4 + 13 + 12 + 15 \equiv 6 \pmod{17},$$

The following is helpful for deciding if a set of numbers is a complete set of representatives.

Proposition 4. Given a set $\mathscr{R} = \{r_1, r_2, \ldots, r_m\}$ of m integers, the following conditions are equivalent:
a) Every integer is congruent modulo m to some r_i in $\mathscr{R}$,
b) For every i, j with $1 \leq i < j \leq m$, $r_i \not\equiv r_j \pmod{m}$.

A complete set of representatives mod m is a set $\mathscr{R}$ of m integers satisfying either a) or b).

Proof. Given a set $\mathscr{R} = \{r_1, r_2, \ldots, r_m\}$ of m integers, the map $r \to [r]_m$ defines a function f from $\mathscr{R}$ to $\mathbb{Z}/m\mathbb{Z}$. Then a) says that f is onto, and b) says that f is one-to-one. Let $f(\mathscr{R}) = \{f(r)|r \text{ in } \mathscr{R}\}$, the image of the function f.
Now if $m = |\mathscr{R}|, |f(\mathscr{R})|$ and $|\mathbb{Z}/m\mathbb{Z}| = m$ denote the cardinalities of the sets $\mathscr{R}, f(\mathscr{R})$ and $\mathbb{Z}/m\mathbb{Z}$, then

$$|\mathscr{R}| \geq |f(\mathscr{R})| \leq |\mathbb{Z}/m\mathbb{Z}|.$$

Also, f is one-to-one if $|\mathscr{R}| = |f(\mathscr{R})|$, and f is onto if $|f(\mathscr{R})| = |\mathbb{Z}/m\mathbb{Z}|$. Since $|\mathscr{R}| = m = |\mathbb{Z}/m\mathbb{Z}|$, it follows that $|\mathscr{R}| = |f(\mathscr{R})|$ iff $|f(\mathscr{R})| = |\mathbb{Z}/m\mathbb{Z}|$, that is, f is one-to-one iff f is onto. So a) and b) are equivalent. $\square$

Exercises.

17. Which of the following sets is a complete set of representatives modulo 7?
  (i) $\{1,3,5,7,9,11,13\}$
  (ii) $\{1,4,7,10,13,16,19\}$
  (iii) $\{1,8,27,64,125,216,343\}$
  (iv) $\{1,-3,9,-27,81,-243,0\}$
  (v) $\{0,1,-2,4,-8,16,-32\}$.

18. Find a complete set of representatives for $\mathbb{Z}/9\mathbb{Z}$ consisting of numbers ≥ 2008.

19. Show that there is no complete set of representatives for $\mathbb{Z}/71\mathbb{Z}$ that includes two of the numbers 1066, 1492 and 1776.

20. Which of the following sets is a complete set of representatives for $\mathbb{Z}/9\mathbb{Z}$?
  (i) $\{1234,4567,8901,-1234,-5677,2534,8654,-1500,-33331\}$
  (ii) $\{-1111,-111,-11,-1,0,1,11,111,1111\}$).

21. For which exponents k is $\{1^k,2^k,3^k,4^k,5^k,6^k,7^k,8^k,9^k,10^k,11^k\}$ a complete set of representatives modulo 11? (Can you generalize your answer to other moduli?)

22. Show that $\{0,2,2^2,2^3,\dots,2^{11},2^{12}\}$ is a complete set of representatives for $\mathbb{Z}/13\mathbb{Z}$.

23. Show that 3 and 5 are primitive roots modulo 7, but 1, 2, 4 and 6 are not.

24. Show that 5 is a primitive root mod 17.

25. Find a primitive root modulo 17 other than 3, 5, 6 or 7.

26. Show that if b is a primitive root modulo 17, then so is $-b$.

27. Let $f(x) = x^4 + 5x^3 + 8x^2 + x + 15$. Compute
  (i) $f(6) \mod 17$,
  (ii) $f(14) \mod 17$,
  (iii) $f(9) \mod 17$.

28. Let $f(x) = x^4 + 6x^3 + 13x^2 + 8x + 7$.
  (i) Find $f(5)$ (mod 17),
  (ii) Find $f(15)$ (mod 17).

29. Set up a "logarithm to the base 5" table modulo 7, let $f(x) = 3x^8 + 5x^4 - 2x^3 + 6$, and, using your table, compute
  (i) $f(4) \mod 7$,
  (ii) $f(3) \mod 7$.

30. Show that if $\{a_1,\dots,a_m\}$ is a complete set of representatives modulo m, and $b_1 \equiv a_1 \pmod{m}, b_2 \equiv a_2 \pmod{m},\dots,b_m \equiv a_m \pmod{m}$, then $\{b_1,\dots,b_m\}$ is a complete set of representatives modulo m.

31. Show that if b is a primitive root modulo p and $c \equiv b \pmod{p}$, then c is a primitive root modulo p.

32. Find all primitive roots b modulo 11 with $1 \le b \le 11$.

33. Show that if b is a primitive root modulo p, then the smallest exponent $e > 0$ so that $b^e \equiv 1 \pmod{p}$ is $e = p - 1$.

34. Show that any m consecutive integers form a complete set of representatives for $\mathbb{Z}/m\mathbb{Z}$.

35. Given $0 < n < m$, show that there exists some t, $n \le t < m$, so that $m - n$ divides t.

36. Suppose $\{a_1, \ldots, a_m\}$ is a complete set of representatives for $\mathbb{Z}/m\mathbb{Z}$. Show that for any integer b,
$$\{a_1 + b, a_2 + b, \ldots, a_m + b\}$$
is a complete set of representatives for $\mathbb{Z}/m\mathbb{Z}$.

37. Suppose $\{a_1, \ldots, a_m\}$ is a complete set of representatives for $\mathbb{Z}/m\mathbb{Z}$. Show:
  (i) If $(b, m) = 1$, then $\{ba_1, \ldots, ba_m\}$ is a complete set of representatives.
  (ii) If $(b, m) > 1$, then $\{ba_1, \ldots, ba_m\}$ is not a complete set of representatives.

38. Let a, b be relatively prime integers with $a > b > 0$. Define the sequence of numbers $s_1, s_2, \ldots, s_k$ by

$$s_1 = a,$$
$$s_2 = a - b,$$
$$s_{k+1} = s_k + a \text{ if } s_k < b, \text{ or}$$
$$= s_k - b \text{ if } s_k \ge b.$$

Show that the numbers $s_1, s_2, \ldots, s_{a+b}$ form a complete set of representatives (mod $a + b$) (see Chapter 3, exercise 66.)
  What if $(a, b) > 1$?

39. Call a set $S = \{a_1, a_2, \ldots, a_n\}$ of distinct integers admissible if S does not include a complete set of representatives modulo any prime.
  (i) Show that $\{0, 2\}$ and $\{1, 3\}$ are admissible, but $\{1, 2\}$ is not.
  (ii) Show that $\{1, 3, 5\}$ is not admissible, but $\{1, 3, 7\}$ is admissible.
  (iii) Show that $\{1, 3, 7, 13\}$ is admissible.
  (iv) Find an admissible set containing ten integers.
  (v)) What is the largest subset of the primes p with $3 \le p \le 43$ that is admissible?
  (vi) Find an admissible set with n integers for any n.

40. Given a, b, c with $(a, b) = 1$, show that there is some m so that $(a + bm, c) = 1$, as follows:
  (i) Show that $c = fg$ where f divides b^k for some $k > 0$ and $(b, g) = 1$.

(ii) Show that $\{a + bn \mid n = 1, 2, \ldots, g\}$ is a complete set of representatives modulo g.

(iii) Show that if $(a + bm, g) = 1$, then $(a + bm, c) = 1$.

41. If $(a, b) = 1$, show that for all $m \geq (a - 1)(b - 1)$, there are integers r, s, both ≥ 0, so that $m = ar + bs$.

E. Units

Given a number a, is there a number b so that $ab = ba = 1$? Whenever there is such a number b, we call a a *unit* and call the number b the *inverse* of a. "The," because there can be at most one inverse. For suppose b and c are two inverses of a, so that $ab = ac = 1$ and also $ba = 1$. Then since $ab = ac$, multiplying by b gives $b(ab) = b(ac)$, then $(ba)b = (ba)c$, then $1 \cdot b = 1 \cdot c$, then $b = c$.

In $\mathbb{Z}$, very few numbers have inverses: in fact, 1 and -1 are the only units in $\mathbb{Z}$: $1 \cdot 1 = 1$, so 1 has an inverse, namely itself; and $-1 \cdot -1 = 1$, so -1 has an inverse, also itself. In order to talk about the inverse of a number such as 2, we have to introduce fractions. The inverse of 2 is $1/2$, but $1/2$ is not an integer. Seeking inverses of natural numbers led to the first expansion of the concept of number beyond the counting numbers in the history of mathematics. Fractions of the form $1/n$ for n a natural number were used by the ancient Egyptians 4000 years ago.

Every fraction except 0 has an inverse: if a/b is any fraction with $a, b \neq 0$, then the inverse of a/b is b/a. So every non-zero element of the rational numbers $\mathbb{Q}$ is a unit of $\mathbb{Q}$.

What about units in $\mathbb{Z}/m\mathbb{Z}$?

To pose the question, we first must decide that $[1]_m$ will play the role of 1 in the definition of unit. This is a reasonable choice. The number 1 is the multiplicative identity of $\mathbb{Z}$, in the sense that for any integer a, $1 \cdot a = a$. Because 1 is the multiplicative identity of $\mathbb{Z}$, $[1]_m$ is a multiplicative identity of $\mathbb{Z}/m\mathbb{Z}$: for any integer a,

$$[1]_m \cdot [a]_m = [1 \cdot a]_m = [a]_m.$$

Moreover, $[1]_m$ is the only multiplicative identity of $\mathbb{Z}/m\mathbb{Z}$. To see this, assume that $[e]_m$ were also a multiplicative identity, then

$$[e]_m \cdot [a]_m = [a]_m$$

for all integers a, so in particular,

$$[e]_m \cdot [1]_m = [1]_m.$$

But since $[1]_m$ is a multiplicative identity,

$$[e]_m \cdot [1]_m = [1]_m \cdot [e]_m = [e]_m.$$

So $[e]_m = [1]_m$.

A unit of $\mathbb{Z}/m\mathbb{Z}$, then, is an element $[a]_m$ for which there is some element $[b]_m$ with $[a]_m[b]_m = [1]_m$.

What are the units of $\mathbb{Z}/m\mathbb{Z}$? Let U_m denote the set of units of $\mathbb{Z}/m\mathbb{Z}$. Consider some examples:

In $\mathbb{Z}/3\mathbb{Z} = \{[0],[1],[2]\}$, the elements with inverses are $[1]$ and $[2] = [-1]$. Each is the inverse of itself. So

$$U_3 = \{[1],[2]\}.$$

In $\mathbb{Z}/5\mathbb{Z} = \{[0],[1],[2],[3],[4]\}$, all elements except $[0]$ are units: $[2]$ and $[3]$ are inverses of each other, because $[2] \cdot [3] = [6] = [1]$, while $[1]$ and $[4] = [-1]$ are their own inverses. Thus

$$U_5 = \{[1],[2],[3],[4]\}.$$

In $\mathbb{Z}/9\mathbb{Z} = \{[0],[1],[2],[3],[4],[5],[6],[7],[8]\}$, we have, besides $[1]$ and $[8] = [-1]$, also the units $[2]$ and $[5]$, which are inverses of each other, and $[4]$ and $[7]$, which are inverses of each other, a total of six units. $[3]$ and $[6]$ are not units. Thus

$$U_9 = \{[1],[2],[4],[5],[7],[8]\}.$$

One (very inefficient!) way to find the units of $\mathbb{Z}/m\mathbb{Z}$ is to write down the multiplication table for $\mathbb{Z}/m\mathbb{Z}$. Then $[a]_m$ is a unit if $[1]_m$ is somewhere in the row of $[a]_m$. But there are better ways to find the units modulo m.

For any modulus m, $[a]_m$ is a unit of $\mathbb{Z}/m\mathbb{Z}$ if and only if there is some number b so that $[a]_m[b]_m = [1]_m$. Translating into congruence notation, $[a]$ is a unit if there is some integer b so that $ab \equiv 1 \pmod{m}$.

We know for which a such a b can be found:

Theorem 5. *In $\mathbb{Z}/m\mathbb{Z}$, $[a]$ is a unit iff a and m are coprime.*

Proof. (Review). Suppose $(a,m) = 1$. Then by Bezout's identity, there are integers r,s with $ar + ms = 1$. (The integers r and s can be found by Euclid's algorithm.) Then $[ar+ms]_m = [1]_m$. But $[ar+ms]_m = [ar]_m = [a]_m[r]_m$. So $[r]_m$ is the inverse of $[a]_m$ in $\mathbb{Z}/m\mathbb{Z}$.

Conversely, if $[a]_m[r]_m = [1]_m$, then $ar \equiv 1 \pmod{m}$, so there exists an integer s so that $ar + ms = 1$, which implies that a and m are coprime. $\square$

Corollary 6. *The number of units of $\mathbb{Z}/m\mathbb{Z}$ is equal to the number of numbers a with $1 \le a \le m$ that are coprime to m.*

Definition. For each $m \ge 2$, $\phi(m)$ denotes the number of numbers a with $1 \le a \le m$ that are coprime to m. The function ϕ is called *Euler's phi function* (or sometimes "Euler's totient").

Thus the number of elements of U_m is $\phi(m)$.

Notice that if $[a]$ and $[a']$ are units, then $[aa']$ is also a unit, because if $[a][b] = [1]$ and $[a'][b'] = [1]$, then $[aa'][bb'] = [1]$. So the set of units U_m is closed under multiplication.

The Primitive Root Theorem, stated in Section D, asserts that if p is prime, then there is some number b so that $\{0, b, b^2, \ldots, b^{p-1}\}$ is a complete set of representatives for $\mathbb{Z}/p\mathbb{Z}$. Then $[b]$ must be a unit, because $[1] = [b^r]$ for some r. Hence all powers of $[b]$ are also units. It follows that the converse of the Primitive Root theorem holds: if there is a number b so that $\{0, b, b^2, \ldots, b^{m-1}\}$ is a complete set of representatives for $\mathbb{Z}/m\mathbb{Z}$, then m must be prime. This is because every congruence class other than $[0]$ is of the form $[b^r]$ for some r, hence is a unit of $\mathbb{Z}/m\mathbb{Z}$, and so every number $<m$ is coprime to m.

Inverses are helpful for solving equations. If we can find the inverse of $[a]_m$ in $\mathbb{Z}/m\mathbb{Z}$, say $[r]_m$, then we can solve the equation $[a]_m X = [c]_m$ for any c: simply let $X = [r]_m[c]_m = [rc]_m$.

For example, in $\mathbb{Z}/17\mathbb{Z}$, the inverse of $[3]_{17}$ is $[6]_{17}$, so we may solve $[3]_{17}X = [11]_{17}$ by multiplying both sides by $[6]_{17}$. Since $[6]_{17}[3]_{17}X = [18]_{17}X = [1]_{17}X = X$, we have:

$$[3]_{17}X = [11]_{17}$$
$$[6]_{17}[3]_{17}X = [6]_{17}[11]_{17}$$
$$X = [6]_{17}[11]_{17} = [6 \cdot 11]_{17} = [66]_{17} = [15]_{17}.$$

Exercises.

42. (i) In $\mathbb{Z}/13\mathbb{Z}$, find the inverses of $[4]$, of $[5]$, of $[7]$.
(ii) In $\mathbb{Z}/13\mathbb{Z}$, solve
(a) $[4]X = [7]$,
(b) $[5]X = [12]$,
(c) $[7]X = [8]$.

43. (i) In $\mathbb{Z}/25\mathbb{Z}$, find the inverse of $[3]$, of $[11]$, of $[23]$.
(ii) In $\mathbb{Z}/25\mathbb{Z}$, solve
(a) $[11]X = [7]$,
(b) $[23]X = [12]$,
(c) $[3]X = [8]$.

44. In $\mathbb{Z}/12\mathbb{Z}$, decide which elements have inverses, and for each element that has an inverse, find the inverse.

45. Same question for $\mathbb{Z}/14\mathbb{Z}$.

46. Same question for $\mathbb{Z}/20\mathbb{Z}$.

47. Show that if $[a]_m = [a']_m$, and a and m are coprime, then a' and m are coprime.

48. Prove that $[a]_{mn}$ is a unit of $\mathbb{Z}/mn\mathbb{Z}$ iff $[a]_m$ is a unit of $\mathbb{Z}/m\mathbb{Z}$ and $[a]_n$ is a unit of $\mathbb{Z}/n\mathbb{Z}$.

Definition. A *complete set of units* modulo m is a set of integers $\{a_1, a_2, \ldots, a_m\}$, so that every unit of $\mathbb{Z}/m\mathbb{Z}$ is represented by exactly one integer in the set.

49. (i) If $p > 2$ is prime, show that $\{1, 2, \ldots, p - 1\}$ is a complete set of units modulo p.

(ii) Show that $\{1, 3, -1, -3\}$ is a complete set of units modulo 10.

(iii) Find a complete set of units modulo 24.

50. Show that if p is prime, then

$$\{1, -1, 2, -2, \ldots, \frac{p-1}{2}, -\frac{p-1}{2}\}$$

is a complete set of units modulo p.

51. Decide whether $\{2, 2^2, 2^3, \ldots, 2^{p-1}\}$ is a complete set of units modulo p, where

(i) $p = 11$

(ii) $p = 23$

(iii) $p = 31$

52. Show that if b is a primitive root modulo p, then

$$\{b, b^2, b^3, \ldots, b^{p-1}\}$$

is a complete set of units modulo p.

53. (i) Suppose $[b]$ is a unit of $\mathbb{Z}/m\mathbb{Z}$. Show that for all $[a]$ and $[a']$ in $\mathbb{Z}/m\mathbb{Z}$, if $[a] \neq [a']$ then $[ba] \neq [ba']$.

(ii) Show that if $\{a_1, a_2, \ldots, a_m\}$ is a complete set of units modulo m, and b is an integer with $(b, m) = 1$, then $\{ba_1, ba_2, \ldots, ba_m\}$ is also a complete set of units modulo m.

54. Let b be any integer. Define the function $f_b : U_m \to \mathbb{Z}/m\mathbb{Z}$ by "multiplication by $[b]_m$":

$$f_b([a]_m) = [b]_m[a]_m.$$

Show that f_b is a one-to-one function if and only if $[b]_m$ is a unit of $\mathbb{Z}/m\mathbb{Z}$.

55. Show that if d divides m, then any complete set of units mod m includes a complete set of units mod d.

F. Solving Equations in $\mathbb{Z}_m$

Consider the equation

$$[6]X = [14]$$

in $\mathbb{Z}/16\mathbb{Z}$. We can't solve this by finding the inverse of $[6]_{16}$ because $[6]_{16}$ is not a unit. But we can still find a solution, in fact two solutions. One solution comes from observing that $[14]_{16} = [30]_{16}$, so the equation becomes

$$[6]X = [30]$$

which has the obvious solution $[5]_{16}$. On the other hand,

$$[14] = [14 - 32] = [-18],$$

so the equation becomes

$$[6]X = [-18]$$

which has the obvious solution $X = [-3] = [13]$.

So if the coefficient of X is not a unit of $\mathbb{Z}/m\mathbb{Z}$, then the solution we found may not be unique.

To completely solve linear equations of the form $aX = b$ in $\mathbb{Z}/m\mathbb{Z}$, we have the following general result:

Proposition 7. *Suppose $X = x_0$ is a solution of the equation $aX = b$. Let $\mathcal{N}$ be the set of all solutions to the equation $aX = 0$. Then every solution to $aX = b$ has the form $X = x_0 + t$ for t in $\mathcal{N}$.*

Proof. First notice that if $ax_0 = b$ and t is a solution of $aX = 0$, then

$$a(x_0 + t) = ax_0 + at = b + 0 = b,$$

so any number of the form $x_0 + t$ for t in $\mathcal{N}$ is a solution of $aX = b$.

Conversely, suppose $ax_0 = b$ and also $az = b$ for some z. Then

$$a(z - x_0) = az - ax_0 = b - b = 0,$$

and so $t = z - x_0$ is in $\mathcal{N}$, the set of all solutions of $aX = 0$, and, of course, $z = x_0 + t = x_0 +$ (an element of $\mathcal{N}$), as claimed.   □

This proposition applies to linear equations in various contexts, such as in linear algebra and differential equations, as well as in modular arithmetic. The equation $aX = b$ is called *nonhomogeneous* if $b \neq 0$, and *homogeneous* if $b = 0$.

To see how it applies here, consider our example above.

Example 2. We saw that one solution of

$$[6]_{16}X = [14]_{16}$$

in $\mathbb{Z}/16\mathbb{Z}$ was $X = [5]_{16}$. To find all solutions of $[6]_{16}X = [14]_{16}$, we need to find all solutions of the homogeneous equation

$$[6]_{16}X = [0]_{16}.$$

Translating this equation into a congruence gives

$$6x \equiv 0 \pmod{16}$$

or

$$6x = 16y.$$

To solve this, divide both sides by 2, the greatest common divisor of 6 and 16, to get

$$3x = 8y.$$

Since 3 and 8 are coprime, 8 must divide x, and so the only solutions of this are $x = 8k$, $y = 3k$ for any integer k. Thus the solutions of the homogeneous equation $[6]_{16}X = [0]_{16}$ are $X = [8k]_{16}$ for any integer k. There are two different congruence classes in $\mathbb{Z}/16\mathbb{Z}$ of that form, $X = [8]_{16}$ and $X = [0]_{16}$. So in the notation of Proposition 7, $\mathcal{N} = \{[0]_{16}, [8]_{16}\}$.

Then the solutions of the original nonhomogeneous equation

$$[6]_{16}X = [14]_{16}$$

are $X = [5]_{16}$ and $X = [5]_{16} + [8]_{16} = [13]_{16} = [-3]_{16}$.

To sum up, then, solving a non-homogeneous equation $aX = b$ in $\mathbb{Z}/m\mathbb{Z}$ involves two separate tasks:

(i) Find some solution (if one exists) of the nonhomogeneous equation $aX = b$;
(ii) Find all solutions of the corresponding homogeneous equation $aX = 0$.

For the second task, we have the following generalization of Example 2:

Proposition 8. *If $d = (a, m)$, then the general solution in $\mathbb{Z}/m\mathbb{Z}$ of $[a]X = [0]$ is*

$$X = [\frac{m}{d}k]$$

for $k = 0, 1, 2, \ldots, d - 1$

Proof. Since $d = (a, m)$, we have

$$a\frac{m}{d} = m\frac{a}{d} \equiv 0 \pmod{m}.$$

So

$$[a][\frac{m}{d}k] = [\frac{amk}{d}] = [m\frac{ak}{d}] = [0].$$

Thus every congruence class of the form $X = [\frac{m}{d}k]$ is a solution of $[a]X = [0]$.

Conversely, if $[a][x] = [0]$, then $ax \equiv 0 \pmod{m}$, so $ax = mt$ for some integer t. Let $d = (a, m)$ and write $a = da_0$, $m = dm_0$. Then a_0 and m_0 are coprime, and dividing $ax = mt$ by d yields

$$a_0 x = m_0 t.$$

Thus m_0 divides $a_0 x$, and since a_0 and m_0 are coprime, it follows that m_0 must divide x. Thus $x = m_0 k$ for any k, and so the solutions of $[a][x] = [0]$ are $[x] = [m_0 k]_m$.

To see how many different solutions we obtain, notice that since $m = m_0 d$ we have

$$m_0 k \equiv m_0 l \pmod{m}$$

iff

$$m_0 k \equiv m_0 l \pmod{m_0 d}$$

iff

$$k \equiv l \pmod{d}$$

(using a cancellation property of Section 5E). Since $0, 1, \ldots, d-1$ is a complete set of representatives modulo d, the solutions of $[a][x] = [0]$ are $[x] = [0], [m_0], [2m_0], \ldots$ $[(d-1)m_0]$. Recalling that $m_0 = \frac{m}{d}$ completes the proof. $\square$

Example 3. Once we find that $X = [4]_{90}$ is a solution in $\mathbb{Z}/90\mathbb{Z}$ of

$$[36]_{90}X = [54]_{90},$$

then to find all solutions, we find all solutions to the corresponding homogeneous equation

$$[36]_{90}X = [0]_{90}.$$

Now $(36, 90) = 18$, so there are 18 solutions of $[36]_{90}X = [54]_{90}$, namely,

$$X = [4]_{90} + [5k]_{90} = [4 + 5k]_{90}$$

for $k = 0, 1, 2, \ldots, 17$.

Example 4. To find all solutions in $\mathbb{Z}/90\mathbb{Z}$ of

$$[7]_{90}X = [54]_{90}$$

we first find some solution. Now $[7]_{90}$ is a unit of $\mathbb{Z}/90\mathbb{Z}$, with inverse $[13]_{90}$ since $13 \cdot 7 = 91$, so we have a solution $X = [13]_{90}[54]_{90} = [702]_{90} = [-18]_{90}$. To find all solutions we find all solutions to the corresponding homogeneous equation

$$[7]_{90}X = [0]_{90}.$$

But since $[7]$ is a unit in $\mathbb{Z}/90\mathbb{Z}$, we may multiply both sides by $[7]^{-1} = [13]$ to get $X = [0]_{90}$. Hence $X = [-18]_{90} = [72]_{90}$ is the only solution of $[7]_{90}X = [54]_{90}$.

Example 5. Let $[a]_m$ be a unit of $\mathbb{Z}/m\mathbb{Z}$. If we apply the proposition to find all solutions in $\mathbb{Z}/m\mathbb{Z}$ of

$$[a]_m X = [1]_m$$

we find that since $(a, m) = 1$, the equation has a unique solution. Thus the inverse of $[a]_m$ in $\mathbb{Z}_m$ is unique.

Section 8A, Exercise 4 has a more general version of this result.

Exercises.

56. In $\mathbb{Z}/20\mathbb{Z}$,
   (i) find some solution to $[12]X = [28]$;
   (ii) find all solutions to the corresponding homogeneous equation $[12]X = [0]$;
   (iii) find all solutions to $[12]X = [28]$.

57. In $\mathbb{Z}/12\mathbb{Z}$,
    (i) find some solution to $[14]X = [18]$;
    (ii) find all solutions to the corresponding homogeneous equation $[14]X = [0]$;
    (iii) find all solutions to $[14]X = [18]$.

58. Explain why $[36]X = [6]$ has no solution in $\mathbb{Z}/45\mathbb{Z}$.

59. Show that if $[a]$ is a unit of $\mathbb{Z}/m\mathbb{Z}$, then
    (i) for any $[b]$, there is a solution of the equation $[a]X = [b]$;
    (ii) the only solution of the homogeneous equation $[a]X = [0]$ is $X = [0]$;
    (iii) the equation $[a]X = [b]$ has a unique solution for every $[b]$.

60. In $\mathbb{Z}/11\mathbb{Z}$, find all solutions of
    (i) $[6]X = [3]$
    (ii) $[8]X = [9]$
    (iii) $[5]X = [9]$.

61. In $\mathbb{Z}/15\mathbb{Z}$, find all solutions of:
    (i) $[36]X = [78]$
    (ii) $[42]X = [57]$
    (iii) $[25]X = [36]$.

62. In $\mathbb{Z}/30\mathbb{Z}$, find all solutions of:
    (i) $[4]X = [18]$
    (ii) $[9]X = [48]$
    (iii) $[10]X = [100]$
    (iv) $[12]X = [8]$
    (v) $[6]X = [2]$.

G. Trial Division

Congruence classes are helpful in thinking about how to factor numbers or to find prime numbers.

Factoring. How do we factor a large number?

If N is the number we wish to factor, the most naive method is to divide N by $2, 3, 4, 5, \ldots$, until we find a number that divides N. If we don't find any number $< \sqrt{N}$ which divides N, then N must be prime, for if $N = ab$, then at least one of a or b must be $< \sqrt{N}$.

This factoring method is called *trial division*.

Trial division is an algorithm that always works: it either finds a factor of N, or proves that N is prime.

However, trial division is slow. As just described, trial division would take $p - 1$ divisions to find the smallest prime divisor p of N, and that makes trial division totally unfeasible on even the fastest computers for numbers N used in cryptography (see Section 10A).

Nonetheless, trial division is useful as a first step in a factoring procedure. Before applying a sophisticated factoring algorithm, it makes sense to use trial division to find and divide out the small prime factors of the number to be factored.

Since trial division is a useful tool, it is worth looking at ways to make the tool a bit more efficient.

Suppose we want to use trial division on N to look for prime divisors $<10^9$. The least efficient way to apply trial division is to start dividing the number N by all numbers from 2 to 10^9. For once we find that 2 does not divide N, then clearly no multiple of 2 will divide N, so it is wasted effort to divide N by any even number >2. Similarly, once we find that 3 does not divide N, then neither will any multiple of 3.

To minimize the number of trial divisions, we could divide only by primes $<10^9$. For if N has a factor $m < 10^9$, then m, hence N, has a prime factor $<10^9$. Thus instead of doing $10^9 - 1$ divisions, we could just divide N by the approximately 50.8 million primes less than 10^9.

However, in order to reduce the number of divisions to 50.8 million, we either have to store the 50.8 million primes, or somehow test each number $<10^9$ for primeness before dividing. The former method is costly in preparation and memory; the latter is costly in time.

A useful compromise for improving the efficiency of trial division without having (or creating) large lists of prime numbers is to divide not by all numbers, but only by numbers that are not obviously composite.

For example, once we know 2 is not a factor of N, we don't need to divide N by even numbers. Once we know 5 is not a factor of N, we don't need to divide N by multiples of 5. Thus we could divide the number N only by 2, 5, and numbers that are coprime to $10 = 2 \cdot 5$. In terms of congruence classes, we could restrict trial division to 2, 5 and numbers in the congruence classes of 1, 3, 7, and 9 (mod 10).

Or we could restrict trial division of N to division by 2, 3, 5 and numbers in the congruence classes modulo $30 = 2 \cdot 3 \cdot 5$ that are coprime to 30, namely, numbers in the congruence classes of 1, 7, 11, 13, 17, 19, 23, and 29 (mod 30). Restricting to division by numbers in those eight congruence classes reduces the number of trial divisions to 8/15 the number needed if we were to trial divide by all odd numbers.

Trial division only by divisors in the congruence classes that are units modulo m is called a *wheel*. The congruence classes from which the trial divisors come are called the *spokes*. In the case of the mod 30 wheel, the wheel has 8 spokes.

It is worth recalling the facts which make the wheel method appropriate for trial division.

Let m be the modulus of the wheel. Modulo m, any two numbers in a given congruence class have the same greatest common divisor with m. (For example, 54 = 144 (mod 30), and (54, 30) = (144, 30) = 6.) Thus in a given congruence class modulo m, either all the numbers are coprime to m, hence units modulo m, hence are not multiples of any prime divisor of m, or all the numbers are not coprime to m.

So if we are searching for prime divisors of N that are coprime to m, then, it suffices to divide N only by numbers in those congruence classes modulo m that consist of units modulo m.

Thus in any wheel modulo m, the number of spokes needed for trial division is $\phi(m)$, the number of units in $\mathbb{Z}/m\mathbb{Z}$.

One suggestion [Wunderlich and Selfridge (1974)] was to use a wheel modulo $30030 = 2 \cdot 3 \cdot 5 \cdot 7 \cdot 11 \cdot 13$. This wheel has $\phi(30030) = 5760$ spokes and does trial division only by the prime divisors of 30030 and by numbers not divisible by 2, 3, 5, 7, 11 or 13. With the wheel modulo 30030, trial division is done by only 19 percent (5760/30030) of all numbers below a given bound, rather than by 50 percent of all numbers if trial division is done by all odd numbers, or 40 percent using the wheel modulo 10.

Exercises.

63. For each of the following numbers, use trial division to find a factor or show that the number is prime:

   (i) 433;
   (ii) 1247;
   (iii) 1261;
   (iv) 2413;

64. Compute the percentage of congruence classes modulo $210 = 2 \cdot 3 \cdot 5 \cdot 7$ that consist of numbers which are not coprime to 210.

65. If n is a number not in the congruence class of 1, 7, 11, 13, 17, 19, 23, and 29 mod 30, show that n must be a multiple of 2, 3, or 5.

66. (Review). Show that if $a \equiv b \pmod{m}$, then $(a, m) = (b, m)$.

67. Write a program to do trial division up to 1000 by the mod 30 wheel with 8 spokes.

68. Suppose your computer had ready access to a table of the 78,498 primes below 1,000,000. Compare the number of divisions needed to test a large number N for division by a number $< 1,000,000$ by the wheel modulo 30030, and by trial division by the 78,498 primes $< 1,000,000$.

Sieves. Trial division is useful as the first step in finding large prime numbers. The idea, in fact, goes back to Eratosthenes (200 B.C.).

To find possible prime numbers in some interval, we discard all the numbers in the interval that are divisible by small primes. Then most of the numbers will be eliminated, and the remaining numbers will be potential primes.

If we seek the prime numbers less than 100, for example, this method works perfectly.

We know the primes <10. For the others <100, write down all the numbers from 1 to 100. If we first cross out all multiples of 2 and 5, we are left with numbers ending in 1, 3, 7, and 9:

$$
\begin{array}{cccc}
1 & 3 & 7 & 9 \\
11 & 13 & 17 & 19 \\
21 & 23 & 27 & 29 \\
31 & 33 & 37 & 39 \\
41 & 43 & 47 & 49 \\
51 & 53 & 57 & 59 \\
61 & 63 & 67 & 69 \\
71 & 73 & 77 & 79 \\
81 & 83 & 87 & 89 \\
91 & 93 & 97 & 99
\end{array}
$$

Then we cross out all multiples of 3 and 7.

$$
\begin{array}{cccc}
11 & 13 & 17 & 19 \\
   & 23 &   & 29 \\
31 &   & 37 &   \\
41 & 43 & 47 &   \\
   & 53 &   & 59 \\
61 &   & 67 &   \\
71 & 73 &   & 79 \\
   & 83 &   & 89 \\
   &   & 97 &
\end{array}
$$

What remains are the numbers <100 that are not multiples of 2, 3, 5, or 7. But these are all prime. This is because any composite number <100 must be divisible by some prime $<\sqrt{100}$.

The same strategy will work, albeit not so perfectly, for much larger numbers. For example, suppose we wish to find primes in the set of numbers between 1194601 and 1194700. By a result in Section 4C, we should expect that the chance that a given randomly chosen number in that interval is prime is around $1/\ln(1194600) \sim 1/13$, hence we should expect something like 7 or 8 primes in that interval.

To find the primes, we trial divide to discard all numbers that are divisible by small primes. To start, we discard the numbers divisible by 2 or 5. We are left with the following set of numbers.

$$
\begin{array}{cccc}
1194601 & 1194603 & 1194607 & 1194609 \\
1194611 & 1194613 & 1194617 & 1194619 \\
1194621 & 1194623 & 1194627 & 1194629 \\
1194631 & 1194633 & 1194637 & 1194639 \\
1194641 & 1194643 & 1194647 & 1194649 \\
1194651 & 1194653 & 1194657 & 1194659 \\
1194661 & 1194663 & 1194667 & 1194669 \\
1194671 & 1194673 & 1194677 & 1194679 \\
1194681 & 1194683 & 1194687 & 1194689 \\
1194691 & 1194693 & 1194697 & 1194699
\end{array}
$$

Next, we discard all multiples of 3: since 1194600 is a multiple of 3, multiples of 3 are those ending in 03, 09, 21, 27, 33, 39, 51, 57, 63, 69, 81, 87, 93, and 99. Then we discard multiples of 7, starting with 1194613: 13, 27, 41, 55, 69, 83, 97; then multiples of 11, starting with 1194600: 11, 77; and multiples of 13, starting with 1194609: 09, 35, 61, 87. This leaves nineteen numbers out of the original 100 that are not multiples of 2, 3, 5, 7, 11, or 13:

1194601		1194607	
		1194617	1194619
	1194623		1194629
1194631		1194637	
	1194643	1194647	1194649
	1194653		1194659
		1194667	
1194671	1194673		1194679
			1194689
1194691			

If we continue trial division by the primes between 17 and 53, we find that 1194607 is a multiple of 17, 1194617 is a multiple of 41, 1194619 is a multiple of 37, 1194643 is a multiple of 23, 1194647 is a multiple of 31, 1194673 is a multiple of 53, 1104689 is a multiple of 23, and 1194691 is a multiple of 31. This leaves eleven numbers:

1194601			
	1194623		1194629
1194631		1194637	
			1194649
	1194653		1194659
		1194667	
1194671			1194679

Continuing to trial divide by primes <100 does not allow us to discard any more of these numbers.

In Chapter 20 we will give a method, not trial division, that will efficiently decide with almost perfect certainty whether a number is prime. For now, if you are really curious about which six of the numbers above are prime, and which five are not, you could continue trial division: for if one of those numbers does not have a divisor less than the square root of 1194699, that is, less than 1094, it must be prime.

Exercises.

69. Find all primes between 200 and 250.

70. Find all primes between 700 and 800.

71. Find the smallest number >120120 which is divisible by no prime <20.

Chapter 7
Rings and Fields

In this chapter we introduce and apply to $\mathbb{Z}/m\mathbb{Z}$ some of the most basic concepts of "abstract" algebra: the concepts of group, ring, field, and ring homomorphism.

A. Axioms

Suppose we wish to find an integer x that solves the equation

$$(3+x)+4 = 9.$$

To solve the equation, we need to use various properties of equality: symmetry (if $a = b$ then $b = a$), transitivity (if $a = b$ and $b = c$, then $a = c$), and "al-jabr" (if $a = b$, then $a+c = b+c$). We also use properties of $\mathbb{Z}$. We'll keep track of these as we solve the equation, one step at a time.

First, we simplify the left hand side. We have

$$3+x = x+3 \text{ (commutativity)}$$

we may add 4 to both sides (al-jabr) to get

$$(3+x)+4 = (x+3)+4.$$

Now

$$(x+3)+4 = x+(3+4) \text{ (associativity)} = x+7.$$

By transitivity of equality, we get

$$(3+x)+4 = x+7,$$

and then, by symmetry and transitivity of equality,

$$x+7 = 9.$$

L.N. Childs, *A Concrete Introduction to Higher Algebra*, Undergraduate Texts in Mathematics, © Springer Science+Business Media LLC 2009

Now we add -7 to both sides:

$$(x+7)+(-7)=9+(-7)=2.$$

The left side becomes

$$(x+7)+(-7)=x+(7+(-7)) \text{ (associativity)}$$

and we know

$$7+(-7)=0 \text{ (property of negatives)}$$

so adding x to both sides,

$$x+(7+(-7))=x+0.$$

By transitivity of equality,

$$(x+7)+-7=x+0.$$

Now $x+0=x$ (property of zero), so by transitivity and symmetry of equality, we finally get

$$x=2.$$

In this (painful) solution of the equation, we used these properties of the integers:

- closure of addition: the sum of two integers is an integer: if a,b are integers, so is $a+b$: thus, if $3+x$ is an integer, so is $(3+x)+4$;
- commutativity of addition: $a+b=b+a$ for all integers a and b: thus, $3+x=x+3$;
- associativity of addition: $(a+b)+c=a+(b+c)$ for all integers a,b,c: thus, $(x+3)+4=x+(3+4)$;
- 0 is an additive identity: for all integers b, $b+0=b$: thus $x+0=x$; and
- any integer has a negative: for any integer b there is an integer $-b$ so that $b+(-b)=0$: thus, $7+(-7)=0$.

These five properties mean that the set $\mathbb{Z}$ of integers with the operation of addition $(+)$, is an *abelian group*. The term "abelian" refers to the condition that addition is commutative.

Now, in the rational numbers $\mathbb{Q}$, let's solve this equation:

$$(3x+1)\cdot\frac{2}{5}=7.$$

We'll use the three properties of equality listed above, and also:
  if $a=b$ then $ac=bc$.

For the left side, we see that

$$(3x+1)\cdot\frac{2}{5} = (3x)\cdot\frac{2}{5}+1\cdot\frac{2}{5} \text{ (distributivity)}$$

$$= (3x)\cdot\frac{2}{5}+\frac{2}{5} \text{ (1 is a multiplicative identity)}$$

$$= 3\cdot\left(x\cdot\frac{2}{5}\right)+\frac{2}{5} \text{ (associativity)}$$

$$= 3\cdot\left(\frac{2}{5}\cdot x\right)+\frac{2}{5} \text{ (commutativity)}$$

$$= \left(3\cdot\frac{2}{5}\right)\cdot x+\frac{2}{5} \text{ (associativity)}$$

$$= \frac{6}{5}\cdot x+\frac{2}{5}.$$

So by symmetry and transitivity of equality, the original equation yields

$$\frac{6}{5}\cdot x+\frac{2}{5}=7.$$

Now we add $\left(-\frac{2}{5}\right)$ to both sides. The right side becomes $7+\left(-\frac{2}{5}\right)=\frac{33}{5}$. Working with the left side, we get

$$\left(\frac{6}{5}\cdot x+\frac{2}{5}\right)+\left(-\frac{2}{5}\right) = \frac{6}{5}\cdot x+\left(\frac{2}{5}+\left(-\frac{2}{5}\right)\right) \text{ (associativity)}$$

$$= \frac{6}{5}\cdot x+0 \text{ (negatives)}$$

$$= \frac{6}{5}\cdot x \text{ (property of 0).}$$

So the equation becomes

$$\frac{6}{5}\cdot x=\frac{33}{5}.$$

Multiplying both sides by $\frac{5}{6}$ gives

$$\frac{5}{6}\cdot\left(\frac{6}{5}\cdot x\right)=\frac{5}{6}\cdot\frac{33}{5}.$$

By associativity and the property of the identity element 1, the left side becomes

$$\left(\frac{5}{6}\cdot\frac{6}{5}\right)\cdot x=1\cdot x=x,$$

so the original equation becomes

$$x=\frac{5}{6}\cdot\frac{33}{5}=\frac{33}{6}.$$

Here, in addition to the abelian group properties of addition used to solve the first equation, we used:

- closure of multiplication, which means, the product of two rational numbers is a rational number: if a and b are rational numbers, then so is $a \cdot b$: thus, $(3x) \cdot \frac{2}{5}$ is a rational number;
- distributivity of multiplication over addition: for all a,b,c in $\mathbb{Q}$, $(a+b) \cdot c = a \cdot c + b \cdot c$: thus, $(3x+1) \cdot \frac{2}{5} = (3x) \cdot \frac{2}{5} + 1 \cdot \frac{2}{5}$;
- associativity of multiplication: for all a,b,c in $\mathbb{Q}$, $(a \cdot b) \cdot c = a \cdot (b \cdot c)$; thus, $(3x) \cdot \frac{2}{5} = 3 \cdot (x \cdot \frac{2}{5})$;
- commutativity of multiplication: for all a,b in $\mathbb{Q}$, $a \cdot b = b \cdot a$; thus, $(x \cdot \frac{2}{5}) = (\frac{2}{5} \cdot x)$;
- 1 is a multiplicative identity: for all a in $\mathbb{Q}$, $1 \cdot a = a$; thus, $1 \cdot \frac{2}{5} = \frac{2}{5}$;
- every rational number except 0 has an inverse: for all $a \neq 0$ in $\mathbb{Q}$, there exists a rational number a^{-1} so that $a \cdot a^{-1} = 1$; thus, $\frac{5}{6} \cdot \frac{6}{5} = 1$.

These properties mean that the set $\mathbb{Q}$ of rational numbers with the operations of addition ($+$) and multiplication ($\cdot$), is a *field*.

The set $\mathbb{Z}$ of integers with addition and multiplication satisfies all of the properties of $\mathbb{Q}$ listed above except the existence of multiplicative inverses. Such a set is called a *commutative ring*.

These properties of numbers seem so natural that you probably lost patience with how tediously we solved those two equations.

But if you have seen calculus in 3-space, consider the set $\mathbb{R}^3$ of vectors in space. There are two operations on $\mathbb{R}^3$, vector addition and the cross product. The three unit vectors in $\mathbb{R}^3$ are $i = (1,0,0), j = (0,1,0)$ and $k = (0,0,1)$, and the cross product of these vectors is given by:

$$i \times i = j \times j = k \times k = 0,$$
$$i \times j = -j \times i = k, j \times k = -k \times j = i, k \times i = -i \times k = j.$$

Every vector (a,b,c) in $\mathbb{R}^3$ is a linear combination of i, j and k: $(a,b,c) = ai + bj + ck$, so we extend the cross product to all vectors in $\mathbb{R}^3$ by distributivity.

Example 1. Using the same steps we used in solving $(3 \cdot x + 1) \cdot 2 = 7$ in $\mathbb{Q}$, let us try to solve the equation

$$((i \times v) + j) \times j = k$$

where $v = (x,y,z) = xi + yj + zk$ is a vector with unknown components x,y,z to be found.

Distributivity works, to give

$$(i \times v) \times j + j \times j = k,$$

and $j \times j = 0$, the zero vector. But then, to work with

$$(i \times v) \times j = k,$$

we can't use commutativity and associativity as we did in $\mathbb{Q}$, because both fail (for example, $(i \times j) \times j \neq i \times (j \times j)$). What to do? In fact, none of the properties pertaining purely to multiplication work with the cross product except closure: no associativity, no commutativity, no identity element, hence, of course, no multiplicative inverses. (Since $v = xi + yj + zk$, we can try to solve for the components of v. Using distributivity, we find that $(i \times v) = -zj + yk$, then $(-zj + yk) \times j = -yi$, so the equation becomes

$$-yi = k,$$

which is impossible.)

So if we have a set of objects on which the operations of addition and multiplication are defined, and we want to solve equations in that set, it is very helpful that the operations satisfy the properties we are accustomed to using with numbers.

Now we codify the definitions we gave above.

To define a group, we start with a set G with an operation $*$. The operation $*$ may be thought of as a function whose domain is all of $G \times G$ (ordered pairs of elements of G) and whose range is G. That means that for every ordered pair (a,b) in $G \times G$, $a * b$ is an element of G. The property that $*$ is defined for all pairs of elements of G is often described by saying that G is *closed* under the operation $*$.

Definition. A set G with an operation $*$ is a *group* if:

(associativity) For all a,b,c in G, $(a * b) * c = a * (b * c)$.

(identity) There exists a special element e in G, called the *identity*, so that for all a in G, $e * a = a * e = a$.

(inverse) For every a in G, there is an element b in G so that $a * b = b * a = e$. The group G is called *abelian* if in addition:

(commutativity) For all a,b in G, $a * b = b * a$.

With the operation $+$, $\mathbb{Z}$ and $\mathbb{Q}$ are abelian groups.

Definition. A *ring (with identity)* is a set R with two operations, $+$ and $\cdot$, and two special elements, 0 and 1, that satisfy:

R with the operation $+$ (addition) is an abelian group with identity 0, called the *zero element* of R. The element b so that $a + b = b + a = 0$ is the *negative* of a.

R with the operation $\cdot$ (multiplication) satisfies the associative property, and the element 1 is the identity element under multiplication.

R with $+$ and $\cdot$ satisfies the distributive laws: for every a,b,c in R, $a(b+c) = (ab) + (ac)$, and $(a+b)c = (ac) + (bc)$.

If in addition, the multiplication $\cdot$ on R satisfies the commutative law: for all a,b in R, $a \cdot b = b \cdot a$, then R is called a *commutative ring*.

Examples: $\mathbb{Z}$, $\mathbb{Q}$, $\mathbb{R}, \mathbb{C}$, and the sets $\mathbb{Z}/m\mathbb{Z}$ of congruence classes of integers mod m are all commutative rings.

Some examples of sets that are not rings are:

the set $\mathbb{N}$ of natural numbers, with the usual operations of $+$ and $\cdot$;

the set $\mathbb{R}_+$ of all nonnegative real numbers with the usual operations of $+$ and $\cdot$;

the set $\mathbb{Z} - \{3\}$ of all integers except 3 with the usual operations of $+$ and $\cdot$; and the set $\mathbb{R}^3$ of all vectors in real three-space, with vector addition, and cross product as multiplication, as we observed above.

The set $\mathbb{Z} - \{3\}$ is not a ring with respect to the usual addition and multiplication in $\mathbb{Z}$ for the reason that $\mathbb{Z} - \{3\}$ is not closed under addition: if a, b are in $\mathbb{Z} - \{3\}$, then $a + b$ need not be in $\mathbb{Z} - \{3\}$. For example, $1 + 2 = 3$: 1 and 2 are in $\mathbb{Z} - \{3\}$, but 3 is not.

If R is a ring, and S is a subset of R that is closed under addition, multiplication, taking negatives, and has 0 and 1, then S is also a ring. To see this, one has to check the properties, associativity of addition, distributivity, etc. But all of them hold in S because S is a subset of R, and all of the operations on S are the same as those on R. So the axioms are valid for S because they are valid for R. When R is a ring and S is a subset of R which is a ring with the operations those of R, we call S a *subring* of R.

Example 2. $\mathbb{Z}$ can be thought of as a subset of $\mathbb{Q}$ by identifying the integer a with the rational number $a/1$. Then $\mathbb{Z}$ is a subring of $\mathbb{Q}$. Similarly, $\mathbb{R}$ is a subring of $\mathbb{C}$. But $\mathbb{Z}/m\mathbb{Z}$ is not a subring of $\mathbb{Z}$, since $\mathbb{Z}/m\mathbb{Z}$ is not a subset of $\mathbb{Z}$. Rather, it is a set of subsets of $\mathbb{Z}$, the congruence classes of $\mathbb{Z}$ modulo m.

We now verify that $\mathbb{Z}/m\mathbb{Z}$ is a commutative ring. We shall write $[a]_m$ as $[a]$ if the modulus m is clear from the context.

Theorem 1. $\mathbb{Z}/m\mathbb{Z}$ *is a commutative ring with identity for every* $m \geq 2$.

Proof (Sketch). We defined addition, multiplication, and subtraction in $\mathbb{Z}/m\mathbb{Z}$ by $[a] + [b] = [a + b]$, $-[a] = [-a]$, $[a] \cdot [b] = [a \cdot b]$. Set $1 = [1], 0 = [0]$. With these definitions, it is easy to show that since $\mathbb{Z}$ is a commutative ring with identity, then so is $\mathbb{Z}/m\mathbb{Z}$. For example, to verify the associative law for multiplication, let a, b, c be any elements of $\mathbb{Z}$, then

$$[a] \cdot ([b] \cdot [c]) = [a] \cdot [b \cdot c]$$
$$= [a \cdot (b \cdot c)]$$
$$= [(a \cdot b) \cdot c] \text{ (since associativity holds in } \mathbb{Z})$$
$$= [a \cdot b] \cdot [c]$$
$$= ([a] \cdot [b]) \cdot [c].$$

The other properties are equally easy to verify.   □

Units. To define a field, it is helpful to look at the set of invertible elements of a commutative ring:

Definition. An element a of a commutative ring R is called a *unit* of R if there exists some b in R so that $a \cdot b = b \cdot a = 1$.

Example 3. In $\mathbb{Z}$ only 1 and -1 are units. In $\mathbb{Q}$ every nonzero rational number is a unit.

The statements "a is a unit" and "a has an inverse" mean the same thing.

We explored the units of $\mathbb{Z}/m\mathbb{Z}$ in Chapter 6E, and found that $[a]_m$ in $\mathbb{Z}/m\mathbb{Z}$ is a unit iff there is an integer b so that $ab \equiv 1 \pmod{m}$, if and only if $(a,m) = 1$.

A useful fact about units is that *units are closed under multiplication*; that is, if a and b are units of R, so is ab. For if a,b are units of a ring R, and a^{-1}, b^{-1} are their inverses, then ab has an inverse also, namely, $b^{-1}a^{-1}$. Thus *the units of a ring R form a group under multiplication.*

We will denote the group of units of R by U_R.

Now we define a field.

Definition. A *field* F is a commutative ring (hence is a set with addition, multiplication, 0, and 1 satisfying all the properties of a commutative ring) with two additional properties:

(inverses) Each $a \neq 0$ in F is a unit.

(non-triviality) F has at least two elements.

Thus F is a field if F is a commutative ring and the group of units U_F contains all elements of F except the zero element 0.

Examples of fields include $\mathbb{Q}, \mathbb{R}, \mathbb{C}$, but not $\mathbb{Z}$. We will determine the m for which $\mathbb{Z}/m\mathbb{Z}$ is a field in Section 7C.

Here are some basic properties of groups and rings:

Proposition 2. *A group has only one identity element.*

Proof. Suppose e and e' are both identity elements. Then $e * e' = e'$ since e is an identity element, and $e * e' = e$ since e' is an identity element. So $e = e'$.   □

Proposition 3. *A ring with identity contains only one zero element and only one identity element.*

Proof. If R is a ring, then R is a group under addition, so has only one zero element by the last proposition. If 1 and $1'$ are two identity elements, then $1 = 1 \cdot 1' = 1'$ as in the proof of the last proposition.   □

Proposition 4. *In a group, an element has only one inverse.*

Proof. Given g in the group, let h and k be inverses for g. Then $h * g = g * h = e$, and $g * k = e$. Then

$$g * h = g * k.$$

Multiplying by h on the left gives:

$$h * (g * h) = h * (g * k)$$
$$(h * g) * h = (h * g) * k \text{ by associativity}$$
$$e * h = e * k \text{ since } h * g = e,$$
$$h = k$$

                □

This proposition implies that in a ring, an element a has only one negative, denoted $-a$, and if b is a unit of a ring with identity, then b has only one inverse, denoted b^{-1}.

Exercises.

1. In a ring with identity, prove that $(-1) \cdot (-1) = 1$.

2. In a ring R with identity, prove that $-(-a) = a$ for all a in R.

3. Show that in a ring R, if $a + b = d$ and $a + c = d$, then $b = c$.

4. Show that in a ring R, if $a \cdot b = b \cdot a = 1$ and $a \cdot c = 1$, then $b = c$.

5. Show that in a ring R, if a has an inverse in R, then there is a unique solution in R to the equation $ax = d$.

6. Prove that for all a, b in a ring R, $(-a)b = -(ab) = a(-b)$

7. Show that if R is a ring, then for every b in R, $b \cdot 0 = 0$.

8. Let G be a group, with operation $*$ and identity element e. Prove left cancellation in G: for all a, b, c in G, if $a * b = a * c$, then $b = c$.

9. Let G be a group, with operation $*$. Prove left solvability in G: for every a and b in G, there is some x in G so that $a * x = b$.

10. Prove: A field F is a commutative ring with identity and with at least two elements, such that for all $a \neq 0$ and b in F, the equation $ax = b$ has a unique solution in F.

11. Suppose F is a field, and a is a nonzero element of F. Show that if r, s are in F and $ar = as$, then $r = s$.

12. Determine which of the axioms for a commutative ring hold and which fail for
(i) $\mathbb{N}$;
(ii) $\mathbb{R}_+$.

13. For a/b and c/d rational numbers, say $a/b \equiv c/d \pmod 1$ if $(a/b) - (c/d)$ is an integer. Call the set of congruence classes mod 1, $\mathbb{Q}/\mathbb{Z}$.
(i) Show that every rational number is congruent (mod 1) to a rational number a/b with $0 \le a/b < 1$.
(ii) Define addition and multiplication in $\mathbb{Q}/\mathbb{Z}$ by working with representatives between 0 and 1 as follows. If $0 < \frac{a}{b} < 1$ and $0 < \frac{c}{d} < 1$ then $\frac{a}{b} \cdot \frac{c}{d} = \frac{ac}{bd}$ and $\frac{a}{b} + \frac{c}{d} = $ the fractional part of $\frac{ad+bc}{bd}$.
Is $\mathbb{Q}/\mathbb{Z}$ a commutative ring? a field? Check the axioms.

14. In $\mathbb{R}^3$ with the operations of vector addition and crossed product:
    (i) can you find two vectors **v** and **w** so that

$$\mathbf{v} \times \mathbf{w} = \mathbf{w} \times \mathbf{v}$$

and $\mathbf{v} \times \mathbf{w}$ is not the zero vector?
    (ii) can you find three vectors **v**, **w** and **y** so that

$$(\mathbf{v} \times \mathbf{w}) \times \mathbf{y} = \mathbf{v} \times (\mathbf{w} \times \mathbf{y})$$

and $(\mathbf{v} \times \mathbf{w}) \times \mathbf{y}$ is not the zero vector?

15. Find the units of $\mathbb{Z}/6\mathbb{Z}$; of $\mathbb{Z}/7\mathbb{Z}$; of $\mathbb{Z}/8\mathbb{Z}$.

16. If m is odd, find the inverse of $[2]_m$ in $\mathbb{Z}/m\mathbb{Z}$.

17. Find the units of $\mathbb{Z}/21\mathbb{Z}$ and verify that they are closed under multiplication.

B. FOIL

In this section we look at some ancient and modern consequences of the axioms of a commutative ring.

    Let's start with an ancient property. Proposition 1 of Book II of Euclid's *Elements* is the distributive law in the generalized form:

$$A(B+C+D+\ldots) = AB+AC+AD+\ldots.$$

Proposition II-5 of Euclid's *Elements* is the property,

$$A^2 - B^2 = (A+B)(A-B).$$

Perhaps it is worthwhile, if tedious, to prove this from the properties of a commutative ring:

$$
\begin{aligned}
(A+B)(A-B) &= (A+B)(A+(-B)) \\
&= A(A+(-B)) + B(A+(-B)) \text{ (distributive law)} \\
&= (A^2 + A(-B)) + (BA + B(-B)) \text{ (distributive law)} \\
&= (A^2 + (-(AB))) + (BA + (-(B^2))) \text{ (by Exercise 6, above)} \\
&= (A^2 + (-AB))) + (AB + (-(B^2))) \text{ (commutativity of multiplication)} \\
&= (A^2 + (-AB + AB)) + (-(B^2)) \text{ (associativity of addition)} \\
&= (A^2 - 0) - B^2 \text{ (definition of negatives)} \\
&= A^2 - B^2 \text{ (definition of zero)}
\end{aligned}
$$

Table 7.1 Ordinary multiplication

	a_1	a_0
	b_1	b_0
	$a_1 b_0$	$a_0 b_0$
$a_1 b_1$	$a_0 b_1$	
$a_1 b_1$	$a_0 b_1 + a_1 b_0$	$a_0 b_0$

The three applications of the distributive law in the first two lines of the proof are equivalent to an application of the mnemonic, "FOIL". To multiply $(a+b)(c+d)$, multiply the First terms, then the Outside terms, then the Inside terms, then the Last terms, and then add them, to get $ac + ad + bc + bd$.

Here is an easy consequence of FOIL:

Proposition 5. *For all a_0, a_1, b_0, b_1 in a commutative ring,*

$$a_1 b_0 + a_0 b_1 = (a_1 b_1 + a_0 b_0) - (a_1 - a_0)(b_1 - b_0).$$

This proposition may seem routine, but in fact it has a rather interesting consequence discovered only in 1962, known as:

Karatsuba multiplication. Consider the usual algorithm for multiplying two 2-digit numbers $a_1 r + a_0$ and $b_1 r + b_0$ in base r, as laid out in Table 7.1.

Ordinary multiplication uses the distributive law, associativity of addition and commutativity of addition and multiplication:

$$(a_1 r + a_0)(b_1 r + b_0) = (a_1 b_1 r^2 + a_1 b_0 r) + (a_0 b_1 r + a_0 b_0)$$
$$= a_1 b_1 r^2 + a_1 b_0 r + a_0 b_1 r + a_0 b_0$$
$$= a_1 b_1 r^2 + (a_1 b_0 + a_0 b_1) r + a_0 b_0$$

and involves four multiplications of digits: $a_1 b_1$, $a_0 b_1$, $a_1 b_0$, and $a_0 b_0$.

Proposition 5 shows that we can replace the middle term $a_0 b_1 + a_1 b_0$ by $(a_1 b_1 + a_0 b_0) - (a_1 - a_0)(b_1 - b_0)$, in which case the algorithm is

$$(a_1 r + a_0)(b_1 r + b_0) = a_1 b_1 r^2 + (a_1 b_0 + a_0 b_1) r + a_0 b_0$$
$$= a_1 b_1 r^2 + (a_1 b_1 r + a_0 b_0 r - (a_1 - a_0)(b_1 - b_0) r) + a_0 b_0$$
$$= (a_0 b_0 r + a_0 b_0) + (a_1 b_1 r^2 + a_1 b_1 r) - (a_1 - a_0)(b_1 - b_0) r,$$

which may be laid out as in Table 7.2.

This way of multiplying is called *Karatsuba multiplication*. Multiplying takes more time than adding (why were logarithms invented in 1614?), so Karatsuba multiplication has an advantage over the usual algorithm for multiplying two two-digit numbers—we only need to do three multiplications of digits:

Table 7.2 Karatsuba multiplication

		a_1	a_0
		b_1	b_0
		a_0b_0	a_0b_0
	a_1b_1	a_1b_1	
		$-(a_1-a_0)(b_1-b_0)$	
a_1b_1		$a_0b_1+a_1b_0$	a_0b_0

$$a_0b_0$$
$$a_1b_1$$
$$(a_1-a_0)(b_1-b_0)$$

rather than four multiplications:

$$a_0b_0$$
$$a_1b_1$$
$$a_1b_0$$
$$a_0b_1$$

The strength of Karatsuba multiplication lies in extending the method to large numbers.

To multiply two numbers of four digits each in base 10, write them as $a_1 \cdot 10^2 + a_0$ and $b_1 \cdot 10^2 + b_0$. Then using Karatsuba,

$$(a_1 \cdot 10^2 + a_0)(b_1 \cdot 10^2 + b_0)$$
$$= (a_1b_1 \cdot 10^4 + (a_1b_1 + a_0b_0 - (a_1-a_0)(b_1-b_0))10^2 + a_0b_0.$$

This involves three multiplications of two-digit numbers:

$$a_0b_0$$
$$a_1b_1$$
$$(a_1-a_0)(b_1-b_0).$$

Each of these multiplications requires three multiplications of single digit numbers, for a total of nine multiplications of single digit numbers. The usual method requires 16 single digit multiplications.

In general, we can show

Proposition 6. *Multiplying two numbers of 2^n digits each in any base r requires 4^n digit multiplications with the usual algorithm, and 3^n digit multiplications with Karatsuba.*

Thus for multiplying two $32 = 2^5$-digit numbers, Karatsuba requires $3^5 = 243$ digit multiplications, rather than $4^5 = 1024$ digit multiplications for the usual algorithm.

Proof. For $n = 1$ we showed that the usual algorithm requires 4 digit multiplications, and Karatsuba requires 3. Suppose the result is true for numbers of 2^k digits. If we have two numbers of 2^{k+1} digits each, write them as $a = a_1 2^k + a_0$ and $b = b_1 2^k + b_0$ where a_1, a_0, b_1, b_0 have at most 2^k digits. Karatsuba computes the product ab using three multiplications of numbers with 2^k digits, instead of four multiplications. If we assume, by induction, that multiplying two numbers of 2^k digits takes 3^k digit multiplications via Karatsuba, and 4^k via the usual algorithm, then to multiply a and b by the usual algorithm requires $4 \cdot 4^k = 4^{k+1}$ digit multiplications, and and by Karatsuba requires $3 \cdot 3^k = 3^{k+1}$ digit multiplications. That proves the proposition by induction. $\square$

Karatsuba multiplication is an example where the desire for fast algorithms for computing led to a fresh look at one of the most basic algorithms in mathematics.

Exercises.

18. Prove Euclid's Proposition 1 in the form

$$A \cdot (B_1 + B_2 + \ldots + B_n) = A \cdot B_1 + A \cdot B_2 + \ldots A \cdot B_n$$

for all $A, B_1, \ldots, B_n$ in a ring, by induction on n.

19. Try Karatsuba multiplication on
  (i) $37 \cdot 56$
  (ii) $3456 \cdot 4528$

20. Try Karatsuba multiplication on $(56, 45)_{60} \cdot (37, 34)_{60}$ (two numbers in base 60, as used by the ancient Babylonians).

C. Zero divisors and $\mathbb{Z}/m\mathbb{Z}$

In $R = \mathbb{Z}$ or any field, the following property holds:

(nzd) For all a, b in R, if $ab = 0$, then $a = 0$ or $b = 0$.

A *nonzero* element a of a ring R for which there is some b, also not zero, with $ab = 0$, is called a *zero divisor*. The terminology is reasonable, for if a, b, c are numbers and $ab = c$, then a is a divisor of c, or a "c-divisor."

A commutative ring, such as $\mathbb{Z}$, which satisfies (nzd) is said to have *no zero divisors*. Thus the ring $\mathbb{Z}$ has no zero divisors.

But there are rings that do have zero divisors, and we have begun to encounter them. To take the smallest example, consider $\mathbb{Z}/4\mathbb{Z}$:

$$[2]_4 \cdot [2]_4 = [4]_4 = [0]_4,$$

the zero element of $\mathbb{Z}/4\mathbb{Z}$. Thus $[2]_4$ is a zero divisor in $\mathbb{Z}/4\mathbb{Z}$, so $\mathbb{Z}/4\mathbb{Z}$ has zero divisors. Or in $\mathbb{Z}/6\mathbb{Z}$,

$$[3]_6 \cdot [4]_6 = [12]_6 = [0]_6,$$

so both $[3]_6$ and $[4]_6$ are zero divisors in $\mathbb{Z}/6\mathbb{Z}$.

If you have encountered matrices, then you probably know that the set M of 2×2 matrices with real entries has an addition and a multiplication on it, and if we set

$$I = \begin{pmatrix} 1 & 0 \\ 0 & 1 \end{pmatrix}, 0 = \begin{pmatrix} 0 & 0 \\ 0 & 0 \end{pmatrix}$$

then I and 0 act as the identity and zero elements, respectively. Then the set M is a ring. But M is not commutative, for

$$\begin{pmatrix} 0 & 1 \\ 0 & 0 \end{pmatrix} \begin{pmatrix} 2 & 3 \\ 0 & 0 \end{pmatrix} = \begin{pmatrix} 0 & 0 \\ 0 & 0 \end{pmatrix},$$

while

$$\begin{pmatrix} 2 & 3 \\ 0 & 0 \end{pmatrix} \begin{pmatrix} 0 & 1 \\ 0 & 0 \end{pmatrix} = \begin{pmatrix} 0 & 2 \\ 0 & 0 \end{pmatrix},$$

and M also has zero divisors, as you can see.

In a commutative ring R, if a and b are not 0 and $ab = 0$, then we'll call b a *complementary zero divisor for a*, or say that a and b are *complementary zero divisors*.

If a is a zero divisor, then a may have many complementary zero divisors. For example, in $\mathbb{Z}/12\mathbb{Z}$, $[6]$ has complementary zero divisors $[2], [4], [6], [8]$ and $[10]$. Any non-zero solution of $[6]_{12}X = [0]_{12}$ is a complementary zero divisor for $[6]_{12}$.

Non-zero divisors can be canceled:

Proposition 7 (Cancellation). *Let R be a commutative ring and suppose $a \neq 0$ in R is not a zero divisor. Then if b, c are in R and $ab = ac$, then $b = c$.*

Proof. From $ab = ac$ we obtain $ab - ac = 0$, hence

$$a(b - c) = 0.$$

Since a is not a zero divisor and $a(b - c) = 0$, we must have $b - c = 0$, and so

$$b = c.$$

$\square$

Corollary 8. *If a is a non-zero divisor of a commutative ring R, then for all b in R, the equation $ax = b$ has at most one solution.*

Proof. If $ax_1 = b$ and $ax_2 = b$, then $ax_1 = ax_2$. By cancellation, $x_1 = x_2$. $\square$

If the commutative ring R has no zero divisors, Corollary 8 applies to all equations $ax = b$ for $a \neq 0$.

A similar result is true for polynomial equations of higher degree, as we'll see in general in a later chapter. But we can do degree 2 here.

Recall that a quadratic equation $x^2 - rx + s = 0$ with r, s in the real numbers $\mathbb{R}$ has at most two roots in $\mathbb{R}$, namely, the two roots given by the quadratic formula: $x = \frac{r \pm b}{2}$ where b is a number whose square is $r^2 - 4s$, if such a number b exists. But in general we have:

Proposition 9. *Let R be a commutative ring. If R has no zero divisors, then for every r, s in R, the equation $x^2 - rx + s = 0$ has at most two solutions in R. On the other hand, if R has non-zero elements a and b such that $ab = 0$ and at least three of $0, a, b$ and $a + b$ are distinct, then the equation $x^2 - (a+b)x = 0$ has at least three roots in R.*

Proof. Suppose a and b are complementary zero divisors in R, so that $a, b \neq 0$ and $ab = 0$. Then it's easy to check that

$$x^2 - (a+b)x = 0$$

has four solutions: $a, b, a+b$, and 0. So if at least three of these are distinct, then the equation has at least three distinct roots.

Conversely, suppose R has no zero divisors, and suppose $x^2 - rx + s = 0$ has two solutions a, b: thus,

$$a^2 - ra + s = 0,$$
$$b^2 - rb + s = 0.$$

Suppose also c is a solution, so that

$$c^2 - rc + s = 0.$$

Subtracting this last equation from each of the previous two, we get

$$r(a - c) = a^2 - c^2 = (a+c)(a-c),$$
$$r(b - c) = b^2 - c^2 = (b+c)(b-c).$$

Since R has no zero divisors, if $c \neq a$ and $c \neq b$, we can cancel $a - c$ from the first equation and $b - c$ from the second, to get

$$r = a + c,$$
$$r = b + c,$$

which implies that $a = b$. Thus if $a \neq b$, then c must equal a or b, and so there cannot be more than two solutions of $x^2 - rx + s = 0$. $\qquad\square$

Our experience with numbers might suggest that a ring having zero divisors is not as "natural" as a ring that does not, because the rings we encounter in courses through calculus: $\mathbb{Z}$, $\mathbb{Q}$, $\mathbb{R}$, $\mathbb{C}$, all have no zero divisors. In fact, $\mathbb{Q}$, $\mathbb{R}$ and $\mathbb{C}$ are fields, $\mathbb{Z}$ is a subring of a field, and:

Proposition 10. *A field has no zero divisors.*

The converse of this proposition is not true: $\mathbb{Z}$ is not a field, and $\mathbb{Z}$ has no zero divisors.

Proposition 10 follows immediately from

Proposition 11. *In a commutative ring R, a unit cannot be a zero divisor.*

Proof. Suppose a is a unit in R. To show that a is not a zero divisor in R, we will show that if $ab = 0$ in F, then b must be 0. Suppose $ab = 0$. Multiply both sides on the left by a^{-1}, to get $a^{-1}(ab) = a^{-1}0 = 0$. Reassociating the left side, we have $(a^{-1}a)b = 0$, hence $1 \cdot b = 0$, hence $b = 0$. □

Now we consider $\mathbb{Z}/m\mathbb{Z}$. Is it a field? Does it have zero divisors? If you check back to the multiplication tables for $\mathbb{Z}/3\mathbb{Z}$ and $\mathbb{Z}/4\mathbb{Z}$, you will observe that $\mathbb{Z}/3\mathbb{Z}$ is in fact a field–every non-zero element has an inverse–whereas in $\mathbb{Z}/4\mathbb{Z}$ property (*nzd*) (no zero divisors) fails, because $[2] \cdot [2] = 0$, and also the property that all the non-zero elements have inverses fails, because $[2]$ has no inverse.

Whether or not $\mathbb{Z}/m\mathbb{Z}$ is a field is easily decided once we recall which elements are units in $\mathbb{Z}/m\mathbb{Z}$.

Theorem 12. *In $\mathbb{Z}/m\mathbb{Z}$,*
(i) $[a]$ is a unit, if $(a,m) = 1$;
(ii) $[a]$ is a zero divisor, if $1 < (a,m) < m$; and
(iii) $[a] = 0$, if m divides a.

Proof. If m divides a, then $a \equiv 0 \pmod{m}$, so $[a] = [0]$. Thus (iii) holds.

To prove (ii): suppose $(m,a) = d$ and $1 < d < m$. Then a is not a multiple of m, so $[a] \neq [0]$. But since d divides m, there is a number e with $1 < e < m$ so that $de = m$. Then $[e] \neq [0]$, while ae is a multiple of $de = m$, so $[a][e] = [ae] = [0]$. Thus $[a]$ is a zero divisor.

We proved (i) in Chapter 6. (Can you recall the proof?) □

Corollary 13. *$\mathbb{Z}/m\mathbb{Z}$ is a field iff m is prime.*

Proof. If $[a]$ is any nonzero element of $\mathbb{Z}/m\mathbb{Z}$, then m doesn't divide a. If m is prime, it follows that $(a,m) = 1$, hence $[a]$ is a unit. Thus every nonzero element of $\mathbb{Z}/m\mathbb{Z}$ is a unit, so $\mathbb{Z}/m\mathbb{Z}$ is a field.

If m is not prime, then $m = a \cdot b$ with $1 < a, b < m$; then $[a][b] = [m] = [0]$, while $[a]$ and $[b]$ are not zero. Thus $\mathbb{Z}/m\mathbb{Z}$ has zero divisors, and so cannot be a field. □

Let p be a prime number. The field $\mathbb{Z}/p\mathbb{Z}$ is so often used in mathematics that it has been given its own Special Roman symbol, just like the integers ($\mathbb{Z}$), rationals ($\mathbb{Q}$), real numbers ($\mathbb{R}$), and complex numbers ($\mathbb{C}$), namely, $\mathbb{F}_p$, "the field of p elements."

We will use $\mathbb{Z}/p\mathbb{Z}$ and $\mathbb{F}_p$ interchangeably. Note, however, that if m is not prime, then $\mathbb{Z}/m\mathbb{Z}$ is not a field, so we do not use the notation $\mathbb{F}_m$ to refer to $\mathbb{Z}/m\mathbb{Z}$ unless m is prime. (In fact, if q is a prime power, like 9 or 16, then there is a field with q

elements, which is essentially unique. The notation $\mathbb{F}_q$ then denotes that field with q elements. $\mathbb{F}_q$ will, of course, not be $\mathbb{Z}/q\mathbb{Z}$ if q is not prime, because $\mathbb{Z}/q\mathbb{Z}$ is not a field.)

We can prove a version of Theorem 6 that is valid for any ring with a finite number of elements.

Theorem 14. *If R is a finite commutative ring with identity, and a is any non-zero element of R, then a is either a unit or a zero divisor.*

Proof. Suppose R has n elements (it does not follow that $R = \mathbb{Z}/n\mathbb{Z}$). Letting $a^s = a \cdot a \cdot \ldots \cdot a$ (s factors) for any natural number s, and $a^0 = 1$, consider the set of elements

$$1 = a^0, a, a^2, a^3, \ldots, a^n$$

This is a set of $n+1$ elements in R, a set of n elements. So two of them must be equal. Suppose $a^r = a^{r+d}$ for some $r \geq 0, d > 0$. Then

$$a^{r+d} - a^r = 0,$$

so

$$a^r(a^d - 1) = 0.$$

Choose r minimal so that $a^r(a^d - 1) = 0$. If $r = 0$, then $a^d - 1 = 0$, so $a(a^{d-1}) = 1$ and a is a unit of R. If $r > 0$, then

$$a^{r-1}(a^d - 1) \neq 0$$

by minimality of r, while

$$a(a^{r-1}(a^d - 1)) = 0$$

Thus a is a zero divisor of R. $\qquad\qquad\qquad\square$

The proof shows that if a is a unit of R, then there is some $d > 0$ with $a^d = 1$. The minimal such $d > 0$ is called the *order* of a. We will study the orders of elements of $\mathbb{Z}/m\mathbb{Z}$ in Chapter 9.

Example 4. In $\mathbb{Z}/5\mathbb{Z}$ the order of $[2]_5$ is 4: $[2] \neq 1, [2]^2 \neq 1, [2]^3 \neq 1$, while $[2]^4 = [16] = 1$. In $\mathbb{Z}/7\mathbb{Z}$, the order of $[2]_7$ is 3, because $[2] \neq 1, [2]^2 \neq 1$, while $[2]^3 = [8] = 1$.

Exercises.

21. For each m with $6 \leq m \leq 13$, how many zero divisors does $\mathbb{Z}/m\mathbb{Z}$ have?

22. Suppose R is a ring with no zero divisors, and S is a subring of R. Show that S has no zero divisors.

23. Suppose R is a ring, and a is a zero divisor in R. Show that the "homogeneous" equation $ax = 0$ in R has more than one solution for x.

24. In a ring R, show that for any b in R, if the "non-homogeneous" equation $ax = b$ has some solution $x = x_0$ in R, then every solution of the equation $ax = b$ is of the form $x = x_0 + t$ where t is a solution of the homogeneous equation $ax = 0$.

25. (For those who have had a course in linear algebra). Let R be the ring of $n \times n$ matrices with real coefficients. Suppose M is a $n \times n$ matrix. Show that the following conditions are equivalent:
  (i) $M \neq 0$ and the rank of M is $<n$.
  (ii) M is a zero divisor in R (on both sides).
  (iii) $M \neq 0$ and for any column vector B, the equation $MX = B$ has either no solutions or infinitely many solutions.

26. In $\mathbb{Z}/18\mathbb{Z}$, show that $[6]_{18}$ is a zero divisor. Find all solutions of the equation $[6]_{18}X = [12]_{18}$.

27. Show that if $[a]$ in $\mathbb{Z}/m\mathbb{Z}$ is a zero divisor, it cannot have an inverse in $\mathbb{Z}/m\mathbb{Z}$.

28. In $\mathbb{Z}/15\mathbb{Z}$, identify the zero divisors, and for each, find all of the complementary zero divisors.

29. In $\mathbb{Z}/16\mathbb{Z}$, identify the zero divisors, and for each, find all of the complementary zero divisors.

30. In $\mathbb{Z}/18\mathbb{Z}$, identify the zero divisors, and for each, find all of the complementary zero divisors.

31. In $\mathbb{Z}/26\mathbb{Z}$, find the inverses of $[9]$, $[11]$, $[17]$, and $[22]$.

32. In $\mathbb{Z}/365\mathbb{Z}$, find, if possible, the inverses of $[53]$, $[73]$, $[93]$, and $[113]$.

33. Let R be a commutative ring with no zero divisors. Suppose z is an element of R satisfying $z^2 = 1$. Show that $z = 1$ or $z = -1$.

34. Find the order of $[3]$ in $\mathbb{Z}/7\mathbb{Z}$.

Here are some new examples of rings:

35. In $\mathbb{F}_3 = \mathbb{Z}/3\mathbb{Z}$, there is no solution of the equation $x^2 = -1$, just as in $\mathbb{R}$. So "invent" a solution, call it i. Then i is a new "number" that satisfies $i^2 = -1$. Consider the set $\mathbb{F}_3[i]$ consisting of all numbers $a + bi$, with a, b in $\mathbb{F}_3$. Add and multiply these numbers as though they were polynomials in i, except whenever you get i^2 replace it by -1.
  (i) Write down the nine elements of $\mathbb{F}_3[i]$.
  (ii) Show that every nonzero element of $\mathbb{F}_3[i]$ has an inverse, so that $\mathbb{F}_3[i]$ is a field.
  (iii) Find the order of $1 + i$.

36. Consider, as in the last exercise, the set $\mathbb{F}_2[i]$ of numbers of the form $a + bi$ where a and b are elements of $\mathbb{F}_2 = \mathbb{Z}/2\mathbb{Z}$. Again, $i^2 = -1$, which in $\mathbb{F}_2$ is the same as 1. Write down the four elements of $\mathbb{F}_2[i]$. Which elements have inverses?

37. Consider the set $\mathbb{Q}[i]$ of numbers $a + bi$ where a and b are in $\mathbb{Q}$. Show that $\mathbb{Q}[i]$ is a field.

D. Homomorphisms

Functions from one set to another are ubiquitous in higher mathematics. Calculus, for example, is almost entirely devoted to the study of functions from the real numbers to the real numbers. Linear algebra is the study of vector spaces and certain kinds of functions (linear transformations) from one vector space to another. So it is not surprising that mathematicians studying algebra should be interested in functions as well. But just as in linear algebra, the functions of interest in algebra have special properties.

We begin with some terminology. Let R, S be two rings. Let f be a function from R to S. Thus R is the *domain* of f, and S is the *range* of f: for each r in R, $f(r)$ is an element of S. To state concisely the domain and range of f, we often write the function as

$$f : R \to S.$$

Related to the range is the *image* of $f : R \to S$, namely, the set of elements s in S such that $s = f(r)$ for some r in R. The image of $f : R \to S$ may or may not be all of S. If the image is all of S, we say that the function f is *onto* S, or is *surjective*.

Among all possible functions $f : R \to S$, we are interested in those functions which "respect" the fact that R, S, as rings, have algebraic operations: $+, \cdot, -$ and special elements $0, 1$. Thus we call $f : R \to S$ a *ring homomorphism*, or, for short, a *homomorphism*, if f satisfies the following properties:

$$(i) \quad f(r + r') = f(r) + f(r') \text{ for all } r, r' \text{ in } R.$$

Here the addition of $f(r)$ and $f(r')$ is the addition in S.

$$(ii) \quad f(r \cdot r') = f(r) \cdot f(r') \text{ for all } r, r' \text{ in } R.$$

Again, the multiplication on the right-hand side of the equation is in S.

$$(iii) \quad f(1) = 1.$$

Here the 1 in $f(1)$ is in R, and the 1 on the right side of the equation is in S. To be perfectly precise, we might better label the identity element of R by 1_R and the identity element of S by 1_S, and write $f(1_R) = 1_S$. But if you recall that f is a function from R to S, then the 1 in the expression "$f(1)$" can only be an element of R, and the 1 on the right side of the equation $f(1) = 1$ can only be in S. So there should be no confusion arising from leaving out the subscripts.

If f satisfies the conditions (i)-(iii), then

$$(iv) \quad f(0) = 0.$$

Here again the left 0 is in R, and the right 0 is in S. This property follows from (i). For given any b in R,

$$f(b) = f(0 + b) = f(0) + f(b);$$

adding $-f(b)$ to both sides gives

$$0 = f(0) + 0 = f(0).$$

Also

(v) $f(-r) = -f(r)$ for any r in R.

To see this, notice that by definition of the negative in S, we have

$$0 = f(0) = f(r + (-r)) = f(r) + f(-r).$$

Since the negative of any element of S is unique, $f(-r) = -f(r)$.

If f is a homomorphism then f also satisfies:

(vi) If a has an inverse a^{-1} in R,

then $f(a)$ has an inverse in S, namely, $f(a^{-1})$.

Thus, if $f : R \rightarrow S$ is a ring homomorphism, then f maps U_R, the group of units of R, into U_S, the group of units of S. (The proof is left as an exercise.)

A ring homomorphism f is one-to-one, or injective, if f is one-to-one as a function, that is, for all a, b in R, if $f(a) = f(b)$ then $a = b$.

Here is a convenient test:

Proposition 15. *A ring homomorphism f is one-to-one if and only if 0 is the only element r of R with $f(r) = 0$.*

Proof. If $r \neq 0$ and $f(r) = 0$, then since $f(0) = 0$, f is not one-to-one; on the other hand, if f is not one-to-one, then there are two different elements a and b of R so that $f(a) = f(b)$. But then $f(a - b) = 0$, and $a - b$ is not the zero element of R. □

Definition. Let $f : R \rightarrow S$ be a homomorphism. The *kernel* of f, written $\ker(f)$, is the set of elements r of R so that $f(r) = 0$. Concisely,

$$\ker(f) = \{r \text{ in } R \mid f(r) = 0\}.$$

The size of the kernel of a homomorphism f describes how far $f : R \rightarrow S$ is from being one-to-one. If $\ker(f) = \{0\}$, then f is one-to-one. In general, we have:

Proposition 16. *Let $f : R \rightarrow S$ be a ring homomorphism and let s be in the image of f. Then $\{r \text{ in } R \mid f(r) = s\}$ is in one-to-one correspondence with $\ker(f)$*

Proof. It is easy to see that if $f(r_0) = s$, then

$$\{r \text{ in } R \mid f(r) = s\} = \{r_0 + k \mid k \text{ in } \ker f\}. \qquad \square$$

Thus if $\ker f$ has m elements, then f is an m-to-one function.

Here is a useful property of fields:

Proposition 17. *Let $f : R \rightarrow S$ be a homomorphism where R is a field and $1 \neq 0$ in S. Then f is one-to-one.*

Proof. Suppose $a \neq 0$ in R. We show $f(a) \neq 0$. Since R is a field, a has an inverse, a^{-1}. Then $1 = f(1) = f(a \cdot a^{-1}) = f(a) \cdot f(a^{-1})$. If $f(a) = 0$. then $1 = 0 \cdot (a^{-1}) = 0$. This is a contradiction, since $1 \neq 0$ in S. Thus the kernel of f contains no element of R except 0, and f is one-to-one. $\square$

Here are two very simple examples of ring homomorphisms.

Example 5. The most trivial examples are the identity homomorphisms. Let R be any commutative ring, and let $i : R \rightarrow R$ be the function defined by $i(r) = r$ for any r in R. Then i is obviously a ring homomorphism and is one-to-one and onto.

Example 6. Let S be a commutative ring and let R be a subset of S which is also a commutative ring with the same addition and multiplication that S has, and such that the identity and zero elements of S are in R. For example, let S be the real numbers and R be the rational numbers. We can then define a homomorphism $i : R \rightarrow S$, the inclusion map, by $i(r) = r$.

The only difference between the two examples are the ranges. In the first example the range is R, while in the second example the range is the ring S. If S is truly larger than R, then the function in the second example is not onto.

These two examples illustrate a difference between functions in algebra and functions in beginning calculus. In calculus, all the functions have the same range, namely, the real numbers (depicted geometrically as the y-axis); in algebra there are many different rings, and hence many different possible ranges. When we introduce a function it is important to specify both the domain and the range of the function as part of the definition of the function.

Homomorphisms with Domain $\mathbb{Z}$. In the rest of this section we find all ring homomorphisms with domain $\mathbb{Z}$.

Example 7. Let R be a commutative ring with identity 1_R. Define a homomorphism $f : \mathbb{Z} \rightarrow R$ as follows:
  $f(0) = 0_R$, the zero element of R. This is required by property (v).
  $f(1) = 1_R$. This is required by property (iii).
  For $k > 1$, define $f(k+1) = f(k) + f(1) = f(k) + 1_R$. This definition of $f(k+1)$ is required by property (i). Then by induction, for any $n > 0$,

$$f(n) = 1_R + 1_R + \ldots + 1_R \quad (n \text{ summands})$$

which we shall write as $n \cdot 1_R$.
  If $n > 0$, then $f(-n) = -f(n) = -(n \cdot 1_R)$, which we'll write as $(-n) \cdot 1_R$.
  Thus for any n in $\mathbb{Z}$, $f(n) = n \cdot 1_R$, and this definition of $f(n)$ is forced by the condition that f be a homomorphism.

Proposition 18. *The function $f : \mathbb{Z} \rightarrow R$ defined by $f(n) = n \cdot 1_R$, is a homomorphism, and is the only ring homomorphism from $\mathbb{Z}$ to R.*

Proof. In the proof write $1_R = 1$. We just showed that if f is a homomorphism from $\mathbb{Z}$ to R, then $f(n) = n \cdot 1_R$, so f is unique if in fact f is a homomorphism. To see that f is a homomorphism, we need to check properties (i)-(iii).

Property (iii) is true by definition.

Property (i) is that for any m, n in $\mathbb{Z}$, $f(m+n) = f(m) + f(n)$, that is,

$$m \cdot 1 + n \cdot 1 = (m+n) \cdot 1.$$

This follows by associativity of addition in R.

Property (ii) is that $f(mn) = f(m)f(n)$, that is,

$$(m \cdot 1) \cdot (n \cdot 1) = (mn) \cdot 1.$$

This is a consequence of the distributive law: if $n > 0$, then

$$n \cdot 1 = (1 + 1 + \ldots + 1) \quad (n \text{ summands}),$$

and so

$$
\begin{aligned}
(m \cdot 1)(n \cdot 1) &= (m \cdot 1)(1 + 1 + \ldots + 1) \\
&= (m \cdot 1)1 + \ldots + (m \cdot 1)1 \\
&= (m \cdot 1) + \ldots + (m \cdot 1) \quad (n \text{ summands}) \\
&= 1 + 1 + \ldots + 1 \quad (mn \text{ times }) \\
&= (mn)1,
\end{aligned}
$$

If $n < 0$, write $n \cdot 1 = (-n) \cdot (-1) = -(-n) \cdot 1$ and again use distributivity. $\qquad \square$

Here are some examples of the homomorphisms defined by Proposition 18.

(1) Let $R = \mathbb{Z}$, then the homomorphism $f : \mathbb{Z} \to \mathbb{Z}$ is defined by $f(n) = n \cdot 1 = n$. So in this case f is the identity function on $\mathbb{Z}$.

(2) Let $R = \mathbb{Q}$, then $f : \mathbb{Z} \to \mathbb{Q} :$ is defined by $f(n) = n \cdot \frac{1}{1} = \frac{n}{1}$, and is the usual embedding of the integers inside the rationals. That is, f is the inclusion map from $\mathbb{Z}$ to $\mathbb{Q}$, and is one-to-one.

(3) Let $f_m : \mathbb{Z} \to \mathbb{Z}/m\mathbb{Z}$ be the function defined by $f_m(a) = [a]_m$. We check the properties (i)-(iii):

(i) $f_m(a+b) = f_m(a) + f_m(b)$? This is the same as $[a+b]_m = [a]_m + [b]_m$. But this is the way we add congruence classes, So (i) holds for f_m.

(ii) $f_m(a \cdot b) = f_m(a) \cdot f_m(b)$? This is the same as $[a \cdot b]_m = [a]_m \cdot [b]_m$. This is the way congruence classes multiply. So (ii) holds for f_m.

(iii) $f_m(1) = 1$? The "1" on the right side is the congruence class of the integer 1, that is, $[1]_m$. Since $f_m(1) = [1]_m$ by definition, (iii) holds. Thus f_m is a homomorphism.

Notice that for $n > 0, f_m(n) = [n] = [1] + [1] + \ldots + [1] = n \cdot [1]$, so f_m is the map defined in Proposition 18.

The functions $f_m : \mathbb{Z} \to \mathbb{Z}/m\mathbb{Z}$ are all onto, that is, for any congruence class $[a]$ in $\mathbb{Z}/m\mathbb{Z}$, there is some integer, namely a, so that $f_m(a) = [a]$. However, f_m is not one-to-one. In fact, any two integers that are congruent (mod m) are mapped to the same class in $\mathbb{Z}/m\mathbb{Z}$ by f_m. The kernel of f_m is the set of integers that are multiples of m. That is, $\ker(f_m) = [0]_m$ thought of as a set of integers, rather than an element of $\mathbb{Z}/m\mathbb{Z}$.

The characteristic of a ring.

Definition. Let $f : \mathbb{Z} \to R$ be the homomorphism defined in Proposition 18. If f is one-to-one then R is said to have *characteristic zero*.

If f is not one-to-one then there is some nonzero integer c in the kernel of f. If $f(c) = 0$, then $f(-c) = -f(c) = -0 = 0$, so there is a natural number in $ker(f)$. Let m be the smallest natural number (> 0) in $\ker(f)$.

Proposition 19. *If $f : \mathbb{Z} \to R$ is a homomorphism and m is the smallest natural number in* $\ker(f)$, *then* $\ker(f)$ *is the set of integers that are multiples of m.*

Proof. If b is in $\ker(f)$, then divide b by m:

$$b = mq + r,$$

where $0 \le r < m$. Applying f to that equation, we have

$$0 = f(b) = f(m)f(q) + f(r) = 0 \cdot f(q) + f(r) = f(r),$$

so r is in $\ker(f)$. But since m is the smallest natural number in $ker(f)$, r must $= 0$. So m divides b.   □

Let $m\mathbb{Z}$ denote the set of all multiples of the natural number m.

Proposition 20. *Let R be a commutative ring with no zero divisors, and $f : \mathbb{Z} \to R$, $f(n) = n \cdot 1$. If $\ker(f) = m\mathbb{Z}$ and $m \ne 0$, then m is prime.*

Proof. If m is not prime, then $m = a \cdot b$ for some a, b with $0 < a < m, 0 < b < m$. Then $f(a) \ne 0, f(b) \ne 0$, but $0 = f(m) = f(ab) = f(a)f(b)$, so R has zero divisors.   □

Definition. If R has no zero divisors, and $f : \mathbb{Z} \to R$ by $f(n) = n \cdot 1$ is not one-to-one, then $ker(f) = p\mathbb{Z}$ where p is a prime number. In that case we say that R has *characteristic p*.

The last proposition implies that any field has either characteristic zero or characteristic p for some prime p.

To rephrase our definition of the characteristic of a field F: add the identity element 1 of F to itself repeatedly, that is, look at $n \cdot 1$ for $n = 1, 2, 3, \ldots$. If $n \cdot 1$ is never $= 0$, then the field has characteristic zero. Otherwise, the smallest positive number n so that $n \cdot 1 = 0$ is prime, and is the characteristic of the field.

Example 8. $\mathbb{F}_9$ has characteristic 3.

More generally, we have

Proposition 21. *If F is a field with a finite number of elements, then F has characteristic p for some prime number p.*

The proof is left as an exercise.

Isomorphisms.

Definition. A ring homomorphism $f : R \to S$ is an *isomorphism* if f is one-to-one and onto. Two rings R and S are *isomorphic* if there is an isomorphism between them.

As a first set of examples of isomorphisms, we have

Proposition 22. *Let R be a commutative ring and let $f : \mathbb{Z} \to R$ be the homomorphism defined by $f(n) = n \cdot 1_R$ for all n in $\mathbb{Z}$. If f is one-to-one, so that R has characteristic zero, then f defines an isomorphism from $\mathbb{Z}$ onto $\{n \cdot 1_R | n \text{ in } \mathbb{Z}\} \subseteq R$.*

Proof. If R has characteristic zero, then f is one-to-one. A function maps onto its image. Thus if f is a one-to-one homomorphism, then f is an isomorphism from its domain to its image. □

Proposition 23 (Homomorphism Theorem). *Let R be a commutative ring and let $f : \mathbb{Z} \to R$ be the homomorphism defined by $f(n) = n \cdot 1_R$ for all n in $\mathbb{Z}$. If f is not one-to-one and $ker(f) \supseteq m\mathbb{Z}$ for some $m \neq 0$ in $\mathbb{Z}$, then f induces a homomorphism $\overline{f}$ from $\mathbb{Z}/m\mathbb{Z}$ onto $\{n \cdot 1_R | n \text{ in } \mathbb{Z}\} \subseteq R$, defined by $\overline{f}([a]_m) = f(a) = a \cdot 1_R$.*

If $ker(f) = m\mathbb{Z}$ then $\overline{f}$ is an isomorphism from $\mathbb{Z}/m\mathbb{Z}$ onto

$$\{n \cdot 1_R | n \text{ in } \mathbb{Z}\} \subseteq R.$$

Proof. If $ker(f)$ contains $m\mathbb{Z}$, then we must show that $\overline{f}$ is a well-defined homomorphism. "Well-defined" relates to the fact that while $\overline{f}$ has as its domain the ring $\mathbb{Z}/m\mathbb{Z}$ of congruence classes of integers, the definition of $\overline{f}$ is given in terms of representatives of congruence classes. So we have to show that if we choose different representatives, the value of $\overline{f}$ is the same.

We defined $\overline{f}([a]_m) = f(a)$. Suppose $[a]_m = [b]_m$. Then $f(b) = f(a)$. For if $[a]_m = [b]_m$, then $a = b + mk$ for some integer k, so $f(a) = f(b + mk) = f(b) + f(mk)$. Since mk is in the kernel of f, $f(b) + f(mk) = f(b)$. So $f(a) = f(b)$ and $\overline{f}$ is well defined.

Then $\overline{f}$ is a homomorphism, because for any a, b in $\mathbb{Z}$,

$$\overline{f}([a] + [b]) = \overline{f}([a + b]) = f(a + b) = f(a) + f(b) = \overline{f}([a]) + \overline{f}([b]).$$

Now suppose $ker(f) = m\mathbb{Z}$. Then $\overline{f}([a]) = 0$ iff $f(a) = 0$, iff a is in $m\mathbb{Z}$, iff $[a]_m = [0]_m$. Thus the kernel of $\overline{f}$ is $\{[0]_m\}$ and so $\overline{f}$ is one-to-one. Since the image of $\overline{f}$ is the same as the image of f, f maps onto $\{n \cdot 1_R | n \text{ in } \mathbb{Z}\} \subseteq R$. □

The following consequences of Propositions 22 and 23 will be useful in later chapters.

Corollary 24. *Let R be a commutative ring with no zero divisors. If R has characteristic zero, then R contains a subring isomorphic to $\mathbb{Z}$. If R has characteristic p, a prime, then R contains a subring isomorphic to $\mathbb{Z}/p\mathbb{Z}$.*

Corollary 25. *If d, m are integers and d divides m, then the homomorphism $f : \mathbb{Z} \to \mathbb{Z}/d\mathbb{Z}$ defined by $f(n) = n \cdot 1$ induces a homomorphism $\overline{f} : \mathbb{Z}/m\mathbb{Z} \to \mathbb{Z}/d\mathbb{Z}$, and a map from U_m, the group of units of $\mathbb{Z}/m\mathbb{Z}$, onto U_d.*

The proof of Corollary 25 is left as an exercise.

Example 9. Corollary 25 implies that there is a homomorphism from $\mathbb{Z}//6\mathbb{Z}$ to $\mathbb{Z}/3\mathbb{Z}$ by $[a]_6 \mapsto [a]_3$. Thus:

$$[1]_6 \mapsto [1]_3$$
$$[2]_6 \mapsto [2]_3$$
$$[3]_6 \mapsto]3]_3 = [0]_3$$
$$[4]_6 \mapsto [4]_3 = [1]_3$$
$$[5]_6 \mapsto [5]_3 = [2]_3$$
$$[0]_6 \mapsto [6]_3 = [0]_3$$

The group of units $\{[1]_6, [5]_6\}$ of $\mathbb{Z}/6\mathbb{Z}$ maps onto the group of units $\{[1]_3, [2]_3\}$ of $\mathbb{Z}/3\mathbb{Z}$.

Exercises.

38. Show that if $f : R \to S$ is a homomorphism, and if a is a unit of R, then $f(a)$ is a unit of S. Show, in fact, that $f(a^{-1}) = f(a)^{-1}$ for any unit a of R.

39. Using the previous exercise, show that the identity function is the only homomorphism from $\mathbb{Q}$ to $\mathbb{Q}$.

40. Let $f : R \to S$, $g : S \to T$ be functions. Let $g \circ f : R \to T$ be the composite of f and g, that is, $(g \circ f)(r) = g(f(r))$. Show that if f and g are homomorphisms, then so is $g \circ f$.

41. Prove Corollary 25 that if d divides m and $\overline{f} : \mathbb{Z}/m\mathbb{Z} \to \mathbb{Z}/d\mathbb{Z}$ is the map defined by $\overline{f}([a]_m) = [a]_d$, then $\overline{f}$ maps the group of units U_m onto the group of units U_d.

42. Show that if R is a ring, the function f from R to $M_2(R)$, the set of 2×2 matrices with entries in R, given by

$$f(r) = \begin{pmatrix} r & 0 \\ 0 & r \end{pmatrix}$$

for any r in R, is a homomorphism.

43. Show that if F is a field of characteristic p for some prime $p > 0$, then for every a in F, $a + a + \ldots + a$ (p times$) = 0$.

44. Show that if F is a field of characteristic 2, then:
(i) $-a = a$ for any a in F;
(ii) for any a, b in F, $(a + b)^2 = a^2 + b^2$.

45. Prove Proposition 21 .

Chapter 8
Matrices and Codes

Many applications of the mathematics developed so far are most conveniently described in terms of matrices and linear algebra. So in this chapter we present a summary of matrix notation and its relationship to systems of linear equations.

Readers with some background in matrix theory will need at most to skim the first four sections of this chapter for notation. Readers for whom matrix theory is new will find our treatment rather terse, and are urged to refer, as needed, to any of the numerous textbooks available on linear algebra and matrices.

One point of this chapter is that the formalism of vectors and matrices makes sense over any commutative ring with identity, and nearly all of the theorems of elementary linear algebra are valid over any field, not just over the real numbers. Thus as soon as a set R is identified as a commutative ring with identity, we can work with matrices and vectors with entries in R, and if we find that R is a field, then the theory of vector spaces and linear transformations will be applicable over R.

The last two sections illustrate the use of matrices over commutative rings in two applications: error-correcting codes, which uses matrices over $\mathbb{Z}/2\mathbb{Z}$, and cryptography, which uses matrices over $\mathbb{Z}/26\mathbb{Z}$. Both applications are historically among the earliest examples of the use of mathematics in their respective areas.

We assume in the first four sections that R is a commutative ring with identity.

A. Matrix Multiplication

This section covers the most basic properties of matrices.

A column vector is a column of elements of R, viz.,

$$\begin{pmatrix} a_1 \\ a_2 \\ \vdots \\ a_n \end{pmatrix}.$$

A row vector is a row of elements of R, viz.,

$$(a_1 \ a_2 \ \cdots \ a_n).$$

An $m \times n$ matrix is a rectangular array of mn elements of R, viz.,

$$\begin{pmatrix} a_{11} & a_{12} & \cdots & a_{1n} \\ a_{21} & a_{22} & \cdots & a_{2n} \\ \vdots & \vdots & & \vdots \\ a_{m1} & a_{m2} & \cdots & a_{mn} \end{pmatrix},$$

which can be thought of as a collection of row vectors placed in a column, or a collection of column vectors laid out in a row. When we say that a matrix is $m \times n$, the first number m is always the number of rows, and the second number n is the number of columns.

Given a row vector with n elements (placed on the left) and a column vector with the same number of elements (placed on the right), we may multiply them to get an element of the ring R:

$$(a_1 \ a_2 \ \cdots \ a_n) \begin{pmatrix} b_1 \\ b_2 \\ \vdots \\ b_n \end{pmatrix} = a_1 b_1 + a_2 b_2 + \ldots + a_n b_n.$$

All we need to know about R for this to make sense is that R has addition and multiplication and addition is associative, so that we can be casual about the order in which we add the terms in the right side of this last equation.

Examples where $R = \mathbb{Z}$ are:

$$(3 \ 2 \ 5) \begin{pmatrix} 1 \\ 2 \\ -1 \end{pmatrix} = 2, \quad (1 \ 2 \ 3) \begin{pmatrix} 2 \\ 2 \\ -2 \end{pmatrix} = 0, \quad (-3 \ 2) \begin{pmatrix} 1 \\ 2 \end{pmatrix} = 1.$$

Given an $m \times n$ matrix $\mathbf{A}$, we can multiply the matrix (placed on the left) with an n-element column vector $\mathbf{X}$ (placed on the right) by thinking of the matrix as a collection of m row vectors, each containing n elements, and doing m multiplications of the row vectors of $\mathbf{A}$ with $\mathbf{X}$. The result, $\mathbf{AX}$, is a column of m elements:

$$\begin{pmatrix} 1 & 2 \\ 2 & 4 \\ 2 & 3 \end{pmatrix} \begin{pmatrix} -1 \\ 2 \end{pmatrix} = \begin{pmatrix} 3 \\ 6 \\ 4 \end{pmatrix}, \quad \begin{pmatrix} 1 & 2 & 3 \\ 0 & 0 & 1 \end{pmatrix} \begin{pmatrix} 2 \\ 0 \\ 1 \end{pmatrix} = \begin{pmatrix} 5 \\ 1 \end{pmatrix}.$$

Given an $m \times n$ matrix $\mathbf{A}$ (on the left) and an $n \times p$ matrix $\mathbf{B}$ (on the right), we can multiply them by thinking of $\mathbf{A}$ as a collection of n-element rows and $\mathbf{B}$ as a collection of n-element columns. The result, $\mathbf{AB}$, is an $m \times p$ matrix whose element in the ith row and jth column is obtained by multiplying the ith row of $\mathbf{A}$ and the

jth column of **B**. Thus in the example

$$\begin{pmatrix} 1 & 2 & 1 \\ 2 & 3 & 0 \end{pmatrix} \begin{pmatrix} 2 & 1 \\ 0 & 1 \\ 1 & 3 \end{pmatrix} = \begin{pmatrix} 3 & 6 \\ 4 & 5 \end{pmatrix},$$

the 3 comes from multiplying

$$\begin{pmatrix} 1 & 2 & 1 \end{pmatrix} \begin{pmatrix} 2 \\ 0 \\ 1 \end{pmatrix},$$

the 6 from

$$\begin{pmatrix} 1 & 2 & 1 \end{pmatrix} \begin{pmatrix} 1 \\ 1 \\ 3 \end{pmatrix},$$

etc. Other examples:

$$\begin{pmatrix} 0 \\ 1 \\ 3 \end{pmatrix} \begin{pmatrix} 1 & 2 & 5 \end{pmatrix} = \begin{pmatrix} 0 & 0 & 0 \\ 1 & 2 & 5 \\ 3 & 6 & 15 \end{pmatrix};$$

$$\begin{pmatrix} 2 & 1 \\ 0 & 1 \\ 1 & 3 \end{pmatrix} \begin{pmatrix} 0 & 1 \\ 4 & 5 \end{pmatrix} = \begin{pmatrix} 4 & 7 \\ 4 & 5 \\ 12 & 16 \end{pmatrix}.$$

Notice that the order in which the matrices are multiplied (i.e., which matrix is on the left and which is on the right) is very important. In the last example,

$$\begin{pmatrix} 0 & 1 \\ 4 & 5 \end{pmatrix} \begin{pmatrix} 2 & 1 \\ 0 & 1 \\ 1 & 3 \end{pmatrix}$$

makes no sense, because it requires multiplying row vectors and column vectors with different numbers of elements. Even when it makes sense to multiply in either order, the results are usually different: compare

$$\begin{pmatrix} 0 \\ 1 \\ 3 \end{pmatrix} \begin{pmatrix} 1 & 2 & 5 \end{pmatrix} = \begin{pmatrix} 0 & 0 & 0 \\ 1 & 2 & 5 \\ 3 & 6 & 15 \end{pmatrix},$$

a 3×3 matrix, with

$$\begin{pmatrix} 1 & 2 & 5 \end{pmatrix} \begin{pmatrix} 0 \\ 1 \\ 3 \end{pmatrix} = \begin{pmatrix} 17 \end{pmatrix},$$

a 1×1 matrix; or compare the two products

$$\begin{pmatrix} 1 & 0 \\ 0 & 0 \end{pmatrix} \begin{pmatrix} 0 & 1 \\ 0 & 0 \end{pmatrix} = \begin{pmatrix} 0 & 1 \\ 0 & 0 \end{pmatrix}$$

and

$$\begin{pmatrix} 0 & 1 \\ 0 & 0 \end{pmatrix} \begin{pmatrix} 1 & 0 \\ 0 & 0 \end{pmatrix} = \begin{pmatrix} 0 & 0 \\ 0 & 0 \end{pmatrix}.$$

One special matrix is the $n \times n$ identity matrix $\mathbf{I}$, whose entries are 1 along the main diagonal (from upper left to lower right) and 0 elsewhere. The matrix $\mathbf{I}$ has the property that for any n-rowed column vector $\mathbf{B}$, hence for any $n \times p$ matrix $\mathbf{B}$, $\mathbf{IB} = \mathbf{B}$. This is easily verified for $n = 2$:

$$\begin{pmatrix} 1 & 0 \\ 0 & 1 \end{pmatrix} \begin{pmatrix} a \\ b \end{pmatrix} = \begin{pmatrix} a \\ b \end{pmatrix}.$$

We can also define addition of vectors and matrices, first for column vectors with equal numbers of components:

$$\begin{pmatrix} a_1 \\ a_2 \\ \vdots \\ a_n \end{pmatrix} + \begin{pmatrix} b_1 \\ b_2 \\ \vdots \\ b_n \end{pmatrix} = \begin{pmatrix} a_1 + b_1 \\ a_2 + b_2 \\ \vdots \\ a_n + b_n \end{pmatrix},$$

then for matrices of the same size, by thinking of them as rows of column vectors:

$$\begin{pmatrix} a_{11} & a_{12} & \cdots & a_{1n} \\ a_{21} & a_{22} & \cdots & a_{2n} \\ \vdots & \vdots & & \vdots \\ a_{m1} & a_{m2} & \cdots & a_{mn} \end{pmatrix} + \begin{pmatrix} b_{11} & b_{12} & \cdots & b_{1n} \\ b_{21} & b_{22} & \cdots & b_{2n} \\ \vdots & \vdots & & \vdots \\ b_{m1} & b_{m2} & \cdots & b_{mn} \end{pmatrix}$$

$$= \begin{pmatrix} a_{11} + b_{11} & a_{12} + b_{12} & \cdots & a_{1n} + b_{1n} \\ a_{21} + b_{21} & a_{22} + b_{22} & \cdots & a_{2n} + b_{2n} \\ \vdots & \vdots & & \vdots \\ a_{m1} + b_{m1} & a_{m2} + b_{m2} & \cdots & a_{mn} + b_{mn} \end{pmatrix}.$$

Note that $\mathbf{A} + \mathbf{B}$ makes sense only if $\mathbf{A}$ and $\mathbf{B}$ have the same size.

Addition of matrices or vectors is associative because addition in R is associative.

If $\mathbf{A}$ is a matrix of any size (in particular, a row or column vector) and s is an element of R, that is, a scalar, then define the matrix $s\mathbf{A}$ to be the element in which each element of $\mathbf{A}$ is multiplied by s. That is,

$$s\mathbf{A} = s \begin{pmatrix} a_{11} & a_{12} & \cdots & a_{1n} \\ a_{21} & a_{22} & \cdots & a_{2n} \\ \vdots & \vdots & & \vdots \\ a_{m1} & a_{m2} & \cdots & a_{mn} \end{pmatrix}$$

$$
= \begin{pmatrix} sa_{11} & sa_{12} & \cdots & sa_{1n} \\ sa_{21} & sa_{22} & \cdots & sa_{2n} \\ \vdots & \vdots & & \vdots \\ sa_{m1} & sa_{m2} & \cdots & sa_{mn} \end{pmatrix}.
$$

For example,

$$
3 \begin{pmatrix} 0 \\ 1 \\ 3 \end{pmatrix} = \begin{pmatrix} 0 \\ 3 \\ 9 \end{pmatrix},
$$

$$
-2 \begin{pmatrix} 0 & 0 & 0 \\ 1 & 2 & 5 \\ 3 & -1 & 15 \end{pmatrix} = \begin{pmatrix} 0 & 0 & 0 \\ -2 & -4 & -10 \\ -6 & 2 & -30 \end{pmatrix}.
$$

B. Linear Equations

Matrices and vectors are a convenient way to describe systems of linear equations.

Suppose given a system of m equations in n unknowns:

$$
\begin{aligned}
a_{11}x_1 + a_{12}x_2 + \cdots + a_{1n}x_n &= b_1, \\
a_{21}x_1 + a_{22}x_2 + \cdots + a_{2n}x_n &= b_2, \\
&\vdots \\
a_{m1}x_1 + a_{m2}x_2 + \cdots + a_{mn}x_n &= b_m,
\end{aligned}
$$

where the elements $a_{11} \ldots, a_{mn}$ and $b_1, \ldots, \ldots b_m$ are elements of the commutative ring R. We call such a system *homogeneous* if $b_1 = b_2 = \ldots = b_m = 0$, and *nonhomogeneous* otherwise.

We can make the two sides into column vectors and write the system as an equality of column vectors,

$$
\begin{pmatrix} a_{11}x_1 + a_{12}x_2 + \cdots + a_{1n}x_n \\ a_{21}x_1 + a_{22}x_2 + \cdots + a_{2n}x_n \\ \vdots \\ a_{m1}x_1 + a_{m2}x_2 + \cdots + a_{mn}x_n \end{pmatrix} = \begin{pmatrix} b_1 \\ b_2 \\ \vdots \\ b_m \end{pmatrix} \tag{8.1}
$$

because two column vectors are equal precisely when their respective components are equal. We can rewrite (8.1) in either of two ways. On the one hand we can use the definition of addition and scalar multiplication of column vectors ($= m \times 1$ matrices) to write equation (8.1) as

$$
x_1 \begin{pmatrix} a_{11} \\ a_{21} \\ \vdots \\ a_{m1} \end{pmatrix} + x_2 \begin{pmatrix} a_{12} \\ a_{22} \\ \vdots \\ a_{m2} \end{pmatrix} + \ldots + x_n \begin{pmatrix} a_{1n} \\ a_{2n} \\ \vdots \\ a_{mn} \end{pmatrix} = \begin{pmatrix} b_1 \\ b_2 \\ \vdots \\ b_m \end{pmatrix}.
$$

This says that to solve the original system is the same as to write the vector

$$\begin{pmatrix} b_1 \\ b_2 \\ \vdots \\ b_m \end{pmatrix}$$

as a linear combination (i.e., a sum of scalar multiples) of the column vectors

$$\begin{pmatrix} a_{11} \\ a_{21} \\ \vdots \\ a_{m1} \end{pmatrix}, \begin{pmatrix} a_{12} \\ a_{22} \\ \vdots \\ a_{m2} \end{pmatrix}, \ldots, \begin{pmatrix} a_{1n} \\ a_{2n} \\ \vdots \\ a_{mn} \end{pmatrix}.$$

On the other hand, we can write down the $m \times n$ matrix whose columns are the vectors we just wrote down, and observe that the left side of (8.1) is the product of that matrix, called the matrix of coefficients of the original system, with a column vector of the x_i's:

$$\begin{pmatrix} a_{11}x_1 + a_{12}x_2 + \ldots + a_{1n}x_n \\ a_{21}x_1 + a_{22}x_2 + \ldots + a_{2n}x_n \\ \vdots \\ a_{m1}x_1 + a_{m2}x_2 + \ldots + a_{mn}x_n \end{pmatrix} = \begin{pmatrix} a_{11} & a_{12} & \cdots & a_{1n} \\ a_{21} & a_{22} & \cdots & a_{2n} \\ \vdots & & & \\ a_{m1} & a_{m2} & \cdots & a_{mn} \end{pmatrix} \begin{pmatrix} x_1 \\ x_2 \\ \vdots \\ x_n \end{pmatrix}.$$

If we set

$$\mathbf{A} = \begin{pmatrix} a_{11} & a_{12} & \cdots & a_{1n} \\ a_{21} & a_{22} & \cdots & a_{2n} \\ \vdots & & & \\ a_{m1} & a_{m2} & \cdots & a_{mn} \end{pmatrix}, \quad \mathbf{X} = \begin{pmatrix} x_1 \\ x_2 \\ \vdots \\ x_n \end{pmatrix}, \quad \mathbf{B} = \begin{pmatrix} b_1 \\ b_2 \\ \vdots \\ b_m \end{pmatrix},$$

then the set of equations can be written in the form $\mathbf{AX} = \mathbf{B}$.

Example 1. The set of equations

$$3x_1 - 2x_2 + x_3 = 4,$$
$$x_1 + x_2 - x_3 = 2,$$
$$x_1 + 3x_3 = 1$$

may be written

$$x_1 \begin{pmatrix} 3 \\ 1 \\ 1 \end{pmatrix} + x_2 \begin{pmatrix} -2 \\ 1 \\ 0 \end{pmatrix} + x_3 \begin{pmatrix} 1 \\ -1 \\ 3 \end{pmatrix} = \begin{pmatrix} 4 \\ 2 \\ 1 \end{pmatrix},$$

or as

$$\begin{pmatrix} 3 & -2 & 1 \\ 1 & 1 & -1 \\ 1 & 0 & 3 \end{pmatrix} \begin{pmatrix} x_1 \\ x_2 \\ x_3 \end{pmatrix} = \begin{pmatrix} 4 \\ 2 \\ 1 \end{pmatrix}. \tag{8.2}$$

Suppose there were an $n \times m$ matrix $\mathbf{C}$ such that $\mathbf{CA} = \mathbf{I}$. If we could find such a $\mathbf{C}$, then $\mathbf{CB} = \mathbf{CAX} = \mathbf{IX} = \mathbf{X}$ would be a solution of the equations. Thus solving equations is closely related to finding inverses of matrices. For example,

$$\begin{pmatrix} 3 & -2 & 1 \\ 1 & 1 & -1 \\ 1 & 0 & 3 \end{pmatrix}$$

turns out to have the inverse

$$\begin{pmatrix} 3/16 & 6/16 & 1/16 \\ -1/4 & 1/2 & 1/4 \\ -1/16 & -1/8 & 5/16 \end{pmatrix},$$

so equation (8.2) has the solution

$$\begin{pmatrix} x_1 \\ x_2 \\ x_3 \end{pmatrix} = \begin{pmatrix} 3/16 & 6/16 & 1/16 \\ -1/4 & 1/2 & 1/4 \\ -1/16 & -1/8 & 5/16 \end{pmatrix} \begin{pmatrix} 4 \\ 2 \\ 1 \end{pmatrix} = \begin{pmatrix} 25/16 \\ 1/4 \\ -3/16 \end{pmatrix}.$$

C. Determinants and Inverses

If $\mathbf{A}$ is an $n \times n$ (square) matrix with entries in the commutative ring R, then the determinant of $\mathbf{A}$ is defined and is an element of R. For 1×1, 2×2 and 3×3 matrices, the determinant of $\mathbf{A}$ is defined as follows:

$$\det(a) = a;$$

$$\det \begin{pmatrix} a & b \\ c & d \end{pmatrix} = ad - bc;$$

$$\det \begin{pmatrix} a_1 & b_1 & c_1 \\ a_2 & b_2 & c_2 \\ a_3 & b_3 & c_3 \end{pmatrix} = a_1 b_2 c_3 + b_1 c_2 a_3 + c_1 a_2 b_3$$

$$- a_1 c_2 b_3 - b_1 a_2 c_3 - c_1 b_2 a_3.$$

If $\mathbf{A}$ is a triangular matrix, that is, a square matrix of the form

$$\begin{pmatrix} a_{11} & 0 & \cdots & 0 \\ a_{21} & a_{22} & \cdots & 0 \\ \vdots & \vdots & \cdots & \vdots \\ a_{n1} & a_{n2} & \cdots & a_{nn} \end{pmatrix}$$

with all entries above the main (upper left to lower right) diagonal equal to zero, then

$$\det(\mathbf{A}) = a_{11}a_{22}\cdots a_{nn}.$$

For nontriangular 4×4 or larger matrices the explicit formula for the determinant is too complicated to write down here and will not be needed in this book.

If $\mathbf{A}$ is an $n \times n$ matrix with entries in R, sometimes $\mathbf{A}$ has an inverse, an $n \times n$ matrix $\mathbf{B}$ such that $\mathbf{AB} = \mathbf{BA} = \mathbf{I}$, the $n \times n$ identity matrix. If $\mathbf{A}$ has an inverse, the inverse is unique and is usually denoted by $\mathbf{A}^{-1}$.

In elementary linear algebra over the real numbers, there is a theorem that states that an $n \times n$ matrix $\mathbf{A}$ has an inverse if and only if $\det(\mathbf{A})$ is not zero. The corresponding theorem for a square matrix over a commutative ring R is: An $n \times n$ matrix $\mathbf{A}$ is invertible if and only if $\det(\mathbf{A})$ is a unit of R. For 2×2 matrices part of this theorem can be seen explicitly, as follows. Suppose

$$\mathbf{A} = \begin{pmatrix} a & b \\ c & d \end{pmatrix}$$

then $\det(\mathbf{A}) = ad - bc$. If $\det(\mathbf{A})$ is a unit of R, with inverse $1/(ad - bc)$, then the inverse $\mathbf{B}$ of $\mathbf{A}$ may be written down explicitly as

$$\mathbf{A}^{-1} = \begin{pmatrix} \frac{d}{ad-bc} & \frac{-b}{ad-bc} \\ \frac{-c}{ad-bc} & \frac{a}{ad-bc} \end{pmatrix}$$

as is easily checked. Analogous formulas (involving cofactors and the classical adjoint of $\mathbf{A}$) are available for $n \times n$ matrices $\mathbf{A}$ for $n > 2$, again with entries in any commutative ring with identity R, but the formulas are complicated and of limited practical value, and we will not present them here.

To solve the matrix equation $\mathbf{AX} = \mathbf{D}$ for $\mathbf{X}$ where $\mathbf{A}$ has an inverse $\mathbf{A}^{-1}$, we can multiply both sides *on the left* by $\mathbf{A}^{-1}$ to get

$$\mathbf{A}^{-1}\mathbf{AX} = \mathbf{A}^{-1}\mathbf{D},$$

hence $\mathbf{X} = \mathbf{A}^{-1}\mathbf{D}$.

The matrix theory thus far presented is sufficient for the applications to codes in Sections E and F.

D. $M_n(R)$. We observed that if $\mathbf{A}$ is an $m \times n$ matrix and $\mathbf{B}$ is an $n \times p$ matrix, then $\mathbf{AB}$ is defined and is an $m \times p$ matrix. If $\mathbf{A}$ and $\mathbf{B}$ are both $m \times n$ matrices, then $\mathbf{A} + \mathbf{B}$ is defined and is an $m \times n$ matrix. Thus if we consider the set $M_n(R)$ of all $n \times n$ (square) matrices with entries in the commutative ring R, then $M_n(R)$ is equipped with both addition and multiplication. In fact:

Theorem 1. *If R is a commutative ring with identity, then $M_n(R)$ is a ring with identity.*

Proof. We sketch the ideas but omit most details.

The axioms for addition follow almost immediately from the fact that R satisfies the same axioms, because addition of matrices is done component by component–an $m \times n$ matrix is just a vector with mn components.

Much harder is associativity of multiplication: $\mathbf{A}(\mathbf{BC}) = (\mathbf{AB})\mathbf{C}$. Perhaps the cleanest way to show associativity is to view a matrix as representing a linear transformation with respect to some bases of vector spaces of appropriate dimensions and show that matrix multiplication corresponds to composition of linear transformations. Then associativity of multiplication follows from the associativity of the composition of three functions, which is obvious. See a linear algebra textbook for details.

For the multiplicative identity, let

$$\mathbf{I} = \begin{pmatrix} 1 & 0 & \cdots & 0 \\ 0 & 1 & \cdot & 0 \\ \vdots & & & \vdots \\ 0 & 0 & \cdots & 1 \end{pmatrix}.$$

Then $\mathbf{I}$ is the multiplicative identity: $\mathbf{AI} = \mathbf{IA} = \mathbf{A}$ for any $n \times n$ matrix $\mathbf{A}$, as we already observed.

To show the two distributivity laws: $\mathbf{A}(\mathbf{B} + \mathbf{C}) = \mathbf{AB} + \mathbf{AC}$; $(\mathbf{A} + \mathbf{B})\mathbf{C} = \mathbf{AC} + \mathbf{BC}$, observe that for the first law, the i-j-th component of the left side is

$$a_{i1}(b_{1j} + c_{1j}) + \ldots + a_{n1}(b_{nj} + c_{nj})$$

and the i-j-th component of the right side is

$$(a_{i1}b_{1j} + \ldots + a_{n1}b_{nj}) + (a_{i1}c_{1j} + \ldots + a_{n1}c_{nj});$$

the respective components are equal because associativity and commutativity of addition and the distributive law holds in R. The same argument works for the second distributive law.

Thus $M_n(R)$ is a ring with identity. $\qquad\qquad\square$

Note that $M_n(R)$ is not a commutative ring if $n \geq 2$. As we indicated earlier, even for square matrices matrix multiplication is rarely commutative.

Our point in introducing $M_n(R)$ is to exhibit a natural collection of examples of noncommutative rings. We will hardly ever use these rings later in the book, but they are of considerable importance in modern algebra.

Exercises.

1. Check associativity of addition for $M_n(R)$.

2. Check commutativity of addition for $M_n(R)$.

3. Check associativity of multiplication for 2×2 matrices.

4. Check distributivity for 2×2 matrices.

5. Show that $M_n(R)$ has zero divisors for each $n \geq 2$, even if R is a field.

6. For each $n \geq 2$, find a nonzero $n \times n$ matrix without an inverse.

7. Show that if R is a field, then each nonzero $n \times n$ matrix is either a unit or a zero divisor. (This requires some matrix theory not presented in this chapter.)

8. The ring $M_2(\mathbb{F}_2)$ has 16 elements. Find all the units of $M_2(\mathbb{F}_2)$. Show that the units are closed under multiplication. Write down the multiplication table for the units. (You can check your answer in Section 11H.)

E. Error-Correcting Codes, I

Error-correcting codes are an application of $\mathbb{F}_2$ and other finite fields that was discovered only around 1948. Our exposition will assume some acquaintance with matrices and vectors.

The problem is the following. Suppose a message consisting of words, that is, blocks of digits, is to be transmitted through a channel to a receiver. If he channel is "noisy" and tends to introduce random errors into what was sent, i.e., change digits, how can the receiver determine what was sent?

The basic idea for the solution is to send messages with redundant data, that is, messages with digits which are repeated, partially repeated, or presented in a certain special format. The receiver can detect or even correct errors in the digits of the message received, by seeing how what was received varies from the format in which the message was known to be originally sent.

Two examples of schemes that detect errors are "casting out 9's", for checking arithmetic operations such as multiplication, and the Luhn formula for checking credit card numbers, both discussed in Chapter 5. These schemes cannot correct errors. For example, if the Luhn formula applied to a 16 digit credit card number shows that there is an error in the number, that error could have arisen with any of the 16 digits, and so there is no way to know which digit to change.

On the other hand, here is a simple example of a scheme that corrects errors.

Sandra wishes to send one of the numbers 0 or 1 through a noisy channel to Rob. Suppose Sandra wishes to send a. She encodes a as follows: she sends out the five-digit word $aaaaa$. Rob's decoding rule is that he thinks the bit A was sent if he receives a word with at least three A's in it. He would be misled only if during the transmission from Sandra to Rob, at least three of the a's sent had been erroneously changed to A's, where $A \neq a$. Thus, for example, if Sandra sends 00000, Rob would assume that 0 was sent unless he receives a five-bit word with three or more 1's in it, such as 10011 or 01111. If two or fewer errors occurred in the transmission of 00000, Rob would correctly determine what Sandra sent.

Error correcting capability is very desirable for data in certain situations.

One situation is where data is coming from a measuring device (such as a space probe) that is continuously transmitting measurements (such as of Jupiter's magnetic field as it travels rapidly through space) and cannot retransmit data, even if the receiver knows they are erroneous. (The channel in this situation would be the space through which the radio waves pass, and the noise would be radio noise, or static).

Another situation is where the transmitting consists of the placing of data into the memory of a computer. The channel here is the memory, which may contain imperfectly manufactured components, and the receiving is the retrieval of the data.

In the two situations just described, the information sent is numerical, and might naturally be in numbers expressed in base 2. Since all of the mathematics tends to be easiest in base 2 also, we shall henceforth assume that we are sending words written in base 2, that is, sequences of 0's and 1's.

Here are two examples in base 2 similar to the examples described above.

Example 2. The parity check code. Sandra wishes to transmit n information digits $abcd \cdots e$. Let

$$f \equiv a+b+c+d+ \ldots +e \quad (\text{mod } 2).$$

She sends $(a,b,c,d,\ldots,e,f)$.

Rob, the receiver, receives $(A,B,C,D,\ldots,E,F)$. If

$$A+B+C+D+ \ldots +E \equiv F \quad (\text{mod } 2),$$

then Rob decides that no error occurred and $(A,B,C,D,\ldots,E,F) = (a,b,c, d,\ldots,e,f)$. Otherwise, Rob decides that an error occurred, but doesn't know where it occurred.

If $A+B+C+D+ \ldots +E \not\equiv F \pmod 2$, then in fact an odd number of errors occurred, while if $\equiv$, then there is an even number of errors: none, or two or more. Rob would be misled if two or more errors occurred in the transmission. If more than one error is extremely unlikely, Rob would have confidence that he decided correctly.

This code, like the Luhn check, detects but does not correct a single error. It is a very efficient code, since $n/(n+1)$ of each code word is information, and only one digit in each word is redundant.

Example 3. The repetition code. Given one information digit a ($a = 0$ or 1), Sandra sends the word of odd length n, $aaa \cdots a$. Rob receives $ABCD \cdots E$, where each of $A,B,\ldots,E$ is 0 or 1. If the number of 1's among $AB \cdots E$ exceeds the number of 0's, Rob decides that $a = 1$; otherwise he decides that $a = 0$. Rob will decode incorrectly only if there are more than $n/2$ errors in $AB \cdots E$.

This code then corrects up to $n/2$ errors, in the sense that if there are fewer than $n/2$ errors then Rob can determine correctly the transmitted word. This code is, however, quite inefficient, for only $1/n$ of each code word is information, and $n - 1$ of the n digits in each word are redundant.

The development of codes has tended to proceed from the assumption that errors are uncommon, so that a desirable code is one that is efficient (that is, the ratio of information digits per word to word length is "large") and capable of correcting errors in a small proportion of the digits of each word.

In the rest of this section we describe two examples of efficient codes constructed using matrices with entries in $\mathbb{F}_2$. These codes are examples of codes described by R.W. Hamming of Bell Telephone Laboratories [Hamming (1950)]. In Chapter 25 we shall describe other codes.

Code I This is an example of a code which corrects one error in words of length 7, where each word has 4 information bits.

We work with elements of $\mathbb{F}_2 = \mathbb{Z}/2\mathbb{Z}$, integers mod 2. We write $[0]_2 = 0, [1]_2 = 1$, so $\mathbb{F}_2 = \{0, 1\}$. Our words are 7-tuples with entries in $\mathbb{F}_2$. Let

$$\mathbf{H} = \begin{pmatrix} 1 & 0 & 1 & 0 & 1 & 0 & 1 \\ 0 & 1 & 1 & 0 & 0 & 1 & 1 \\ 0 & 0 & 0 & 1 & 1 & 1 & 1 \end{pmatrix}$$

This matrix $\mathbf{H}$ has the pleasant property that for r, s, t in $\mathbb{F}_2$ not all zero,

$$\begin{pmatrix} r \\ s \\ t \end{pmatrix}$$

is the $(tsr)_2$-th column of $\mathbf{H}$. Thus the sixth column is $\begin{pmatrix} 0 \\ 1 \\ 1 \end{pmatrix}$, and reading the digits from bottom to top, $(110)_2 = 6$. In particular, all columns of $\mathbf{H}$ are different, an essential fact.

Suppose Sandra wishes to send the word $\mathbf{W} = \begin{pmatrix} a \\ b \\ c \\ d \end{pmatrix}$, where a, b, c, d are in $\mathbb{F}_2$.

Call $\mathbf{W}$ the information word. Sandra forms the vector

$$\mathbf{C} = \begin{pmatrix} x \\ y \\ a \\ z \\ b \\ c \\ d \end{pmatrix}$$

with x, y, z chosen so that $\mathbf{HC} = \mathbf{0}$, that is, so that (in $\mathbb{F}_2$)

$$x + a + b + d = 0,$$
$$y + a + c + d = 0,$$
$$z + b + c + d = 0.$$

Here (x, y, z) is the redundant part of the word. Sandra can find the numbers x, y, z from a, b, c, d quickly using the equation $\mathbf{HC} = \mathbf{0}$, or she can find $\mathbf{C}$ by multiplying $\mathbf{W}$ by a 7×4 matrix $\mathbf{G}$ obtained by solving for x, y, z, a, b, c, d in terms of a, b, c, d:

$$
\begin{pmatrix} x \\ y \\ a \\ z \\ b \\ c \\ d \end{pmatrix} = \begin{pmatrix} 1 & 1 & 0 & 1 \\ 1 & 0 & 1 & 1 \\ 1 & 0 & 0 & 0 \\ 0 & 1 & 1 & 1 \\ 0 & 1 & 0 & 0 \\ 0 & 0 & 1 & 0 \\ 0 & 0 & 0 & 1 \end{pmatrix} \begin{pmatrix} a \\ b \\ c \\ d \end{pmatrix}.
$$

The vector $\mathbf{C}$, made up of the information word $\mathbf{W} = (a, b, c, d)$ and the redundancy (x, y, z), is the coded word. Sandra transmits the vector $\mathbf{C}$ to Rob over a possibly noisy channel.

Suppose Rob receives

$$
\mathbf{R} = \begin{pmatrix} x' \\ y' \\ a' \\ z' \\ b' \\ c' \\ d' \end{pmatrix}.
$$

Case 0. Suppose $\mathbf{R} = \mathbf{C}$. Then $\mathbf{HR} = \mathbf{0}$, because $\mathbf{HC} = \mathbf{0}$.

Case 1. Suppose one component of $\mathbf{C}$ was changed in the transmission, so that $\mathbf{R}$ differs from $\mathbf{C}$ in at most one component. Then $\mathbf{R} - \mathbf{C} = \mathbf{E}$ is a column vector with all entries 0 except for a 1 in the component where the error occurred. Then $\mathbf{HE}$ is the column of $\mathbf{H}$ corresponding to the location of the 1 in the vector $\mathbf{E}$. When Rob computes $\mathbf{HR}$, he gets:

$$\mathbf{HR} = \mathbf{HC} + \mathbf{HE}$$
$$= \mathbf{0} + \mathbf{HE}$$
$$= \text{(the column of } \mathbf{H} \text{ corresponding to where the 1 is in } \mathbf{E}).$$

Thus, if there is one error, Rob can determine where the error is by examining $\mathbf{HR}$; once he knows $\mathbf{E}$, he knows $\mathbf{R} - \mathbf{E} = \mathbf{C}$, the word that Sandra transmitted.

Case 2. If $\mathbf{R}$ differs from $\mathbf{C}$ in two or more entries, then

$$\mathbf{HR} = \mathbf{HC} + \mathbf{HE}$$
$$= \mathbf{0} + \text{(sum of two or more columns of } \mathbf{H}).$$

Since the sum of two or more columns of $\mathbf{H}$ is either $\mathbf{0}$ or a column of $\mathbf{H}$, Rob will decode inaccurately because he is assuming that no errors or one error occurred.

Thus with this code, Rob is capable of correcting exactly one error. If Rob can confidently assume that at most one error occurred in the transmission of a word, then he will be able to confidently determine what was sent. He will be misled whenever more than one error occurs in a given word. (If p, the probability of an error in any given digit, is $p = .1$, and the probability of an error in some digit is independent of the probability of an error in any other digit, then the probability of at most one error in a word is $e = (1 - .1)^7 + 7(1 - .1)^6(.1) = .85$, so there is a 15 percent chance that Rob will be misled on each word. If $p = .01, e = .998$, so there is a 0.2 percent chance that he will be misled).

To illustrate the decoding of Code I with some examples, suppose Rob receives

$$\mathbf{R} = \begin{pmatrix} 1 \\ 0 \\ 1 \\ 1 \\ 0 \\ 1 \\ 1 \end{pmatrix}. \text{ Then } \mathbf{HR} = \begin{pmatrix} 1 & 0 & 1 & 0 & 1 & 0 & 1 \\ 0 & 1 & 1 & 0 & 0 & 1 & 1 \\ 0 & 0 & 0 & 1 & 1 & 1 & 1 \end{pmatrix} \begin{pmatrix} 1 \\ 0 \\ 1 \\ 1 \\ 0 \\ 1 \\ 1 \end{pmatrix} = \begin{pmatrix} 1 \\ 1 \\ 1 \end{pmatrix},$$

so assuming one error, it must be the last digit, and so Rob will assume that Sandra sent

$$\mathbf{C} = \begin{pmatrix} 1 \\ 0 \\ 1 \\ 1 \\ 0 \\ 1 \\ 0 \end{pmatrix}.$$

If $\mathbf{R} = \begin{pmatrix} 1 \\ 1 \\ 1 \\ 0 \\ 0 \\ 1 \\ 0 \end{pmatrix}$, then $\mathbf{HR} = \begin{pmatrix} 0 \\ 1 \\ 1 \end{pmatrix}$, so $\mathbf{C} = \begin{pmatrix} 1 \\ 1 \\ 1 \\ 0 \\ 0 \\ 0 \\ 0 \end{pmatrix}$.

If $\mathbf{R} = \begin{pmatrix} 0 \\ 0 \\ 0 \\ 1 \\ 0 \\ 0 \\ 1 \end{pmatrix}$, then $\mathbf{HR} = \begin{pmatrix} 1 \\ 1 \\ 0 \end{pmatrix}$, so $\mathbf{C} = \begin{pmatrix} 0 \\ 0 \\ 1 \\ 1 \\ 0 \\ 0 \\ 1 \end{pmatrix}$.

If $\mathbf{R} = \begin{pmatrix} 0 \\ 1 \\ 0 \\ 1 \\ 0 \\ 1 \\ 0 \end{pmatrix}$ then $\mathbf{HR} = \begin{pmatrix} 0 \\ 0 \\ 0 \end{pmatrix}$, so $\mathbf{C} = \mathbf{R}$.

Code II A modification of Code I will enable Rob to detect the presence of two errors, as well as to correct one error. Let

$$\mathbf{H} = \begin{pmatrix} 1 & 1 & 1 & 1 & 1 & 1 & 1 & 1 \\ 0 & 1 & 0 & 1 & 0 & 1 & 0 & 1 \\ 0 & 0 & 1 & 1 & 0 & 0 & 1 & 1 \\ 0 & 0 & 0 & 0 & 1 & 1 & 1 & 1 \end{pmatrix},$$

the matrix of Code I with an additional column of zeros on the left and then a row of 1's on the top. Sandra wishes to send Rob the information word $\mathbf{W} = (a,b,c,d)$. To add redundancy, she constructs

$$\mathbf{C} = \begin{pmatrix} w \\ x \\ y \\ a \\ z \\ b \\ c \\ d \end{pmatrix}$$

with $\mathbf{HC} = \mathbf{0}$. Then x, y, z satisfy equations (1) of Code I, and w satisfies

$$0 = w + x + y + z + a + b + c + d.$$

Adding this equation to equations (1) of Code I yields the simpler equation defining w:

$$w + a + b + c = 0.$$

Sandra transmits the vector $\mathbf{C}$. (She can derive $\mathbf{C}$ from $\mathbf{W}$ by multiplying $\mathbf{W}$ by an 8×4 matrix $\mathbf{G}$ as in Code I).

Suppose Rob receives $\mathbf{R}$. He computes $\mathbf{HR}$.

Case 0. If no errors occurred, then $\mathbf{R} = \mathbf{C}$ and $\mathbf{HR} = \mathbf{0}$.

Case 1. If one error occurred, then $\mathbf{R} - \mathbf{C} = \mathbf{E}$ has one nonzero entry, one error. Then $\mathbf{HR} = \mathbf{HE}$ since $\mathbf{HC} = \mathbf{0}$, and $\mathbf{HE}$ is the column of $\mathbf{H}$ corresponding to where the error occurred. Since all columns of $\mathbf{H}$ are distinct, Rob can locate and correct the error by identifying $\mathbf{HR}$ with a column of $\mathbf{H}$ and correcting the corresponding entry of $\mathbf{R}$.

Case 2. If two errors occurred, then $\mathbf{R} - \mathbf{C} = \mathbf{E}$ has two nonzero entries. Then $\mathbf{HR} = \mathbf{HE}$ is the sum of two columns of $\mathbf{H}$. Rob cannot determine which two columns of $\mathbf{H}$ make up the sum. For example, here are two sums of columns of $\mathbf{H}$:

$$\begin{pmatrix} 1 \\ 1 \\ 1 \\ 0 \end{pmatrix} + \begin{pmatrix} 1 \\ 1 \\ 0 \\ 1 \end{pmatrix} = \begin{pmatrix} 0 \\ 0 \\ 1 \\ 1 \end{pmatrix} = \begin{pmatrix} 1 \\ 0 \\ 1 \\ 0 \end{pmatrix} + \begin{pmatrix} 1 \\ 0 \\ 0 \\ 1 \end{pmatrix}.$$

But what he knows is that $\mathbf{HE}$ is not a column of $\mathbf{H}$, since the sum of two columns of $\mathbf{H}$ is non-zero but always has top entry $= 0$, while every column of $\mathbf{H}$ has top entry $= 1$. So Rob knows that at least two errors occurred.

Code II, then, is a code that corrects one error and detects two errors in words of length 8 with 4 information digits. Rob will be misled only if there are 3 or more errors.

Exercises.

9. Here is a collection of received words which were transmitted after being encoded with Code II. For each word assume there are 0, 1, or 2 errors. Decode each word.

$$\begin{pmatrix} 1 \\ 0 \\ 1 \\ 1 \\ 0 \\ 1 \\ 1 \\ 1 \end{pmatrix}, \begin{pmatrix} 1 \\ 1 \\ 1 \\ 1 \\ 0 \\ 0 \\ 0 \\ 1 \end{pmatrix}, \begin{pmatrix} 1 \\ 0 \\ 1 \\ 1 \\ 1 \\ 0 \\ 0 \\ 1 \end{pmatrix}, \begin{pmatrix} 0 \\ 1 \\ 0 \\ 1 \\ 1 \\ 0 \\ 1 \\ 1 \end{pmatrix}$$

10. What is the maximum allowable probability of error is a typical digit in order that Code II can be used with probability .999 that the receiver will not be misled (i.e., 3 or more errors occur) in a single word?

11. Define a code, analogous to Code II, that uses a 5×16 matrix H, and sends out binary words of length 16 (of which 11 are information digits) such that the receiver can correct one error and detect two errors.

12. In Code II, do there exist received words which the receiver can determine with certainty have at least 3 errors?

13. In Code I, we found $\mathbf{C}$ by multiplying the information word $\mathbf{W}$ by the 7×4 matrix $\mathbf{G}$.

Suppose instead of using $\mathbf{H}$, we used

$$\mathbf{H_0} = \begin{pmatrix} 1 & 0 & 0 & 1 & 1 & 0 & 1 \\ 0 & 1 & 0 & 1 & 0 & 1 & 1 \\ 0 & 0 & 1 & 0 & 1 & 1 & 1 \end{pmatrix} = (\mathbf{I}, \mathbf{P}).$$

where $\mathbf{I}$ is the 3×3 identity matrix. Describe the corresponding encoding matrix $\mathbf{G_0}$ in terms of $\mathbf{P}$.

14. Find the encoding matrix $\mathbf{G}$ for Code II.

15. If $\mathbf{C_1}$ and $\mathbf{C_2}$ are two 8-tuples, let $d(\mathbf{C_1} - \mathbf{C_2})$, the distance between $\mathbf{C_1}$ and $\mathbf{C_2}$, be the number of 1's in $\mathbf{C_1} - \mathbf{C_2}$. Can you determine the minimum distance between two coded words in Code II? Do you see any relationship between the minimum distance and the error correcting ability of the code?

F. Hill Cryptosystems

Cryptography is the study of techniques to ensure secure communication in the presence of a malicious adversary.

The problem is that two parties, say Alice and Bob, wish to communicate over an insecure channel, such as via radio, a cell phone or the internet, and want to ensure that messages between them are private and authentic. Private means that no third party, say Eve (an eavesdropper), can comprehend their messages; authentic means that when Bob receives a message purportedly from Alice, he can be confident that only Alice could have sent the message.

To try to achieve the desired security, Alice encrypts her message m, called the *plaintext*, by transforming it into a encrypted message or *ciphertext*, c, by some function f: that is, $c = f(m)$. The encrypted message c is transmitted to Bob, who decrypts it via the inverse g of the function f to get the original message: $m = g(c)$. The function f and its inverse g are known only to Alice and Bob. If Eve wished to impersonate Alice, Eve would need to know the encrypting function f. If Eve intercepts a message from Alice to Bob, Eve would have c, the ciphertext, but in order to obtain the plaintext, Eve would need to know the decrypting function g.

Cryptography seeks to find encrypting and decrypting functions that are sufficiently secure, and, at the same time, studies ways to crack proposed cryptosystems with given encrypting and decrypting functions. It looks at the problem from both Alice and Bob's point of view, and from Eve's.

Historically, the problem of security of communication was most acute in wartime. However, in the present day, the need for secure information goes far beyond uses in warfare. The rapid growth of electronic commerce depends on the integrity of communication over the internet. For this reason, cryptosystems are important tools for business, and cryptography has become an important research area of computer science and applied mathematics.

In this book we will look at several cryptosystems, all based on modular arithmetic and related algebra and number theory.

For our first example, in this section we look at a cryptographic protocol published by Lester Hill (1931). While never widely used, and known to be insecure, the Hill scheme is historically significant because it represented the first public systematic use of mathematics in the design of a cryptosystem.

The encrypting and decrypting functions work on numbers. So we need to translate English messages into sequences of numbers. One way to do so is to count the letters A to Z to make a correspondence between the letters and the numbers from 1 to 26:

$$A\ B\ C\ D\cdots\ \ J\cdots\ \ O\cdots\ T\cdots\ Y\ Z,$$
$$1\ 2\ 3\ 4\ \cdots\ 10\ \cdots\ 15\ \cdots\ 20\ \cdots\ 25\ 26.$$

In the Hill cryptosystem, plaintext messages are vectors of congruence classes in $\mathbb{Z}/26\mathbb{Z}$. We view $\mathbb{Z}/26\mathbb{Z}$ as the numbers from 1 to 26 with arithmetic modulo 26.

For encrypting n-tuples of numbers, we choose an encrypting matrix, an invertible $n \times n$ matrix $\mathbf{A}$ with entries in $\mathbb{Z}/26\mathbb{Z}$. To encrypt, we take a plaintext word $\mathbf{w}$, that is, an n-tuple of elements of $\mathbb{Z}/26\mathbb{Z}$, write $\mathbf{w}$ as a column vector, and encrypt $\mathbf{w}$ by multiplying it by the $n \times n$ matrix $\mathbf{A}$ (on the left) to get the ciphertext, $\mathbf{c} = \mathbf{Aw}$, another n-tuple of elements of $\mathbb{Z}/26\mathbb{Z}$. In short, the encrypting function is multiplication by the matrix $\mathbf{A}$.

If $\mathbf{A}$ is an invertible matrix, that is, if there exists a matrix $\mathbf{B}$ with entries in $\mathbb{Z}/26\mathbb{Z}$ so that $\mathbf{BA} = \mathbf{I}$, the $n \times n$ identity matrix, then multiplying an n-tuple by $\mathbf{B}$ is the inverse of the function that multiplies an n-tuple by $\mathbf{A}$. As noted earlier in this chapter, $\mathbf{A}$ will have such an inverse $\mathbf{B}$ if and only if the determinant of $\mathbf{A}$ is a unit of $\mathbb{Z}/26\mathbb{Z}$. So the decrypting function is multiplication by $\mathbf{B}$, the inverse of $\mathbf{A}$. If $\mathbf{c} = \mathbf{Aw}$ is a ciphertext, then $\mathbf{w} = \mathbf{Bc}$ is the plaintext.

Suppose Alice wants to send Bob the plaintext message:

BUYXENRON

where X replaces the space between words.

Alice writes the message as a sequence of elements of $\mathbb{Z}/26\mathbb{Z}$ using the counting correspondence above, to get:

$$2, 21, 25, 24, 5, 14, 18, 15, 14.$$

To communicate, Alice and Bob need to agree in advance on a private encrypting key: an invertible $n \times n$ matrix $\mathbf{A}$. Alice will use $\mathbf{A}$ to encrypt, and Bob will use $\mathbf{B} = \mathbf{A}^{-1}$, the inverse of $\mathbf{A}$, to decrypt. The matrix $\mathbf{A}$ must be kept secret, because if Eve knew $\mathbf{A}$, she could easily find the inverse of $\mathbf{A}$ and decrypt any encrypted message Alice sent to Bob.

We illustrate the Hill codes by encrypting and decrypting this message using matrices $\mathbf{A}$ of various sizes.

Codes of Size 1×1. A 1×1 invertible matrix is just a unit of $\mathbb{Z}/26\mathbb{Z}$. For an example, take the element 5 of $\mathbb{Z}/26\mathbb{Z}$: its inverse in $\mathbb{Z}/26\mathbb{Z}$ is 21, since $5 \cdot 21 = 105 \equiv 1 \pmod{26}$.

To encrypt the plaintext, Alice multiplies each number in the message by 5 (mod 26) to get

$$10, 1, 21, 16, 25, 18, 12, 23, 18.$$

She sends the ciphertext either as that sequence of numbers or as the corresponding sequence of letters, JAUPYRLWR.

Bob receives the ciphertext, puts it into a sequence of numbers if needed, and decrypts by multiplying each of the resulting numbers by 21 (mod 26). Since 21 is the inverse of 5 modulo 26, , Bob ends up with the sequence of numbers corresponding to the plaintext, the original message.

For example, Bob would multiply the first number, 10, in the encrypted message, by 21, then find $2 = 10 \cdot 21 \bmod 26$. So the first letter of the original message is B. Bob would then do the same for the other numbers in the encrypted message.

In this 1×1 code, each letter corresponds to a single number or letter. Thus wherever it occurs in the original message, N is always replaced by 18 or R in the encrypted message. Also, different letters in the original message become different numbers or letters in the encrypted message. Codes with those properties are easy enough to crack that they show up as cryptogram puzzles in daily newspapers.

But for $n > 1$ these codes become somewhat more difficult to crack.

Codes of Size 2×2. Alice breaks up the enumerated plaintext

$$2, 21, 25, 24, 5, 14, 18, 15, 14$$

into a sequence of 2-tuples that she puts into vectors with two components:

$$\binom{2}{21}, \binom{25}{24}, \binom{5}{14}, \binom{18}{15}, \binom{14}{24},$$

where to finish the last vector, she adds $X = 24$ to the end of the message.

Alice and Bob agree on a secret invertible 2×2 matrix $\mathbf{A}$. To encrypt, Alice multiplies each vector by $\mathbf{A}$.

Example 4. Suppose

$$\mathbf{A} = \begin{pmatrix} 9 & 1 \\ 1 & 3 \end{pmatrix}.$$

Then $\det \mathbf{A} = 1$, and

$$\mathbf{A}^{-1} = \begin{pmatrix} 3 & -1 \\ -1 & 9 \end{pmatrix}.$$

The encrypted message Alice constructs by multiplying each vector by $\mathbf{A}$ is the sequence of vectors:

$$\begin{pmatrix} 9 & 1 \\ 1 & 3 \end{pmatrix} \begin{pmatrix} 2 \\ 21 \end{pmatrix} = \begin{pmatrix} 13 \\ 13 \end{pmatrix},$$

$$\begin{pmatrix} 9 & 1 \\ 1 & 3 \end{pmatrix} \begin{pmatrix} 25 \\ 24 \end{pmatrix} = \begin{pmatrix} 15 \\ 19 \end{pmatrix},$$

$$\begin{pmatrix} 9 & 1 \\ 1 & 3 \end{pmatrix} \begin{pmatrix} 5 \\ 14 \end{pmatrix} = \begin{pmatrix} 7 \\ 21 \end{pmatrix},$$

$$\begin{pmatrix} 9 & 1 \\ 1 & 3 \end{pmatrix} \begin{pmatrix} 18 \\ 15 \end{pmatrix} = \begin{pmatrix} 21 \\ 11 \end{pmatrix},$$

$$\begin{pmatrix} 9 & 1 \\ 1 & 3 \end{pmatrix} \begin{pmatrix} 24 \\ 14 \end{pmatrix} = \begin{pmatrix} 20 \\ 8 \end{pmatrix},$$

The resulting sequence of numbers is

$$13, 13, 15, 19, 7, 21, 21, 11, 20, 8.$$

If we replace the numbers by the corresponding letters, we get

MMOSGUUKTH

Notice that, for example, X in the plaintext is replaced by S or H in the ciphertext depending on its location in the plaintext, and the two letters U in the ciphertext correspond to E and N in the plaintext. In this code, only pairs of letters are set to the same thing, and then only if they both begin at an odd, or both at an even location in the message. So this encryption is apparently much more difficult for Eve to decrypt.

Bob would take the encrypted message, put it back into a sequence of vectors, and multiply each vector by A^{-1}. Since $A^{-1} \cdot A = I$, he will end up with the original set of 2-tuples and finally, the original plaintext, using the counting correspondence.

Example 5. If we use the matrix

$$A = \begin{pmatrix} 2 & -3 \\ 3 & 1 \end{pmatrix}$$

whose determinant is 11, an invertible element of $\mathbb{Z}/26\mathbb{Z}$, then (see Section C, above)

$$A^{-1} = \begin{pmatrix} -7 & 5 \\ -5 & 12 \end{pmatrix}$$

Applying this matrix A to the 2-tuples (4), Alice would send Bob the encrypted message

$$19, 1, 4, 21, 20, 3, 17, 17, 8, 14 = \text{SADUTCQQHN},$$

which Bob can decipher using A^{-1}.

If higher security is needed we can use larger matrices.

Codes of Size 3×3.

Example 6. Alice breaks up the message into words of length 3

$$\begin{pmatrix} 2 \\ 21 \\ 25 \end{pmatrix}, \begin{pmatrix} 24 \\ 5 \\ 14 \end{pmatrix}, \begin{pmatrix} 18 \\ 15 \\ 14 \end{pmatrix}.$$

Alice and Bob agree on a 3×3 matrix with invertible determinant, such as

$$C = \begin{pmatrix} 2 & 3 & 5 \\ 5 & 11 & 2 \\ 1 & 2 & 2 \end{pmatrix}.$$

The determinant of $\mathbf{C}$ is 7, a unit modulo 26, and so $\mathbf{C}$ is invertible. Its inverse turns out to be

$$\mathbf{C}^{-1} = \begin{pmatrix} 10 & 8 & -7 \\ 10 & 11 & 3 \\ 11 & 11 & 1 \end{pmatrix}.$$

Encrypting the original message BUYXENRON with the matrix $\mathbf{C}$, Alice gets the ciphertext

$$10,5, 16, 3, 21, 10, 21, 23, 24 = \text{JEPCUJUWX},$$

which Bob can decrypt using $\mathbf{C}^{-1}$.

It is evident how one could increase security by using larger matrices.

There is nothing special about $\mathbb{Z}/26\mathbb{Z}$ in all we have done. We might prefer to add three symbols and work in $\mathbb{Z}/29\mathbb{Z}$, a field. Then we could use the counting correspondence as before to translate from letters to elements of $\mathbb{Z}/29\mathbb{Z}$, and let 27, 28, 29 denote ".", "?", and "_" (space). Then

BUY ENRON??

becomes

$$2,21,25,29,5,14,18,15,14,28,28$$

The Hill codes have been thoroughly analyzed by cryptanalysts. Konheim (1981) describes how one might recover the coding matrix $\mathbf{A}$ assuming one has enough pairs of plaintext and corresponding ciphertext. It turns out that if $\mathbf{A}$ is $n \times n$ then not many more than n pairs of plain- and ciphertext will usually suffice to determine $\mathbf{A}$ and crack the code. In "real life" situations, one often finds that such pairs can be obtained. (They certainly were during World War II, and having such pairs available aided cryptanalysts immensely . See Kahn (1967) and Hodges (1983)).

Exercises.

16. Encrypt and decrypt HAPPYXBIRTHDAY using the 1×1 code, multiplying by 7 modulo 26.

17. Encrypt and decrypt HAPPYXBIRTHDAY using the 2×2 code of Example 4.

18. Decrypt MXGWGCCCUKMQNGRC using the code of Example 5.

19. (i) In Example 6, verify that $\mathbf{C}^{-1}$ is as claimed.
(ii) encrypt and decrypt HAPPYXBIRTHDAY using the 3×3 code of Example 6.

20. Do Example 5 (2×2 case) with the same matrix $\mathbf{A}$, except think of $\mathbf{A}$ as having entries in $\mathbb{Z}/29\mathbb{Z}$, and illustrate the example by encoding and decoding the message

BUY ENRON??

Note: $\mathbf{A}^{-1}$ will be different.

Part III
Congruences and Groups

Chapter 9
Fermat's and Euler's Theorems

Pierre de Fermat (1601–1665) was a public official in the French city of Toulouse, and in his spare time was one of the greatest mathematicians of the seventeenth century. Fermat's "little theorem", as generalized by Euler a century later, is perhaps the first theorem in what is now known as group theory. It also has some remarkably interesting applications. We present Fermat's theorem and related theory in this chapter and some applications in the next.

A. Orders of Elements

The mathematics in this chapter starts from the observation that if we take powers of a number a: $1 = a^0, a, a^2, a^3, \ldots$, then eventually two of the powers will be congruent modulo m. For example, modulo 7, the powers of 2:

$$1, 2, 4, 8, 16, 32, 64, \ldots,$$

are congruent to

$$1, 2, 4, 1, 2, 4, 1, \ldots.$$

Modulo 10, the powers of 2 are congruent to

$$1, 2, 4, 8, 6, 2, 4, \ldots.$$

Modulo 12, the powers of 2 are congruent to

$$1, 2, 4, 8, 4, 8, 4, \ldots.$$

The explanation for what is happening is simple. There are exactly m congruence classes modulo m. If we look at the powers of a: $1, a, a^2, a^3, \ldots, a^m$, then since there are $m + 1$ powers and m congruence classes, at least two of the powers must be in the same congruence class. (We used this idea in the proof of Theorem 14 of Section 7C).

L.N. Childs, *A Concrete Introduction to Higher Algebra*, Undergraduate Texts in Mathematics, © Springer Science+Business Media LLC 2009

Suppose $a^r \equiv a^s \pmod{m}$ for some $r \geq 0, s > r$. Then $a^{r+k} \equiv a^{s+k} \pmod{m}$ for every $k \geq 0$. So from a^s on the powers of a modulo m repeat earlier powers of a.

For example, $2^5 \equiv 2 \pmod{10}$. So $2^6 \equiv 2^2$, $2^7 \equiv 2^3$, etc. Pictorially, we have

$$
\begin{array}{ccc}
2 & \rightarrow & 4 \\
\uparrow & & \downarrow \\
6 & \leftarrow & 8
\end{array}
$$

where the arrows mean "multiply by 2 (mod 10)."

Sometimes there is a positive power of a congruent to 1 $\pmod{m}$, sometimes not. In the examples above $2^3 \equiv 1 \pmod{7}$, while $2^s \not\equiv 1 \pmod{10}$ for all $s > 0$. If $a^s \equiv 1 \pmod{m}$ for some $s > 0$, then a must be a unit modulo m, which means that a and m must be coprime. The converse is also true:

Proposition 1. *If a and m are coprime, then $a^t \equiv 1$ (mod m) for some t, $1 \leq t < m$.*

Proof. To get $t < m$ instead of $t \leq m$, we refine the above discussion slightly.

Since a and m are coprime, m does not divide a^s for any s, and so the m numbers $1, a, a^2, \ldots, a^{m-1}$ all belong to the $m - 1$ congruence classes other than the congruence class of 0. So two of the numbers must be in the same congruence class: that is, there exist numbers s and t with $s \geq 0$ and $0 < t < m - 1$ so that $a^s \equiv a^{s+t} \pmod{m}$. Now since a and m are coprime, we can cancel the common factor a^s from both sides of the congruence to get $1 \equiv a^t \pmod{m}$. $\square$

In the notation of congruence classes, Proposition 1 states that if $[a]$ is a unit of $\mathbb{Z}/m\mathbb{Z}$, then $[a]^t = 1$ for some t with $0 < t < m$.

Definition. Let $m \geq 2$ and a be any integer coprime to m. The *order of a modulo m* is the smallest positive integer e so that $a^e \equiv 1 \pmod{m}$. In terms of congruence classes, the *order* of $[a]$ in $\mathbb{Z}/m\mathbb{Z}$ is the smallest $e > 0$ so that $[a]^e = 1$. In terms of divisibility, the order of $a \bmod m$ is the smallest $e > 0$ so that m divides $a^e - 1$.

Example 1. The order of 2 modulo 7 is 3, because $2^3 \equiv 1 \pmod{7}$, while $2^1 \not\equiv 1$ and $2^2 \not\equiv 1 \pmod{7}$.

Notice that in showing that e is the order of a modulo m, two things must be checked:

(i) $a^e \equiv 1 \pmod{m}$; and

(ii) for $1 \leq s < e$, $a^s \not\equiv 1 \pmod{m}$.

Thus the notion of order is similar to that of least common multiple. Recall that m is the least common multiple of a and b if:

(i) m is a common multiple of a and b; and

(ii) no number $<m$ is a common multiple of a and b.

Since ab is a common multiple of a and b, there is a least common multiple of a and b by well ordering.

Similarly, we know by Proposition 1 that if a is a unit modulo m, then there is some positive exponent t so that $a^t \equiv 1 \pmod{m}$, hence by well ordering there is a least positive exponent so that $a^t \equiv 1 \pmod{m}$. That least positive exponent is the order of a modulo m.

We found the order of $[2]$ in $\mathbb{Z}/7\mathbb{Z}$; let us find the orders of the other nonzero elements of $\mathbb{Z}/7\mathbb{Z}$, by direct computation.

The order of $[1]$ is 1.
The order of $[2]$ is 3, for $[2]^1 = [2], [2]^2 = [4], [2]^3 = [1]$.
The order of $[3]$ is 6, for $[3]^1 = [3], [3]^2 = [2], [3]^3 = [6], [3]^4 = [4], [3]^5 = [5], [3]^6 = [1]$.
The order of $[4]$ is 3, for $[4]^1 = [4], [4]^2 = [2], [4]^3 = [1]$.
The order of $[5]$ is 6, for $[5]^1 = [5], [5]^2 = [4], [5]^3 = [6], [5]^4 = [2], [5]^5 = [3], [5]^6 = [1]$.
The order of $[6]$ is 2, for $[6]^1 = [6], [6]^2 = [1]$.

In tabular form, here are the results:

element	order
[1]	1
[2]	3
[3]	6
[4]	3
[5]	6
[6]	2

We found that the least common multiple of a and b not only is $\leq$ any positive common multiple, but in fact divides any common multiple (Chapter 4, Proposition 7). The same is true for the order of a modulo m.

Proposition 2. *If e is the order of a modulo m, and $a^f \equiv 1 \pmod{m}$, then e divides f.*

Proof. We have $a^e \equiv 1 \pmod{m}$ and $a^f \equiv 1 \pmod{m}$. Divide e into f : $f = eq + r$, with $0 \leq r < e$. Then $a^f = (a^e)^q \cdot a^r$. Modulo m, this becomes $1 \equiv (1)^q \cdot a^r$, so $a^r \equiv 1 \pmod{m}$. But $r < e$ and e is the least positive number with $a^e \equiv 1 \pmod{m}$. So $r = 0$, and e divides f. □

The next result would have saved us some effort in computing the orders of the elements of $\mathbb{Z}/7\mathbb{Z}$.

Proposition 3. *If a has order e modulo m and $d > 0$, then the order of a^d modulo m is $e/(d,e)$, where (d,e) is the greatest common divisor of d and e.*

Proof. Recall that for numbers d, e, the least common multiple $[d,e]$ satisfies

$$\frac{[d,e]}{d} = \frac{e}{(d,e)}.$$

Since e divides $[d,e]$ and $a^e \equiv 1 \pmod{m}$, we have

$$a^{[d,e]} \equiv 1 \pmod{m}.$$

It follows that

$$(a^d)^{\frac{[d,e]}{d}} \equiv (a^d)^{\frac{e}{(d,e)}} \equiv 1 \pmod{m}.$$

To show that $\frac{[d,e]}{d}$ is the order of a^d, suppose $(a^d)^s \equiv 1 \pmod{m}$ for $s > 0$. Then $a^{ds} \equiv 1 \pmod{m}$, so by Proposition 2, e divides ds. So ds is a common multiple of e and d, so $ds \geq [d,e]$, hence $s \geq \frac{[d,e]}{d}$. So the order of a^d is $[d,e]/d$.   □

To apply Proposition 3 to the orders of elements of $\mathbb{Z}/7\mathbb{Z}$, suppose we find that the order of $[5]$ is 6. Then:
Since $[4] = [5]^2$, the order of $[4]$ is $6/(6, 2) = 6/2 = 3$.
Since $[6] = [5]^3$, the order of $[6]$ is $6/(6, 3) = 6/3 = 2$.
Since $[2] = [5]^4$, the order of $[2]$ is $6/(6, 4) = 6/2 = 3$.
Since $[3] = [5]^5$, the order of $[3]$ is $6/(6, 5) = 6/1 = 6$.
Since $[1] = [5]^6$, the order of $[1]$ is $6/(6, 6) = 6/6 = 1$.

Exercises.

1. Find the orders of the nonzero elements of $\mathbb{Z}/5\mathbb{Z}$.

2. Find the orders of the units of $\mathbb{Z}/9\mathbb{Z}$.

3. Find the order of $[2]$ in $\mathbb{Z}/m\mathbb{Z}$ where:
(i) $m = 11$;
(ii) $m = 17$;
(iii) $m = 31$;
(iv) $m = 9$;
(v) $m = 14$.

4. Find the orders of the nonzero elements of $\mathbb{Z}/11\mathbb{Z}$. (Hint: Start with $[2]$.)

5. Find the orders of the nonzero elements of $\mathbb{Z}/13\mathbb{Z}$.

6. Find the orders of the nonzero elements of $\mathbb{Z}/17\mathbb{Z}$.

7. Find the orders of the invertible elements of $\mathbb{Z}/24\mathbb{Z}$.

8. Let r and s be coprime numbers >2, and suppose the order of a modulo r is d, and the order of a modulo s is e. Let $m = rs$. Show that the order of a modulo m is the least common multiple of d and e.

9. Using the last exercise, find the order of 2 $\pmod{77}$.

10. Find the order of $2^{10} \pmod{77}$.

11. Find the order of $[32]$ in $\mathbb{Z}/17\mathbb{Z}$.

12. Prove Proposition 2 from Proposition 3.

13. Modulo 163, 3 has order 162. What is the order of 3^{26}? Of 3^{27}?

14. Find the order of $[2^{24}]$ in $\mathbb{Z}/m\mathbb{Z}$, where:
  (i) $m = 11$;
  (ii) $m = 17$;
  (iii) $m = 31$.

15. Modulo 83, $2^{41} \equiv 82$. Find the order of 2 (mod 83).

16. (i) Find the order of 3 (mod 14).
  (ii) Find the least nonnegative integer in the congruence class $[59^{110}]_{14}$.

B. Fermat's Theorem

From the exercises you may have noticed that if p is a prime, and a is any number not divisible by p, then the order of a modulo p divides $p - 1$. (Or notice the table of orders for $\mathbb{Z}/7\mathbb{Z}$.) If so, you have recognized a theorem which was discovered by Fermat in 1640. Fermat never made public a proof of the theorem, and the first proof was given by Euler, a century later. We will give three different proofs of Fermat's theorem in this book.

Fermat's theorem may be expressed in various ways. In terms of congruence:

Theorem 4 (Fermat's Theorem). *If p is a prime and a is an integer not divisible by p, then*
$$a^{p-1} \equiv 1 \pmod{p}.$$

In terms of congruence classes, Fermat's Theorem reads:

If p is prime and $[a]_p$ is a unit of $\mathbb{Z}/p\mathbb{Z}$, then $[a]_p^{p-1} = [1]_p$.

In terms of divisibility:

If p is prime and a is coprime to p, then p divides $a^{p-1} - 1$.

This last version was the one that Fermat stated. The concept of congruence did not arise until 160 years after Fermat's discovery.

A useful way to visualize the first proof we will give for Fermat's theorem is to look at that portion of the multiplication table for multiplication modulo p that does not involve the number 0. We illustrate with $p = 7$:

·	1	2	3	4	5	6
1	1	2	3	4	5	6
2	2	4	6	1	3	5
3	3	6	2	5	1	4
4	4	1	5	2	6	3
5	5	3	1	6	4	2
6	6	5	4	3	2	1

For each a, $1 \leq a \leq 6$, the row starting with a contains the entries

$$a \cdot 1, a \cdot 2, \ldots, a \cdot 6 \pmod{7}$$

from left to right.

Denote by $a \cdot U$ the set of entries in the row associated to multiplication by a. Then

$$U = 1 \cdot U = \{1, 2, 3, 4, 5, 6\},$$

while, for example, $3 \cdot U$ is the set

$$3 \cdot U = \{3 \cdot 1, 3 \cdot 2, 3 \cdot 3, 3 \cdot 4, 3 \cdot 5, 3 \cdot 6\} \pmod{7}$$
$$= \{3, 6, 2, 5, 1, 4\} \pmod{7}$$

Now when we multiply the elements of the set $3 \cdot U$ together, on the one hand we get

$$(3 \cdot 1) \cdot (3 \cdot 2) \cdot (3 \cdot 3) \cdot (3 \cdot 4) \cdot (3 \cdot 5) \cdot (3 \cdot 6) = 3^6 \cdot 1 \cdot 2 \cdot 3 \cdot 4 \cdot 5 \cdot 6,$$

while on the other hand, the elements of $3 \cdot U$ are congruent modulo 7 to the numbers 3, 6, 2, 5, 1, 4, which is just a rearrangement of the numbers 1, 2, 3, 4, 5, 6. So the product of the elements of $3 \cdot U$ is congruent modulo 7 to $1 \cdot 2 \cdot 3 \cdot 4 \cdot 5 \cdot 6$. Thus

$$3^6 \cdot 1 \cdot 2 \cdot 3 \cdot 4 \cdot 5 \cdot 6 \equiv 1 \cdot 2 \cdot 3 \cdot 4 \cdot 5 \cdot 6 \pmod{7}.$$

Cancelling $1 \cdot 2 \cdot 3 \cdot 4 \cdot 5 \cdot 6$ from both sides, we get $3^6 \equiv 1 \pmod{7}$, which is Fermat's theorem.

To prove Fermat's theorem in general uses the same idea. We prove the congruence class version.

Proof. Write down the multiplication table for $\mathbb{Z}/p\mathbb{Z}$ (we omit brackets $[\ ]_p$ in the table):

·	1	2	3	$\cdots$	$p-1$
1	1	2	3	$\cdots$	$p-1$
2	2	$2 \cdot 2$	$2 \cdot 3$	$\cdots$	$2(p-1)$
3	3	$3 \cdot 2$	$3 \cdot 3$	$\cdots$	$3(p-1)$
$\vdots$	$\vdots$	$\vdots$	$\vdots$		$\vdots$
a	a	$a \cdot 2$	$a \cdot 3$	$\cdots$	$a(p-1)$
$\vdots$	$\vdots$	$\vdots$	$\vdots$		$\vdots$
$p-1$	$p-1$	$(p-1) \cdot 2$	$(p-1) \cdot 3$	$\cdots$	$(p-1)(p-1)$

For any $[a] \neq [0]$, let $a \cdot U$ denote the set of nonzero elements in the row corresponding to multiplication by $[a]$, that is, the congruence classes

$$[a \cdot 1], [a \cdot 2], \dots, [a \cdot (p-1)].$$

The set $a \cdot U$ is then the same as the set $1 \cdot U = U$, except for the ordering of the elements. To see this, let $[a] \neq [0]$, then $[a]$ is a unit of $\mathbb{Z}/p\mathbb{Z}$. Let $[b]$ be the inverse of $[a]$. If $[m]$ is any nonzero element of $\mathbb{Z}/p\mathbb{Z}$, then $[m] = [1][m] = [a][b][m] = [a][(bm)]$, so $[m]$ is in the set $a \cdot U$. Thus the set U, consisting of all the nonzero elements of $\mathbb{Z}/p\mathbb{Z}$, is a subset of $a \cdot U$. But U and $a \cdot U$ both contain $p-1$ elements. So $U = a \cdot U$.

Now the product of the elements of $a \cdot U$ is

$$[a \cdot 1][a \cdot 2][a \cdot 3] \dots [a(p-1)] = [a^{p-1}][1 \cdot 2 \cdot 3 \cdot \dots \cdot (p-1)]$$

while the product of the elements of U is

$$[1][2][3] \dots [p-1] = [1 \cdot 2 \cdot 3 \cdot \dots \cdot (p-1)].$$

Since $U = a \cdot U$, the products of the elements in U and in $a \cdot U$ are equal:

$$[1 \cdot 2 \cdot 3 \cdot \dots \cdot (p-1)] = [a^{p-1}][1 \cdot 2 \cdot 3 \cdot \dots \cdot (p-1)].$$

Canceling the element $[1 \cdot 2 \cdot 3 \cdot \dots \cdot (p-1)]$, a unit of $\mathbb{Z}/p\mathbb{Z}$, gives $[a^{p-1}] = [1]$. $\square$

Proposition 5. *If p is prime and a is not divisible by p, then the order of a modulo p divides $p-1$.*

This follows immediately from Fermat's theorem and Proposition 2 (Section A).

Applying Proposition 5 shortens the process of finding the order of $[a]$ in $\mathbb{Z}/p\mathbb{Z}$. For example, consider the order of $[7]$ in $\mathbb{Z}/11\mathbb{Z}$. By Proposition 5 , the order divides 10, so the order can only be 1, 2, 5 or 10. We find that $[7]^1 - [7]$, $[7]^2 - [5]$ and $[7]^5 = [7][5][5] = [7][3] = [21] \neq [1]$. So the order of $[7]$ must be 10.

Fermat's theorem also gives a way of describing the inverse of an invertible element of $\mathbb{Z}/p\mathbb{Z}$, p prime. If a is any integer with $(a, p) = 1$, then $[a]^{p-1} = 1$, so $[a] \cdot [a]^{p-2} = 1$. Thus $[a]^{p-2}$ is the inverse of $[a]$. For example, the inverse of $[4]$ in $\mathbb{Z}/7\mathbb{Z}$ is $[4]^5 = [1024]$. (Since $1024 \equiv 2 \pmod 7$, $[1024] = [2]$).

Exercises.

17. Verify Fermat's theorem for $[5]_{11}$ in $\mathbb{Z}/11\mathbb{Z}$.

18. Verify Fermat's theorem for $[2]_{13}$ in $\mathbb{Z}/13\mathbb{Z}$. Then verify Fermat's theorem for all $[a]_{13}$ with $(a, 13) = 1$ (use Proposition 3).

19. Find $2^9 \bmod 11$ and verify that 2^9 is the inverse of 2 modulo 11.

20. Find the order of $[3]$ in $\mathbb{Z}/23\mathbb{Z}$. (Hint: Use Proposition 5).

21. Without any significant computations, explain why the order of 7 modulo 167 is at least 80.

22. Show that the order of 10 (mod 83) is at least 30.

23. Is there an element of order 15 in $\mathbb{Z}/97\mathbb{Z}$? If so, find it.

24. Find the order of every nonzero element of $\mathbb{Z}/19\mathbb{Z}$.

25. Find the least nonnegative residue of 2^{47} (mod 23).

26. Prove that if p is prime, then for every number a, divisible by p or not, $a^p \equiv a$ (mod p).

27. Show that $n^5/5 + n^3/3 + 7n/15$ is an integer for every n.

28. Show that for every integer n, $n^9 + 2n^7 + 3n^3 + 4n$ is divisible by 5.

29. Show that if 7 does not divide n, then 7 divides $n^{12} - 1$.

30. Show that $n^{13} - n$ is divisible by 2, 3, 5, 7, and 13 for all n.

31. Show that for every n, $n^{111} \equiv n$ (mod 11).

32. Show that $23^{560} \equiv 1$ (mod 561).

33. Show that for every prime p, the product of the elements of $U(1)$ is $[(p-1)!] = [-1]$.

34. Find the least nonnegative residue of 3^{255} (mod 29).

35. Let $m = 2^{15} - 1 = 32767$. Show that
   (i) The order of 2 modulo m is 15.
   (ii) 15 does not divide $m - 1$.
   Why does that imply that m is not prime?

36. (i) Show that 9 is the order of 2 (mod 511).
   (ii) Show that if 511 were prime, then 9 would have to divide 510; since that is not so, 511 must be composite.

37. Let p be a prime ≥ 7. Show that p divides at least one of the numbers in the set

$$\{1, 11, 111, 1111, 11111, \ldots\}.$$

(These numbers are called *repunits*.)

38. Let n be any number of at most e digits. Let p be a prime ≥ 7. Show that there is some number r so that p divides

$$n + n \cdot 10^e + n \cdot 10^{2e} + \ldots + n \cdot 10^{re}$$

39. Let $m = 2^{2^e} + 1$, the e-th Fermat number.

(i) Show that 2 has order 2^{e+1} (mod m).

(ii) Let p be a prime divisor of m. Show that $2^{2^e} \equiv -1$ (mod p), hence the order of 2 modulo p is 2^{e+1}.

(iii) Using Fermat's theorem, show that any prime divisor p of m satisfies $p - 1 = k \cdot 2^{e+1}$ for some k, hence $p = 1 + k \cdot 2^{e+1}$.

(This result of Euler is helpful in looking for prime factors of Fermat numbers. Lucas subsequently showed that any prime factor of m must be of the form $p = 1 + k \cdot 2^{e+2}$).

C. Euler's Theorem

Proposition 1 of Section A showed that if m is a modulus and a any integer with $(a, m) = 1$, then there is some t with $a^t = 1$ (mod m). If m is prime, Fermat's theorem asserts that we can choose $t = m - 1$. When m is composite, Euler's theorem gives an appropriate t.

Recall from Section 6E that for $m \geq 2$, we defined $\phi(m)$ to be the number of units of $\mathbb{Z}/m\mathbb{Z}$. The function ϕ is called *Euler's phi-function*. The number of units of $\mathbb{Z}/m\mathbb{Z}$ is equal to the number of numbers r with $1 \leq r \leq m$ that are coprime to m.

Here is Euler's theorem expressed in the language of congruence classes:

Theorem 6 (Euler's Theorem). *For every unit* $[a]$ *of* $\mathbb{Z}/m\mathbb{Z}$,

$$[a]^{\phi(m)} = [1].$$

Here is the theorem expressed in terms of congruences:

For every integer a coprime to m, $a^{\phi(m)} \equiv 1$ (mod m).

Example 2. In $\mathbb{Z}/14\mathbb{Z}$, the units are the classes of $1, 3, 5, 9, 11$, and 13. So $\phi(14) = 6$, and for every integer a coprime to 14, $a^6 \equiv 1$ (mod 14). For example, $3^6 = 27^2 \equiv (-1)^2 = 1$ (mod 14).

Euler's theorem can be proved in the same way Fermat's theorem was proved. The only refinement is to let U be the set of units modulo m, and $a \cdot U$ denote the set U of units each multiplied by a for $[a]$ a unit of $\mathbb{Z}/m\mathbb{Z}$. Then $a \cdot U$ consists of the entries in the row of the multiplication table for the units of $\mathbb{Z}/m\mathbb{Z}$ corresponding to multiplication by a. For example, consider the units in the multiplication table for $\mathbb{Z}/8\mathbb{Z}$. (We omit $[\ ]_8$ in the table).

$\cdot$	1	3	5	7		$\cdot$	1	3	5	7
1	$1 \cdot 1$	$1 \cdot 3$	$1 \cdot 5$	$1 \cdot 7$		1	1	3	5	7
3	$3 \cdot 1$	$3 \cdot 3$	$3 \cdot 5$	$3 \cdot 7$	=	3	3	1	7	5
5	$5 \cdot 1$	$5 \cdot 3$	$5 \cdot 5$	$5 \cdot 7$		5	5	7	1	3
7	$7 \cdot 1$	$7 \cdot 3$	$7 \cdot 5$	$7 \cdot 7$		7	7	5	3	1

Then
$$3 \cdot U = \{[3] \cdot [1], [3] \cdot [3], [3] \cdot [5], [3] \cdot [7]\}$$
$$= \{[3], [1], [7], [5]\}$$
$$= \{[1], [3], [5], [7]\} = U$$

and so the products of the elements of $3 \cdot U$ and of U are equal:

$$[3]^4 \cdot [1] \cdot [3] \cdot [5] \cdot [7] = [1] \cdot [3] \cdot [5] \cdot [7].$$

Canceling $[1] \cdot [3] \cdot [5] \cdot [7]$ from both sides gives

$$[3]^4 = [1].$$

Having an appropriate set $a \cdot U$, the proof of Euler's theorem is exactly the same as that of Fermat's theorem.

Proof. Let $U = \{u_1, u_2, \ldots, u_{\phi(m)}\}$ be the units of $\mathbb{Z}/m\mathbb{Z}$ and a be any unit of $\mathbb{Z}/m\mathbb{Z}$. Since U is closed under multiplication, au_i is in U for every i. Let $a \cdot U$ be the set

$$a \cdot U = \{au_1, au_2, \ldots, au_{\phi(m)}\}.$$

Then the sets $a \cdot U$ and U are equal, for the function $f_a : U \to a \cdot U$ defined by $f_a(u) = au$ is a one-to-one function with inverse $f_{a^{-1}} : a \cdot U \to U$ defined by multiplication by a^{-1}. Thus $au_1, au_2, \ldots, au_{\phi(m)}$ are $\phi(m)$ different units of $\mathbb{Z}/m\mathbb{Z}$, hence $a \cdot U = U$.

Since $U = a \cdot U$, the product of all the elements of $a \cdot U$ and of U are equal, that is,

$$au_1 \cdot au_2 \cdots au_{\phi(m)} = u_1 \cdot u_2 \cdots u_{\phi(m)}.$$

Since $u_1, \ldots, u_{\phi(m)}$ are units, we can cancel them from the two sides of the equation, leaving
$$a^{\phi(m)} = 1.$$
$\square$

Notice that Fermat's theorem is a special case of Euler's theorem. If m is prime, then $\phi(m) = m - 1$. Even the proofs are the same, for if m is prime, then the $a \cdot U$ used in the proof of Euler's theorem is the same as the $a \cdot U$ used in the proof of Fermat's theorem.

One question remaining with Euler's theorem is the number $\phi(m)$, Euler's phi function. In order to use Euler's theorem, we need to know $\phi(m)$. Even for fairly small numbers m, the description of $\phi(m)$ as the number of units of $\mathbb{Z}/m\mathbb{Z}$, or as the number of numbers r with $1 \le r \le m$ that are coprime to m, is not all that helpful for computing $\phi(m)$. For example, what is $\phi(60)$?

Understanding $\phi(m)$, given m, is a problem of some current interest, as will become clear in Section 10A. But $\phi(m)$ is easy to compute if we can factor m into products of prime powers:

Proposition 7. *a) If p is prime, then $\phi(p) = p - 1$;*
b) If p is prime, then for all $e > 0$, $\phi(p^e) = p^{e-1}(p - 1)$;
c) If a and b are coprime, then $\phi(ab) = \phi(a)\phi(b)$.

Proofs are sketched in the exercises. We will also give a proof of part c) in Section 12D using the Chinese Remainder Theorem.

Exercises.

40. Verify Euler's theorem for $[1], [3], [7]$, and $[9]$ in $\mathbb{Z}/10\mathbb{Z}$.

41. Find a^5 (mod 7) for $a = 1, 2, 3, 4, 5, 6$, and verify that $a^5 \cdot a \equiv 1$ (mod 7).

42. Verify that $2^{\phi(21)} \equiv 1$ (mod 21).

43. Find 48^{322} mod 25.

44. Find 40^{322} mod 21.

45. Prove that for any n, 33 divides $n^{101} - n$.

46. Find the order of 7 modulo 172.

47. Observe that $2^{10} = 1024 \equiv -1$ mod 25. Find the order of 2 modulo 25.

As with Fermat's theorem, Euler's theorem can be used to find the inverse of a unit modulo m. Since

$$a \cdot a^{\phi(m)-1} \equiv 1 \quad (\text{mod } m),$$

the inverse of a modulo m is congruent to $a^{\phi(m)-1}$.

48. Verify that 5^{11} mod $26 = 21$, the inverse of 5 modulo 26.

49. Describe the inverse of 5 modulo 18 as a positive power of 5 mod 18.

50. Prove that if a and m are coprime and $f \equiv 1$ (mod $\phi(m)$), then

$$a^f \equiv a \quad (\text{mod } m).$$

51. Prove that (i) $\phi(p) = p - 1$ if p is prime;
(ii) $\phi(p^n) = p^n - p^{n-1}$ if p is prime.

52. Verify that $\phi(ab) = \phi(a)\phi(b)$ for a and $b =$
(i) 3 and 5;
(ii) 4 and 7;
(iii) 5 and 6.

53. Prove that $\phi(15) = \phi(3)\phi(5)$ as follows:
(i) Write down the numbers <15 in 5 columns of 3 numbers, as follows:

$$\begin{vmatrix} 1 & 4 & 7 & 10 & 13 \\ 2 & 5 & 8 & 11 & 14 \\ 3 & 6 & 9 & 12 & 15 \end{vmatrix}.$$

Show that the numbers coprime to 15 lie only in the rows of numbers all of which are coprime to 3. Show that each row is a complete set of representatives modulo 5, hence each row has $\phi(5)$ numbers coprime to 5.

(ii) Conclude that $\phi(15) = \phi(3)\phi(5)$.

54. Generalize the last exercise to prove that $\phi(ab) = \phi(a)\phi(b)$ for any a and b with $(a,b) = 1$.

We will give an alternate proof of this last exercise in Chapter 12.

55. Verify that $\phi(4)\phi(6) < \phi(24)$.

56. Prove that for every prime p and for every $r, s \geq 1$,

$$\phi(p^r)\phi(p^s) < \phi(p^{r+s}).$$

57. Prove that for all numbers $a, b \geq 1$, if $(a,b) > 1$, then $\phi(a)\phi(b) < \phi(ab)$. (Hint: try induction on the number of primes dividing (a,b)).

58. Define $\phi(1) = 1$ and compute $\sum_{d\mid n} \phi(d)$ for:
   (i) $n = 16$;
   (ii) $n = 15$;
   (iii) $n = 45$;

59. Show that for every $n \geq 1$ and every divisor d of n, the number of elements $[a]$ of $\mathbb{Z}/n\mathbb{Z}$ so that $(a,n) = d$ is $\phi(n/d)$.

60. Prove that $\sum_{d\mid n} \phi(d) = n$ for every $n > 0$.

61. We can prove Euler's theorem in terms of congruence, rather than congruence classes, as follows. Given a modulus m, we say that a set of integers $a_1, a_2, \ldots, a_r$ is a *complete set of units mod m* if every integer a coprime to m is congruent to exactly one of the numbers $a_1, a_2, \ldots, a_r$ (see Section 6E). Otherwise stated, $a_1, a_2, \ldots, a_r$ is a complete set of units mod m if $\{[a_1], [a_2], \ldots, [a_r]\}$ is the set of units of $\mathbb{Z}/m\mathbb{Z}$.

(i) Show that $r = \phi(m)$.

(ii) Show that if $a_1, a_2, \ldots, a_r$ is a complete set of units mod m, and b is any integer coprime to m, then $ba_1, ba_2, \ldots, ba_r$ is a compete set of units mod m. Illustrate with $m = 12$ and $b = 17$.

(iii) Show that if $a_1, a_2, \ldots, a_r$ and $c_1, c_2, \ldots, c_r$ are both complete sets of units mod m, then $a_1 \cdot a_2 \cdot \ldots \cdot a_r \equiv c_1 \cdot c_2 \cdot \ldots \cdot c_r \pmod{m}$. Illustrate with $m = 12$ and the two complete sets of units of the previous exercise.

(iv) Use parts (i) - (iii) to prove that $b^{\phi(m)} \equiv 1 \pmod{m}$ for any number b coprime to m.

62. Prove that if $q > 2$ is odd and the order of 2 (mod q) is $q - 1$, then q is prime. Why is this not contradicted by the fact that $2^{340} \equiv 1 \pmod{341}$?

D. Repeating Decimals

This section describes a connection between the decimal expansion of a fraction and the order of 10 modulo the denominator.

Everyone learns how to find the decimal expansion of a fraction. For example, to expand 1/7 into a decimal fraction, divide 7 into 10, (with quotient 1), multiply the remainder (3) by 10, divide the result (30) by 7 (with quotient 4 and remainder 2), etc.:

$$10 = 7 \cdot 1 + 3$$
$$30 = 7 \cdot 4 + 2$$
$$20 = 7 \cdot 2 + 6$$
$$60 = 7 \cdot 8 + 4$$
$$40 = 7 \cdot 5 + 5$$
$$50 = 7 \cdot 7 + 1$$
$$\vdots$$

Then divide each equation by 7 and 10 to get

$$\frac{1}{7} = \frac{1}{10} + \frac{3}{70}$$
$$\frac{3}{7} = \frac{4}{10} + \frac{2}{70}$$
$$\frac{2}{7} = \frac{2}{10} + \frac{6}{70}$$
$$\frac{6}{7} = \frac{8}{10} + \frac{4}{70}$$
$$\frac{4}{7} = \frac{5}{10} + \frac{5}{70}$$
$$\frac{5}{7} = \frac{7}{10} + \frac{1}{70}$$
$$\vdots$$

Successively substituting, we find that

$$\frac{1}{7} = \frac{1}{10} + \frac{4}{10^2} + \frac{2}{10^3} + \frac{8}{10^4} + \frac{5}{10^5} + \ldots = 0.14285714\ldots.$$

The quotients become the numerators, or the digits in the decimal expansion of 1/7. This process is laid out efficiently by long division of $1.00000\ldots$ by 7.

Here are some other examples:

$$\frac{1}{3} = 0.3333\ldots$$
$$\frac{1}{5} = 0.20000\ldots$$

$$\frac{3}{7} = 0.4285714285714\ldots$$

$$\frac{4}{13} = 0.3076923076923\ldots$$

$$\frac{7}{24} = 0.291666\ldots$$

$$\frac{59}{148} = 0.398648648\ldots$$

All of these examples are *eventually repeating* decimal expansions–after a few initial digits, the digits cycle. To describe such expansions efficiently, we can put a bar over the repeating digits, as follows:

$$\frac{1}{3} = 0.\overline{3}$$

$$\frac{1}{5} = 0.2\overline{0} = 0.2$$

$$\frac{3}{7} = 0.\overline{428571}$$

$$\frac{4}{13} = 0.\overline{307692}$$

$$\frac{7}{24} = 0.291\overline{6}$$

$$\frac{59}{148} = 0.39\overline{864}.$$

A decimal expansion is *terminating* if every digit from some point on is equal to zero. A terminating expansion is a special case of an eventually repeating expansion.

If the decimal expansion of a/b is not terminating, then the *period* of the fraction a/b, or of its decimal expansion, is the number of digits under the bar, that is, the least number $d > 0$ so that the decimal expansion of a/b eventually repeats every d digits. For example, the period of $\frac{1}{3}$ is 1; the period of $\frac{3}{7}$ is 6; the period of $\frac{59}{148}$ is 3.

The period of a fraction relates to Euler's theorem. In fact, we have

Theorem 8. *Let t be a number coprime to 10, and let $\frac{u}{t}$ be a reduced fraction (that is, u and t are coprime). Then the period of $\frac{u}{t}$ is equal to the order of 10 modulo t.*

To see that this result is plausible, we observe that:

10 has order 1 modulo 3 (since $10 \equiv 1 \pmod{3}$) and the period of $\frac{1}{3} = 0.\overline{3}$ is 1;

10 has order 6 modulo 7 (since $10 \equiv 3 \pmod{7}$ and 3 is a primitive root modulo 7) and the period of $\frac{5}{7} = 0.\overline{714285}$ is 6;

10 has order 2 modulo 11 (since $10 \equiv -1 \pmod{11}$) and the period of $\frac{3}{11} = 0.\overline{27}$ is 2;

10 has order 6 modulo 13 (since $10 \equiv -3 \pmod{13}$ and $(-3)^3 = -27 \equiv -1$ (mod 13)) and the period of $\frac{4}{13} = 0.\overline{307692}$ is 6.

Proof. Assume $u < t$ and $(u,t) = 1$, so that u/t is a reduced proper fraction.

Let e be the order of 10 modulo t. Then

$$10^e \equiv 1 \pmod{t}$$

and so

$$10^e - 1 = tk > uk.$$

This means that uk has at most e decimal digits. Now using the geometric series

$$\frac{1}{1-x} = 1 + x + x^2 + \dots,$$

which converges for every real number x with $|x| < 1$, and setting $x = 1/10^e$, we have

$$\frac{u}{t} = \frac{uk}{tk} = \frac{uk}{10^e - 1}$$

$$= \frac{uk}{10^e}\left(\frac{1}{1 - \frac{1}{10^e}}\right)$$

$$= \frac{uk}{10^e}\left(1 + \frac{1}{10^e} + \frac{1}{10^{2e}} + \frac{1}{10^{3e}} + \dots\right)$$

$$= \frac{uk}{10^e} + \frac{uk}{10^{2e}} + \frac{uk}{10^{3e}} + \dots,$$

a decimal expansion that repeats every e digits. If we write uk as usual in base 10 as

$$uk = a_{e-1}a_{e-2}\dots a_1 a_0,$$

then

$$\frac{u}{t} = 0.\overline{a_{e-1}a_{e-2}\dots a_1 a_0}.$$

Conversely, suppose $\frac{u}{t}$ repeats every d digits. Then there is some number s of at most d digits so that

$$\frac{u}{t} = \frac{s}{10^d} + \frac{s}{10^{2d}} + \frac{s}{10^{3d}} + \dots$$

$$= \frac{s}{10^d}\left(1 + \frac{1}{10^d} + \frac{1}{10^{2d}} + \dots\right)$$

$$= \frac{s}{10^d}\left(\frac{1}{1 - \frac{1}{10^d}}\right)$$

$$= \frac{s}{10^d - 1}$$

Multiplying by both denominators gives

$$u(10^d - 1) = ts.$$

Since $(u,t) = 1$ it follows that t divides $10^d - 1$, hence $10^d \equiv 1 \pmod{t}$. Thus the order e of 10 modulo t divides d.

We showed above that if e is the order of 10 modulo t, then u/t repeats every e digits. Since the period of $\frac{u}{t}$ is the least number $d > 0$ so that $\frac{u}{t}$ repeats every d digits, it follows that the period of $\frac{u}{t}$ is equal to e. $\square$

Observe that the period of a reduced fraction $\frac{u}{t}$ with $(t,10) = 1$ doesn't depend on the numerator u at all. For example, all proper fractions with denominator 37 have period 3, such as

$$\frac{5}{37} = 0.\overline{135}$$

$$\frac{14}{37} = 0.\overline{378}$$

$$\frac{22}{37} = 0.\overline{594}$$

$$\frac{36}{37} = 0.\overline{972}$$

because the order of 10 modulo 37 is 3: $10^3 = 1 + 999 = 1 + 27 \cdot 37$.

Theorem 8 explains the expansions of fractions where the denominator is co-prime to 10. The situation for general fractions, such as $\frac{59}{148}$, reduces to that for fractions with denominators coprime to 10 by the following result.

Proposition 9. *For any reduced fraction $\frac{a}{b}$ there is some $g \geq 0$ and integers q, u and t so that*

$$\frac{10^g a}{b} = q + \frac{u}{t}$$

where $u < t, (t,u) = 1$ and $(t,10) = 1$. Hence $\frac{a}{b}$ is eventually periodic with period equal to the order of 10 modulo t.

For example, $\frac{7}{24}$ may be multipliplied by 10^3 to yield

$$\frac{7000}{24} = 291 + \frac{16}{24} = 291 + \frac{2}{3},$$

hence

$$\frac{7}{24} = \frac{291}{1000} + \left(\frac{1}{1000}\right)\frac{2}{3}.$$

Proof. Given $\frac{a}{b}$ with $(a,b) = 1$, factor the highest powers possible of 2 and 5 from b, to get $b = 2^e 5^f t$ where t is coprime to 2, 5, and a. Let g be the larger of the two exponents e and f, and multiply $\frac{a}{b}$ by 10^g. Then divide $10^g a$ by b to get the integer and fractional part of $\frac{10^g a}{b}$. If

$$10^g a = bq + r$$

with $0 \leq r \leq b$, then

$$\frac{10^g a}{b} = q + \frac{r}{b}.$$

Reducing $\frac{r}{b}$ to lowest terms, we obtain a fraction with denominator t. For

$$r = 10^g a - bq = 10^g a - 2^e 5^f tq,$$

so

$$\frac{r}{b} = \frac{10^g a - 2^e 5^f tq}{2^e 5^f t}.$$

Since $g \geq e$ and $\geq f$, we can cancel 2^e and 5^f to get

$$\frac{r}{b} = \frac{2^{g-e} 5^{g-f} a - tq}{t} = \frac{u}{t}.$$

Then t is coprime to $u = 2^{g-e} 5^{g-f} a - tq$ since t is coprime to $2^{g-e} 5^{g-f} a$, so $\frac{u}{t}$ is reduced. Then

$$\frac{10^g a}{b} = q + \frac{u}{t},$$

so

$$\frac{a}{b} = \frac{q}{10^g} + \frac{1}{10^g}\left(\frac{u}{t}\right).$$

Expanding $\frac{u}{t}$ and substuting gives the decimal expansion of $\frac{a}{b}$, and the period of $\frac{a}{b}$ is the same as the period of $\frac{u}{t}$.

$\square$

Corollary 10. *If $\frac{a}{b}$ is a reduced fraction and $b = 2^e 5^f t$ with $(t, 10) = 1$, then the period of $\frac{a}{b}$ divides $\phi(t)$, the order of the group of units of $\mathbb{Z}/t\mathbb{Z}$.*

Proof. The period of $\frac{a}{b}$ is the order of $10 \pmod{t}$ by Theorem 8. We know $10^{\phi(t)} \equiv 1 \pmod{t}$ by Euler's Theorem, and so the period of $\frac{a}{b}$ divides $\phi(t)$. $\square$

Theorem 8 may be applied either to find orders or to find periods.

We can use Theorem 8 to find the order of 10 modulo 21, by finding the period of 1/21.

We can find the period of 1/21 by computing the decimal expansion of 1/21 by long division. If we do so, we find that $1/21 = 0.\overline{047619}$, and so the order of 10 modulo 21 is 6.

Long division also yields the least non-negative residues modulo 21 of the powers of 10. For spreading out the long division,

$$10 = 21 \cdot 0 + 10, \text{ so } 10 \equiv 10 \pmod{21}$$
$$10 \cdot 10 = 100 = 21 \cdot 4 + 16, \text{ so } 10^2 \equiv 16 \pmod{21}$$
$$10^3 \equiv 10 \cdot 16 = 160 = 21 \cdot 7 + 13, \text{ so } 10^3 \equiv 13 \pmod{21}$$
$$10^4 \equiv 10 \cdot 13 = 130 = 21 \cdot 6 + 4, \text{ so } 10^4 \equiv 4 \pmod{21}$$
$$10^5 \equiv 10 \cdot 4 = 40 = 21 \cdot 1 + 19, \text{ so } 10^5 \equiv 19 \pmod{21}$$
$$10^6 \equiv 10 \cdot 19 = 190 = 21 \cdot 9 + 1, \text{ so } 10^6 \equiv 1 \pmod{21}$$

The powers of 10 modulo 21 are the remainders at each stage of the long division.

We can also use Theorem 8 to find the periods of fractions without actually computing the decimal expansion. For example, to find the period of 33/83, we can instead find the order of 10 modulo 83, a prime number. That order must be 1, 2,

41 or 82 by Proposition 5. The order of 10 is not 1 or 2. To check 41, we use the technique of Section F, below:

$$10^2 \equiv 17 \quad (\text{mod } 83)$$
$$10^4 \equiv 17^2 = 289 \equiv 40 \quad (\text{mod } 83)$$
$$10^8 \equiv 40^2 = 1600 \equiv -60 \equiv 23 \quad (\text{mod } 83)$$
$$10^{16} \equiv 23^2 = 529 \equiv 31 \quad (\text{mod } 83)$$
$$10^{32} \equiv 31^2 = 961 \equiv 48 \quad (\text{mod } 83)$$

and so

$$10^{41} = 10^{32} \cdot 10^8 \cdot 10 \equiv 48 \cdot 23 \cdot 10 \equiv 25 \cdot 10 = 250 \equiv 1 \quad (\text{mod } 83).$$

Therefore the order of 10 modulo 83 is 41. So the period of 33/83 is also 41.

All of this works in any base (radix) b. In that more general setting, Theorem 8 becomes

Let b be a base and let $(t,b) = 1$. Then the period of $\frac{1}{t}$ in base b equals the order of b modulo t, and divides $\phi(t)$.

Exercises.

63. Find the period of $\frac{1}{17}$.

64. Find the period of $\frac{11}{47}$.

65. Find the period of $\frac{17}{143}$.

66. Find all prime numbers t so that the expansion of $1/t$ in base 10 has period 3.

67. (i) Show that 10 has order 5 modulo 11111.

(ii) Show that if a prime p divides 11111, then 5 must divide $p - 1$, hence $p = 1 + 5k$ for some k, and k must be even.

(iii) Show that if 11111 is composite, then 11111 must be divisible by a prime <110.

(iv) Factor 11111 or show that 11111 is prime.

68. Find all prime numbers t so that the expansion of $1/t$ in base 10 has period 5.

69. Find all numbers t so that the expansion of $1/t$ in base 10 has period 4.

70. Find all numbers t so that $\phi(t)$ divides 100. Then show that for every b coprime to t, the period of $1/t$ in base b divides 100.

E. The Binomial Theorem and Fermat's Theorem

In this section we give a proof of Fermat's theorem that uses the Binomial Theorem modulo a prime number p. Recall from Section 2D that the Binomial Theorem is:

$$(x+y)^n = x^n + \binom{n}{1}x^{n-1}y + \ldots + \binom{n}{r}x^r y^{n-r} + \ldots + y^n,$$

where the coefficients are integers and satisfy

$$\binom{n}{r} = \frac{n!}{r!(n-r)!}.$$

In preparation for the proof, we observe:

Proposition 11. *If p is prime, then p divides $\binom{p}{r}$ for all r, $0 < r < p$.*

Proof. For p prime, $\binom{p}{r} = \frac{p!}{r!(p-r)!}$. Since $\binom{p}{r}$ is an integer, $r!(p-r)!$ divides $p!$. For $1 \le r \le p-1$, the prime p does not divide $r!$ and does not divide $(p-r)!$, so p and $r!(p-r)!$ are coprime. So $r!(p-r)!$ divides $(p-1)!$ and $\binom{p}{r} = p[\frac{(p-1)!}{r!(p-r)!}]$ is an integer multiple of p. $\qquad\square$

Proposition 12. *If p is prime, then $(x+y)^p \equiv x^p + y^p \pmod{p}$ for all integers x and y.*

Proof. Expand $(x+y)^p$ by the Binomial Theorem. By Proposition 11, the prime p divides $\binom{p}{r}$ for $1 \le r \le p-1$, and so modulo p, the only terms with non-zero coefficients are the first and last:

$$(x+y)^p = x^p + y^p \pmod{p}.$$

$\qquad\square$

Using Proposition 12, we can prove Fermat's Theorem by induction.

Theorem 13 (Fermat's Theorem). *If p is a prime, then every integer a satisfies the congruence $a^p \equiv a \pmod{p}$.*

If a is coprime to p, then canceling a gives Fermat's Theorem in its original form: $a^{p-1} \equiv 1 \pmod{p}$.

Proof. We prove the theorem for all $a > 0$ by induction on a. For $a = 1$ it is clear. Suppose a is an integer ≥ 1 and $a^p \equiv a \pmod{p}$. Then

$$(a+1)^p \equiv a^p + 1^p \pmod{p}$$

by Proposition 12, and this

$$\equiv a+1 \pmod{p}$$

by the induction assumption. So Fermat's Theorem is true for all $a \geq 0$. If b is any integer, then $b \equiv a \pmod{p}$ for some positive integer a. Then $b^p \equiv a^p \equiv a \equiv b \pmod{p}$. $\qquad\square$

The proof of Fermat's Theorem in Section B extended easily to a proof of Euler's theorem. With a bit more effort we can also extend the new proof:

Theorem 14 (Euler's Theorem). *If a and m are coprime integers, $m \geq 2$, then $a^{\phi(m)} \equiv 1 \pmod{m}$.*

Proof. Write $m = p_1^{e_1} p_2^{e_2} \ldots p_g^{e_g}$, a product of powers of distinct primes. It suffices to show that
$$a^{\phi(m)} \equiv 1 \pmod{p_i^{e_i}}$$
for each i. Now by Proposition 7 of Section C, above,
$$\phi(m) = \phi(p_1^{e_1})\phi(p_2^{e_2}) \ldots \phi(p_g^{e_g}).$$

If we show that
$$a^{\phi(p_i^{e_i})} \equiv 1 \pmod{p_i^{e_i}}$$
for each i, then
$$a^{\phi(m)} \equiv 1 \pmod{p_i^{e_i}}$$
for all i, and so $a^{\phi(m)} - 1$ is a common multiple of $p_1^{e_1}, \ldots, p_g^{e_g}$. Then $a^{\phi(m)} - 1$ is divisible by m, the least common multiple of $p_1^{e_1}, \ldots, p_g^{e_g}$. Hence
$$a^{\phi(m)} \equiv 1 \pmod{m}.$$

So let p be any of the p_i's. We will show that for every a coprime to p,
$$a^{\phi(p^e)} \equiv 1 \pmod{p^e}$$
that is,
$$a^{p^{e-1}(p-1)} \equiv 1 \pmod{p^e},$$
which we will prove by induction on e.

The case $e = 1$ is Fermat's Theorem, and gives us that
$$a^{p-1} = 1 + ps_1$$
for some integer s_1. To do the case $e = 2$, we have
$$a^{p(p-1)} = (1 + ps)^p$$
$$= 1 + \binom{p}{1}ps + \binom{p}{2}p^2s^2 + \ldots + p^p s^p$$
$$= 1 + p^2 s_2$$

for some integer s_2 by Proposition 12. So

$$a^{p(p-1)} \equiv 1 \quad (\text{mod } p^2).$$

and the case $e = 2$ is true. Similarly, if $e > 2$ and we assume by induction that

$$a^{p^{e-2}(p-1)} \equiv 1 \quad (\text{mod } p^{e-1}),$$

then

$$a^{p^{e-2}(p-1)} = 1 + p^{e-1}s_{e-1}$$

for some integer s_{e-1}. So

$$a^{p^{e-1}(p-1)} = (1 + p^{e-1}s_{e-1})^p$$
$$= 1 + p^e s_e$$

for some integer s_e, just as in the case $e = 2$. So the prime power case is done by induction on e, completing the proof. $\qquad \square$

This last proof of Euler's Theorem yields

Proposition 15. *Let* $m = p_1 p_2 \cdots p_g$ *be a product of distinct prime numbers (m is "squarefree"). Let*

$$\lambda(m) = lcm(p_1 - 1, p_2 - 1, \ldots, p_g - 1).$$

Then for every integer a and every number k,

$$a^{\lambda(m)k+1} \equiv a \quad (\text{mod } m).$$

Proof. As in the proof of Theorem 14, it suffices to show that $a^{\lambda(m)k+1} \equiv a$ (mod p_i) for each prime p_i dividing m. If p_i divides a, this is clear. Otherwise, a is a unit modulo p_i. Now $\lambda(m)$ is a multiple of $p_i - 1$, so $\lambda(m)k = (p_i - 1)l_i$ for some number l_i. Hence by Fermat's Theorem,

$$a^{\lambda(m)k+1} \equiv (a^{p_i-1})^{l_i} a \equiv a \quad (\text{mod } p_i).$$

$\qquad \square$

Since $\phi(m)$ is a multiple of $\lambda(m)$, we have

Corollary 16. *If m is a squarefree integer, then for every integer a and every number k,*

$$a^{\phi(m)k+1} \equiv a \quad (\text{mod } m).$$

These last two results will be useful in Section 10A.

The exponent $\lambda(m)$ will reappear in Section 19A as the *exponent* of the abelian group U_m of units of $\mathbb{Z}/m\mathbb{Z}$.

The Frobenius map. Proposition 12 has useful applications to finite fields. We first extend Proposition 12 from integers modulo p, a prime number, to elements of arbitrary rings of characteristic p (defined in Section 7D):

Theorem 17. *If R is a commutative ring of characteristic p and a, b are elements of R, then $(a+b)^p = a^p + b^p$.*

Proof. By the Binomial Theorem,

$$(y+z)^p = y^p + \binom{p}{1} y^{p-1} z + \binom{p}{2} y^{p-2} z^2 + \ldots + z^p$$

for any indeterminates y and z. Set $y = a, z = b$ with a, b in R. Now if $r \neq 0$ or p, $\binom{p}{r} = pq$ for some integer q by Proposition 11, so

$$\binom{p}{r} a^{p-r} b^r = pq a^{p-r} b^r = 0.$$

Thus in R, $(a+b)^p = a^p + b^p$.   □

Let R be a ring of characteristic p. Then the pth power map f_p, which takes a in R to a^p, is a ring homomorphism from R to R (see Section 7D) because

$$f_p(a+b) = (a+b)^p = a^p + b^p = f_p(a) + f_p(b)$$

by Theorem 17, and

$$f_p(ab) = (ab)^p = a^p b^p = f_p(a) f_p(b);$$

$$f_p(0) = 0; f_p(1) = 1.$$

The map f_p is called the *Frobenius map*.

Proposition 18. *If F is a finite field of characteristic p, then f_p is a one-to-one function from F onto F.*

A homomorphism from a ring R to itself which is one-to-one and onto is called an *automorphism* of R.

Proof. To show that f_p is an automorphism of F we need to show that f_p is one-to-one and onto. Since F has a finite number of elements, if f_p is one-to-one, then f_p must be onto. But f_p is one-to-one because it is a nonzero homomorphism whose domain is a field–see Proposition 17 of Section 7D.   □

Corollary 19. *If R is a ring of characteristic p, then for all a, b in R and every $n > 0$,*

$$(a+b)^{p^n} = a^{p^n} + b^{p^n}.$$

Proof. Let $f_{p^n} : R \to R$ be the function defined by $f_{p^n}(a) = a^{p^n}$ for any a in R. Then f_{p^n} is the composition of f_p with itself n times:

$$f_{p^n}(a) = f_p(f_p(\cdots(a)\cdots)).$$

Since the composition of homomorphisms is a homomorphism (an easy exercise), we have that

$$(a+b)^{p^n} = f_{p^n}(a+b) = f_{p^n}(a) + f_{p^n}(b) = a^{p^n} + b^{p^n}.$$

□

Exercises.

71. Let $m = 15$, then $\lambda(m) = 4$. Verify that for every number a,

$$a^5 \equiv a \pmod{m}.$$

72. Let $m = 41 \cdot 11 = 451$. Verify that

$$11^{\lambda(m)+1} \equiv 11 \pmod{m}.$$

73. Show that $\lambda(m) < \phi(m)$ for every odd composite number m.

74. Find examples of p, q primes > 10 so that $\lambda(pq) = p - 1$.

75. Let a, b, c be integers and p a prime. Show that

$$(a+b+c)^p \equiv a^p + b^p + c^p \pmod{p}.$$

Generalize.

76. Find integers u, b so that

$$(a+b)^4 \not\equiv a^4 + b^4 \pmod{4}.$$

77. Show that for all integers a, b and every $n > 0$,

$$(a+b)^n \equiv a^n + b^n \pmod{2}.$$

F. Finding High Powers Modulo m

For finding inverses by Euler's theorem and for other applications, we often need to find the least nonnegative residue of a high power of a number modulo m.

For example, one way to find the inverse of 87 modulo 179 is as $87^{177} \bmod 179$.

But if we put 87^{177} into a calculator, it will either choke or give us something like "$1.972\,E + 343$", which is useless for discovering that 107 is the inverse of 87 modulo 179.

To find 87^{177} modulo 179, it is helpful to write the exponent in base 2 and then find the result using a sequence of squarings modulo 179. We first find that $179 = 128 + 32 + 16 + 1$. Then we compute

$$87$$
$$87^2 \equiv 51 \quad (\text{mod } 179),$$
$$87^4 \equiv 51^2 \equiv 95 \quad (\text{mod } 179),$$
$$87^8 \equiv 95^2 \equiv 75 \quad (\text{mod } 179),$$
$$87^{16} \equiv 75^2 \equiv 76 \quad (\text{mod } 179),$$
$$87^{32} \equiv 76^2 \equiv 48 \quad (\text{mod } 179),$$
$$87^{64} \equiv 48^2 \equiv 156 \quad (\text{mod } 179),$$
$$87^{128} \equiv 156^2 \equiv 171 \quad (\text{mod } 179).$$

Since $179 = 128 + 32 + 16 + 1$, we have

$$87^{179} = 87^{128+32+16+1}$$
$$= 87^{128} \cdot 87^{32} \cdot 87^{16} \cdot 87$$
$$\equiv (171)(48)(76)(87)$$
$$\equiv 107 \quad (\text{mod } 179).$$

An efficient way to do the computations is as follows: write the exponent, 177, in base 2: $177 = (10110001)_2$. Then write down that base 2 number with an S inserted in the spaces between adjacent digits:

$$1S0S1S1S0S0S0S1.$$

Now replace each 1 by X and erase each 0, to get

$$X\,SSX\,SX\,SSSSX.$$

Beginning with the number 1, view X and S, from left to right, as operations to compute $a^{177} \pmod{m}$, as follows: X means, multiply the result by a and reduce modulo m; and S means, square the result and reduce modulo m. If we do not reduce modulo m, we would get:

$$
\begin{array}{ccccccccccc}
X & & S & & S & & X & & S & & \\
1 & \to & a & \to & a^2 & \to & a^4 & \to & a^5 & \to & a^{10} \\
X & & S & & S & & S & & S & & X \\
\to a^{11} & \to & a^{22} & \to & a^{44} & \to & a^{88} & \to & a^{176} & \to & a^{177}
\end{array}
$$

If we reduce modulo m at each step, we get the least nonnegative residue of a^{101} (mod m) at the end. Thus

$$X: \quad 1 \cdot 87 \equiv 87 \quad (\text{mod } 179)$$
$$S: \quad 87 \cdot 87 \equiv 51 \quad (\text{mod } 179)$$
$$S: \quad 51 \cdot 51 \equiv 95 \quad (\text{mod } 179)$$
$$X: \quad 95 \cdot 87 \equiv 31 \quad (\text{mod } 179)$$
$$S: \quad 31 \cdot 31 \equiv 66 \quad (\text{mod } 179)$$
$$X: \quad 66 \cdot 87 \equiv 14 \quad (\text{mod } 179)$$
$$S: \quad 14 \cdot 14 \equiv 17 \quad (\text{mod } 179)$$
$$S: \quad 17 \cdot 17 \equiv 110 \quad (\text{mod } 179)$$
$$S: \quad 110 \cdot 110 \equiv 107 \quad (\text{mod } 179)$$
$$S: \quad 107 \cdot 107 \equiv 172 \quad (\text{mod } 179)$$
$$X: \quad 172 \cdot 87 \equiv 107 \quad (\text{mod } 179).$$

So

$$87^{177} \equiv 107 \quad (\text{mod } 179).$$

Exercises.

78. Find the least nonnegative residue (mod 34) of 12^{87}.

79. Find the least nonnegative number a congruent to 2^{69} (mod 71). Verify that $2a \equiv 1$ (mod 71).

80. Find the least nonnegative number a congruent to 5^{69} (mod 71). Verify that $5a \equiv 1$ (mod 71).

81. Find the least nonnegative number a congruent to 3^{340} (mod 341).

82. Find the least nonnegative number a congruent to 5^{1728} (mod 1729).

83. Find the least nonnegative residue (mod 101) of 18^{77}.

84. (i) Find the least nonnegative number a congruent to $2^{1194648}$ (mod 1194649) Could 1194649 be prime?

(ii) Find the least nonnegative number a congruent to $3^{1194648}$ (mod 1194649) Is 1194649 prime?

85. Let $m = 252601$. Suppose we discover that

$$3^{126300} \equiv 67772 \quad (\text{mod } 252601)$$
$$3^{252600} \equiv 1 \quad (\text{mod } 252601)$$

Is then 252601 prime? composite? Or can we not decide for sure from the information given?

86. Show how to adapt Russian Peasant Arithmetic (Chapter 2, Exercise 29) with multiplication replacing addition and squaring replacing multiplying by 2, to efficiently find a^e and a^e mod m for any numbers a, e and m.

G. Modular Multiplication

When we find a^e mod m as in Section F, every time we perform an operation (multiplication, squaring), we immediately reduce the result modulo m to bring the result back to a number $<m$. (If we don't, the size of the numbers can become unmanageably large.)

For example, suppose the modulus $m = 179$ and we square 107 to get $107^2 = 11449$. To find its least non-negative residue modulo m, we divide 179 into 11449 and take the remainder.

But long division is the only algorithm in classical arithmetic that is not automatic.

Consider dividing 179 into 11449. We look for the first digit of the quotient. Since 17 is bigger than 11, we can't guess the first digit by dividing the first digit of the divisor into the first digit or two of the dividend. So we start guessing with 9:

$$179 \cdot 9 = 1611;$$
$$179 \cdot 8 = 1432;$$
$$179 \cdot 7 = 1253;$$
$$179 \cdot 6 = 1074$$

and 1074 is less than 1144. So the first digit is 6.

We subtract 10740 from 11449 and get 709. Now we guess the next digit. How many times does 179 go into 709? We try the first digit idea: since 1 goes into 7, 7 times, we start with 7:

$$179 \cdot 7 = 1253,$$
$$179 \cdot 6 = 1074,$$
$$179 \cdot 5 = 895,$$
$$179 \cdot 4 = 716,$$
$$179 \cdot 3 = 537,$$

and 537 is less than 709. So the second digit is 3, and the remainder is $709 - 537 = 172$.

Hence 11449 mod 179 = 172, and so $107 \cdot 107$ mod 179 is 172.

To find the digits of the quotient, we needed to guess, and trial divide, nine times.

Evidently, we can learn how to do long division by guessing. But for programming a computer, it could be helpful to find a systematic way to find the least non-negative residue of a number without the trial dividing that is part of the long division algorithm.

We present a method, due to P. Montgomery in 1985, which replaces the long division by several multiplications. Here is how it works.

Given the modulus m, we choose a base, or radix $r > m$ such that m is coprime to r, and such that finding the least non-negative residue of any number modulo r is easy. For example, if we are working with numbers written in the usual decimal notation and the modulus m is coprime to 10, then we can choose r to be a power of

10. For then the least non-negative residue of a number is just the rightmost digits of the number.

For example, if $r = 1000$ then $324,554,217$ modulo 1000 is 217, while 11449 modulo 1000 is 449.

In our example, if $m = 179$, then we can choose $r = 1000$.

For many applications, such as cryptography, the assumption that the modulus m is coprime to 10 will always hold.

Precomputation. Given the modulus m and the base $r > m$, we first precompute some constants for the algorithm. Since m and r are coprime, we can find numbers r' and m' so that $r'r - m'm = 1$, or

$$r'r = 1 + m'm,$$

where $0 < r' < m$ and $0 < m' < r$. Note that r' is the inverse of r modulo m.

We also find the least non-negative residue w of $r^2 \bmod m$. The constants r', m' and w are used in the algorithm.

The algorithm. Now let b be a number $<mr$. We want to find $b \bmod m$.

We do it in two parts.

For the first part we find $br' \bmod m$, as follows.

First, let $s = bm' \bmod r$. (That, recall, is easy to do.) Then, multiplying by m yields

$$sm = bm'm \pmod{mr},$$

and since $s < r$, then $sm < rm$ and sm is the least non-negative residue of $b'bm$ modulo mr. Then

$$b + sm \equiv b + bm'm = b(1 + m'm) = br'r \pmod{mr},$$

so $b + sm$ is a multiple of r. Divide the congruence

$$b + sm \equiv br'r \pmod{mr}$$

by r (again, easy to do), to get $z = (b + sm)/r$. Then

$$z \equiv br' \bmod m.$$

We also have that

$$z < 2m.$$

To see this, recall that $b < mr$ by assumption, and $sm < mr$. So $rz = b + sm < 2mr$, hence $z < 2m$.

The least non-negative residue c of $br' \bmod m$ is then either z, if $z < m$, or $z - m$, if $m \le z < 2m$.

For the second part of the algorithm, multiply c and w, where w is the least non-negative residue of r^2 modulo m that we precomputed earlier. Then $wc < m^2 < mr$, and

$$wc \equiv r^2 br' \equiv br \pmod{m}.$$

If we then repeat the first part on wc instead of b, we will end up with a number $d < m$ so that

$$d \equiv wcr' \equiv brr' \equiv b \pmod{m},$$

and so d is the least non-negative residue of b modulo m.

In outline, to find $b \bmod m$ for $b < mr$:

- find $s = bm' \bmod r$,
- compute $z = (b + sm)/r$. Then $z < 2m$.
- determine c where $c = z$ if $z < m$ and $c = z - m$ if $z \geq m$.
- find $s' = wcm' \bmod r$,
- compute $z' = (wc + s'm)/r$. Then $z' < 2m$.
- determine d where $d = z'$ if $z' < m$ and $d = z' - m$ if $z' \geq m$.

Then $d = b \bmod m$.

Example 3. Let $m = 179$ and choose the radix $r = 1000$. For the precomputation, we find that $179 \cdot 581 + 1 = 1000 \cdot 104$, and $r^2 = 1000^2 \equiv 106 \pmod{179}$. So

$$r' = 104,$$
$$m' = 581,$$
$$w = 106.$$

For the algorithm itself, let $b = 107 \cdot 107 = 11449$. We want to find b modulo $m = 179$.

First, we find

$$s \equiv bm' = 11449 \cdot 581 \bmod 1000.$$

We can find s efficiently by first reducing $b = 11449 \bmod 1000$ to get 449, then multiplying 449 by $m' = 581$ to get 260869, then reducing 260869 modulo 1000 to get

$$s = 869.$$

Then $sm = 869 \cdot 179 = 155551$, the least non-negative residue of $bm'm$ modulo mr. So

$$b + sm = 11449 + 155551 = 167000$$
$$\equiv b + bm'm = b(1 + m'm) = br'r \pmod{rm}$$

is a multiple of $r = 1000$. So

$$z = (b + sm)/r = 167000/1000 = 167.$$

Then $z = 167$ satisfies

$$167 \equiv 11449 \cdot 104 = br' \bmod 179.$$

Since $167 < 179$, we have
$$c = 167.$$

Now we multiply $c = 167$ by the least non-negative residue $w = 106$ of r^2 to get

$$wc = 167 \cdot 106 = 17702 \equiv br \quad (\text{mod } m).$$

We find $wc \cdot m' = 17702 \cdot 581 = 10284862$, then

$$s' = wcm' \bmod 1000 = 862.$$

Then $s'm = 862 \cdot 179 = 154298$, and

$$wc + s'm = 17702 + 154298 = 172000$$
$$\equiv wc + wcm'm = wc(1 + m'm) = wcr'r \quad (\text{mod } rm),$$

a multiple of $r = 1000$. So
$$z' = 172.$$

Since $172 < 179$, we have
$$d = 172,$$

the least non-negative residue of $b = 11449$ modulo 179.

To sum up, once we set up Montgomery's algorithm for a particular modulus m and radix r by precomputing m', r' and w, the algorithm finds the least non-negative residue of any number $b < mr$, replacing long division by m with five multiplications of numbers $<m$ and five divisions by r. The guessing or trial division that can arise in long division is eliminated.

The Montgomery algorithm has been called the most efficient algorithm available for modular multiplication. It is being built into circuitry designed to do fast modular multiplication of numbers of sizes up to 2^{2048} (numbers of up to 616 digits). We'll see some applications of modular multiplication of large numbers in later chapters.

The original algorithm appeared in Montgomery (1985).

Exercises.

87. Use Montgomery's algorithm to find
  (i) $132 \cdot 89 \bmod 179$,
  (ii) $167 \cdot 148 \bmod 179$.

88. Set up Montgomery's algorithm for $m = 267$. Use it to find $167 \cdot 239 \bmod 267$.

89. Try this "pick a number" puzzle on a friend:
  Pick something you know your friend does at least one day per week. Ask her:
  "Write down how many days last week you did [that thing]? Don't show it to me."
Call the secret number m. (m should be a number with $1 \le m \le 7$).

Tell her to do the following:

- Take her secret number, add it to 42, call the result t.
- Take the units digit of t, multiply it by 7, take the units digit of the result, multiply that by 7, add the result to t, then divide by 5. Call the result u.
- Then take the units digit of u, multiply it by 7, take the units digit of the result, multiply that by 7, add the result to u, then divide by 10.

Then tell her that the number she computed was the number of days last week she did [that thing].

(i) If she says you're wrong, can you accuse her of making an error in her computations?

(ii) Try to explain to her why it works (if it does!).

(iii) Write up the result of your trial.

90. Make up your own "pick a number" puzzle based on Montgomery's algorithm.

Chapter 10
Applications of Euler's Theorem

The applications of Fermat's and Euler's Theorems in this chapter are to cryptography and to the study of large numbers.

A. RSA cryptography

Euler's theorem is the key result behind the widely used RSA cryptosystem, developed by R.L. Rivest, A. Shamir, and L. Adleman (1977).

Suppose two parties, call them Alice and Bob, wish to send messages back and forth to each other, and want them to be incomprehensible to a third party, say Eve. The idea is to encrypt each message, to transform the plaintext message into a message that would be unreadable except to the intended receiver. Even if the encrypted message is broadcast publicly, or sent over the internet, Eve, reading the encrypted message, should not be able to determine in a reasonable amount of time what the original message is.

Any message in words can be translated into a sequence of numbers by replacing the letters of the message by numbers in some agreed-upon way. For example, we could count the alphabet and replace each letter by the corresponding two-digit number:

$$A \leftrightarrow 01; B \leftrightarrow 02; ...; M \leftrightarrow 13; ...; Z \leftrightarrow 26; \text{ (space)} \leftrightarrow 00.$$

Thus the word I LOVE YOU would become 10001215220500251521 (Or we could use the ASCII numbering of characters.) In this way a message is translated into a sequence of decimal digits. We'll assume hereafter that every message is a sequence of numbers.

We first describe how the method works, and then explain why it is effective.

How RSA works. For Alice to send Bob a message, Bob chooses two different large primes p and q that he keeps secret, and sets $m = pq$. He chooses an encrypting exponent e coprime to $\phi(m) = (p-1)(q-1)$. Then Bob finds a number d so that

$$ed \equiv 1 \quad (\text{mod } \phi(m)).$$

Then d is the inverse of e modulo $\phi(m)$. Bob can find d by solving the equation

$$ex + \phi(m)y = 1$$

for x. Since e and $\phi(m)$ are coprime, Bob can solve this equation efficiently by the extended Euclidean Algorithm (Bezout's Identity, Section 3D). Thus

$$ed = 1 + \phi(m)k$$

for some k.

Bob keeps d secret but broadcasts m and e to Alice.

Alice has a message that consists of a sequence of numerical words. Each word is a number w that is smaller than m. To encrypt the word w, Alice computes

$$c = w^e \bmod m.$$

That is, Alice finds the number $c < m$ that is congruent to w^e modulo m. She broadcasts the encrypted word c to Bob.

Bob computes
$$w' = c^d \bmod m.$$

Then w' will be the original word w of Alice. For since $ed \equiv 1 \pmod{\phi(m)}$, we have
$$w' \equiv c^d \equiv (w^e)^d = w^{1+k\phi(m)} \equiv w \pmod{m}$$

for some integer k, where the last congruence follows from Corollary 16 of Chapter 9. Since both w and w' are numbers less than m, then $w = w'$.

We illustrate the computations with some small unrealistic examples where the modulus is prime, rather than a product of two primes.

Example 1. Let the modulus $m = 101$, a prime. Then $\phi(101) = 100$. Suppose Bob chooses the encrypting exponent $e = 13$. Then

$$13 \cdot 77 \equiv 1 \quad (\text{mod } 100)),$$

so 77 is the decrypting exponent. Bob broadcasts $m = 101$ and $e = 13$ to Alice.

Suppose Alice wants to send the message HELLO, which she translates into five two-digit words (the plaintext) as 08, 05, 12, 12, 15. Then she encrypts the message by raising each word to the exponent 13 modulo 101 as follows:

$$8^{13} \equiv 18 \quad (\text{mod } 101)$$
$$5^{13} \equiv 56 \quad (\text{mod } 101)$$
$$12^{13} \equiv 53 \quad (\text{mod } 101)$$
$$15^{13} \equiv 7 \quad (\text{mod } 101).$$

Then the coded message, or ciphertext, is 18, 56, 53, 53, 07. Alice sends those five encrypted words to Bob. Bob decodes the words by raising each word to the exponent 77 modulo 101:

$$18^{77} \equiv 8 \quad (\text{mod } 101),$$

$$56^{77} \equiv 5 \quad (\text{mod } 101),$$

$$53^{77} \equiv 12 \quad (\text{mod } 101),$$

$$7^{77} \equiv 15 \quad (\text{mod } 101),$$

By doing so, Bob recovers the sequence 08, 05, 12, 12, 15, the plaintext message.

Example 2. Let $m = 2803$, a prime. Let $e = 113$. Since $\phi(2803) = 2802 = 2 \cdot 3 \cdot 467$, e is coprime to $\phi(m)$. Using Bezout's identity, we find that

$$113 \cdot 1463 \equiv 1 \quad (\text{mod } 2802),$$

and so the decoding exponent $d = 1463$. Bob chooses e, computes d, and broadcasts m and e to Alice.

Alice wishes to send the message GO, or 0715. (Note that 2803 is larger than any numerical word corresponding to a two-letter message.) To encode the plaintext message 0715, Alice finds the least nonnegative residue of 715^{113} (mod 2803). (This can be done efficiently by writing the exponent in base 2: $113 = 64 + 32 + 16 + 1$ and successively squaring 715 modulo 2803, organizing the computation as shown in Section 9F.) She finds that

$$715^{113} \equiv 708 \quad (\text{mod } 2803).$$

Thus the encrypted message, or ciphertext, is 708.

Alice broadcasts $c = 708$ to Bob.

Bob takes the ciphertext 708, and finds the least nonnegative residue mod 2803 of 708^{1463}. The resulting calculation yields 715, which translates back into the message GO.

In practice, the modulus is much larger. To set up the cryptosystem, Bob finds two large primes p and q of approximately the same number of digits. Suppose the modulus $m = pq$ has $r + 1$ digits for some r. Bob chooses e, the encrypting exponent, coprime to $\phi(m) = (p-1)(q-1)$, and determines d, his private decrypting exponent, to satisfy $de \equiv 1 \pmod{\phi(m)}$.

Bob broadcasts m and e, but keeps $p, q, \phi(m)$ and d secret.

Alice splits up her (numerical) message for Bob into words consisting of r digit numbers. She encrypts by raising each word w to the eth power mod m. The resulting sequence of numbers is the ciphertext. Alice broadcasts the ciphertext to Bob.

Bob knows d and m, so can decrypt the ciphertext by taking each encrypted word c from Alice, and computing c^d mod m. The resulting sequence of words will be Alice's original numerical message.

Why is RSA effective? Suppose Eve eavesdrops on the messages between Alice and Bob. Then she knows m and e and each encrypted word c. To read each word w, all Eve needs to do is to find the e-th root of c modulo m. This is easily done if she can find d, the decrypting exponent. But that is the problem. To find d, Eve would need to solve the congruence $ed \equiv 1 \pmod{\phi(m)}$ for d. But to do so, she needs to know what $\phi(m)$ is.

Now $\phi(m)$ is the number of numbers $<m$ that are coprime to m. To find $\phi(m)$ by counting is out of the question if m is sufficiently large. To find $\phi(m)$, Eve would need to know how to factor m. But factoring large numbers into products of primes is a hard problem.

Thus the key to the effectiveness of the RSA cryptosystem is that if Bob chooses m to be a product of two large primes, then someone like Eve, who knows m, but not its factorization, will be unable to determine the factorization (and hence $\phi(m)$, and hence the decrypting exponent d) in a reasonable amount of time.

How large should m be? The standard changes over time with improvements in computer power and in factoring algorithms. For example, in 1999, a modulus of 155 decimal digits would have been fairly secure. That year, a number m of 155 digits, offered as a challenge by RSA Laboratories, was successfully factored into a product of two 77 digit primes in seven months.

In 2006, a modulus of 230 digits was considered secure.

RSA Laboratories (2004), which markets software with RSA cryptography commercially, recommended moduli of 1024 binary bits for use through 2010, but for longer term security (through 2030) recommends moduli of 2048 binary bits. For 1024 binary bits, the modulus should be a product of two 154 digit primes [http://www.rsa.com/rsalabs/node.asp?id=2004]. (For current information, their website, www.rsa.com/rsalabs/ is a useful source of information. It also offers challenge moduli of increasing size with cash prizes for their factorization).

On the other hand, if the factorization of m is known then $\phi(m)$ can be found instantly and the decoding exponent d can be found by Euclid's algorithm in a few seconds.

Thus the effectiveness of the RSA cryptosystem ultimately lies in the fact that factoring large numbers into products of primes is an inefficient computational process.

Even if factoring becomes more efficient, or in fact is more efficient than is publicly known, code users can (presumably) always stay ahead of the state of the art of factoring by choosing m to be a product of two primes that are sufficiently large.

Signatures. RSA codes have a "signature" feature that makes them particularly useful.

To illustrate how the feature works, suppose Bob is a stock broker on Wall Street, and Alice is a wealthy client, reclining with her wireless laptop on a beach in the Caribbean. Alice wants to send buy and sell orders to Bob. Bob wants to be certain that when he receives an order from Alice, it is authentic.

One way to handle these problems is for Alice to set up an RSA cryptosystem (m_A, e_A), and Bob to set up a different RSA system (m_B, e_B). Both could be published. But only Bob would know the secret decrypting exponent d_B for his system, and only Alice would know the secret decrypting exponent d_A for her system.

To send an order to broker Bob, client Alice would encrypt her order twice: first by using the pair (m_A, d_A), then using the pair (m_B, e_B). Bob would receive the encrypted order, and first decrypt it using the secret d_B, then using the public e_A. Since Alice encrypted the message using the secret exponent d_A, which broker Bob was able to decrypt, Bob would know that only client Alice could have sent the message. Since only Bob knows the secret exponent d_B, Alice would know that only Bob could decrypt the message. Thus both Alice and Bob are assured of the authenticity and secrecy of the message, and communication between them is secure. (But see Sections 12C and 20D for potential pitfalls of a careless application of this scheme).

Exercises.

1. Encode the message BUY using the code of Example 1.

2. Encode the message SELL using the code of Example 2.

3. Let $m = 29, e = 5$. Encode the message HOLD.

4. Let $m = 29$. Suppose you choose $e = 4$, encode SELL and send it to Merrill. How would Merrill decode the message?

5. Let $m = 3337, e = 11, d = 1171$. Encode and decode the message NO.

6. Let $m = 3501697, e = 17, d = 1440269$. Encode and decode (not by hand!) the message 5552 0307 4562 1587.

7. We observed above that if the factorization of the modulus m is known, then $\phi(m)$ is easy to compute. The converse is also true. Suppose m is a product of two unknown prime numbers p and q, and suppose m and $\phi(m)$ are known. Show that p and q can be found as the roots of an appropriate quadratic equation.

8. Let p, q be distinct odd primes and let $\lambda(m) = [p - 1, q - 1]$, the least common multiple of $p - 1$ and $q - 1$. Suppose we set up an RSA cryptosystem with a modulus $m = pq$ and an encrypting exponent e coprime to $p - 1$ and $q - 1$. Show that we can use d' for the decrypting exponent, where d' is a solution of $ex \equiv 1 \pmod{\lambda(m)}$.

B. Pseudoprimes

It is often of interest to know whether a given number is prime. For example, in Chapter 7 we observed that $\mathbb{Z}/n\mathbb{Z}$ is a field iff n is prime. In the last section we saw how large primes are used in the construction of cryptographic codes.

Given a number n, how can we decide if n is prime?

The most naive approach to this question is to treat it as a special case of the problem of factoring n. Namely, try to factor n, for example, by trial division. If we succeed, n is not prime; otherwise, n is prime.

But this naive approach has its drawbacks: as we saw in Section 6G, trial division is hopelessly inefficient for large numbers n. Furthermore, if n happens in fact to be prime, trial division is a particularly inefficient method to demonstrate the primeness of n, as we have already observed, because we would have to verify that n is not divisible by each prime $< \sqrt{n}$.

For this reason, mathematicians have sought other ways to try to show that a number n is prime, or to test n for primeness, without having to use trial division.

One of the simplest tests involves Fermat's theorem.

Fermat's theorem says that if n is a prime number, then for any integer a relatively prime to n, n divides $a^{n-1} - 1$. The contrapositive of Fermat's theorem is the following:

If a is some number coprime to n so that $a^{n-1} - 1$ is not divisible by n, then n is not prime.

A special case is

The 2-pseudoprime test. If $2^{m-1} \not\equiv 1 \pmod{m}$, then m is composite.

The 2-pseudoprime test is a compositeness test. We can show, for example, that 9 is not prime, by observing that $2^8 \equiv 4 \neq 1 \pmod 9$ and using Fermat's theorem: if 9 were prime, then 2^8 would be congruent to 1 $\pmod 9$: since it isn't, 9 can't be prime.

But we can't conclude that 561 is prime by determining that $2^{560} \equiv 1 \pmod{561}$ (which is true), because in fact 561 is not prime–it is clearly divisible by 3.

But as a test for compositeness of a number m, seeing if $2^{m-1} \equiv 1 \pmod m$ can be done rather quickly, using the technique of section 9F, and is surprisingly effective.

Example 3. Let us look for primes in the set of numbers between 1194601 and 1194700 (inclusive).

In Section 6G, we eliminated 89 of the 100 numbers in this set by showing they were divisible by some prime < 53. The remaining numbers are:

$$1194601$$
$$1194623 \qquad 1194629$$
$$1194631 \qquad\qquad 1194637$$
$$1194649$$
$$1194653 \qquad 1194659$$
$$1194667$$
$$1194671 \qquad\qquad 1194679$$

Every prime between 1194600 and 1194700 must be among these eleven numbers. To test these numbers, we try the 2-pseudoprime test. We find that $2^{m-1} \equiv 1$

(mod m) for $m = 1194601, 1194631, 1194649, 1194659, 1194667, 1194671,$ and $1194679.$ However:

$$2^{1194622} \equiv 965745 \quad (\text{mod } 1194623)$$
$$2^{1194628} \equiv 506389 \quad (\text{mod } 1194629)$$
$$2^{1194636} \equiv 1031720 \quad (\text{mod } 1194637)$$
$$2^{1194652} \equiv 553181 \quad (\text{mod } 1194653).$$

Thus 1194623, 1194629, 1194637, and 1194653 are composite. This leaves seven potential primes out of the original 100 numbers.

It turns out that of those seven numbers, six are prime. The only one which is not is $1194649 = 1093 \cdot 1093$. (The factorizations of the candidates that failed the 2-pseudoprime test are:

$$1194623 = 509 \cdot 2347$$
$$1194629 = 269 \cdot 4441,$$
$$1194637 = 241 \cdot 4957,$$
$$1194653 = 521 \cdot 2293).$$

If a number m is prime, then m passes the 2-pseudoprime test, but as we saw with 1194649, some composite numbers m also pass the 2-pseudoprime test. They are called 2-pseudoprimes:

Definition. A number m is a *2-pseudoprime* if:
(i) m is composite; and
(ii) $2^{m-1} \equiv 1 \pmod{m}$.

It turns out that 2-pseudoprimes are much less common than primes. How scarce are they? Here are some counts:

There are 168 primes <1000, but only three 2-pseudoprimes: $341 = 11 \cdot 31$, $561 = 3 \cdot 11 \cdot 17$, and $645 = 3 \cdot 5 \cdot 43$.

There are 5,761,455 primes under 100,000,000, but only 2057 2-pseudoprimes. There are 882,206,716 primes less than $2 \cdot 10^{10}$, compared with 19,685 2-pseudoprimes less than $2 \cdot 10^{10}$. [Pomerance, Selfridge, and Wagstaff (1980)].

Based on these counts and a comparison of the number of 2-pseudoprimes $<m$ with the number of primes $<m$ for large numbers m, it is evident that if you pick randomly a large number n and verify that n divides $2^{n-1} \equiv 1 \pmod{n}$, then it is highly probable that n is prime. For this reason, the mathematician Henri Cohen has called a number n with $2^{n-1} \equiv 1 \pmod{n}$ an "industrial grade prime."

However, even if a randomly chosen number that satisfies the 2-pseudoprime test is likely to be prime, that conclusion does not necessarily apply to special types of numbers. In the rest of this section we will look at two famous special sets of numbers, both of which are sets of 2-pseudoprimes.

I. Mersenne Numbers. Marin Mersenne (1588–1648) was a French cleric and mathematician who corresponded with Descartes, Fermat, and other mathematicians of the time. Like most mathematicians in the 17th century, he was very familiar with Euclid's Elements and other ancient Greek mathematics, so he knew of Euclid's interest in primes of the form $2^p - 1$. Mersenne conjectured in 1644 that $2^p - 1$ was prime for the following p: 2, 3, 5, 7, 13, 17, 19, 31, 67, 127, and 257, and composite for the other primes $p \leq 257$. This conjecture turned out to be quite inaccurate above 31 ($2^p - 1$ is prime for $p = 61, 89, 107$, and 127, and composite for the other primes p with $31 < p \leq 257$).

However, other Mersenne numbers, numbers of the form $2^p - 1$ for p prime and $p > 257$ have been found that are prime. This set of Mersenne primes includes some very large primes. In fact, for 75 years (from 1876 to 1950) the largest known prime was the Mersenne prime

$$2^{127} - 1 = 170141183460469231731687303715884105727.$$

[Zagier (1977)]. In 1951 a larger prime was found that was not a Mersenne prime, but by 1952 a larger Mersenne prime was found, and ever since, the largest known prime has been a Mersenne prime. In the summer of 2008, the 45th and 46th Mersenne primes were found, the first known primes with more than ten million digits [GIMPS (2008)].

There are several reasons why Mersenne numbers have been singled out for primality testing. One is because if we seek a prime number that is "almost" a power, then we are forced to look only at Mersenne numbers:

Proposition 1. *If a number of the form $a^n - 1$ is prime, then $a = 2$ and n is prime.*

The proof is an easy exercise, below.

Another reason for inquiring about Mersenne numbers goes back to the ancient Greeks. The Pythagoreans (c. 500 BC), who were fascinated by numbers, discovered that the number 6 has the property that 6 is the sum of its proper divisors:

$$6 = 1 + 2 + 3.$$

They called such a number *perfect*, and sought other numbers with the same property. They found 28: $28 = 1 + 2 + 4 + 7 + 14$. By the time of Euclid, they had learned, and Euclid proved, the following theorem:

Proposition 2. *If $2^n - 1$ is a prime number, then $m = 2^{n-1}(2^n - 1)$ is a perfect number.*

The proof is left as an exercise, below.

This theorem, known to Mersenne and his correspondents, made the quest for Mersenne primes of particular interest to any mathematician with an interest in classical Greek mathematics.

The full story on perfect numbers is not known.

Euler, a century after Mersenne, proved a partial converse of Euclid's theorem, namely, that if m is an even perfect number, then $m = 2^{p-1}(2^p - 1)$ where $2^p - 1$ is a Mersenne prime.

Euler's theorem left open the question of whether or not there exist any odd perfect numbers. None are known, and knowledgeable number theorists tend to believe that no odd perfect numbers can exist. But no one has yet (as of 2008) proved that none exist. Among the known facts are that any odd perfect number must be divisible by at least 9 distinct primes, have at least 75 prime factors, and be greater than 10^{300}. For references, see http://mathworld.wolfram.com/OddPerfectNumber.html.

Mersenne numbers are at least plausible candidates for being primes, because they are always 2-pseudoprimes:

Proposition 3. *If n passes the 2-pseudoprime test, then $2^n - 1$ does also. Thus if n is prime, $2^n - 1$ is either prime or a 2-pseudoprime.*

The proof is left as an exercise, below.

Fermat's theorem is of assistance in finding non-trivial factors of Mersenne numbers if they exist.

For example, consider trying to factor $2^{37} - 1$. Suppose p is a prime divisor of $2^{37} - 1$. Then $2^{37} \equiv 1 \pmod{p}$, and since 37 is prime, 37 must be the order of 2 modulo p. Thus by Fermat's theorem, 37 divides $p - 1$, that is, $p - 1 = 37k$ for some k. It follows that any prime divisor of $2^{37} - 1$ must be of the form $p = 1 + 37k$. Evidently we can assume that k is even, that is, p is a prime of the form $p = 1 + 74h$, for $h = 1, 2, 3, \ldots$.

Such an analysis drastically reduces the number of trial divisions needed to decide whether or not $2^{37} - 1$ is prime. So we try:

$h = 1, p = 75$–not prime;

$h = 2, p = 149$–prime, but not a divisor of $2^{37} - 1$;

$h = 3, p = 223$–prime, and a divisor of $2^{37} - 1$.

This computation was in fact performed by Fermat, as an application of his theorem [Weil (1984), p. 57].

For another factorization of a Mersenne number, see section C below.

II. Fermat Numbers. Another interesting collection of 2-pseudoprimes is the Fermat numbers. The nth Fermat number is $F_n = 2^{2^n} + 1$. The Fermat numbers for $n = 0, 1, 2, 3, 4$ are 3, 5, 17, 257, and 66537, all primes.

As with Mersenne numbers, the focus on Fermat numbers is partly based on the fact that if a number of the form $2^a + 1$ is prime, then a must be of the form $a = 2^n$ for some n.

We showed in section 4C that for $m \neq n$, F_m and F_n are coprime, hence yield infinitely many primes as factors of Fermat numbers.

Fermat conjectured that all Fermat numbers are prime. This conjecture may have been suggested by:

Proposition 4. *For any n, F_n passes the 2-pseudoprime test.*

The proof is an exercise.

Euler observed (see Weil (1984), p. 58) that, as with the Mersenne numbers, Fermat's theorem considerably restricts the form of the primes p that may divide a Fermat number. Euler's example was $F_5 = 2^{32} + 1$, the fifth Fermat number. If p divides F_5, then, since $2^{32} \equiv -1 \pmod{p}$, we have $2^{64} \equiv 1 \pmod{p}$ but $2^r \not\equiv 1 \pmod{p}$ for r dividing 64 and $r < 64$. So the order of 2 mod p is 64. Thus 64 divides $p - 1$, that is,

$$p = 1 + 64k$$

for some k.

Euler then proceeded to test F_5 by trial division by primes of the form $p = 1 + 64k$, $k = 1, 2, 3...$, and found that with $k = 10, p = 641$ divides F_5. In this way, Euler refuted Fermat's conjecture.

This same idea has been used to help factor larger Fermat numbers [Brent and Pollard (1981), p. 629, line 8].

The state of Fermat's conjecture as of 2007 is that there is no number $n > 4$ for which F_n is known to be prime. A website that keeps track of current work on factoring Fermat numbers is W. Keller's website, www.prothsearch.net/fermat.html.

III. Carmichael numbers. If we want to see if a number m is prime using trial division, we would not test m by just dividing by 2–we would try dividing m by many primes.

Similarly, there is no reason to test a number n for primeness by using just the 2-pseudoprime test. Fermat's theorem states that if a is *any* number not a multiple of m, and m is prime, then $a^{m-1} - 1$ is divisible by m.

Thus instead of checking to see if $2^{m-1} \equiv 1 \pmod{m}$, we could pick any number $a < m$ and see if $a^{m-1} \equiv 1 \pmod{m}$. Call this the *a-pseudoprime test*.

If m fails an a-pseudoprime test, that is, if

$$a^{m-1} \not\equiv 1 \pmod{m},$$

for a single number $a < m$, then m is composite.

Definition. A number m is an *a-pseudoprime* if m is composite and $a^{m-1} \equiv 1 \pmod{m}$.

We observed that 2-pseudoprimes are fairly rare, and the same is true of a-pseudoprimes for any a. So it is plausible that numbers that are simultaneously 2- and 3-pseudoprimes are rarer still.

For example, neither of the 2-pseudoprimes 341 and 645 is a 3-pseudoprime: the latter because 3 divides 645. Neither is 1194649, the 2-pseudoprime we found between 1194601 and 1194700. In fact,

$$3^{194648} \equiv 341017 \pmod{1194649}$$

so 1194649 is not prime because it fails the 3-pseudoprime test.

More systematically, we could treat Fermat's theorem like trial division: take a number m, and check to see if

$$a^{m-1} \equiv 1 \pmod{m}$$

for many numbers a.

Very few numbers would pass these a-pseudoprime tests and not be prime. For example, here are the a-pseudoprimes between 50 and 999 for various prime numbers a:

2-pseudoprimes: 341, 561, 645.
3-pseudoprimes: 91, 121, 286, 671, 703, 949.
5-pseudoprimes: 124, 217, 561, 781.
7-pseudoprimes: 325, 561, 703, 817.
11-pseudoprimes: 133, 190, 259, 305, 481, 645, 703, 793.
13-pseudoprimes: 105, 231, 244, 276, 357, 427, 561.
17-pseudoprimes: 145, 261, 781.
19-pseudoprimes: 153, 169, 343, 561, 637, 889, 905, 906.
23-pseudoprimes: 154, 165, 169, 265, 341, 385, 451, 481, 553, 561, 638, 956.
29-pseudoprimes: 105, 231, 268, 341, 364, 469, 481, 561, 651, 793, 871.

Evidently, numbers that are a-pseudoprimes for more than one number a in this list are uncommon. However, looking over this list, you will perhaps notice that $561 = 3 \cdot 11 \cdot 17$ appears on all the lists except for 3, 11, and 17, the three factors of 561. That is, for all primes $a \leq 29$ that are coprime to 561, 561 is an a-pseudoprime.

It happens that 561 is an example of a composite number m with the property that for every number a coprime to m,

$$a^{m-1} \equiv 1 \pmod{m}.$$

Such a number is called a Carmichael number.

Definition. A composite odd number m is a *Carmichael number* if m is an a-pseudoprime for every a coprime to m.

To confirm what we observed with $n = 561$:

Proposition 5. $561 = 3 \cdot 11 \cdot 17$ *is a Carmichael number.*

Proof. First note that $561 = 3 \cdot 11 \cdot 17$ is composite. We want to show that for any a coprime to 561, $a^{560} \equiv 1 \pmod{561}$. To do so, it suffices to show that

$$a^{560} \equiv 1 \pmod 3,$$

$$a^{560} \equiv 1 \pmod{11},$$

and

$$a^{560} \equiv 1 \pmod{17}.$$

Now by Fermat's theorem, $a^2 \equiv 1 \pmod 3$, $a^{10} \equiv 1 \pmod{11}$ and $a^{16} \equiv 1 \pmod{17}$ for every number a coprime to 3, 11, and 17. Since 560 is a multiple of 2, 10 and 16, it follows that each of the displayed congruences is true. $\square$

Carmichael numbers are quite rare. The twelve Carmichael numbers less than 50,000 are 561, 1105, 1729, 2465, 2821, 6601, 8911, 10585, 15841, 29341, 41041, and 46657. Among the 2051 2-pseudoprimes less than 100,000,000, only 252 are Carmichael numbers. It was proved only in 1992 that there are infinitely many Carmichael numbers.

The existence of Carmichael numbers dashes the hope that Fermat's Theorem alone can be used as a primality test. If m is a Carmichael number, then no amount of a-pseudoprime testing will reveal that m is composite (unless we are lucky enough to pick a not coprime to m).

Chapter 20 has much more on Carmichael numbers and presents an extension of the a-pseudoprime test that is effective on Carmichael numbers.

Exercises.

9. C.P. Snow describes a meeting between the English mathematician G.H. Hardy and the Indian genius Ramanujan, when the latter was terminally ill with tuberculosis. Hardy started the conversation with, "I thought the number of my taxicab was 1729. It seemed to me rather a dull number." To which Ramanujan replied, "No, Hardy! It is a very interesting number. It is the smallest number expressible as the sum of two cubes in two different ways." (Hardy (1969), page 37). (Reprinted with Permission of Cambridge University Press).

(i) Find two pairs (a,b) of integers with $0 < a < b$ so that $1729 = a^3 + b^3$.

(ii) Verify that 1729 is a Carmichael number.

10. Verify that 341 is a 2-pseudoprime.

11. Verify that 91 is a 3-pseudoprime.

12. Verify that 645 is a 2-pseudoprime.

13. Prove Proposition 1, that if $a^n - 1$ is prime, then $a = 2$ and n is prime.

14. Prove Proposition 5, Euclid's Theorem on even perfect numbers.

15. Prove Proposition 3, that if n passes the 2-pseudoprime test, then so does $2^n - 1$.

16. Show that there are infinitely many 2-pseudoprimes.

17. (i) Show that every prime factor of the Mersenne number $2^{11} - 1$ is congruent to 1 modulo 22.

(ii) Find a prime factor of the Mersenne number $2^{11} - 1$.

18. Using a computer, find a prime factor of the Mersenne number $2^{23} - 1$.

19. Prove that if a number of the form $2^a + 1$ is prime, then $a = 2^n$ for some n.

20. Prove Proposition 4, that every Fermat number passes the 2-pseudoprime test.

21. Let p be an odd prime and b a primitive root modulo p.

(i) Show that b has order $p-1$ or order $p(p-1)$ modulo p^2.

(ii) Show that there is a unique k with $0 \leq k \leq p-1$ such that $b + kp$ has order $p-1$ modulo p^2.

(iii) Conclude that if p is any odd prime, then there is a number b so that every unit modulo p^2 is congruent modulo p^2 to a power of b.

22. Let $p = 2^q - 1$ be a Mersenne prime. Show that $2^p \not\equiv 1 \pmod{p^2}$.

23. Show that $20^7 \equiv 1 \pmod{551}$. Conclude that 551 is not a 20-pseudoprime, hence is composite.

24. (i) Show that if m is an a-pseudoprime for $a = 2$ and $a = 3$, then m is an a-pseudoprime for $a = 6$.

(ii) Find a number m so that m is a 6-pseudoprime but not a 2-pseudoprime or a 3-pseudoprime.

25. Show that for each m, the set

$$\{[a]_m \text{ in } \mathbb{Z}/m\mathbb{Z} \,|\, a^{m-1} \equiv 1 \pmod{m}\}$$

is an abelian group under multiplication of congruence classes (see Section 8B).

26. Let $n \equiv 5 \pmod 6$. Show that n is not a perfect number, as follows:

(i) Show that n is not a square.

(ii) Let $\sigma(n)$ be the sum of the divisors of n (including 1 and n). Show that

$$\sigma(n) = \sum_{d < \sqrt{n}, d|n} d + \frac{n}{d}.$$

(iii) Show that for each divisor $d < \sqrt{n}$,

$$d \cdot \frac{n}{d} = n = -1 \pmod 3,$$

hence

$$d + \frac{n}{d} \equiv 0 \pmod 3.$$

(iv) Show that $\sigma(n) \equiv 0 \pmod 3$. Hence $\sigma(n) \neq 2n$ and n is not a perfect number. [Holdener (2002)].

C. The Pollard $p - 1$ Factoring Algorithm

Fermat's theorem is the basis of the Pollard $p - 1$ factoring algorithm, discovered by J. M. Pollard (1974). The algorithm is effective for finding a prime factor p of a number N when $p - 1$ is a product of small primes.

To make more precise the idea that $p - 1$ is a product of small primes, we use the following terminology:

Definition. Let k be a number ≥ 2. A number m is *k-smooth* if every prime divisor of m is $<k$.

Some examples:
The only 2-smooth numbers are powers of 2;
If we factor the numbers from 180 to 190, we find:

$$180 = 2^2 \cdot 3^2 \cdot 5$$
$$181 \text{ is prime}$$
$$182 = 2 \cdot 7 \cdot 13$$
$$183 = 3 \cdot 61$$
$$184 = 2^3 \cdot 23$$
$$185 = 5 \cdot 37$$
$$186 = 2 \cdot 3 \cdot 31$$
$$187 = 11 \cdot 17$$
$$188 = 2^2 \cdot 47$$
$$189 = 3^3 \cdot 7$$
$$190 = 2 \cdot 5 \cdot 19.$$

Thus 180 is 5-smooth, 180 and 189 are 7-smooth, while 180, 182, 187, 189 and 190 are 19-smooth. Evidently, the larger k is, the more k-smooth numbers there are.

Here is the idea behind the Pollard $p-1$ algorithm.
We wish to factor the number N.
Suppose p is a prime factor of N. If we take some base a, Fermat's Theorem says that $a^{p-1} \equiv 1 \pmod{p}$, or equivalently, p divides $a^{p-1} - 1$. Then, since p also divides N, it follows that the greatest common divisor of N and $a^{p-1} - 1$ is a non-trivial divisor of N.

But of course, we don't know p, so we can't compute a^{p-1}.

But if $a^{p-1} \equiv 1 \pmod{p}$, then $a^B \equiv 1 \pmod{p}$ for every number B that is a multiple of $p-1$. If we find a number B that is divisible by $p-1$ for many primes p, then $a^B \equiv 1 \pmod{p}$ for many primes. If one of those primes also divides N, then

$$(a^B - 1, N) > 1.$$

In the Pollard $p-1$ algorithm the exponent B is chosen sufficiently large, so that if p is a prime divisor of N and $p-1$ is k-smooth for small k, then $p-1$ divides B.

Pollard $p-1$ algorithm. Pick some base number a, and some smoothness bound k.
For each prime $q \leq k$, let e be the exponent such that

$$E_q = q^e \leq N < q^{e+1},$$

and set

$$B_k = \prod_{q \text{ prime, } q \leq k} E_q.$$

Then B_k is the product of all prime powers q^e where $q \leq k$ and $q^e \leq N < q^{e+1}$.

For example, let $N = 323$ and $k = 7$. Then $E_2 = 2^8, E_3 = 3^5, E_5 = 5^3, E_7 = 7^2$, and $B_7 = 2^8 3^5 5^3 7^2$.

If p is a prime so that $p - 1$ is k-smooth, then $p - 1$ will divide B_k, and so by Fermat's Theorem,

$$a^{B_k} \equiv 1 \pmod{p}$$

and p divides $a^{B_k} - 1$.

We compute the greatest common divisor of $a^{B_k} - 1$ and N. If p also divides N, then p divides $(N, a^{B_k} - 1)$, so $(N, a^{B_k} - 1) > 1$ and is a factor of N.

Note that both parts of the procedure, raising to a high power modulo N and finding the greatest common divisor of two numbers, can be done by fast algorithms–Euclid's algorithm for the latter, and the squaring algorithm of Section 9F for the former.

Let's try this algorithm out on a couple of numbers that didn't factor by trial division in Chapter 6.

Example 4. Let $N = 1194653$. Let $a = 2$, let $B = B_{13}$ and compute a^B. Here,

$$B = 2^{20} \cdot 3^{12} \cdot 5^8 \cdot 7^7 \cdot 11^5 \cdot 13^5,$$

the product of all prime powers of primes $q \leq 13$ where the exponent of q is defined by

$$q^e \leq N < q^{e+1}.$$

(For example, $5^8 \leq N < 5^9$.) We compute the least non-negative residue:

$$2^B \equiv 56790 \pmod{N},$$

and find that

$$(2^B - 1, N) = (56790 - 1, N) = (56789, 1194653) = 521,$$

a prime divisor of N. The reason we found the divisor 521 of N is that

$$520 = 2^3 \cdot 5 \cdot 13$$

so that $521 - 1 = 520$ is 13-smooth, hence 520 divides B and so

$$2^B \equiv 1 \pmod{521}.$$

Example 5. Let $N = 1194637$. With the same $B = B_{13}$ as in the last example,

$$2^B \equiv 1010514 \pmod{N}$$

and
$$(2^B - 1, N) = (1010514 - 1, N) = (1010513, 1194637) = 241,$$

a prime divisor of N. Here
$$240 = 2^4 \cdot 3 \cdot 5$$

so that 241 is a prime divisor of N such that $241 - 1 = 240$ is 13-smooth, hence divides B.

For this last example, we could have found 241 by using $k = 5$. In that case, $B_5 = 2^{20} \cdot 3^{12} \cdot 5^8$. Setting $B = B_5$, we find that

$$2^B \equiv 301733 \pmod{N}$$

and
$$(2^B - 1, N) = (301732, N) = 241.$$

But $k = 5$ is too small for the number in the previous example:

Example 6. Again let $N = 1194653$. Let $k = 5$ and $B = B_5 = 2^{20} \cdot 3^{12} \cdot 5^8$. We find that
$$2^B \equiv 43666 \pmod{N}$$

and
$$(2^B - 1, N) = (43666 - 1, N) = (43665, 1194653) = 1.$$

The algorithm fails here because $N = 1194653$ has no prime divisor p so that $p - 1$ is 5-smooth.

In practice, we don't need to specify in advance a particular smoothness bound. If a given bound k fails and there is more time, just choose a larger bound k'. Then $B_{k'} = B_k F$ where F is a product of powers of primes q with $k < q \leq k'$, and

$$2^{B_{k'}} = \left(2^{B_k}\right)^F.$$

So to change from k to k' we just continue the computation we did for 2^{B_k} to get $2^{B_{k'}}$.

For example, with $N = 1194653$,

$$2^{B_{13}} = 2^{\left(2^{20} 3^{12} 5^8 7^7 11^5 13^5\right)} = \left(2^{\left(2^{20} 3^{12} 5^8\right)}\right)^{\left(7^7 11^5 13^5\right)} = \left(2^{B_5}\right)^F$$

where $F = 7^7 11^5 13^5$.

Given a large number N, is it reasonable to hope that N has a prime divisor p so that $p - 1$ is k-smooth for small k?

For example, suppose we have a six-digit number N which is known to be composite. How large must k be to give a reasonable chance of factoring N?

If $N < 10^6$, then N will be divisible by at least one prime $p < 1000$. There are 167 odd primes under 1000. Of these,

There are 4 primes p so that $p-1$ is 2-smooth (3, 5, 17 and 257);

There are 17 primes p so that $p-1$ is 3-smooth (the above, and 7, 13, 97, 193, 769, 19, 37, 73, 577, 109, 433, 163, 487);

There are 33 primes p so that $p-1$ is 5-smooth (the above, and also 11, 31, 41, 61, 101, 151, 181, 241, 251, 271, 401, 541, 601, 641, 751 and 811).

So suppose we do the Pollard $p-1$ algorithm with $k=5$, and suppose we just want to find prime divisors of N that are <1000. We can use $B = 2^9 \cdot 3^6 \cdot 5^4$, the product of the highest powers of 2, 3, and 5 that are less than 1000. Then for every prime $p < 1000$ so that $p-1$ is 5-smooth, $p-1$ will divide B. If any of those 33 primes divides our number N, then for any base a, $a^B - 1$ and N will be divisible by p. Thus just one computation checks to see if any of those 33 primes <1000 are factors of N. One computation checks 20 percent of the primes <1000 all at once.

Increasing the smoothness bound k will capture more primes. We find:

For 75 of the 167 odd primes $p < 1000$, $p-1$ is 13-smooth,

For 107 of the 167, $p-1$ is 31-smooth. Thus if we are doing Pollard $p-1$ on a random six-digit number N and we choose $k=31$, then if N is divisible by any of the 107 primes p so that $p-1$ is 31-smooth, then $(a^B - 1, N) > 1$.

On the other hand, the worst possible prime $p < 1000$ for the algorithm is 983. Since $983 - 1 = 2 \cdot 491$ and 491 is prime, $983 - 1$ is not k-smooth for any $k < 491$.

The next example illustrates a phenomenon that can occur with the Pollard $p-1$ algorithm:

Example 7. Let $N = 125561$. Suppose we compute $2^{B_q} \pmod{N}$ for $q = 13$, where $B = B_{13} = 2^{16} \cdot 3^{10} \cdot 5^7 \cdot 7^6 \cdot 11^4 \cdot 13^4$ (B is divisible by all prime powers $<N$ where each prime is <13.) We find that

$$2^B \equiv 1 \pmod{N}$$

hence $N = 125561$ divides $2^B - 1$ and

$$(2^B - 1, N) = N.$$

The algorithm appears to fail in this example because the greatest common divisor of $2^{B_{13}} - 1$ and N is N itself. When this happens, it implies that the order of 2 divides B_{13} modulo every prime divisor of N.

If we were to compute the greatest common divisor with $k = 7$ rather than waiting until $k = 13$, then we would not have reached the point we did in the last example.

Example 8. For the same N, $B_7 = 2^{16} \cdot 3^{10} \cdot 5^7 \cdot 7^6$, and

$$2^B \equiv 124839 \pmod{N}.$$

Computing the greatest common divisor at this point yields

$$(2^B - 1, N) = (124838, 125561) = 241,$$

a prime divisor of N. We found the prime divisor $p = 241$ of 125561 because $p - 1 = 240$ is 7-smooth. But $125561 = 241 \cdot 521$ and for $p = 521$, the other factor of 125561, $p - 1 = 520 = 8 \cdot 5 \cdot 13$ is 13-smooth but not 7-smooth.

An alternative way to deal with the phenomenon that $a^B \equiv 1 \pmod{N}$ is to view N as like an a-pseudoprime. See Chapter 20 or Exercise 31 below.

Remark. Note that in most of our examples, the exponents of the primes q in the prime factorization of B_k were much larger than needed. We assumed that B_k was the least common multiple of all prime powers l^e where $l < k$ and $l^e < N$. Alternative suggestions for B_k are

(i) let B_k be the smallest number divisible by l^e for all primes $l < k$, but where $l^e < M$ with M much less than N (in the last example, we chose $M = \sqrt{N}$), or

(ii) let $B_k = k!$, or

(iii) let B_k be the least common multiple of $2, 3, 4, \ldots, k$.

Each alternative gives a much smaller exponent B_k for a given N and k, but leaves open the possibility that a factor p of N has the property that $p - 1$ is k-smooth but $p - 1$ does not divide B_k. For an example of the use of a smaller exponent, see the subsection, "Factoring Mersenne numbers", below. Thus the implementation of the Pollard $p - 1$ algorithm raises issues of balancing the size of B_k for a given k against the size of k in order to obtain the greatest likelihood of success in the time available. These issues go beyond the scope of this book.

A great deal of work has been done studying smoothness of numbers. A result of Dickman (1930) implies that for large N, the number of $\sqrt{N}$-smooth numbers $< N$ is approximately $N/4$, while the number of $N^{1/4}$-smooth numbers $< N$ is around $N/200$. Thus for a number N of around 10^{12}, roughly $1/4$ of the numbers less than N are 10^6-smooth, but only around $1/200$ of the numbers less than N are 1000-smooth. See Granville (2004) for an extended discussion of smooth numbers and their uses.

RSA security. A situation where the issue of success or failure of the Pollard $p - 1$ algorithm is of importance is in connection with RSA cryptography.

Recall from Section A above that in an RSA code, the modulus $m = p_1 p_2$ is a product of two large primes, p_1 and p_2, whose secrecy is essential for the code to be secure. Since m is typically made public, m should be impossible to factor in a reasonable amount of time.

In view of the Pollard $p - 1$ factoring algorithm, it is important that if p is any prime dividing m, then $p - 1$ should not be k-smooth for any small k.

One way to insure that the Pollard $p - 1$ algorithm is maximally ineffective for factoring the modulus m is to choose the prime divisors p of m so that $p - 1 = 2q$ for q a prime number. Then the smallest k so that $p - 1$ is k-smooth is $k = q$, a number nearly as large as p. It would take much less time to trial divide m by numbers $< \sqrt{m}$ than to do the Pollard $p - 1$ algorithm to find a q-smooth prime factor of m.

A prime number p with the property that $p - 1 = 2q$ with q prime is sometimes called a *safeprime*.

Of the 167 odd primes <1000, 25 are safeprimes. The largest safeprime <1000 is 983. See Section 12C, Exercise 29 on finding small safeprimes.

A prime number q so that $2q + 1 = p$ is also prime is called a *Sophie Germain prime*, after the early 19th century French mathematician Sophie Germain. Around 1825 she proved that if q is a prime so that $2q + 1$ is also prime, then $x^q + y^q = z^q$ has no solutions with x, y and z natural numbers all not $\equiv 0$ modulo q (the "First Case" of Fermat's Last Theorem).

At this writing, it is unknown whether or not there are infinitely many Sophie Germain primes, although there is a conjectured estimate that the number of Sophie Germain primes $<m$ is approximately

$$\frac{1.32m}{(\ln m)^2}.$$

See Ribenboim (1980), p. 328.

Factoring Mersenne numbers. For p a prime number, $N = 2^p - 1$ is a Mersenne number. As we saw in section B, among the Mersenne numbers are the largest known prime numbers. However, most Mersenne numbers are composite.

To find a prime factor of a Mersenne number, it is helpful to apply Fermat's theorem in the following way. Recall that if q is a prime divisor of $N = 2^p - 1$, then

$$2^p \equiv 1 \pmod{N},$$

therefore

$$2^p \equiv 1 \pmod{q},$$

and so, since p is prime, p must be the order of 2 modulo q. Now

$$2^{q-1} \equiv 1 \pmod{q}.$$

and therefore p must divide $q - 1$. This means that q has the form

$$q = 1 + mp$$

for some m.

Using this fact, we can adapt the Pollard $p - 1$ method to try to factor N as follows. Suppose there is a prime divisor q of N, necessarily of the form $q = 1 + pm$, where m is k-smooth for some small k. Let $B = B_k$ be the exponent corresponding to m and k in the Pollard $p - 1$ algorithm. That is, B is the smallest number divisible by all primes l^e where $l \le k$ and $l^e \le m$. Then m divides B, and so $q - 1 = pm$ divides pB. Thus, by Fermat's Theorem, for every $a < q$,

$$a^{pB} \equiv 1 \pmod{q}$$

and so

$$(a^{pB} - 1, N) \ge q.$$

(We need to choose $a \neq 2$, for

$$2^{pB} = (2^p)^B \equiv 1 \pmod{N}$$

since $N = 2^p - 1$, and so the Pollard $p - 1$ algorithm will fail).

Example 9. Let $N = 2^{163} - 1$, a number of 50 digits. We look for a prime factor $q = 1 + 163m$ where m is 9-smooth. Rather than choosing B to guarantee picking up any prime factor p so that $p - 1$ is 9-smooth, we choose one of the alternative smaller exponents noted in the remark following Example 8, above.

For the second alternative, we set $B = 9!$ and find that

$$(5^{163B} - 1, N) = 704161,$$

a factor of N. The reason 704161 showed up was that 704161 is prime and

$$704161 - 1 = 704160 = 163m$$

where

$$m = 2^5 \cdot 3^3 \cdot 5$$

is 9-smooth and divides B.

For the first alternative choice of B, we set $B = 2^3 \cdot 3 \cdot 5 \cdot 7$, but then $m = 2^5 \cdot 3^3 \cdot 5$ is not a divisor of B, and

$$(3^{163B} - 1, N) = 1,$$

so the algorithm would fail.

It turns out that $N = 2^{163} - 1$ is a product of five primes. For the other four primes, the corresponding m's are k-smooth for minimal $k = 461, 6037, 10061$ and 22392890561.

The GIMPS website gives a 27 digit prime factor q of

$$N = 2^{2944999} - 1,$$

found using the Pollard $p - 1$ algorithm. The Pollard $p - 1$ algorithm was successful in this case because $q - 1 = 2944999m$ where m is 69061-smooth.

Exercises.

27. Try factoring the following numbers by the Pollard $p - 1$ algorithm with $k = 3$:
   (i) 1679;
   (ii) 1739;
   (iii) 2231.

28. Try factoring the following numbers by the Pollard $p - 1$ algorithm with $k = 3$:
   (i) 44329;
   (ii) 42919;

29. Try factoring the following numbers by the Pollard $p - 1$ algorithm with $k = 3$:
(i) 202211
(ii) 218663
(iii) 222559

30. Suppose you apply the Pollard $p - 1$ algorithm to a number N. You pick a smoothness bound k and a base a, and find that for the a you chose,

$$a^B \equiv 1 \pmod{N}.$$

Does that imply that every prime factor p of N has the property that $p - 1$ is k-smooth? If so, explain why; if not, give an example.

31. If you try to factor 7081 by the Pollard $p - 1$ algorithm with $k = 3$, you will find that $B = 2^{12} \cdot 3^8 = 26873856$, and (choosing $a = 5$)

$$5^B \equiv 1 \pmod{7081}.$$

This means that 7081 divides $5^B - 1$. What to do? One strategy is to let $s = 3^8$ and find $5^s \pmod{7081}$. We find that

$$5^s \equiv 5704 \pmod{7081}.$$

Then we proceed by repeatedly squaring 5^s modulo 7081:

$$5^s \equiv 5704;$$
$$5^{2s} \equiv 5502;$$
$$5^{4s} \equiv 729;$$
$$5^{8s} \equiv 366;$$
$$5^{16s} \equiv 6498;$$
$$5^{32s} \equiv 1.$$

(We must get $5^{2^r s} \equiv 1$ for some $r \leq 12$).
  (i) Show that then 7081 cannot be prime because

$$7081 \text{ divides } (6498^2 - 1) = (6498 + 1)(6498 - 1).$$

but doesn't divide $6498 + 1$ or $6498 - 1$.
  (ii) Show that the greatest common divisor of $6498 - 1$ and 7081 is a non-trivial divisor of 7081.

32. The number 220459 is a product of two primes p and q such that $p - 1$ and $q - 1$ are both 7-smooth. Factor 220459 by a strategy similar to that of the last problem.

Chapter 11
Groups

In this chapter we reintroduce groups, and show that the mathematics involving orders, Fermat's and Euler's Theorems is part of elementary group theory.

A. Groups of Units and Euler's Theorem

Let us begin by reviewing the proof of Euler's theorem from Chapter 9.

Let $U = U_m$ be the set of units of $\mathbb{Z}/m\mathbb{Z}$, and let $\phi(m)$ be the number of elements of U. In proving that for a in U, $a^{\phi(m)} = 1$, we took the set U,

$$U = \{u_1, u_2, \ldots, u_{\phi(m)}\}$$

and multiplied each of the elements of U by the unit a to get the set

$$aU = \{au_1, au_2, \ldots, au_{\phi(m)}\}.$$

We found that the sets aU and U were the same, except for the ordering of the elements. Hence the product of the elements of U:

$$u_1 u_2 \cdots u_{\phi(m)}$$

and of aU:

$$au_1 au_2 \cdots au_{\phi(m)}.$$

are the same:

$$u_1 u_2 \cdots u_{\phi(m)} = au_1 au_2 \cdots au_{\phi(m)}.$$

Canceling the common factor $u_1 u_2 \cdots u_{\phi(m)}$ from both sides gives Euler's theorem: $a^{\phi(m)} = 1$.

How did we show that the set aU and the set U are the same? We needed to show for every a in U:

L.N. Childs, *A Concrete Introduction to Higher Algebra*, Undergraduate Texts in Mathematics, © Springer Science+Business Media LLC 2009

(1) The set aU is a subset of U: that is, for any units a and u, au is a unit; this is the property that U is *closed* under multiplication.

(2) All the elements of aU are different: that is, for any units u, v and a, if $au = av$, then $u = v$; call this the *cancellation property* for U.

(3) Every element of U is in aU: that is, for any unit u, there is a unit v so that $u = av$; call this the *solvability property* for U.

These properties, together with the property that multiplication of elements of U is associative, mean that U, a set with multiplication, is a group. We recall the definition from Section 7A:

Definition. A *group* is a set G together with an operation

$$* : G \times G \to G$$

(this means that for any two elements a, b of G, $a * b$ is in G) that satisfies the following properties:

associativity: for any three elements a, b, c of G, $a * (b * c) = (a * b)c$;

cancellation: for any three elements a, b, c of G, if $a * b = a * c$ or if $b * a = c * a$, then $b = c$;

solvability: for any elements a, b of G, there is an element c in G with $a * c = b$, and an element d in G with $d * a = b$;

existence of identity: there is an element e of G so that for all a in G, $e * a = a * e = a$;

and

existence of inverses: for any a in G, there is an element b in G so that $a * b = b * a = e$.

It is easy to see that the set U_m of units of $\mathbb{Z}/m\mathbb{Z}$ is a group, where the operation $*$ is multiplication; in fact U_m also satisfies the property

commutativity: for all a, b in U_m, $a * b = b * a$.

A group satisfying the commutative law is called an *abelian group*, after the Norwegian mathematician N. H. Abel (1802–1829).

The properties just listed which characterize a group are redundant: a set G with an associative multiplication is a group if it either has the identity and inverse properties, or has the cancellation and solvability properties.

Nowadays a group is customarily defined as a set with an associative operation that has the identity and inverse properties, as in Section 7A. In 1893, the standard definition of a group was that of a set with an associative operation that satisfies the solvability and cancellation properties (see Van der Waerden (1985), p. 154). The two definitions are equivalent. The proof of equivalence is not difficult and is left as an exercise.

Among the many examples of groups are two collections of examples arising from rings:

A ring R itself is an abelian group under addition. The identity element e is 0. When we refer to R as just a group under addition, rather than as a ring, we sometimes call it "the additive group of R".

The set of units of a ring R is a group under (i.e., with the operation of) multiplication. If R is a commutative ring, then the set of units of R is an abelian group. Here the identity element e is 1.

Euler's Theorem is a theorem about the group of units of $\mathbb{Z}/m\mathbb{Z}$. The proof of Euler's theorem above is really the proof of a theorem about abelian groups with a finite number of elements. To see this, we reprove Euler's theorem in that setting. In what follows we will write the operation on the group as multiplication, and often omit the $\cdot$, so that ba means $b \cdot a$.

Theorem 1 (Abstract Fermat Theorem). *Let G be an abelian group with n elements. Then for any a in G, $a^n = e$, the identity element of G.*

Here a^n denotes a multiplied by itself n times: $a \cdot a \cdot \ldots \cdot a$.

Proof. Let $u_1, u_2, \ldots, u_n$ be the elements of G, and let a be any element of G. Consider the set

$$aG = \{au_1, au_2, \ldots, au_n\}.$$

Then aG is exactly the same as the set G: all the elements of G are distinct, by the cancellation property, and every element of G is in aG by the solvability property. Thus $aG = G$.

To finish the proof, we simply multiply all the elements of aG together, and all the elements of G together, equate the two products, and cancel the common factors. We'll be left with $a^n = e$.

But if we are to be careful, we need to know that these manipulations of the elements of G work. For this, we need two consequences of the axioms:

Generalized Associativity. If G is a group, so that $a(bc) = (ab)c$ for all a, b, c in G, then for every $n > 3$ all possible ways of associating the product of every n elements of G are equal.

For example, with $n = 5$, $(a(bc))(de) = ((ab)(cd))e = a(b(c(de))) = \ldots$.

Generalized associativity means that when we see a product $abcde$ we are free to associate it in any way we want. The resulting product will not depend on how we did it. For this reason we can without confusion omit parentheses entirely.

Generalized Commutativity. If G is an abelian group, so that $ab = ba$ for all a, b in G, then for every $n > 2$, all possible ways of multiplying n elements $a_1, \ldots, a_n$ of G, regardless of order, give the same element of G.

For example, $abcde = edcba = acedb = \ldots$ (where we have omitted parentheses by generalized associativity).

The proofs of Generalized Associativity and Generalized Commutativity can be done by induction: see exercises 4 and 5, below.

The properties of Generalized Associativity and Generalized Commutativity permit the manipulations in the remainder of the proof of the abstract Fermat theorem. Since the set

$$G = \{u_1, \ldots, u_n\}$$

is the same as the set

$$aG = \{au_1, \ldots, au_n\},$$

the products of all the elements in each of the two sets are the same (by Generalized Commutativity):

$$u_1 u_2 \cdot \ldots \cdot u_n = au_1 au_2 \cdot \ldots \cdot au_n$$

(where we can omit parentheses by Generalized Associativity). We rearrange the right side, using Generalized Commutativity:

$$au_1 au_2 \cdot \ldots \cdot au_n = (aa \cdot \ldots \cdot a)(u_1 u_2 \cdot \ldots \cdot u_n)$$

so

$$u_1 u_2 \cdot \ldots \cdot u_n = (aa \cdot \ldots \cdot a)(u_1 u_2 \cdot \ldots \cdot u_n)$$

By cancellation,

$$e = aa \cdots a = a^n,$$

which was to be proved.   □

This abstract Fermat theorem is an abstraction of Euler's, and in turn, Fermat's theorem. If G is the group U_m of units of $\mathbb{Z}/m\mathbb{Z}$ for m any number ≥ 2, we have Euler's theorem. If G is the group U_p of units of $\mathbb{Z}/p\mathbb{Z}$ for p a prime, we have the original Fermat theorem.

The abstract Fermat theorem also holds for any finite group of units of any commutative ring. Other than $\mathbb{Z}/m\mathbb{Z}$ the only example we've seen so far is $\mathbb{F}_9$, the set of elements of the form $a + bi$, where a, b are in $\mathbb{Z}/3\mathbb{Z}$ and $i^2 = -1$ (see Section 7C, Exercise 35). Later in the book we will introduce many more examples of finite commutative rings–see Chapter 23.

Exercises.

1. Prove that if G is a set with an associative operation that has the identity and inverse properties, then it has the cancellation and solvability properties.

2. Prove that if G is a set with an associative operation that has the solvability and cancellation properties, than it has the identity and inverse properties.

3. Write down all possible ways of associating the product $abcd$, where a, b, c, d are elements of a group G, and show, using the associative law, that the products are all equal.

4. Let $a_1, a_2, \ldots, a_n$ be elements of a group G. Prove generalized associativity for $a_1, a_2, \ldots, a_n$ by showing that every product u of $a_1, a_2, \ldots, a_n$, associated in any way, equals

$$a_1(a_2(a_3(\ldots(a_{n-1}u_n)\ldots))).$$

(Hint: Assume by induction that the result is true for any product of $n-1$ elements of G).

5. Let G be an abelian group, assuming generalized associativity, prove generalized commutativity for G by induction on n, by showing that for every set of n elements $\{a_1, a_2, \ldots, a_n\}$ of G, every product of $a_1, \ldots, a_n$ in any order is equal to $a_1 \cdot a_2 \cdots a_n$.

6. Prove that in a group, the identity element is unique.

7. Using cancellation, prove that in a group, every element has a unique inverse.

8. Define an operation on the set $\mathbb{N}$ of natural numbers (>0) by $a*b = [a,b]$, the least common multiple of a and b.
    (i) Show that this operation is associative and commutative.
    (ii) Find an identity element for $\mathbb{N}$ under this operation.
    (iii) Which elements of $\mathbb{N}$ have inverses?
    (iv) When is it possible to solve the equation $a*x = b$?
    (v) If $a*b = a*c$, does it follow that $b = c$?

9. In $\mathbb{F}_9$, show that $(1-i)$ has order 8, and so every unit of $\mathbb{F}_9$ is a power of $(1-i)$. Why does this imply that every unit has order dividing 8?

B. Subgroups

Let G be a group with operation $*$ and identity element e, and where the inverse of an element a is denoted a'. A *subgroup* H of G is a nonempty subset of G with two properties:
    (i) if a,b are in H, then $a*b$ is in H; and
    (ii) if a is in H, so is a'.
    In other words, H is a subset of G which is closed under products and inverses.
    Here are some examples of groups and subgroups that you have perhaps already seen:
    1. Let $G = \mathbb{Z}$, the integers, where the operation $*$ is $+$, the identity element is 0, and the inverse of an element is its negation. Let m be any nonnegative integer, and let $H = m\mathbb{Z}$, the set of all multiples of m. Then H is a subgroup of G. For it is easy to see that if a,b are two multiples of m, then so is $a+b$, and if a is a multiple of m, so is $-a$. (Note that H is the congruence class (mod m) of zero).

    2. Let $G = \mathbb{Z}$ again, and let $H = \mathbb{N}$, the set of natural numbers. Then H is not a subgroup of G. For while the sum of two natural numbers is a natural number, the negation of a natural number is not a natural number, so is not in $\mathbb{N}$. Since $\mathbb{N}$ is not closed under negations, $\mathbb{N}$ is not a subgroup of G.

3. Let G be the group under multiplication of the non-zero complex numbers. For each number k, let $U(k)$ be the group

$$U(k) = \{\alpha \in \mathbb{C} | \alpha^k = 1\}.$$

The subgroup $U(k)$ is the group of *kth roots of unity* in $\mathbb{C}$, and is a subgroup of G. For example,

$$U(2) = \{1, -1\};$$
$$U(4) = \{1, i, -1, -i\};$$
$$U(3) = \{\frac{-1 + \sqrt{-3}}{2}, \frac{-1 + \sqrt{-3}}{2}, 1\};$$
$$U(8) = \{1, i, -1, -i, \frac{1 + i}{\sqrt{2}}, \frac{-1 + i}{\sqrt{2}}, \frac{1 - i}{\sqrt{2}}, \frac{-1 - i}{\sqrt{2}}\}.$$

Then $U(2)$ is a subgroup of $U(4)$, which in turn is a subgroup of $U(8)$.

4. Let G be set of nonzero real numbers with the operation being multiplication. Let $H = U(2)$ be the subset consisting of 1 and -1. Then H is a subgroup of G.

5. Let $G = U_m$, the group of units of $\mathbb{Z}/m\mathbb{Z}$ under multiplication. For any k, let

$$U_m(k) = \{\alpha \in U_m | \alpha^k = 1\}.$$

Then $U_m(k)$ is a subgroup of U_m, the group of *kth roots of unity* in U_m. By Euler's Theorem, $U_m(\phi(m)) = U_m$: every unit is a $\phi(m)$th root of unity in U_m.

6. If G is a vector space over some field, then G is an abelian group under +. Any subspace H of G is a subgroup of G.

7. Two "trivial" subgroups of any group G are the group G itself, and the subgroup consisting only of the identity element of G. So a "non-trivial" subgroup means a subgroup other than the trivial subgroups.

Definition. Let G be a group with operation $*$, identity e and inverse $^{-1}$. Fix an element a of G. The *cyclic subgroup generated by* a is the set H of elements of G of the form a^n for all integers n. Here a^0 denotes the identity element e, a^n for $n > 0$ denotes $a * a * \cdots * a$ (n factors), and a^{-n} for $n > 0$ denotes $a^{-1} * a^{-1} * \ldots a^{-1}$ (n factors).

The cyclic subgroup of G generated by a is denoted by $\langle a \rangle$.

A group G is *cyclic* if $G = \langle b \rangle$ for some b in G.

It is easy to see that for a in G, the set $\langle a \rangle$ is closed under products and inverses and is a subgroup of G.

For example, if $G = \mathbb{Z}$, with $*$ being $+$, then the cyclic subgroup $\langle m \rangle$ generated by the integer m is $m\mathbb{Z}$, the set of all integers rm, where r is any element of $\mathbb{Z}$. This is because any integer in $\langle m \rangle$ is obtained by adding either m to itself or $-m$ to itself. For example, if $s > 0$, then $sm = m + m + \ldots + m$ (s terms). This example is the same as example 1, above.

The cyclic subgroup $\langle 1 \rangle$ of $\mathbb{Z}$ generated by 1 is all of $\mathbb{Z}$. So $\mathbb{Z}$ is a cyclic group.

For G a finite group (with operation $*$ and identity e), the cyclic subgroup generated by an element a is just the set of all positive powers of a.

To see this, recall that the *order* of an element a of G, if it exists, is the smallest exponent $n > 0$ so that $a^n = e$, the identity of G.

Proposition 2. *Suppose G is a finite group with n elements. Every element a of G has an order, and the order d of a is $\leq n$. If a has order d, then the cyclic subgroup $\langle a \rangle$ of G generated by a has d elements:*

$$\langle a \rangle = \{a, a^2, \ldots, a^d\}.$$

Hence the order of a is equal to the number of elements in $\langle a \rangle$.

Proof. The proof that the order of a is $\leq n$ is an argument that we gave in Section 9A. In brief, the elements $a, a^2, \ldots, a^{n+1}$ cannot all be different since G has only n elements. Hence $a^r = a^{r+t}$ for some r, t with $1 \leq r < r + t \leq n + 1$, hence $t \leq n$. Cancelling a^r yields $a^t = e$. Since $t \leq n$, the order d of a must be $\leq n$.

Let d be the order of a, and let

$$A = \{a, a^2, \ldots, a^{d-1}, a^d\},$$

where $a^d = a^0 = e$. For every $l > 0$, $l = dq + r$ with $0 \leq r < d$. Then

$$a^l = a^{dq}a^r = e^q a^r = a^r.$$

So A contains every positive power of a. In particular, A is closed under multiplication. Also, for each r with $1 \leq r < d$, $a^r a^{d-r} = a^d = e$. So A is closed under taking inverses. So A is a subgroup of G containing a.

Since A is closed under products and inverses, A contains every positive or negative power of a, so $\langle a \rangle = A = \{a, a^2, \ldots, a^d\}$.

Since d is the order of a, we are left only with showing that the elements $a, a^2, a^3, \ldots, a^d$ of $\langle a \rangle$ are all different. So suppose $a^s = a^{s+k}$ where $1 \leq s < s + k \leq d$. Then, cancelling a^s, we have $e = a^k$. But $1 \leq k < d$. Hence this last equation violates the assumption that d is the order of a. Thus all the elements in the set $\langle a \rangle$ are different. That is, the number of elements in the subgroup $\langle a \rangle$ generated by a is equal to the order of a, as we wished to show. $\square$

Example 1. If $G = U_{13}$, the group of units of $\mathbb{Z}/13\mathbb{Z}$, then the cyclic subgroup $\langle [3] \rangle$ generated by $[3]$ has three elements: $[3], [9]$ and $[27] = [1]$. The cyclic subgroup $\langle [5] \rangle$ generated by $[5]$ has four elements: $[5], [25] = [-1], [-5] = [8]$ and $[1]$.

If $\langle a \rangle$ is a cyclic subgroup of G of order $m = rs$, then

$$\langle a^r \rangle = \{a^r, a^{2r}, \ldots, a^{sr}\}$$

is a cyclic subgroup of $\langle a \rangle$, hence a cyclic subgroup of G, of order s.

Example 2. If $G = U_{13}$, then $\langle [2] \rangle$ has order 12, so is all of U_{13}. Then, for example,

$$\langle [2^3] \rangle = \{[2^3] = [8], [2^6] = [64] = [-1], [2^9] = [-8], [2^{12}] = [1]\},$$

and

$$\langle [2^4] \rangle = \langle [3] \rangle.$$

Most groups have subgroups other than the trivial subgroups:

Proposition 3. *If G is an abelian group then G has a non-trivial subgroup unless the order n of G is 1 or a prime p.*

Proof. For each $a \neq e$ in G, consider the cyclic subgroup $\langle a \rangle$ generated by a. If $\langle a \rangle \neq G$, then $\langle a \rangle$ is a non-trivial subgroup of G. If $\langle a \rangle = G$ and $n = rs$ with $r, s > 1$, then $\langle a^r \rangle$ has order s, so is a proper subgroup of G. Thus if G has no non-trivial proper subgroups, then the order of G is 1 or a prime. $\qquad\square$

We defined a subgroup of a group G to be a subset of G that is closed under the operation and also under inverses. In case G is a finite group, we may omit this last condition:

Proposition 4. *Let G be a group with operation $*$. If G is finite, then a non-empty subset H of G is a subgroup of G if and only if H is closed under $*$.*

The proof is essentially included in the proof of Proposition 2 so is left as an exercise.

Exercises.

10. Prove Proposition 4.

11. Let $G = U_{19}$.
   (i) Find the cyclic subgroup of G generated by $[7]$;
   (ii) Find the cyclic subgroup of G generated by $[12]$;
   (iii) Find the cyclic subgroup of G generated by $[8]$.

12. Let $G = U_{21}$.
   (i) Find the cyclic subgroup of G generated by $[10]$;
   (ii) Find the cyclic subgroup of G generated by $[2]$;
   (iii) Find the cyclic subgroup of G generated by $[8]$.

13. In $G = U_{16}$ find a subgroup H with 4 elements that is not the cyclic subgroup generated by some element of G.

14. Let $G = \mathbb{Z}/n\mathbb{Z}$ with operation $+$. Show that the cyclic subgroup H generated by $[b]_n$ in $\mathbb{Z}/n\mathbb{Z}$ is all of G if and only if $(b, n) = 1$.

C. Cosets and Lagrange's Theorem

In Section A we proved the abstract Fermat theorem: if G is an abelian group with n elements, then for any element a of G, the order of a divides n. In Proposition 2 of the last section, we showed that if a is any element of G, then the number of elements in the subgroup $\langle a \rangle$ generated by a is equal to the order of a, and hence the number of elements of $\langle a \rangle$ divides n.

In this section we will generalize this result to show that if G is a finite group and H is any subgroup of G, then the number of elements of H is a divisor of the number of elements of G. This famous result is called Lagrange's theorem. Euler's and Fermat's theorems are easy consequences of Lagrange's theorem.

The proof we will give will work for any finite group, abelian or not. (The final section of this chapter gives some examples of nonabelian groups.)

In order to prove Lagrange's theorem, we need to generalize the notion of congruence classes.

Definition. Let G be a group with operation $*$, and H a subgroup. For any b in G, the *left coset of b*, denoted $b * H$, is the set of elements $b * h$, where h runs through all elements of H. In symbols,

$$b * H = \{b * h \mid h \text{ in } H\}.$$

Example 3. Let $G = \mathbb{Z}$ (the operation is +), $H = 2\mathbb{Z}$. Then the coset $1 + 2\mathbb{Z}$ is the set of integers of the form $1 + 2k$ where k runs through all elements of $\mathbb{Z}$. Thus $1 + 2\mathbb{Z}$ is the set of all integers congruent to 1 (mod 2) (the odd integers). We have called that set the congruence class of 1 (mod 2) and called it $[1]_2$. Similarly, the coset $0 + 2\mathbb{Z}$ is just the set of elements in the subgroup $2\mathbb{Z}$, that is, the set of multiples of 2 (the even integers), which we called $[0]_2$.

Any integer is either even or odd, so is either in $0 + 2\mathbb{Z}$ or in $1 + 2\mathbb{Z}$. So there are two cosets of the subgroup $2\mathbb{Z}$ in $\mathbb{Z}$, every integer is in one of the two cosets, and the cosets have no elements in common (no integer is both even and odd).

Example 4. More generally, let $G = \mathbb{Z}$, $H = m\mathbb{Z}$ for some $m > 1$, the modulus. If a is any integer, then $a + m\mathbb{Z}$, the coset of a, is the set of integers of the form $a + mk$ for k any integer, that is, the set of integers congruent to a (mod m), i.e., the congruence class $[a]_m$. Then the coset $a + m\mathbb{Z}$ is equal to the coset $b + m\mathbb{Z}$ iff a is congruent to b (mod m). There are m cosets, namely, $0 + m\mathbb{Z}, 1 + m\mathbb{Z}, 2 + m\mathbb{Z}, \ldots, (m-1) + m\mathbb{Z}$. This is because any integer is congruent (mod m) to (exactly) one of the numbers $0, 1, 2, \ldots, m-1$.

Just as with congruence classes, we can prove, quite generally, that

Proposition 5. *Let H be a subgroup of a group G. Two left cosets are either disjoint or equal.*

Proof. (See also Exercise 17.) Write the group operation as $*$. Suppose $a * H$ and $b * H$ have some element in common. Let c be such an element. Then $c = a * h = b * k$ for some h, k in H.

We show $a * H$ is contained in $b * H$. We know $a * h = b * k$ is in $b * H$. Let $a * h'$ be any element of $a * H$. Then we can find some t in H so that $h * t = h'$, since H is a group. But then $a * h' = a * h * t = b * k * t$, an element of $b * H$. So every element of $a * H$ is in $b * H$.

The same argument shows that $b * H$ is contained in $a * H$. Thus if $a * H$ and $b * H$ have an element in common, they are equal. $\square$

For counting, the following proposition is quite useful:

Proposition 6. *If $a * H$ is any coset of H, then the number of elements in $a * H$ is equal to the number of elements in H.*

Proof. To show that two sets have the same number of elements ("the same cardinality"), the idea is to define a one-to-one, onto function ("a one-to-one correspondence") from one set to the other.

To prove the proposition, define a function T from H to $a * H$ by the rule, $T(h) = a * h$. Thus T is the function, "operate on the left by a." Then T is a one-to-one correspondence between H and $a * H$. To see this most easily, observe that we can define an inverse function S from $a * H$ to H, namely, "operate on the left by a^{-1}." Then

$$S(a * h) = a^{-1} * (a * h) = (a^{-1} * a) * h = e * h = h.$$

So the composition $S \circ T$ is the identity function on H. Similarly, $T \circ S$ is the identity function on $a * H$. So T and S define a one-to-one correspondence between H and $a * H$. That proves the proposition. $\square$

Now we can prove Lagrange's theorem.

Theorem 7 (Lagrange's Theorem). *Let G be a finite group, H a subgroup of G. Then the number of elements of H divides the number of elements of G.*

Proof. Let G have n elements, and H have r elements. Write G as a disjoint union of left cosets:

$$G = (a_1 * H) \cup (a_2 * H) \cup \ldots \cup (a_s * H).$$

We can do this as follows: every b in G is in the left coset $b * H$. So we let $b_1, b_2, \ldots, b_g$ be the elements of G, then

$$G = (b_1 * H) \cup (b_2 * II) \cup \ldots \cup (b_g * H).$$

Unless H contains only one element, there will be cosets in this union that are equal. So starting with the coset $b_2 * H$, look at each coset $b_{k+1} * H$ to see if it has an element in common with one of the earlier cosets $b_1 * H, \ldots, b_k * H$. If so, then $b_{k+1} * H$ is equal to the coset it has an element in common with. So toss $b_{k+1} * H$ out. Once we toss out all the duplicates, we're left with G as the disjoint union of the remaining cosets. Call the non-duplicative cosets $a_1 * H, a_2 * H, \ldots, a_s * H$. Then

$$G = (a_1 * H) \cup (a_2 * H) \cup \ldots \cup (a_s * H).$$

Now we count the elements of G.

We see that n, the number of elements of G, is equal to the number of elements in the coset $a_1 * H$ plus the number of elements of $a_2 * H$ plus ... plus the number of elements of $a_s * H$. But Proposition 6 tells us that every coset in the disjoint union has the same number of elements, namely m, the number of elements in H. Thus if G has n elements and s cosets, then $n = ms$. To state this formula in words, the number of elements in G is equal to the number of elements in H times the number of left cosets of H in G.

That completes the proof of Lagrange's theorem. □

Corollary 8. *For every element b of a finite group G, the order of b divides the number of elements of G.*

Proof. Let $H = \langle b \rangle$ be the subgroup of G generated by b. Then the order of b is the number of elements of H by Proposition 2. The corollary then follows immediately from Lagrange's theorem. □

Corollary 9. *Euler's theorem.*

Proof. Let $G = U_m$, the group (under multiplication) of units of $\mathbb{Z}/m\mathbb{Z}$, and let a be any number coprime to m. Then the order d of $[a]_m$ is the number of elements of the subgroup $\langle a \rangle$ of U_m. Hence d divides the number of elements of U_m, namely $\phi(m)$. If $\phi(m) = ds$ for some number s, then $[a]^{\phi(m)} = [a]^{ds} = [1]^s = [1]$; hence, in congruence notation, $a^{\phi(m)} \equiv 1 \pmod{m}$. □

The usual terminology is that the number of elements in a finite group G is called the *order* of G. Lagrange's theorem says that if H is a subgroup of a finite group G, then the order of H divides the order of G. The number of cosets of H in G is called the *index* of H in G. Thus:

$$(\text{order of } H) \times (\text{index of } H \text{ in } G) = (\text{order of } G).$$

If a is an element of G, then the order of a = the order of the subgroup $\langle a \rangle$ generated by a, by Proposition 2, Section B. Thus the two notions of order, for an element and for a group, are compatible.

Exercises.

15. Let $G = U_{19}$. For each of the following subgroups, write down the cosets of the subgroup, and verify Lagrange's theorem in each case.
  (i) $\langle [7] \rangle$;
  (ii) $\langle [12] \rangle$;
  (iii) $\langle [5] \rangle$.

16. Let $G = U_{21}$. Does it make sense to write down the coset of the subgroup of G generated by $[7]$?

17. Prove that if H is a subgroup of G, then the relation,

$$a \sim b \text{ if } a * H = b * H$$

is an equivalence relation. Why does this imply Proposition 5?

18. (i) Let U be the group of units of $\mathbb{Z}/m\mathbb{Z}$. Then U is a subgroup of itself. For every unit $[a]$ of U, show that the coset $[a]U$ is equal to U.

(ii) Show that (when m is prime) the first half of the proof of Fermat's Theorem in Section 9B consists of verifying the statement of part i).

D. A Probabilistic Primality Test

The proof of Lagrange's theorem yields information on Fermat's theorem as a primality test.

Let m be a natural number >2, and let U_m be the group of units of $\mathbb{Z}/m\mathbb{Z}$. Then U_m is an abelian group containing $\phi(m)$ elements. Let

$$U_m(m-1) = \{[a] \text{ in } U_m \mid [a]^{m-1} = [1]\},$$

the group of $(m-1)$-st roots of unity in $\mathbb{Z}/m\mathbb{Z}$. Then

$$U_m(m-1) = \{[a] \text{ in } U_m \mid m \text{ passes the } a\text{-pseudoprime test}\}.$$

Proposition 10. $U_m(m-1)$ *is a subgroup of* U_m.

Proof. By Proposition 3 it suffices to show that if $[a]$ and $[b]$ are in $U_m(m-1)$, so is $[a][b] = [ab]$. But if $[a]^{m-1} = [1]$ and $[b]^{m-1} = [1]$, then

$$[ab]^{m-1} = ([a][b])^{m-1} = [a]^{m-1}[b]^{m-1} = [1].$$

$\square$

Since $U_m(m-1)$ is a subgroup of U_m either $U_m(m-1) = U_m$ or $U_m(m-1) \neq U_m$ (obviously). The case $U_m(m-1) = U_m$ always occurs if m is prime, by Fermat's theorem. A Carmichael number is a composite number m for which $U_m(m-1) = U_m$ (see Section 10B).

If $U_m(m-1) \neq U_m$ and f is the number of elements of $U_m(m-1)$, then among the $s = \phi(m)/f$ cosets of $U_m(m-1)$ in U_m only the f elements in the coset $U_m(m-1)$ itself satisfy $[a]^{m-1} = [1]$, while $(s-1)f$ elements $[a]$ of U_m do not satisfy $[a]^{m-1} = [1]$. Since $U_m(m-1) \neq U_m$, there are at least two cosets, so $s \geq 2$ and $(s-1)f \geq f$. Thus we have

Proposition 11. *If m is not prime and not a Carmichael number, then m will pass the a-pseudoprime test for at most half of the numbers a, $1 \leq a \leq m$.*

This fact has practical significance for testing a number for primeness. Suppose we have a number m which we wish to test for primeness. Pick, say, 20 numbers a, $1 < a < m$, at random, and subject m to the a-pseudoprime test for each a.

- If m is prime, m will pass all of the a-pseudoprime tests.
- If m is Carmichael and all a are chosen coprime to m, then m will pass all of the a-pseudoprime tests.
- If m is composite and not Carmichael, then the chance that m passes the a-pseudoprime test for any single randomly chosen a is at most $1/2$. So the chance that m passes the a-pseudoprime test for all 20 randomly chosen numbers a is less than $1/2^{20}$, or less than one in a million.

So, provided we are not so unlucky to have selected a Carmichael number (or the use we have for m requires only that m be prime or Carmichael), this is a good probabilistic primality test, in the sense that we have less than one chance in a million of being wrong if a number passes our 20 a-pseudoprime tests and we conclude that m is prime or Carmichael.

In fact, for most composite numbers, the index s of $U_m(m-1)$ in U_m is much greater than 2. To see this, we have

Proposition 12. *If $d = (m-1, \phi(m))$, then $U_m(m-1) = U_m(d)$.*

The proof is an application of Bezout's Identity and is left as an exercise.

Example 5. Let $m = 51$. Then $\phi(51) = \phi(17)\phi(3) = 16 \cdot 2 = 32$, and $(50, 32) = 2$. So $U_{51}(50) = U_{51}(2)$: every unit mod 51 whose 50-th power equals 1 has its square equal to 1. The only elements of order <2 in U_{51} are $[1], [-1], [16]$ and $[-16]$. So $U_{51}(50)$ has order 4 and the index of $U_{51}(50)$ in U_{51} is 8.

In Chapter 20 we will give a strengthened version of the a-pseudoprime test that with high probability will distinguish between primes and all composite numbers, even Carmichael numbers.

Exercises.

19. Show that $U_{21}(20) = U_{21}(2) = \{[1], [-1], [8], [-8]\}$.

20. Determine the order f of $U_m(m-1)$ and compare it with $\phi(m)$ if
(i) $m = 9$;
(ii) $m = 20$;
(iii) $m = 25$;
(iv) $m = 75$.

21. In U_m, suppose $[b]^{m-1} = [c]$. Show that every element $[a]$ in the coset $[b]U_m$ satisfies $[a]^{m-1} = [c]$.

22. (i) Show that $(50, 32) = 2$ and that $50 \cdot 41 = 2 + 32 \cdot 64$.

   (ii) Show that if $a^{50} \equiv 1 \pmod{51}$, then $a^2 \equiv 1 \pmod{51}$, and conversely.

   (iii) Show that $U_{51}(50) = U_{51}(2)$ (c.f. Example 5).

23. Prove Proposition 12.

24. Suppose n is a number so that $n - 1$ is prime (e.g., $n = 30, 60, 48$). Show that n is an a-pseudoprime for a coprime to n, iff $a = 1 \pmod{n}$.

25. Let p be an odd prime so that $m = 2p + 1$ is the product of two primes (e.g., $p = 19, 43, 47$). Show that

$$U_m(m - 1) = \{[a] \text{ in } U_m \mid [a]^2 = 1\} = U_m(2).$$

It follows that $U_m(m - 1)$ has four elements (see Section 12A, Exercise 33).

E. Cosets and Equations

In this section we look at cosets that arise in solving equations.

   In Section 6F we studied the question of finding all solutions of the equation

$$[a]_m X = [b]_m$$

in $\mathbb{Z}/m\mathbb{Z}$. If the greatest common divisor of a and m divides b, then we can find a solution $X = [x_0]_m$ (a "particular solution"), for example, by reformulating the problem to that of finding an integer solution of $ax + my = b$, solvable by the extended Euclidean algorithm. Then all other solutions are obtained by adding to that particular solution the solutions of the corresponding homogeneous equation

$$[a]_m X = [0]_m.$$

The general solution to the homogeneous equation is

$$X = [\frac{m}{(a, m)} k]_m$$

for $k = 1, \ldots, (a, m)$. So the general solution to the original, non-homogeneous equation is

$$X = [x_0 + \frac{m}{(a, m)} k]_m, \quad k = 1, \ldots, (a, m).$$

Here is a generalization of this result for any abelian group:

Proposition 13. *Let G be an abelian group with operation multiplication and identity* 1. *Let*

$$G(n) = \{h \in G \mid h^n = 1\},$$

the set of solutions in G of the equation $x^n = 1$. Let c be in G. If there is some b in G so that $b^n = c$, then the set of solutions to the equation $x^n = c$ is the coset

$$bG(n) = \{bh | h \text{ in } G(n)\}.$$

Since 1 is the identity element of G, the equation $x^n = 1$ is "homogeneous". If $c \neq 1$, then $x^n = c$ is "non-homogeneous". The notion of homogeneous can be characterized by the idea that the set of solutions of a homogeneous equation is a subgroup of G.

Proof. If $b^n = c$, then for all h in $G(n)$, $(bh)^n = b^n h^n = c \cdot 1 = c$. So every element of $bG(n)$ is a solution of the equation $x^n = c$. Conversely, if $s^n = c$ for some s in G, then

$$(b^{-1}s)^n = b^{-n}s^n = c^{-1}c = 1,$$

so $b^{-1}s = h$ is in $G(n)$. Thus $s = bh$ is in $bG(n)$. □

Example 6. Let $G = U_{29}$, let

$$G(7) = \{a \in U_{29} | a^7 = 1\}.$$

Since $6^7 \equiv -1 \pmod{29}$, we have

$$\{b \in U_{29} | b^7 = -1\} = [6]_{29} G(7).$$

Example 7. Let $G = \mathbb{R}^2$, real vectors in the plane, a group under addition of vectors, and let

$$L = \{(x,y) \text{ in } \mathbb{R}^2 | 3x + 2y = 8\}.$$

Let

$$N = \{(x,y) \text{ in } \mathbb{R}^2 | 3x + 2y = 0\},$$

the set of solutions of the corresponding homogeneous equation. Then N is a subspace of $\mathbb{R}^2$, hence is a subgroup of $\mathbb{R}^2$. Let $\bar{r} = (2,1)$, then $\bar{r}$ is in L since $3 \cdot 2 + 2 \cdot 1 = 8$. The set of all solutions to $3x + 2y = 8$ is then the coset $L = \bar{r} + N$. For if (u,v) is in N, then $3u + 2v = 0$, so

$$\bar{r} + (u,v) = (2+u, 1+v)$$

satisfies

$$3(2+u) + 2(1+v) = (6+2) + (3u+2v) = 8+0 = 8.$$

Geometrically, L is the line through the point $(2,1)$ with slope $m = -3/2$, and N is the line through the origin with slope $-3/2$, hence is parallel to L.

This example generalizes greatly:

Example 8. Let A be any $m \times n$ matrix, then the set of vectors $\bar{v}$ with $A\bar{v} = 0$ is the null space N of A, a subgroup of $\mathbb{R}^n$. For any vector $\bar{b}$ in $\mathbb{R}^m$, if $\bar{u}$ is some vector in $\mathbb{R}^n$ such that $A\bar{u} = \bar{b}$, then the set of all solutions of $A\bar{x} = \bar{b}$ is the coset $\bar{u} + N$.

This last example is why we chose the letter N for the solutions to the homogeneous equation in the examples above.

Example 9. In U_{17}, the group of units of $\mathbb{Z}/17\mathbb{Z}$, let

$$
\begin{aligned}
U_{17}(8) &= \{[a] \in U_{17} | [a]^8 = [1]\} \\
&= \{[1],[2],[4],[8],[9],[13],[15],[16]\} \\
&= \langle[2]\rangle.
\end{aligned}
$$

Then $[3]^8 = [81]^2 = [13]^2 = [169] = [-1]$, so

$$\{[a] \in U_{17} | [a]^8 = [-1]\}$$

is the coset

$$[3]U_{17}(8) = \{[3],[5],[6],[7],[10],[11],[12],[14]\}.$$

Note that

$$U_{17} = U_{17}(8) \cup [3]U_{17}(8),$$

the disjoint union of the cyclic subgroup $\langle 2 \rangle$ generated by $[2]$ and the coset $[3]\langle[2]\rangle$. So every $[a]$ in U_{17} satisfies either $x^8 = [1]$ or $x^8 = [-1]$.

Example 10. In U_{21}, the group of square roots of 1 is

$$U_{21}(2) = \{[1],[-1],[8],[-8]\}.$$

Now $2^2 = 4$, so the solutions in U_{21} of $x^2 = 4$ is the coset

$$[2]U_{21}(2) = \{[2],[-2],[5],[-5]\}.$$

Similarly, $4^2 = 16$, and so

$$\{[a] \in U_{21} | [a]^2 = [16]\} = [4]U_{21}(2) = \{[4].[-4],[10].[-10]\}.$$

Note that the cosets $U_{21}(2), [2]U_{21}(2)$ and $[4]U_{21}(2)$ are disjoint and their union is all of U_{21}. Thus the only squares in U_{21} are $[1],[4]$ and $[16]$.

Exercises.

26. The group of units U_{91} is the disjoint union of $U_{91}(90)$ and $[2]U_{91}(90)$, where

$$[2]U_{91}(90) = \{[a] \in U_{91} | [a]^{90} = [64]\}.$$

Assuming that fact, how many elements are there in $U_{91}(90)$?

27. (i) Find the four elements of $U_{35}(2)$.
(ii) Find the set of solutions of $x^2 \equiv 14 \pmod{35}$. Identify the set as a coset.

28. Let $p > 10$ and $q - p + 2$ be primes, and let $m - pq$.
(i) Show that $1, p+1, -1$ and $-(p+1)$ are the four solutions of $x^2 \equiv 1 \pmod{m}$.
(ii) Find the coset of $U_m(2)$ consisting of solutions to $x^2 \equiv 9 \pmod{m}$.

29. (i) Find the three 10th powers in U_{31}.
(ii) Show that $U_{31}(10) = \langle -2 \rangle$.
(iii) Show how the three cosets of $U_{31}(10)$ correspond to the three 10th powers in U_{31}.

30. Consider the error-correcting Code I from Section 8E. The set $\mathscr{C}$ of code words is the set of vectors C in $\mathbb{F}_2^7$ so that $HC = 0$. Let

$$W = \begin{pmatrix} 1 \\ 1 \\ 0 \end{pmatrix}$$

Show that if $HR = W$, then the set of vectors X in $\mathbb{F}_2^7$ so that $HX = W$ is the coset $R + \mathscr{C}$ in $\mathbb{F}_2^7$.

F. Homomorphisms

Let G and H be groups, where we denote the operation on G by $*$ and the identity by e_G, and similarly for H.

Definition. A function $f : G \to H$ with domain G and range H is called a *group homomorphism* if

$$f(g_1 * g_2) - f(g_1) * f(g_2) \text{ for all } g_1, g_2 \text{ in } G,$$

and

$$f(e_G) = e_H.$$

If $f : G \to H$ is a homomorphism, and if g in G has inverse g^{-1} so that $g * g^{-1} = e_G$, then $f(g^{-1}) = f(g)^{-1}$, the inverse of $f(g)$. This follows because

$$e_H = f(e_G) = f(g * g^{-1}) = f(g) * f(g^{-1}),$$

and the inverse of $f(g)$ is unique.

We have implicitly seen many examples of group homomorphisms.

Example 11. If R is a ring and we forget that R has a multiplication, then with the operation $+$ and the identity element 0, R is an abelian group, which we call the *additive group* of R.

If $f : R \to S$ is a ring homomorphism, then f is a group homomorphism from the additive group of R to the additive group of S.

Example 12. Also associated with a ring R is $U(R)$, the group of units, or invertible elements, of R. If $f : R \to S$ is a ring homomorphism, then restricting f to the subset $U(R)$ of R yields a group homomorphism from $U(R)$ to $U(S)$, because f takes the identity element 1 of R to the identity element of S, and $f(ab) = f(a)f(b)$.

Other examples of group homomorphisms:

Example 13. Let H be a subgroup of a group G, then the inclusion map $i : H \to G$, which takes an element of H and views it as in G, is a group homomorphism.

Example 14. If $\{e\}$ is the group with one element, then the only possible function from a group G to $\{e\}$ is a homomorphism, called the zero homomorphism. More generally, if G' is any group, and $f : G \to G'$ is the function which takes every element of G to the identity element of G', then f is a homomorphism, also called the zero homomorphism.

Example 15. Let G be the additive group of $\mathbb{Z}/m\mathbb{Z}$, and let α be an element of $\mathbb{Z}/m\mathbb{Z}$. Define a function $L_\alpha : \mathbb{Z}/m\mathbb{Z} \to \mathbb{Z}/m\mathbb{Z}$ by $L_\alpha(\beta) = \alpha\beta$. Then L_α is a group homomorphism, for $L_\alpha(0) = 0$, and $L_\alpha(\beta + \gamma) = L_\alpha(\beta) + L_\alpha(\gamma)$, by the distributive law. Note that L_α is not a ring homomorphism unless $\alpha = 1$, because $L_\alpha(\beta\gamma)$ is not equal to $L_\alpha(\beta)L_\alpha(\gamma)$. If $\alpha = 0$, then L_0 is the zero homomorphism.

Example 16. Let $G = U_m$, the group of units of $\mathbb{Z}/m\mathbb{Z}$. Then for every number r, the function

$$f_r : U_m \to U_m$$

defined by $f_r([a]) = [a]^r$ is a group homomorphism.

Example 17. A vector space is an abelian group under addition, and linear transformations are group homomorphisms.

A group homomorphism $f : G \to G'$ is *one-to-one* if it is one-to-one as a function. That is, for g, h in G, if $f(g) = f(h)$, then $g = h$. As with ring homomorphisms we can test for a homomorphism f to be one-to-one by looking at the *kernel* of f,

$$\ker f = \{g \text{ in } G \mid f(g) = e\}.$$

Proposition 14. *Let $f : G \to G'$ be a group homomorphism. Then $\ker f$ is a subgroup of G. The map f is one-to-one iff $\ker f = \{e\}$, the subgroup of G consisting of just the identity element of G.*

The proof is an exercise.

Example 18. Let $G = U_m$, the group of units of $\mathbb{Z}/m\mathbb{Z}$. Let $f_r : U_m \to U_m$ be the homomorphism defined by $f_r(\alpha) = \alpha^r$, as in Example 14. Then the kernel of f_r,

$$\ker f_r = \{\alpha \text{ in } U_m \mid \alpha^r = 1\} = U_m(r),$$

is the subgroup of r-th roots of unity in U_m. See Example 5 of Section B.

Definition. Let $f : G \to H$ be a group homomorphism. The *image* of f is the set $f(G)$ of elements of H that have the form $f(g)$ for some g in G.

The image of a group homomorphism $f : G \to H$ is a subgroup of H (an exercise).

Example 19. Let $f : R \to S$ be a ring homomorphism. If we view f as a group homomorphism of additive groups, the kernel of f is the same as the kernel of f viewed as a ring homomorphism. Thus, for example, if $f : \mathbb{Z} \to \mathbb{Z}/m\mathbb{Z}$ is the ring homomorphism defined by $f(a) = [a]_m$, then, viewed as a group homomorphism of additive groups, $\ker f = m\mathbb{Z}$, the set of multiples of m.

Example 20. Let $L_\alpha : \mathbb{Z}/m\mathbb{Z} \to \mathbb{Z}/m\mathbb{Z}$ be the homomorphism, "multiply by α," of Example 14. Then $\ker L_\alpha = \{\beta \text{ in } \mathbb{Z}/m\mathbb{Z} \mid \alpha\beta = 0\}$. Thus $\ker L_\alpha = 0$ if and only if α is not a zero divisor in $\mathbb{Z}/m\mathbb{Z}$.

Definition. A group homomorphism $f : G \to G'$ is an *isomorphism* if f is one-to-one and onto.

Proposition 15. *If $f : R \to S$ is an isomorphism of rings, then f restricts to an isomorphism $f : U(R) \to U(S)$ from the group of units of R to the group of units of S.*

Proof. If u is a unit of R, then $f(u)$ is a unit of S. Suppose t is a unit of S. Let r be the unique element in R so that $f(r) = t$, (r is unique because f is one-to-one) and let r' in R be so that $f(r') = t^{-1}$. Then $f(rr') = f(r)f(r') = t \cdot t^{-1} = 1$. But $f(1) = 1$ and f, being an isomorphism, is one-to-one. Thus $rr' = 1$, and r is a unit of R. Thus f is a one-to-one, onto function between the set of units of R and the set of units of S.   □

We shall see some examples of isomorphisms of groups in the next section, and an application of Proposition 14 in Section 12D.

Exercises.

31. Verify that a ring homomorphism $f : R \to S$ yields a group homomorphism from $U(R)$ to $U(S)$.

32. Prove Proposition 14.

33. Find $\ker L_\alpha$ (Example 19) if $L_\alpha : \mathbb{Z}/m\mathbb{Z} \to \mathbb{Z}/m\mathbb{Z}$, where
   (i) $m = 10$ and $\alpha = [2]$;
   (ii) $m = 11$ and $\alpha = [3]$; and
   (iii) $m = 12$ and $\alpha = [4]$.

34. Show that if r and $\phi(m)$ are coprime then $f_r : U_m \to U_m$ (Example 15) is one-to-one.

35. The image of a group homomorphism $f : G \to H$ is

$$f(G) = \{h \text{ in } H \mid h = f(g) \text{ for some } g \text{ in } G\}.$$

Show that the image of a group homomorphism $f : G \to H$ is a subgroup of H.

36. Show that if $f_1 : G \to H$ and $f_2 : H \to K$ are group homomorphisms, then the composition $f_2 \circ f_1 : G \to K$, defined by $f_2 \circ f_1(g) = f_2(f_1(g))$, is a group homomorphism.

37. Show that if $f : G \to H$ is an isomorphism of groups, then the inverse function $f^{-1} : H \to G$, defined by $f^{-1}(f(g)) = g$, is a group homomorphism.

G. Quotient Groups

Let G be a group and H a subgroup of G. We've observed in earlier sections of this chapter the role that the left cosets of H in G play in proving Lagrange's Theorem, in providing information about primality testing via Fermat's Theorem, and in interpreting solutions of equations in groups.

But the most familiar example of cosets, namely the cosets of $m\mathbb{Z}$ in $\mathbb{Z}$, are more than just sets. Those cosets, the congruence classes modulo m, form the elements of the commutative ring $\mathbb{Z}/m\mathbb{Z}$. Hence, in particular, they form a group under addition, where the addition of congruence classes is induced from the addition on $\mathbb{Z}$.

In this section we observe that if G is any abelian group (with operation $*$) and H any subgroup, then the left cosets of H in G also form an abelian group, where the operation $*$ on cosets is induced from the operation on G. By analogy with $\mathbb{Z}/m\mathbb{Z}$, we denote the group of left cosets of H in G by G/H.

To show that the set G/H of left cosets is a group, we need to explain the group operation on G/H.

Recall how we defined addition on $\mathbb{Z}/m\mathbb{Z}$.

Let $[a]_m$ and $[b]_m$ be two congruence classes. Then

$$[a]_m + [b]_m = [a+b]_m.$$

In words, the sum of the congruence class of a and the congruence class of b is the congruence class of $a + b$.

When we defined addition of congruence classes in this way in Chapter 6, we were concerned that the addition was "well-defined". By this, we meant that in defining the addition of $[a]_m$ and $[b]_m$, we were using particular representatives of the congruence classes, namely a and b, to determine the sum $[a+b]_m$ of the two congruence classes. What if we chose different representatives?

What we found was that

If $a \equiv a'$ (mod m) and $b \equiv b'$ (mod m) then $a + b \equiv a' + b'$ (mod m).

Thus if a' is in the congruence class $[a]_m$, and b' is in the congruence class $[b]_m$, then $a' + b'$ is in the congruence class $[a + b]_m$.

To look at it another way, if we interpret $[a]_m + [b]_m$ to be the set of all integers of the form $a' + b'$ where a' is in $[a]_m$ and b' is in $[b]_m$, then that set $[a]_m + [b]_m$ is the same set as the congruence class $[a + b]_m$.

Now the congruence class $[a]_m$ is the left coset $a + m\mathbb{Z}$, and $[b]_m = b + m\mathbb{Z}$. In fact, the notation $a + m\mathbb{Z}$ is really a more suggestive notation for the congruence class of a modulo m than $[a]_m$ is, because $[a]_m$ is the set of all integers of the form $a + mk$ for all k in $\mathbb{Z}$. The notation $a + m\mathbb{Z}$ describes this set well.

Using the notation $a + m\mathbb{Z}$ for $[a]_m$, what we showed in Chapter 6 is that if G is the group $\mathbb{Z}$ under addition, and H is the subgroup $m\mathbb{Z}$ of all multiples of m in $\mathbb{Z}$, then the operation on $G/H = \mathbb{Z}/m\mathbb{Z}$ defined by

$$(a + m\mathbb{Z}) + (b + m\mathbb{Z}) = (a + b) + m\mathbb{Z}$$

is well-defined.

Having done so, then it is clear that addition in $\mathbb{Z}/m\mathbb{Z}$ is associative and commutative because addition in $\mathbb{Z}$ is associative and commutative. Also, $0 + m\mathbb{Z}$, the coset consisting of the subgroup $m\mathbb{Z}$ itself, is the identity element of the group $\mathbb{Z}/m\mathbb{Z}$, and every element of $\mathbb{Z}/m\mathbb{Z}$ has an inverse (negative), namely, for any integer a, the inverse of $a + m\mathbb{Z}$ is $-a + m\mathbb{Z}$.

Hence the set $\mathbb{Z}/m\mathbb{Z}$ of left cosets of $m\mathbb{Z}$ in $\mathbb{Z}$ is an abelian group, where the addition on left cosets is induced from that on $\mathbb{Z}$. "Induced" means that we can determine how cosets add by adding representatives in $\mathbb{Z}$.

Now let G be any abelian group, with operation $*$ and identity e and let H be a subgroup. Denote by $a * H$ the left coset of the element a of G. Let G/H be the set of left cosets of H in G. We define an operation on G/H, induced by $*$ on G, as follows: For every a and b in G, define the product of $a * H$ and $b * H$ by

$$(a * H) * (b * H) = (a * b) * H.$$

We want to show this product is well defined.

Proposition 16. *Let G be an abelian group and H a subgroup of G. Suppose $a * H = a_1 * H$ and $b * H = b_1 * H$. Then $(a * b) * H = (a_1 * b_1) * H$.*

Proof. It suffices to show that if we pick any element out of the coset $a * H$, and any element out of the coset $b * H$, and multiply them in G, we get an element of the coset $(a * b) * H$. We show this. Let $a_1 = a * h$ for some h in H, and $b_1 = b * h_1$ for some h_1 in H. If we take the product $a_1 * b_1 = (a * h) * (b * h_1)$, then, since G is abelian, we can rearrange the product:

$$(a * h) * (b * h_1) = (a * b) * (h * h_1).$$

Since $h * h_1$ is in H, the result is in $(a * b) * H$. □

Here is an application of Proposition 16.

Example 21. We show that if p is an odd prime, then

$$((\frac{p-1}{2})!)^2 \equiv 1 \text{ or } -1 \pmod{p}.$$

To do this, we let $G = U_p$ and $N = \{1, -1\}$ (where we write $[a]_p$ as a for convenience). Then

$$(p-r)N = (-r)N = rN.$$

Repeatedly using Proposition 16, we have

$$(p-1)!N = (1 \cdot 2 \cdots \frac{p-1}{2}) \cdot ((p-1) \cdot (p-2) \cdots (p - \frac{p-1}{2}))N$$

$$= (1N \cdot 2N \cdots \frac{p-1}{2}N) \cdot ((p-1)N \cdot (p-2)N \cdots (p - \frac{p-1}{2})N)$$

$$= (1N \cdot 2N \cdots \frac{p-1}{2}N) \cdot (1)N \cdot (2)N \cdots (\frac{p-1}{2})N)$$

$$= (\frac{p-1}{2})!N \cdot (\frac{p-1}{2})!N = ((\frac{p-1}{2})!)^2 N.$$

But $(p-1)!$ is in N by Wilson's Theorem (Exercises 12 and 13 of Chapter 14, or Exercise 47, below), so $((\frac{p-1}{2})!)^2$ is in N, and hence

$$((\frac{p-1}{2})!)^2 \equiv 1 \text{ or } -1 \pmod{p},$$

as claimed.

Once we see that the operation $*$ on cosets is well-defined, associativity and commutativity follow easily since they hold in G, the coset $e * H$ is the identity element of G/H, and the inverse of $a * H$ is $a' * H$ where a' is the inverse of a in G. Thus:

Theorem 17. *If G is an abelian group and H is a subgroup of G, then the set of left cosets G/H is an abelian group, with the operation on cosets induced by the operation on G.*

The group G/H is called a *quotient group*.
Here are some examples.

Example 22. Let $G = U_{21}$, the group of units modulo 21, a group of order $\phi(21) = 12$. Let $N = U_{21}(2)$ be the subgroup consisting of the square roots of 1 in $\mathbb{Z}/21\mathbb{Z}$, that is, the units modulo 21 whose square is $[1]_{21}$. Then

$$N = \{[1], [-1], [8], [-8]\}.$$

Then G/N consists of three cosets. From Example 8 we found that

$$G/N = \{N, [2]N, [4]N\}.$$

Since $([2]N)^3 = [8]N = N$, the group G/N consists of all multiples of the coset $[2]N$. Thus G/N is a cyclic group of order 3.

Example 23. Let G be a finite abelian group (written multiplicatively) of even order $2m$. Let $f_2 : G \to G$ be the squaring function, $f_2(g) = g^2$. Let H be the image of G. Then G/H is an abelian group, and every coset of H in G has order 1 or 2. For if aH is a coset, then $(aH)^2 = (aH)(aH) = a^2H$ is the identity coset H because a^2 is in H.

Here is an example. Let $G = U_{20}$. Then G has order $\phi(20) = 8$. Let $H = f_2(G)$, the set of squares in U_{20}. Then $H = \{[1], [9]\}$, and

$$G/H = \{H, [3]H, [11]H, [13]H\}.$$

It's easy to verify that the square of each element of G/H is the identity element of G/H. For example, the coset $[3]H = \{[3], [7]\}$, and

$$[3]H \cdot [3]H = H$$

(as can be verified by multiplying $[3]$ or $[7]$ by $[3]$ or $[7]$ and verifying that the product is in $H = \{[1], [9]\}$).

As a nice application of quotient groups, we can prove

Theorem 18 (Cauchy). *If G is a finite abelian group of order n, and p is a prime divisor of n, then G has an element of order p.*

Proof. We do it by induction on n. Assume the group operation is multiplication, with identity e. The result is true if $n = p$ is prime, because if G has order p and $a \neq e$, then a must have order p.

If n is not prime, then by Proposition 3, G has a proper subgroup H of order $m > 1$, a divisor of n. If p divides m, then by induction, H, and therefore G, has an element of order p. If p does not divide m, then let G/H be the quotient group, of order s. By Lagrange's Theorem, $ms = n$, so if p doesn't divide m, p must divide s, and so, by induction, G/H has an element aH of order p.

Then a is not in H, but $(aH)^p = H$, the identity element of G/H. Thus $a^p = b$ in H. Since b is in H and H has order m, we have $b^m = (a^p)^m = (a^m)^p = e$. If we show that $a^m \neq e$, then a^m is an element of order p in G, and the proof is done.

But if $a^m = e$, then $(aH)^m = eH = H$, and also $(aH)^p = H$. Since p doesn't divide m, $pr + ms = 1$ for some integers r, s, so

$$aH = (aH)^{pr+ms} = (aH)^{pr}(aH)^{ms} = H$$

and so aH is the identity element of G/H, contradicting the assumption that aH had order p in G/H. Hence $a^m \neq e$ and a^m is an element of order p in G. $\square$

Once we have quotient groups, we can give the Fundamental Homomorphism Theorem for abelian groups:

Theorem 19. *Let G, H be abelian groups and $f : G \to H$ be a group homomorphism with kernel $K \subseteq G$ and image $f(G) \subseteq H$. Then f induces an isomorphism*

$$\overline{f} : G/K \to f(G)$$

by $\overline{f}(g * K) = f(g)$.

Proof. We want to show that $\overline{f}$ is well-defined, which means that if $g * K = g_1 * K$, then $f(g) = f(g_1)$. But if $g * K = g_1 * K$, then $g = g_1 * k$ for some k in $K = \ker(f)$. So

$$f(g) = f(g_1 * k) = f(g_1) * f(k) = f(g')$$

since $f(k) = e$, the identity of H. So $\overline{f}$ is well-defined.

Then $\overline{f}$ is a group homomorphism because f is, and $\overline{f}$ is onto $f(G)$ because f is. To show that $\overline{f}$ is one-to-one, suppose $\overline{f}(g * K) = e$, then $f(g) = e$, so g is in $\ker(f) = K$. So $g * K = K$ is the identity element of G/K. Thus $\overline{f}$ is an isomorphism from G/K onto $f(G)$. $\square$

Here is an application of these ideas.

Example 24. Let p be an odd prime, U_p the units of $\mathbb{Z}/p\mathbb{Z}$, and let $f : U_p \to U_p$ be the squaring function, $f([a]) = [a]^2 = [a^2]$. Let Q be the image of f. Then

$$Q = \{[a^2] : [a] \text{ in } U_p\}$$

is the set of squares in U_p. The kernel of f is

$$N = \{[a] \text{ in } U_p : [a]^2 = [1]\} = \{[a] \text{ in } U_p : a^2 \equiv 1 \pmod{p}\}.$$

Since p is prime, $N = \{[1], [-1]\}$. The Fundamental Homomorphism Theorem says that

$$U_p/N \cong Q.$$

The order of U_p/N is the number of cosets of N in U_p. By Lagrange's Theorem, this is equal to the order of U_p divided by the order of N. So the order of Q is $(p-1)/2$.

We can now show that the product of any two non-squares in U_p is a square. To do so, consider U_p/Q. By Lagrange's Theorem, U_p/Q has order 2. So there are two cosets, namely Q, the set of squares in U_p, and the other coset, which consists of non-squares in U_p. For every non-square a in U_p, aQ is that other coset.

Now if a is any non-square, then $aQ \cdot aQ = a^2Q = Q$, since a^2 is in Q. If b is any other non-square, then $aQ = bQ$, so $Q = aQ \cdot aQ = aQ \cdot bQ = abQ$ since multiplication of cosets is well-defined, by Proposition 16. Thus ab is in Q, that is, ab is a square in U_p. Thus the product of every two non-squares in U_p is a square in U_p.

We'll give a different proof of this fact in Chapter 21, using the existence of a primitive root.

We remark that Proposition 16 and Theorem 17 are often not true if the group G is not abelian. The equation

$$(a*h)*(b*h_1) = (a*b)*(h*h_1)$$

that showed that the $*$ operation on cosets is well-defined, used the assumption that $*$ is commutative. For a general group G and a subgroup H, the condition that the operation on the cosets of H in G is well-defined is equivalent to the condition that for every a in G, the left coset $a*G$ and the right coset $G*a$ are equal. A subgroup H of G with that property is called a *normal* subgroup of G. Every subgroup of an abelian group is normal, but nearly every non-abelian group has subgroups that are not normal. (The only finite exceptions are groups $Q \times A$ where Q is the quaternion group of order 8 and A is abelian, as Rotman ((2000), p. 153) points out.)

Exercises.

38. Let $G = U_{16}$ be the group of units modulo 16, a group of order 8, and let $J = \{[1], [-1]\}$, a subgroup. Then G/J is a group of order 4. Show that every element of G/J is a power of the coset $[3]H$.

39. Let $G = U_{16}$ be the group of units modulo 16. Let $H = \{[1], [9]\}$, a subgroup. Then G/H is a group of order 4. Show that the square of every element of G/H is equal to the identity.

40. Let $G = U_{45}$ be the group of units modulo 45, a group of order $\phi(45) = 24$. Then $[4]_{45}$ has order 6. Let H be the subgroup generated by $[4]_{45}$. Find the orders of the four elements of G/H.

41. Let $G = U_{45}$ be the group of units modulo 45, a group of order $\phi(45) = 24$. Then $[11]_{45}$ has order 6. Let H be the subgroup generated by $[11]_{45}$. Find the orders of the four elements of G/H.

42. Let $G - U_{p^2}$ be the group of units modulo p^2, where p is an odd prime. Then G is a group of order $\phi(p^2) = p(p-1)$. The element $[1+p]_{p^2}$ has order p. Let H be the subgroup generated by $[1+p]$. Show that G/H is a cyclic group, generated by the coset of any primitive root modulo p.

43. Let $G = U_p$ be the group of units modulo p, where p is an odd prime. Let H be the subgroup generated by $[-1]_p$. Show that G/H is generated by the coset of any primitive element modulo p.

44. Let $G = U_p$ be the group of units modulo p, where p is an odd prime. Let H be the subgroup generated by $[-1]_p$. Show that G/H consists of the cosets of the congruence classes of $1, 2, \ldots, \frac{p-1}{2}$.

45. Show that the product of the elements of $U_7/\{[1], [-1]\}$ is equal to the identity element.

46. Show that the square of the product of the elements of $U_{13}/\{[1], [-1]\}$ is equal to the identity element.

47. Prove that if p is prime, $G = U_p$ and $N = \{1, -1\}$, then $(p-1)!$ is in N, as follows.

(i) Show that every number r with $1 \le r \le p-1$ has a unique inverse r' modulo p with $1 \le r' \le p-1$.

(ii) Show that

$$1 \cdot 2 \cdot 3 \cdots p - 1 \equiv 1' \cdot 2' \cdot 3' \cdots (p-1)' \pmod{p}.$$

(iii) (Gauss's 2nd grade trick) Show that

$$1 \equiv (1 \cdot 2 \cdot 3 \cdots p - 1)(1' \cdot 2' \cdot 3' \cdots (p-1)')$$
$$\equiv (1 \cdot 2 \cdot 3 \cdots p - 1)(1 \cdot 2 \cdot 3 \cdots p - 1)$$
$$\equiv ((p-1)!)^2 \pmod{p}.$$

(iv) To finish the proof, recall that $x^2 \equiv 1 \pmod{p}$, p prime, has only the solutions $x \equiv 1$ or -1.

48. Find $(\frac{p-1}{2})!$ and its square modulo p when $p = 5, 7, 11, 17, 23, \ldots$. Can you guess any pattern for the values modulo p?

H. Some Nonabelian Groups

In this section we give an example of a group which is not abelian. This is a group with 6 elements, and we shall present it in three guises, as $\mathfrak{S}_3$, as $GL_2(\mathbb{F}_2)$ and as D_3, each of which suggests a generalization to a large class of nonabelian groups.

The symmetric group. The group $\mathfrak{S}_3$ is the group of permutations on three symbols. If we denote the symbols by $\{a, b, c\}$, then a permutation is a one-to-one function from the set $\{a, b, c\}$ to itself. The group operation is composition of functions. There are six permutations of $\{a, b, c\}$, namely:

$\iota : (a, b, c) \to (a, b, c)$, the identity permutation,

$\rho : (a, b, c) \to (b, c, a)$.

(The notation means that $\rho(a) = b, \rho(b) = c$, and $\rho(c) = a$.)

$\rho^2 : (a, b, c) \to (c, a, b)$, (i.e., $\rho^2(a) = \rho(b) = c, \rho^2(b) = \rho(c) = a, \rho^2(c) = \rho(a) = b$),

$\tau : (a, b, c) \to (b, a, c)$,

$\tau\rho : (a, b, c) \to (a, c, b)$ (i.e., $\tau\rho(a) = \tau(b) = a$, etc.),

$\tau\rho^2 : (a, b, c) \to (c, b, a)$.

It is easy to verify that $\rho^3 = \iota, \tau^2 = \iota, \rho\tau = \tau\rho^2$ and $\rho^2\tau = \tau\rho$.

Here is the group table for $\mathfrak{S}_3$:

	ι	ρ	ρ^2	τ	$\tau\rho$	$\tau\rho^2$
ι	ι	ρ	ρ^2	τ	$\tau\rho$	$\tau\rho^2$
ρ	ρ	ρ^2	ι	$\tau\rho^2$	τ	$\tau\rho$
ρ^2	ρ^2	ι	ρ	$\tau\rho$	$\tau\rho^2$	τ
τ	τ	$\tau\rho$	$\tau\rho^2$	ι	ρ	ρ^2
$\tau\rho$	$\tau\rho$	$\tau\rho^2$	τ	ρ^2	ι	ρ
$\tau\rho^2$	$\tau\rho^2$	τ	$\tau\rho$	ρ	ρ^2	ι

The group S_3 is called the *symmetric group* on three symbols.

The general linear group. Denote by $GL_2(\mathbb{F}_2)$ the group of units of the ring of 2×2 matrices with entries in $\mathbb{F}_2 = \{0, 1\}$. The group operation is multiplication. (See Chapter 8 for a review of matrices.) Thus $GL_2(\mathbb{F}_2)$ consists of all matrices $\begin{pmatrix} a & b \\ c & d \end{pmatrix}$ such that

$$\det \begin{pmatrix} a & b \\ c & d \end{pmatrix} = ad - bc \neq 0.$$

Since $ad - bc = 0$ or 1, we may find all elements of $GL_2(\mathbb{F}_2)$ by finding all a, b, c, d in $\mathbb{F}_2$ with $ad - bc = 1$. But $ad - bc = 1$ iff $ad = 1$ and $bc = 0$, or $ad = 0$ and $bc = 1$. So we get six solutions, three with $a = d = 1$, three with $b = c = 1$. Thus $GL_2(\mathbb{F}_2)$ has the elements

$$I = \begin{pmatrix} 1 & 0 \\ 0 & 1 \end{pmatrix}, \qquad T = \begin{pmatrix} 0 & 1 \\ 1 & 0 \end{pmatrix}$$

$$R = \begin{pmatrix} 0 & 1 \\ 1 & 1 \end{pmatrix}, \qquad TR = \begin{pmatrix} 1 & 1 \\ 0 & 1 \end{pmatrix}$$

$$R^2 = \begin{pmatrix} 1 & 1 \\ 1 & 0 \end{pmatrix}, \qquad TR^2 = \begin{pmatrix} 1 & 0 \\ 1 & 1 \end{pmatrix}.$$

Then $R^3 = I, T^2 = I, RT = TR^2, R^2T = TR$, and $GL_2(\mathbb{F}_2)$ has the following multiplication table:

	I	R	R^2	T	TR	TR^2
I	I	R	R^2	T	TR	TR^2
R	R	R^2	I	TR^2	T	TR
R^2	R^2	I	R	TR	TR^2	T
T	T	TR	TR^2	I	R	R^2
TR	TR	TR^2	T	R^2	I	R
TR^2	TR^2	T	TR	R	R^2	I

This table looks just like the table for $\mathfrak{S}_3$ (replacing ι by I, ρ by R, τ by T). The resemblance is not an accident. Set

$$a = \begin{pmatrix} 1 \\ 0 \end{pmatrix}, b = \begin{pmatrix} 0 \\ 1 \end{pmatrix}, c = \begin{pmatrix} 1 \\ 1 \end{pmatrix},$$

then the elements of $GL_2(\mathbb{F}_2)$ act on $\{a,b,c\}$ by left multiplication. For example,

$$Ra = \begin{pmatrix} 0 & 1 \\ 1 & 1 \end{pmatrix}\begin{pmatrix} 1 \\ 0 \end{pmatrix} = \begin{pmatrix} 0 \\ 1 \end{pmatrix} = b;$$

$$Rb = \begin{pmatrix} 0 & 1 \\ 1 & 1 \end{pmatrix}\begin{pmatrix} 0 \\ 1 \end{pmatrix} = \begin{pmatrix} 1 \\ 1 \end{pmatrix} = c;$$

$$Rc = \begin{pmatrix} 0 & 1 \\ 1 & 1 \end{pmatrix}\begin{pmatrix} 1 \\ 1 \end{pmatrix} = \begin{pmatrix} 1 \\ 0 \end{pmatrix} = a;$$

So R acts like ρ. Similarly, we can see how each of the other elements of $GL_2(\mathbb{F}_2)$ acts as a permutation of the set $\{a,b,c\}$. By identifying the permutation, we obtain an isomorphism of groups

$$f : GL_2(\mathbb{F}_2) \to \mathfrak{S}_3$$

where the map f takes a matrix and determines the permutation of $\{a,b,c\}$ that the matrix yields. It's not hard to see that $f(1) = \iota$, $f(R) = \rho$, $f(R^2) = \rho^2$, $f(T) = \tau$, $f(TR) = \tau\rho$, and $f(TR^2) = \tau\rho^2$.

The dihedral group. Consider an equilateral triangle cut out of a flat board of uniform thickness. We label the vertices of the triangle by a,b,c, as follows:

A rigid motion of the triangle consists of removing the triangle from its hole in the board, and placing it back in the hole, either right side up or upside down, in some fashion. The group D_3 consists of all rigid motions of the triangle. A rigid motion followed by another rigid is a rigid motion. Thus iteration of rigid motions defines the group operation. There are six rigid motions:

i, the identity;

r, obtained by rotating the triangle $120°$ counterclockwise;

r^2, obtained by rotating the triangle $240°$ counterclockwise;

t, obtained by flipping the triangle across the axis through the vertex c and the midpoint of the edge ab;

tr, obtained by first doing r, then t, and

tr^2, obtained by first doing r^2, then t.

If we let D_3 denote this set of rigid motions, then

$$D_3 = \{i, r, r^2, t, tr, tr^2\}.$$

Notice that the six rigid motions give all of the six distinct permutations of the three vertices $\{a,b,c\}$. For example, tr is the permutation $(a,b,c) \rightarrow (a,c,b)$. Thus D_3 can also be viewed as the set of permutations of the vertices a,b,c. More explicitly, we can define an isomorphism of groups

$$g : D_3 \rightarrow \mathfrak{S}_3$$

by $g(\iota) = \iota, g(r) = \rho, g(r^2) = \rho^2, g(t) = \tau, g(tr) = \tau\rho, g(tr^2) = \tau\rho^2$.

Each of the three versions of this non-abelian group of six elements can be generalized in a distinctive way:

I. If $\{a_1, a_2, \ldots, a_n\}$ is a set of n symbols, then the set $\mathfrak{S}_n$ of all permutations of $a_1, a_2, \ldots, a_n$ is a (nonabelian) group with $n!$ elements, called the symmetric group. The group operation is composition of permutations. This infinite collection of finite groups is of particular importance, since, as we shall see shortly, every finite group may be viewed as a subgroup of some $\mathfrak{S}_n$.

II. If $\mathbb{F}_p$ is the field of p elements, p any prime, then $GL_n(\mathbb{F}_p)$ denotes the group of invertible $n \times n$ matrices with entries in the field $\mathbb{F}_p$, called the general linear group. The set of $n \times n$ matrices with entries in $\mathbb{F}_p$ has p^{n^2} elements, so $GL_n(\mathbb{F}_p)$ is somewhat smaller. This is a doubly infinite collection of finite groups, since both n and p can vary over infinite sets.

III. Let D_k be the group of rigid motions of the regular k-gon. Then any element of D_k can be obtained by iterating in some order two particular motions: a rotation of $360/k$ degrees, and a reflection, or flip, across an axis passing through the center of the k-gon and a vertex. D_k is a non-abelian group with $2k$ elements, called the dihedral group.

We can see that $GL_n(\mathbb{F}_p)$ may be viewed as a subgroup of $\mathfrak{S}_k$, where $k = p^n - 1$. Each element of $GL_n(\mathbb{F}_p)$ is an $n \times n$ matrix which acts as a function from the set V of non-zero n-entry column vectors to V. An invertible matrix is a one-to-one function from V to itself, that is, a permutation of V. So $GL_n(\mathbb{F}_p)$ may be viewed as a subgroup of the group of permutations of V. V is a set with $p^n - 1$ vectors.

Conversely, $\mathfrak{S}_n$ may be viewed as a subgroup of $GL_n(\mathbb{F}_p)$ for every prime p, by viewing an element π of $\mathfrak{S}_n$ as a permutation of the standard basis of $\mathbb{F}_p^n$, and associating to π the matrix of π relative to the standard basis.

Also, D_k may be viewed as a subgroup of the group of permutations of the vertices of the k-gon, because any element of D_k is completely determined by where the vertices of the k-gon end up. So D_k may be viewed as a subgroup of $\mathfrak{S}_k$.

More generally, we have:

Theorem 20. *Cayley's Theorem. Let G be a group with n elements. There is a one-to-one homomorphism L from G to $\mathfrak{S}_n$.*

Proof. Let $G = \{a_1, a_2, \ldots, a_n\}$, and for b in G, let L_b be the function from G to G defined by $L_b(a_i) = b * a_i$. Then by cancellation and solvability, L_b is a permutation of the set G. If e is the identity element of G, then L_e is the identity permutation. The function L is a group homomorphism, that is, for any a, b, c in G,

$$L_{b*c}(a) = (b*c)*a = b*(c*a) = L_b(L_c(a)) = (L_b \circ L_c)(a),$$

so $L_{b*c} = L_b \circ L_c$. Finally, if $L_b = L_c$, then for every a in G, $b*a = c*a$; by cancellation, we get $b = c$. So the function L is a one-to-one homomorphism.   □

Notice that if G has n elements, $\mathfrak{S}_n$ has $n!$ elements. So $\mathfrak{S}_n$ is larger than G for all $n > 2$, and the image of G is a proper subgroup of $\mathfrak{S}_k$ (that is, is not all of $\mathfrak{S}_n$).

The theory of groups has become an extremely rich and deep subject with important applications to molecular chemistry and quantum physics, as well as to many areas of advanced mathematics. If you browse in almost any college or university library you will find numerous books on group theory. One particularly nice treatment of group theory is in Herstein (1975), the source of the proof of Cauchy's Theorem, Theorem 18.

We will not pursue the theory of non-abelian groups further in this book. See Chapter 19 for more on cyclic groups.

Exercises.

49. Describe a group homomorphism from U_8 to $\mathfrak{S}_4$.

50. Describe a group homomorphism from U_5 to $\mathfrak{S}_4$.

51. How many elements are there in $GL_2(\mathbb{F}_3)$?

52. Describe a group homomorphism from U_4 to $\mathfrak{S}_4$. (Hint: One way is to let U_4 act via multiplication on all of $\mathbb{Z}/4\mathbb{Z}$, not just on U_4.)

53. In D_3, show that if H is the subgroup generated by the reflection t, then there is an element g of D_3 so that the left coset gH is not a right coset of H in D_3. Hence multiplication of left cosets by $aH \cdot bH = abH$ is not well-defined.

Chapter 12
The Chinese Remainder Theorem

In this chapter we study systems of two or more linear congruences. When the moduli are pairwise coprime, the main theorem is known as the Chinese Remainder Theorem, because special cases of the theorem were known to the ancient Chinese. In modern algebra the Chinese Remainder Theorem is a powerful tool in a variety of applications, as we shall see.

A. The Chinese Remainder Theorem

Two congruences. Here is the result for two linear congruences.

Theorem 1. *Let m and n be natural numbers > 1 (the moduli) and a, b be any integers. Then there is a solution $x = x_0$ of*

$$x \equiv a \pmod{m}$$
$$x \equiv b \pmod{n},$$

if and only if the greatest common divisor of m and n divides $b - a$. If $x = x_0$ is a solution, then the set of integers x that satisfy the two congruences is the same as the set of x that satisfy

$$x \equiv x_0 \pmod{[m,n]}$$

where $[m,n]$ is the least common multiple of m and n.

Before proving the theorem, we look at three examples.

Example 1. Consider the pair of congruences

$$x \equiv 11 \pmod{74}$$
$$x \equiv 13 \pmod{63}.$$

L.N. Childs, *A Concrete Introduction to Higher Algebra*, Undergraduate Texts in Mathematics, © Springer Science+Business Media LLC 2009

If x is a solution, then

$$x = 11 + 74r$$

for some integer r, and

$$x = 13 + 63s$$

for some integer s. Setting the two expressions for x equal to each other, we obtain

$$11 + 74r = 13 + 63s$$

or, collecting the constants,

$$74r - 63s = 2.$$

This equation is solvable using the extended Euclidean algorithm, since the greatest common divisor of 74 and 63, namely 1, divides 2. In fact, Euclid's algorithm for 74 and 63 is:

$$74 = 63 + 11$$
$$63 = 11 \cdot 6 - 3$$
$$11 = 3 \cdot 3 + 2$$

Using the row operation approach we have

	coeff. of 74	coeff. of 63
74	1	0
63	0	1
11	1	−1
66	6	6
3	6	−7
9	18	−21
2	−17	20.

The last line says that

$$2 = 74 \cdot (-17) + 63 \cdot 20.$$

Since we wish to find r, s so that $74r - 63s = 2$, we set $r = -17, s = -20$. Since $x = 13 + 63s$, we find $x = 13 - 63 \cdot 20 = 13 - 1260 = -1247$ is a solution of the congruences.

As with a single linear congruence, once we find a particular solution $x = -1247$ of

$$x \equiv 11 \pmod{74}$$
$$x \equiv 13 \pmod{63}.$$

we can find the general solution by taking the particular solution $x = -1247$ and adding to it the general solution to the homogeneous system of congruences,

$$x \equiv 0 \pmod{74}$$
$$x \equiv 0 \pmod{63}.$$

Any integer x that solves this homogeneous system must be a multiple of 74 and a multiple of 63, hence a common multiple of 74 and 63. Since 74 and 63 are coprime, the least common multiple $[74, 63]$ of 74 and 63 is $74 \cdot 63 = 4662$. To a particular solution of the original (non-homogeneous) congruences, like $x = -1247$, then, we can add any multiple of 4662 and get another solution. Expressed concisely, the set of all solutions to the set of congruences

$$x \equiv 11 \pmod{74}$$
$$x \equiv 13 \pmod{63}.$$

is the set of all integers x so that

$$x = -1247 + 4662k$$

for some integer k. This is the same as the set of integers x that satisfy the congruence

$$x \equiv -1247 \pmod{4662}.$$

Example 2. We seek the smallest non-negative solution of

$$x \equiv 2 \pmod{24}$$
$$x \equiv 8 \pmod{39}$$

We set $x = 2 + 24r = 8 + 39s$, hence $24r - 39s = 6$. Since $(24, 39) = 3$ and 3 divides 6, there is a solution. After some calculations we find that $39 \cdot 2 - 24 \cdot 3 = 6$, and so $x = 2 + 24(-3) = -70$. The general solution is $x = -70 + [24, 39]k$ for k any integer. Since $[24, 39] = 312$, we can write the solutions as $x = -70 + 312k$ for every integer k, or as

$$x \equiv -70 \pmod{312}.$$

In particular, $x = -70 + 312 = 242$ is the smallest positive solution.

Example 3. Consider

$$x \equiv 5 \pmod{20}$$
$$x \equiv 15 \pmod{16}.$$

We set $x = 5 + 20r = 15 + 16s$ and collect the constant terms to one side:

$$20r - 16s = 10.$$

But in this example, the greatest common divisor of 20 and 16 is 4, and 4 does not divide 10. So there are no integers r and s that solve the equation $5 + 20r = 15 + 16s$, and so there is no solution of the pair of congruences.

The proof of Theorem 1 follows the method of the examples.

Proof. We suppose x is a solution to the two congruences

$$x \equiv a \pmod{m}$$
$$x \equiv b \pmod{n}.$$

Since x is a solution to the first congruence, $x = a + my$ for some integer y. Since x is a solution to the second congruence, $x = b + nz$ for some integer z. Equating the two expressions for x yields the following equation in y and z:

$$a + my = b + nz$$

or

$$my - nz = b - a.$$

Now we follow the method of Section 3D for these equations. The result we obtained there was:

- if the greatest common divisor of m and n doesn't divide $b - a$, then there are no integers y, z so that $my - nz = b - a$ (for if d is the greatest common divisor of m and n, then d divides $my - nz$, hence must divide $b - a$). Thus there is no integer x that solves the original pair of congruences.
- if $d = (m, n)$ divides $b - a$, so that $b - a = qd$, then we can use Bezout's identity to solve the equation as follows: we find integers t and w so that $mt + nw = d$, then $m(tq) + n(wq) = b - a$. Hence, setting $y = tq$, we find that $x = a + mtq$ is a solution of the original pair of congruences.

These results prove the first part of the Proposition.

For the second part, suppose x_0 and x_1 are solutions of the pair of congruences. Then $x_1 - x_0$ is a solution of the "homogeneous" pair of congruences

$$x \equiv 0 \quad (\mod m)$$
$$x \equiv 0 \quad (\mod n).$$

That means $x_1 - x_0$ is a common multiple of m and n, and hence is a multiple of the least common multiple $[m, n]$. Hence $x_1 - x_0 = [m, n]k$ for some k, and so

$$x_1 = x_0 + [m, n]k$$

for some k.

Conversely, if x_0 is a solution to the pair of congruences and x satisfies $x \equiv x_0$ $(\mod [m, n])$ for some integer k, then $x \equiv x_0$ $(\mod m)$ and $x \equiv x_0$ $(\mod n)$, and so x is also a solution to the original pair of congruences.

The set of integers x of the form $x = x_0 + [m, n]k$ may be described as the set of integers such that

$$x \equiv x_0 \quad (\mod [m, n]).$$

$\square$

As an immediate corollary of the theorem, we obtain the Chinese Remainder Theorem for two congruences:

Corollary 2 (The Chinese Remainder Theorem). *Let m and n be coprime natural numbers > 1 (the moduli) and a, b be any integers. Then there is a solution $x = x_0$ of*

$$x \equiv a \quad (\text{mod } m)$$
$$x \equiv b \quad (\text{mod } n).$$

The set of integers x that satisfy the two congruences is the same as the set of x that satisfy

$$x \equiv x_0 \quad (\text{mod } mn).$$

If we have a pair of congruences where one of the moduli is small, we can reduce the Bezout's identity calculations by solving a single congruence modulo the smaller of the two moduli.

Example 4. Consider

$$x \equiv 38 \quad (\text{mod } 60)$$
$$x \equiv 7 \quad (\text{mod } 11).$$

Then $x = 38 + 60r = 7 + 11s$ for some integers r, s. To find x we don't need to find both r and s in the equation

$$38 + 60r = 7 + 11s,$$

rather, just one of them. So instead of approaching the equation as a Bezout type problem, we look at the equation as a congruence modulo the smaller of the original moduli:

$$38 + 60r \equiv 7 \quad (\text{mod } 11).$$

Reducing 38 and 60 modulo 11 yields

$$5 + 5r \equiv 7 \quad (\text{mod } 11)$$

or

$$5r \equiv 2 \quad (\text{mod } 11).$$

Since the inverse of 5 modulo 11 is 9, we multiply the last congruence by 9 to obtain

$$r \equiv 9 \cdot 5r \equiv 9 \cdot 2 \equiv 7 \quad (\text{mod } 11).$$

Then $x = 38 + 7 \cdot 60 = 458$ is a solution to the original congruences. Since the least common multiple of 11 and 60 is 660, the general solution is

$$x \equiv 458 \quad (\text{mod } 660).$$

Three or more congruences. The key to solving systems of more than two simultaneous congruences is the observation that we can express the set of integers that solve two simultaneous congruences as the set of integers that satisfy one congruence.

Example 5. We find all solutions to

$$x \equiv 2 \quad (\text{mod } 12)$$
$$x \equiv 8 \quad (\text{mod } 10)$$
$$x \equiv 9 \quad (\text{mod } 13).$$

We first solve the first two: we find x of the form $x = 2 + 12r = 8 + 10s$. It's easy enough to see that $x = 38$ is a solution. Since $[12, 10] = 60$, the general solution of the first two congruences is $x = 38 + 60k$ for k any integer. Thus to solve the three congruences is the same as to solve

$$x \equiv 38 \pmod{60}$$
$$x \equiv 9 \pmod{13}.$$

This pair of congruences has the property that the modulus 13 in the third of the original congruences is smaller than the modulus 60 arising from the first two original congruences. Hence the congruence method of the previous example is helpful.

So to find a t so that $x = 38 + 60t = 9 + 13u$ for some u, we set up the congruence

$$38 + 60t \equiv 9 \pmod{13}$$

which reduces modulo 13 to

$$-1 - 5t \equiv 9 \pmod{13}$$

or

$$-5t \equiv 10 \pmod{13}.$$

Thus

$$t \equiv -2 \equiv 11 \pmod{13},$$

hence $x = 38 + 60(11) = 698$. The general solution to the original three congruences is then

$$x \equiv 698 \pmod{780}$$

since $[10, 12, 13] = 60 \cdot 13 = 780$.

If we have a system of n congruences in which the moduli are pairwise coprime (not as in this last example), there is always a solution and the solution is unique modulo the product of the moduli. Thus we have:

Theorem 3 (Chinese Remainder Theorem). *Let $m_1, m_2, \ldots, m_n$ be pairwise coprime natural numbers > 1 (the moduli), and $a_1, a_2, \ldots a_n$ be any integers. Then there is a solution of the set of simultaneous congruences*

$$x \equiv a_1 \pmod{m_1}$$
$$x \equiv a_2 \pmod{m_2}$$
$$\vdots$$
$$x \equiv a_n \pmod{m_n}.$$

If x_0 is a solution, then the set of all solutions is the set of integers congruent to x_0 modulo $M = m_1 m_2 \cdot \ldots \cdot m_n$.

Proof. The proof is by induction on n. The case for two congruences is the corollary above.

For $n > 2$ we assume that any set of $n - 1$ congruences whose moduli are pairwise coprime has a solution. Suppose we have a set of n congruences as in the statement of the theorem. We use the theorem for two congruences to replace the first two congruences by a single congruence, of the form

$$x \equiv x_0 \pmod{m_1 m_2}.$$

Then to show that there is a solution of the original set of n congruences, we need to show that there is a solution for the set of $n - 1$ congruences consisting of all but the first two of the n original congruences, together with the congruence

$$x \equiv x_0 \pmod{m_1 m_2}.$$

To apply the induction hypothesis, the only thing we need to observe is that the new last modulus, $m_1 m_2$, has the property that $m_1 m_2$ and m_j are coprime for $j = 3, \ldots, n$. But we saw in Chapter 4 that if $(m_j, m_1) = 1$ and $(m_j, m_2) = 1$ then $(m_j, m_1 m_2) = 1$. Thus the set of $n - 1$ congruences has a solution by the induction hypothesis, and that solution will be a solution of the original n congruences.

The last statement of the theorem is left as an exercise. □

For a first application of the Chinese Remainder Theorem, we look at single linear congruences to composite moduli.

Example 6. Suppose we wish to solve

$$36x \equiv 29 \pmod{85}.$$

Now $85 = 17 \cdot 5$ and 17 and 5 are coprime, so this congruence is equivalent to the two congruences

$$36x \equiv 29 \pmod 5$$
$$36x \equiv 29 \pmod{17}.$$

To solve the original congruence, we'll solve this new system.

Now the first congruence, $36x \equiv 29 \pmod 5$, is equivalent to

$$x \equiv 4 \pmod 5,$$

while the second, $36x \equiv 29 \pmod{17}$, is equivalent to

$$2x \equiv 12 \pmod{17}$$

or

$$x \equiv 6 \pmod{17}$$

So we need to solve the pair of congruences

$$x \equiv 4 \pmod 5$$
$$x \equiv 6 \pmod{17}.$$

These yield
$$x = 4 + 5r = 6 + 17s.$$

Setting this up as a congruence modulo 5 gives
$$17s \equiv -2 \pmod 5$$

or
$$2s \equiv -2 \equiv 8 \pmod 5$$

which has a solution $s = 4$, hence $x = 6 + 17 \cdot 4 = 74$.

Since $36 \cdot 74 \equiv 29 \pmod 5$ and $36 \cdot 74 \equiv 29 \pmod{17}$, we obtain the solution $x = 74$ of the original congruence
$$36x \equiv 29 \pmod{85}.$$

See Section 12C for a nontrivial extension of this strategy.

The strategy of Example 6 can be applied to systems of the form
$$ax \equiv b \pmod m$$
$$cx \equiv d \pmod n.$$

Example 7. To solve
$$11x \equiv 13 \pmod{20}$$
$$9x \equiv 17 \pmod{25},$$

we first solve $11x \equiv 13 \pmod{20}$, for example, by observing that $11 \cdot 11 = 121 \equiv 1 \pmod{20}$, so
$$x = 13 \cdot 11 \equiv 143 \equiv 3 \pmod{20}.$$

Then we solve $9x \equiv 17 \pmod{25}$, for example, by observing that $9 \cdot 11 \equiv -1 \pmod{25}$, so
$$x \equiv -17 \cdot 11 \equiv 8 \cdot 11 \equiv 88 \equiv 13 \pmod{25}.$$

So the original system is equivalent to
$$x \equiv 3 \pmod{20}$$
$$x \equiv 13 \pmod{25}.$$

Or, having found that the first congruence is equivalent to $x \equiv 3 \pmod{20}$, we can substitute $x = 3 + 20k$ into the second congruence to get
$$9(3 + 20k) \equiv 17 \pmod{25}$$

and simplify to get
$$5k \equiv -10 \pmod{25},$$

which has a solution $k = 3$, $x = 3 + 20 \cdot 3 = 63$. Once we find one solution, then, since $[25, 20] = 100$, the general solution is
$$x \equiv 63 \pmod{100}.$$

Exercises.

1. Find the smallest positive solution, if any, of

$$x \equiv 9 \quad (\text{mod } 16)$$
$$x \equiv 17 \quad (\text{mod } 28).$$

2. Find the smallest positive solution, if any, of

$$x \equiv 24 \quad (\text{mod } 66)$$
$$x \equiv 9 \quad (\text{mod } 48).$$

3. Find the smallest positive solution, if any, of

$$x \equiv 10 \quad (\text{mod } 15)$$
$$x \equiv 17 \quad (\text{mod } 28).$$

4. Write the set of solutions of

$$x \equiv 5 \quad (\text{mod } 24)$$
$$x \equiv 17 \quad (\text{mod } 18),$$

if any, as the solutions to a single congruence.

5. Write the set of solutions of

$$x \equiv 23 \quad (\text{mod } 36)$$
$$x \equiv 3 \quad (\text{mod } 8),$$

if any, as the solutions to a single congruence.

6. Write the set of solutions of

$$x \equiv 17 \quad (\text{mod } 30)$$
$$x \equiv 7 \quad (\text{mod } 40),$$

if any, as the solutions to a single congruence.

7. Find all solutions of

$$x \equiv 2 \quad (\text{mod } 12)$$
$$x \equiv 8 \quad (\text{mod } 10)$$
$$x \equiv 10 \quad (\text{mod } 14).$$

Write the set of solutions, if any, as the solutions to a single congruence.

8. Find all solutions of

$$x \equiv 5 \quad (\text{mod } 14)$$
$$x \equiv 7 \quad (\text{mod } 8)$$
$$x \equiv 13 \quad (\text{mod } 18).$$

Write the set of solutions, if any, as the solutions to a single congruence.

9. Find all solutions of

$$x \equiv 2 \quad (\text{mod } 12)$$
$$x \equiv 16 \quad (\text{mod } 25)$$
$$x \equiv 16 \quad (\text{mod } 35).$$

Write the set of solutions, if any, as the solutions to a single congruence.

10. Show that the system

$$x \equiv 1 \quad (\text{mod } m)$$
$$x \equiv 0 \quad (\text{mod } n)$$
$$x \equiv 0 \quad (\text{mod } q),$$

has a solution if and only if m and nq are coprime.

11. Find the solution closest to 0 of

$$x \equiv 1 \quad (\text{mod } 11)$$
$$x \equiv 0 \quad (\text{mod } 25)$$
$$x \equiv 0 \quad (\text{mod } 32).$$

12. A battalion of 208 men went off to a tough battle, and when the unwounded survivors regrouped after the encounter, their commanding officer was so shaken by the battle and the casualties that he could not count them. So he had them form groups of three, and found that there was one left over. He had them form groups of ten, and there were six left over. He had them form groups of seven, and there were three left over. How many unwounded survivors were there?

13. The West End Athletic Club has around 60 members. Many of the members volunteered to participate in all three events in the club's three-day annual competition with the East End A.C. across town. The competition involved soccer on one day, softball the next day, and volleyball on the final day. When the participating West Enders grouped themselves into 11 person soccer teams, there were 2 participants left over. When they grouped themselves into 9 person softball teams, there was one left over. When they grouped themselves into 6 person volleyball teams, there were 4 left over. How many West End participants were there?

14. A 19 person youth group sold boxes of cookies, at a profit of one dollar per box, and agreed to share the profits equally. When the boxes were sold out, their advisor collected the money, paid the supplier and planned to distribute the profits among the 19 members at the next meeting. She found that after dividing the dollar bills into 19 piles, there were 3 dollars left over. Then she was informed that one member had quit the group. So she redivided the dollar bills into 18 piles and found that there were 10 left over. Just before the next meeting, another member quit. So she redivided the dollar bills into 17 piles and found that there were no bills left over. If no member sold more than 15 boxes of cookies, how many boxes of cookies did the group sell?

15. Find numbers t, u, v so that $33t + 2 = 20u + 13 = 29v + 1$.

16. Find the smallest positive solution to the classical Chinese problem (Yih-hing, 707 A. D.),

$$x \equiv 1 \quad (\text{mod } 2)$$
$$x \equiv 2 \quad (\text{mod } 3)$$
$$x \equiv 5 \quad (\text{mod } 6)$$
$$x \equiv 5 \quad (\text{mod } 12)$$

17. (from Su Shu Chiu Chang, 1247 AD [Joseph (2000)]: Three thieves, A, B, C, entered a rice shop and stole three vessels, X, Y and Z, of equal size filled to the brim with rice, but whose exact capacity was not known. When the thieves were caught and the vessels recovered, it was found that all that was left in Vessels X, Y and Z were 1 ko, 14 ko and 1 ko, respectively. The captured thieves confessed that they did not know the exact quantities that they had stolen. But A said he had used a "horse ladle" (capacity 19 ko) and taken the rice from Vessel X. B confesses to using his wooden shoe (capacity 17 ko) to take rice from Vessel Y. C admitted he had used a bowl (capacity 12 ko) to take the rice from Vessel Z. What was the total amount of rice stolen?

18. From Brahmagupta (c. 625 AD) [Van der Waerden (1983)]: What number, divided by 6, has a remainder of 5, and by 5, a remainder of 4, and by 4, a remainder of 3, and by 3, a remainder of 2?

19. (from Sun Tsu Suan Ching, 4th century AD) [Van der Waerden (1983)]: There is an unknown number of objects. When counted by threes, the remainder is 2; when counted by fives, the remainder is 3; and when counted by sevens, the remainder is 2. How many objects are there?

20. Find the smallest number $a > 0$ that is congruent to 1 modulo 2, 3, 4, 5 and 6 and is a multiple of 7.

21. Use the method of Example 6 to solve $81x \equiv 11 \pmod{100}$.

22. Use the method of Example 6 to solve $83x \equiv 100 \pmod{143}$. Also solve the problem by setting up the equation $83x + 143y = 100$ and solving it using Bezout's identity methods. Which method do you prefer?

23. Use the method of Example 6 to solve $23x \equiv 1 \pmod{504}$. Also solve the problem by Bezout's identity methods.

24. From the Aryabhatiya (498 AD): Find the smallest number x if $8x$ divided by 29 gives a remainder of 4, and $17x$ divided by 45 gives a remainder of 7.

25. Find all solutions of

$$8x \equiv 2 \quad (\text{mod } 18)$$
$$9x \equiv 28 \quad (\text{mod } 30)$$

26. Find all solutions of

$$7x \equiv 2 \quad (\text{mod } 17)$$
$$13x \equiv 21 \quad (\text{mod } 22)$$

27. Find all solutions of

$$11x \equiv 13 \quad (\text{mod } 16)$$
$$13x \equiv 9 \quad (\text{mod } 28)$$

28. Show that there are 12 pairs of numbers (a_1, a_2) with

$$0 \leq a_1 < 4, 0 \leq a_2 < 6$$

so that

$$x \equiv a_1 \quad (\text{mod } 4)$$
$$x \equiv a_2 \quad (\text{mod } 6)$$

has a solution.

29. Show that there are $[m_1, m_2]$ pairs of numbers (a_1, a_2) with

$$0 \leq a_1 < m_1, 0 \leq a_2 < m_2$$

so that

$$x \equiv a_1 \quad (\text{mod } m_1)$$
$$x \equiv a_2 \quad (\text{mod } m_2)$$

has a solution.

30. Generalize the last exercise to more than two congruences.

31. Let $(a, m) = 1$ and $(b, n) = 1$. Show that if

$$c \equiv a \quad (\text{mod } m)$$
$$c \equiv b \quad (\text{mod } n),$$

then $(c, mn) = 1$.

32. A famous theorem of Dirichlet states that a and m are coprime, then there are infinitely many prime numbers p so that $p \equiv a \pmod{m}$. Show that if a_1 and m_1 are coprime, a_2 and m_2 are coprime, and the pair of congruences

$$x \equiv a_1 \quad (\text{mod } m_1)$$
$$x \equiv a_2 \quad (\text{mod } m_2)$$

has a solution, then there are infinitely many primes p so that $x = p$ is a solution of the pair of congruences.

33. (i) Prove that if $m = rs$ with r, s coprime, then $x^2 \equiv 1 \pmod{m}$ has at least four solutions.
   (ii) If r and s are primes, $r \neq s$, then $x^2 \equiv 1 \pmod{m}$ has exactly four solutions.

B. Another Solution Method

In this section we give an alternate method for solving a system of n congruences when the moduli are pairwise coprime. The method is useful for solving several systems of congruences all involving the same moduli.

The idea of this method is that we solve a collection of special systems and then obtain a solution of the original congruence as a linear combination of the solutions of the special systems. Here are some examples.

Example 8. Consider the pair of congruences

$$x \equiv 15 \pmod{20}$$
$$x \equiv 3 \pmod{17}.$$

We know there is a solution, since 20 and 17 are coprime. To solve this system we first solve the two systems

$$x \equiv 1 \pmod{20}$$
$$x \equiv 0 \pmod{17}$$

and

$$x \equiv 0 \pmod{20}$$
$$x \equiv 1 \pmod{17}.$$

In the first system, $x = e_1$ is a solution if

$$e_1 = 1 + 20r = 17s.$$

Making this a congruence modulo 20, we obtain

$$17s \equiv 1 \pmod{20}.$$

Since

$$17 \equiv -3 \pmod{20}$$

and the inverse of 3 modulo 20 is 7, we can set $s = -7$, hence $e_1 = -119$. Similarly, in the second system, $x = e_2$ is a solution if

$$e_2 = 20t = 1 + 17u.$$

Making this a congruence modulo 20, we obtain

$$17u \equiv -1 \pmod{20}.$$

Multiplying this congruence by -1 yields

$$3u \equiv 1 \pmod{20},$$

so we can set $u = 7$ and $e_2 = 120$.

Having found e_1 and e_2, we may find a solution x_0 of the original system by setting

$$x_0 = 15e_1 + 3e_2 = 15 \cdot (-119) + 3 \cdot 120 = -1425.$$

(Check: modulo 20, $x_0 \equiv 15 \cdot (-119) \equiv 15 \cdot 1 = 15$, and modulo 17, $x_0 \equiv 3 \cdot 120 \equiv 3 \cdot 1 = 3$, as desired.)

As before, since $[20, 17] = 20 \cdot 17 = 340$, the general solution is $x \equiv -1425$ (mod 340) and the smallest positive solution is $x = -1425 + 340 \cdot 5 = 275$.

Note that the strategy of finding e_1 and e_2 first will not work if the moduli are not coprime. For example, if we tried to solve the system

$$x \equiv 1 \quad (\text{mod } 20)$$
$$x \equiv 0 \quad (\text{mod } 18)$$

we would get that $x = 1 + 20r = 18s$. But the equation $20r - 18s = 1$ has no solution since 20 and 18 are not coprime.

Example 9. Suppose now we wish to solve

$$x \equiv 8 \quad (\text{mod } 20)$$
$$x \equiv 11 \quad (\text{mod } 17).$$

We've done all the work: knowing e_1 and e_2 for these moduli from the previous example, we simply set

$$x_0 = 8e_1 + 11e_2 = 8 \cdot -119 + 11 \cdot 120 = 368.$$

The general solution is then

$$x \equiv 368 \quad (\text{mod } 340)$$

and the smallest positive solution is $x = 28$.

If we wish to solve

$$x \equiv 9 \quad (\text{mod } 20)$$
$$x \equiv 13 \quad (\text{mod } 17),$$

we obtain $x_0 = 9e_1 + 13e_2 = 9 \cdot (-119) + 13 \cdot 120 = -1071 + 1560 = 489$; the smallest positive solution is then $x = 489 - 340 = 149$.

Example 10. We solve

$$x \equiv 2 \quad (\text{mod } 13)$$
$$x \equiv 8 \quad (\text{mod } 10)$$
$$x \equiv 7 \quad (\text{mod } 11).$$

To do this, we solve the three systems with the same moduli:

$$e_1 \equiv 1 \quad (\text{mod } 13)$$
$$e_1 \equiv 0 \quad (\text{mod } 10)$$
$$e_1 \equiv 0 \quad (\text{mod } 11),$$

and
$$e_2 \equiv 0 \quad (\bmod\ 13)$$
$$e_2 \equiv 1 \quad (\bmod\ 10)$$
$$e_2 \equiv 0 \quad (\bmod\ 11),$$

and
$$e_3 \equiv 0 \quad (\bmod\ 13)$$
$$e_3 \equiv 0 \quad (\bmod\ 10)$$
$$e_3 \equiv 1 \quad (\bmod\ 11).$$

These systems are not as onerous as they might seem at first sight, because they all reduce immediately to systems of two congruences, namely
$$e_1 \equiv 1 \quad (\bmod\ 13)$$
$$e_1 \equiv 0 \quad (\bmod\ 110),$$

and
$$e_2 \equiv 0 \quad (\bmod\ 143)$$
$$e_2 \equiv 1 \quad (\bmod\ 10),$$

and
$$e_3 \equiv 0 \quad (\bmod\ 130)$$
$$e_3 \equiv 1 \quad (\bmod\ 11).$$

For the first, $e_1 = 1 + 13r = 110s$, hence s is the inverse of 110 modulo 13. Since $110 \equiv 6 \ (\bmod\ 13)$, $s = -2$ and $e_1 = -220$.

For the second, $e_2 = 1 + 10t = 143u$, hence u is the inverse of 143 modulo 10, hence $u = -3$ and $e_2 = -429$.

For the third, $e_3 = 1 + 11v = 130w$, hence w is the inverse of 130 modulo 11. Since $130 \equiv -2 \ (\bmod\ 11)$, $w = 5$ and $e_3 = 650$.

Having found e_1, e_2 and e_3, a solution of the original congruences
$$x \equiv 2 \quad (\bmod\ 13)$$
$$x \equiv 8 \quad (\bmod\ 10)$$
$$x \equiv 7 \quad (\bmod\ 11).$$

is
$$x_0 = 2e_1 + 8e_2 + 7e_3 = 2 \cdot (-220) + 8 \cdot (-429) + 7 \cdot 650 = 678.$$

The general solution to the original congruences is
$$x \equiv 678 \quad (\bmod\ 1430)$$

since $10 \cdot 11 \cdot 13 = 1430$.

If we wish now to solve
$$x \equiv 5 \quad (\bmod\ 13)$$
$$x \equiv 3 \quad (\bmod\ 10)$$
$$x \equiv 6 \quad (\bmod\ 11),$$

a solution is

$$x_0 = 5e_1 + 3e_2 + 6e_3 = 5 \cdot (-220) + 3 \cdot (-429) + 6 \cdot 650$$
$$= -1100 - 1287 + 3900 = 1513,$$

and the smallest positive solution is $x = 1513 - 1430 = 83$.

If we are interested in solving only a single system of n congruences to pairwise prime moduli, the method of section 12A involves solving $n - 1$ systems of two congruences, while the method of this section involves solving n systems of two congruences. An additional advantage of the method of 12A is that it deals with systems in which the moduli are not pairwise coprime. The advantage of the method of this section is that it is much faster for solving more than one system of n congruences to the same moduli.

Babylonian multiplication. Here is a fanciful application of the method of this section.

Imagine that you have been transported to a society like ancient Babylonia, where "paper" consists of heavy clay tablets, and numbers are in base 60. To multiply numbers like

$$(35, 43, 52) = 35 \cdot 60^2 + 43 \cdot 60 + 52$$

and

$$(14, 2, 47) = 14 \cdot 60^2 + 2 \cdot 60 + 47,$$

by the usual multiplication algorithm, it would appear that you need to either memorize the base 60 multiplication table, containing $\frac{59 \cdot 60}{2} = 1770$ products, or write the table on a clay tablet that is much too heavy to carry. So what do you do? Use the Chinese Remainder Theorem.

First observe that $5 \cdot 8 \cdot 9 \cdot 11 = 3960 > 59 \cdot 59$, and 5, 8, 9 and 11 are pairwise coprime. So we find e_5 satisfying

$$e_5 \equiv 1 \quad (\text{mod } 5)$$
$$e_5 \equiv 0 \quad (\text{mod } 8 \cdot 9 \cdot 11);$$

e_8 satisfying

$$e_8 \equiv 1 \quad (\text{mod } 8)$$
$$e_8 \equiv 0 \quad (\text{mod } 5 \cdot 9 \cdot 11);$$

e_9 satisfying

$$e_9 \equiv 1 \quad (\text{mod } 9)$$
$$e_9 \equiv 0 \quad (\text{mod } 5 \cdot 8 \cdot 11);$$

and e_{11} satisfying

$$e_{11} \equiv 1 \quad (\text{mod } 11)$$
$$e_{11} \equiv 0 \quad (\text{mod } 5 \cdot 8 \cdot 9).$$

We find that $e_5 = -1584$, $e_8 = -495$, $e_9 = -440$ and $e_{11} = -1440$.

To multiply $52 \cdot 47$, we find that

$$52 \cdot 47 \equiv 2 \cdot 2 \equiv -1 \quad (\text{mod } 5)$$
$$52 \cdot 47 \equiv 4 \cdot -1 \equiv 4 \quad (\text{mod } 8)$$
$$52 \cdot 47 \equiv -2 \cdot 2 \equiv -4 \quad (\text{mod } 9)$$
$$52 \cdot 47 \equiv -3 \cdot 3 \equiv 2 \quad (\text{mod } 11).$$

Then modulo 3960,

$$52 \cdot 47 \equiv (-1)e_5 + 4e_8 + (-4)e_9 + 2e_{11}$$
$$\equiv 1584 - 1980 + 1760 - 2880 \equiv -1516 \equiv 2444 \quad (\text{mod } 3960).$$

Since $52 \cdot 47 < 3960$ and $52 \cdot 47 \equiv 2444$ (mod 3960), we must have $52 \cdot 47 = 2444$.

We can set up a table of all the products that can arise in computations like this.

Since every number modulo 5 is congruent to 0, 1 or 2 or their negatives, our table only needs e_5 and $2e_5$. Similarly, for modulo 8, 9 and 11 our table only needs $e_8, 2e_8, 3e_8$ and $4e_8$; $e_9, 2e_9, 3e_9$ and $4e_9$; and $e_{11}, 2e_{11}, 3e_{11}, 4e_{11}$ and $5e_{11}$, all modulo 3960:

$\cdot$	e_5	e_8	e_9	e_{11}
1	-1584	-495	-440	-1440
2	792	-990	-880	1080
3		-1485	-1320	-360
4		-1980	-1760	-1800
5				720

For example, $3e_9$ is congruent modulo 3960 to -1320; while $4e_{11}$ is congruent modulo 3960 to -1800.

Example 11. We use the table to find $43 \cdot 47$.

We observe that

$$43 \cdot 47 \equiv 1 \quad (\text{mod } 5)$$
$$\equiv -3 \quad (\text{mod } 8)$$
$$\equiv -4 \quad (\text{mod } 9)$$
$$\equiv -3 \quad (\text{mod } 11),$$

so

$$43 \cdot 47 \equiv 1 \cdot e_5 - 3 \cdot e_8 - 4 \cdot e_9 - 3 \cdot e_{11} \quad (\text{mod } 3960)$$

and we read these terms off the table to get

$$43 \cdot 47 \equiv -1584 + 1485 + 1760 + 360 \equiv 2021 \quad (\text{mod } 3960).$$

Since $43 \cdot 47 < 3960$, $43 \cdot 47 = 2021$.

In this way we can solve our Babylonian "memory" problem by creating a clay tablet with just that small table of 15 numbers on it. With that table we can multiply any two numbers <60 using congruences to the moduli 5, 8, 9 and 11, and addition.

Exercises.

34. (i) Find e_1 so that
$$e_1 \equiv 1 \quad (\text{mod } 21)$$
$$e_1 \equiv 0 \quad (\text{mod } 25).$$

and e_2 so that
$$e_2 \equiv 0 \quad (\text{mod } 21)$$
$$e_2 \equiv 1 \quad (\text{mod } 25).$$

(ii) Find the smallest positive solution of

$$x \equiv 3 \quad (\text{mod } 21)$$
$$x \equiv 17 \quad (\text{mod } 25).$$

(iii) Find the smallest positive solution of

$$x \equiv 17 \quad (\text{mod } 21)$$
$$x \equiv -8 \quad (\text{mod } 25).$$

35. (i) Find e_1 so that
$$e_1 \equiv 1 \quad (\text{mod } 11)$$
$$e_1 \equiv 0 \quad (\text{mod } 13)$$
$$e_1 \equiv 0 \quad (\text{mod } 15),$$

find e_2 so that
$$e_2 \equiv 0 \quad (\text{mod } 11)$$
$$e_2 \equiv 1 \quad (\text{mod } 13)$$
$$e_2 \equiv 0 \quad (\text{mod } 15),$$

and e_3 so that
$$e_3 \equiv 0 \quad (\text{mod } 11)$$
$$e_3 \equiv 0 \quad (\text{mod } 13)$$
$$e_3 x \equiv 1 \quad (\text{mod } 15).$$

(ii) Find the smallest positive solution of

$$x \equiv 3 \quad (\text{mod } 11)$$
$$x \equiv 5 \quad (\text{mod } 13)$$
$$x \equiv 8 \quad (\text{mod } 15).$$

(iii) Find all solutions of

$$x \equiv 9 \quad (\text{mod } 11)$$
$$x \equiv 2 \quad (\text{mod } 13)$$
$$x \equiv -7 \quad (\text{mod } 15).$$

36. Multiply the base 60 numbers (39, 26) and (56, 44) using the table as in Example 11 to multiply digits.

37. Verify that $e_9 = -440$ is a solution of

$$e_9 \equiv 1 \quad (\text{mod } 9)$$
$$e_9 \equiv 0 \quad (\text{mod } 5 \cdot 8 \cdot 11).$$

38. Design an adaptation of Example 11 to multiply numbers in base 1000. Which moduli will give the smallest possible number of table entries?

C. Some Applications to RSA Cryptography

RSA Decrypting. This application refers to the RSA cryptosystem described in section 10A.

Suppose Bob develops an RSA cryptosystem for Alice to use to send messages to Bob. Recall that Bob does this as follows: he finds two large primes p and q and sets $m = pq$ (the modulus). He picks an encoding exponent e that is coprime to $\phi(m) = (p-1)(q-1)$, finds a decoding exponent d satisfying $ed \equiv 1 \pmod{\phi(m)}$, and sends m and e to Alice. To make computations easy for Alice, he chooses e to be a small number (such as $e = 3$, or 7).

To send the message w to Bob, Alice computes $c = w^e$ modulo m and sends Bob c. To determine w, Bob must compute c^d modulo m. But c is going to be a number of almost the same number of digits as m, and since e is small, d will have almost the same number of digits as m. Thus determining c^d modulo m takes a bit of effort.

But Bob has the advantage that he knows that $m = pq$. So he can proceed as follows:

(i) Compute $c_1 \equiv c^d \pmod{p}$ and $c_2 \equiv c^d \pmod{q}$ where c_1 and c_2 are as small as possible.

(ii) Find y so that

$$y \equiv c_1 \quad (\text{mod } p)$$
$$y \equiv c_2 \quad (\text{mod } q).$$

Then

$$y \equiv c^d \quad (\text{mod } p)$$
$$y \equiv c^d \quad (\text{mod } q),$$

so

$$y \equiv c^d \quad (\text{mod } pq).$$

and $pq = m$. If we choose $0 < y < m$, then since Alice's original word w satisfies

$$w \equiv c^d \quad (\text{mod } m).$$

and $0 < w < m$, we must have $y = w$.

Example 12. To illustrate how this works, suppose the modulus $m = 187 = 11 \cdot 17$, the encoding exponent is $e = 3$ and Alice wants to send $w = 127$ to Bob. Alice encodes w to get $c = 127^3 \equiv 172 \pmod{187}$, and sends c to Bob. The decoding exponent is $d = 107$, so Bob needs to find $c^d = 172^{107} \pmod{187}$. So he computes

$$172^{107} \pmod{11}$$

and

$$172^{107} \pmod{17}.$$

Now $172 \equiv 7 \pmod{11}$, so $172^{107} \equiv 7^{107}$, and this is congruent to 7^7, since $7^{10} \equiv 1 \pmod{11}$ by Fermat's Theorem. One can check easily that $7^7 \equiv 6 \pmod{11}$.

Also, $172 \equiv 2 \pmod{17}$, so again using Fermat's Theorem, $172^{107} \equiv 2^{107} \equiv 2^{11} \pmod{17}$. But $2^4 \equiv -1 \pmod{17}$, so $2^{11} \equiv 2^3 = 8 \pmod{17}$. Thus $w \equiv c^d = 172^{107} \pmod{187}$ satisfies

$$w \equiv 6 \pmod{11}$$
$$w \equiv 8 \pmod{17}.$$

One checks that $w = 127$ as follows:

$$w = 8 + 17r = 6 + 11s,$$

so

$$17r \equiv -2 \pmod{11},$$

hence

$$r \equiv -4 \pmod{11}$$

and so $w \equiv 8 + 17(-4) = -60 \pmod{187}$. Since $0 < w < 187$, $w = -60 + 187 = 127$.

It has been estimated that decrypting using the Chinese Remainder Theorem in this way takes somewhere between 1/4 and 1/3 of the time needed to compute c^d modulo m directly. Note however, that only someone who knows the factorization of the modulus m can use this method. That's why, if Bob designed the code, then the exponent used by Alice should be small to minimize her computations, since she cannot use the Chinese Remainder Theorem.

Common encoding exponents. Suppose Alice, a financial advisor, has three clients, Bill, Bob and Brian, each of whom has his own modulus, m_1, m_2 and m_3 respectively. Alice wants to send privileged information about a particular stock to each of them. For convenience, Alice always uses the encoding exponent $e = 3$. So Alice sends the message w to each of them, as follows: To Bill she sends $c_1 \equiv w^3 \pmod{m_1}$. To Bob she sends $c_2 \equiv w^3 \pmod{m_2}$. To Brian she sends $c_3 \equiv w^3 \pmod{m_3}$. Eve (perhaps an agent looking for violations of insider trading laws) intercepts c_1, c_2, c_3 and knows m_1, m_2, m_3 and $e = 3$. She doesn't know w or w^3, but she knows that

$$w^3 \equiv c_1 \quad (\text{mod } m_1)$$
$$w^3 \equiv c_2 \quad (\text{mod } m_2)$$
$$w^3 \equiv c_3 \quad (\text{mod } m_3).$$

So she solves

$$t \equiv c_1 \quad (\text{mod } m_1)$$
$$t \equiv c_2 \quad (\text{mod } m_2)$$
$$t \equiv c_3 \quad (\text{mod } m_3)$$

for some number $t < m_1 m_2 m_3$. Then

$$t \equiv w^3 \quad (\text{mod } m_1 m_2 m_3).$$

But $w < m_i$ for $i = 1, 2, 3$, so $w^3 < m_1 m_2 m_3$. Thus $t = w^3$.

Once Eve finds t, she can simply compute the cube root of t to find the message w.

The moral of this example is that one should not send the same message with the same small encoding exponent to many different recipients.

Safeprimes. A prime number p is called a safeprime if $p = 2q + 1$ where q is also prime. Examples: p = 5, 7, 11, 23, ...

Safeprimes are useful primes for constructing moduli for RSA codes, because there are some factoring algorithms, such as the Pollard $p - 1$ algorithm, that factor large numbers m more efficiently when one of the prime factors p of m has the property that $p - 1$ is a product of only small primes. Any odd prime p has the property that the largest possible prime factor of $p - 1$ is $(p - 1)/2$. Hence a safeprime is a prime p so that $p - 1$ has a prime factor that is as large as possible. An RSA modulus that is a product of safeprimes will be maximally effective in thwarting factoring algorithms such as the Pollard $p - 1$ algorithm (Section 10C).

Safeprimes must satisfy certain congruences, as notes in the following exercises.

Exercises.

39. Show that if $p > 20$ is a safeprime, then
(i) $p \equiv 2 \pmod 3$
(ii) $p \equiv 3 \pmod 4$
(iii) $p \equiv 2, 3$ or $4 \pmod 5$
(iv) List the three congruence classes modulo 60 that can contain safeprimes.
(v) Find six safeprimes > 60.

40. Show that for every odd prime l, a safeprime $p > l$ must satisfy

$$p \not\equiv 0 \text{ or } 1 \pmod l.$$

41. List the congruence classes modulo 84 that can contain safeprimes.

42. A prime p is *special* if $p = 2p_1 + 1$ is a safeprime and also $p_1 = 2p_2 + 1$ is a safeprime.

(i) Show that a special prime must be congruent to 7 modulo 8;

(ii) Show that for each odd prime l, a special prime p with $p_2 > l$ cannot be congruent to 0, 1 or 3 modulo l. What are the corresponding conditions on p_2?

43. Find the least non-negative residue of

$$95^{1002} \quad (\mathrm{mod}\ 217).$$

44. Find the least non-negative residue of

$$100^{246} \quad (\mathrm{mod}\ 247).$$

(Note: $247 = 13 \cdot 19$.)

45. Suppose you know $m = 17 \cdot 23 = 391$, Alice's exponent is 3, and she sends you the encrypted word $c = 21$. The decrypting exponent is 235. Find w.

46. Eve knows that Alice sent the same plaintext message w to Bob with $m = 17, e = 3$, to Bill with $m = 23$ and $e = 3$, and to Brent with $m = 31, e = 3$. She intercepts the encrypted messages:

$$c_1 \equiv 11 \quad (\mathrm{mod}\ 17)$$
$$c_2 \equiv 3 \quad (\mathrm{mod}\ 23)$$
$$c_3 \equiv 23 \quad (\mathrm{mod}\ 31).$$

What is w?

D. Homomorphisms and Euler's ϕ-Function

Recall from Section 7D that if R, S are rings, a function $f : R \to S$ is a ring homomorphism if

$$f(r_1 + r_2) = f(r_1) + f(r_2)$$
$$f(r_1 r_2) = f(r_1) f(r_2)$$
$$f(1) = 1.$$

The kernel of a ring homomorphism $f : R \to S$ is the set

$$\ker f = \{r \text{ in } R | f(r) = 0\}.$$

We showed that a ring homomorphism $f : \mathbb{Z} \to S$ is uniquely determined by $f(1)$, for then $f(n) = f(1) + \ldots + f(1)$ (n summands) in S if $n > 0$, and $f(-n) = -f(n)$.

What about ring homomorphisms from $\mathbb{Z}/m\mathbb{Z}$ to S?

It turns out that every such ring homomorphism arises from a homomorphism from $\mathbb{Z}$ to S.

To see this, suppose

$$g : \mathbb{Z}/m\mathbb{Z} \to S$$

is a ring homomorphism. Let

$$\gamma_m : \mathbb{Z} \to \mathbb{Z}/m\mathbb{Z}$$

be the homomorphism that takes r in $\mathbb{Z}$ to the congruence class $[r]_m$ in $\mathbb{Z}/m\mathbb{Z}$. Then the composition of γ_m followed by g is a homomorphism f from $\mathbb{Z}$ to S: f is defined by

$$f(r) = g\gamma_m(r) = g([r]_m).$$

Moreover, if $r = mk$, then $f(mk) = g([mk]_m) = g([0]_m) = 0$. So the kernel of f contains $m\mathbb{Z}$.

To sum up:

Proposition 4. *Every ring homomorphism $g : \mathbb{Z}/m\mathbb{Z} \to S$ lifts to a ring homomorphism $f : \mathbb{Z} \to S$ so that $m\mathbb{Z} \subseteq \ker(f)$.*

The converse is useful.

Proposition 5 (Homomorphism Theorem). *Let S be a commutative ring and let $f : \mathbb{Z} \to S$ be the homomorphism defined by $f(n) = n \cdot 1_S$ for all n in $\mathbb{Z}$. If f is not one-to-one and $\ker(f) \supseteq m\mathbb{Z}$ for some $m \neq 0$ in $\mathbb{Z}$, then f induces a homomorphism $\overline{f}$ from $\mathbb{Z}/m\mathbb{Z}$ onto $\{n \cdot 1_S | n$ in $\mathbb{Z}\}$, defined by $\overline{f}([a]_m) = f(a) = a \cdot 1_S$.*

If $\ker(f) = m\mathbb{Z}$ then $\overline{f}$ is an isomorphism from $\mathbb{Z}/m\mathbb{Z}$ onto

$$\{n \cdot 1_S \text{ in } \mathbb{Z}\} \subseteq S.$$

We proved this as Proposition 23 of Section 7D.

For a useful application of the Homomorphism Theorem, we introduce products of rings.

Products of rings. Let R, S be two sets. The product of R and S, written $R \times S$, is the set of ordered pairs (r, s) where r is in R, s in S. The notion of a set of ordered pairs should be familiar from analytic geometry. Assigning coordinates to points in the plane gives a one-to-one correspondence between points in the plane and the set $\mathbb{R} \times \mathbb{R}$ of ordered pairs of real numbers.

Suppose R and S are not just sets, but are commutative rings. Then the product $R \times S$ can be made into a commutative ring via coordinatewise operations, as follows:

$$(r, s) + (r', s') = (r + r', s + s'),$$

$$(r, s) \cdot (r', s') = (rr', ss'),$$

$$-(r, s) = (-r, -s).$$

The operations on $R \times S$ are defined by using the operations of R and S in the respective coordinates.

The zero and multiplicative identity elements are

$$0 = (0,0),$$
$$1 = (1,1).$$

With these definitions it is easy to see that if R and S are commutative rings, then $R \times S$ is a commutative ring.

If R and S are rings with a finite number of elements, say R has m elements and S has n elements, then $R \times S$ has mn elements.

Example 13. $\mathbb{Z}/2\mathbb{Z} \times \mathbb{Z}/3\mathbb{Z}$ has six elements. Here is its multiplication table:

$\cdot$	$(0,0)$	$(1,1)$	$(0,2)$	$(1,0)$	$(0,1)$	$(1,2)$
$(0,0)$	$(0,0)$	$(0,0)$	$(0,0)$	$(0,0)$	$(0,0)$	$(0,0)$
$(1,1)$	$(0,0)$	$(1,1)$	$(0,2)$	$(1,0)$	$(0,1)$	$(1,2)$
$(0,2)$	$(0,0)$	$(0,2)$	$(0,1)$	$(0,0)$	$(0,2)$	$(0,1)$
$(1,0)$	$(0,0)$	$(1,0)$	$(0,0)$	$(1,0)$	$(0,0)$	$(1,0)$
$(0,1)$	$(0,0)$	$(0,1)$	$(0,2)$	$(0,0)$	$(0,1)$	$(0,2)$
$(1,2)$	$(0,0)$	$(1,2)$	$(0,1)$	$(1,0)$	$(0,2)$	$(1,1)$

Evidently, $(0,0)$ acts as the zero element, and $(1,1)$ as the identity element.

We can find the units and zero divisors of a product of rings:

Proposition 6. *(i) (a,b) is a unit of $R \times S$ iff a is a unit of R and b is a unit of S.*

(ii) (a,b) in $R \times S$ is a zero divisor iff $(a,b) \neq (0,0)$ and either a is zero or a zero divisor of R, or b is zero or a zero divisor of S.

Proof. Part (i) is easy. To prove (ii), suppose $(a,b) \neq (0,0)$. If a is a zero divisor, and a' is non-zero in R with $aa' = 0$, then for every b in S, $(a,b)(a',0) = (0,0)$, so (a,b) is a zero divisor. If $a = 0$, then $(0,b)(1,0) = (0,0)$, so $(0,b)$ is a zero divisor. Similarly if b is either a zero divisor or zero.

Conversely, if (a,b) is a zero divisor, then $(a,b)(a',b') = (0,0)$ for some $(a',b') \neq (0,0)$ in $R \times S$. Then $aa' = 0$ in R, and $bb' = 0$ in S. Either $a' \neq 0$ or $b' \neq 0$. If $a' \neq 0$, then $a = 0$ or is a zero divisor; if $b' \neq 0$ then $b = 0$ or is zero divisor. That proves (ii). □

The Homomorphism Theorem yields the following description of $\mathbb{Z}/m\mathbb{Z}$ when m is a product of two coprime numbers r, s

Theorem 7. *Let $m = rs$ where r and s are coprime natural numbers ≥ 2. Then there is an isomorphism of rings*

$$\psi : \mathbb{Z}/m\mathbb{Z} \rightarrow \mathbb{Z}/r\mathbb{Z} \times \mathbb{Z}/s\mathbb{Z}$$

given by $\psi([a]_m) = ([a]_r, [a]_s)$.

For example, this theorem says that the ring $\mathbb{Z}/2\mathbb{Z} \times \mathbb{Z}/3\mathbb{Z}$ looks just like the ring $\mathbb{Z}/6\mathbb{Z}$. In fact, the map ψ in this case works as follows: here $[a]$ means $[a]_6$, and (a,b) means $([a]_2, [b]_3)$.

$$\psi([0]) = (0,0),$$
$$\psi([1]) = (1,1),$$
$$\psi([2]) = (2,2) = (0,2),$$
$$\psi([3]) = (3,3) = (1,0),$$
$$\psi([4]) = (4,4) = (0,1),$$
$$\psi([5]) = (5,5) = (1,2).$$

Thus the two units, $[1]$ and $[5]$, of $\mathbb{Z}/6\mathbb{Z}$ correspond under the isomorphism ψ to the two units $(1,1)$ and $(1,2)$ of $\mathbb{Z}/2\mathbb{Z} \times \mathbb{Z}/3\mathbb{Z}$; the zero divisors $[2], [3]$ and $[4]$ of $\mathbb{Z}/6\mathbb{Z}$ correspond to the zero divisors $(0,2), (1,0)$ and $(0,1)$ of $\mathbb{Z}/2\mathbb{Z} \times \mathbb{Z}/3\mathbb{Z}$.

The proof of Theorem 2 relates to the Chinese Remainder Theorem.

Proof. Let $m = rs$ and let

$$\phi : \mathbb{Z} \to \mathbb{Z}/r\mathbb{Z} \times \mathbb{Z}/s\mathbb{Z}$$

by $\phi(a) = ([a]_r, [a]_s) = a([1]_r, [1]_s)$. If $a = mk$, then $([mk]_r, [mk]_s) = 0$, since $m = rs$. So mk is in the kernel of ϕ for every k.

By the Homomorphism Theorem we get an induced homomorphism

$$\psi : \mathbb{Z}/m\mathbb{Z} \to \mathbb{Z}/r\mathbb{Z} \times \mathbb{Z}/s\mathbb{Z}$$

by $\psi([a]_m) = ([a]_r, [a]_s)$. To show that ψ is one-to-one, we look at the kernel of ψ, namely, the set of $[a]_m$ so that $\psi([a]_m) = 0$ in $\mathbb{Z}/r\mathbb{Z} \times \mathbb{Z}/s\mathbb{Z}$. Now $\psi([a]_m) = 0$ if and only if $[a]_r = 0$ and $[a]_s = 0$; that is, r divides a and s divides a. But since r and s are coprime, that implies that m divides a, so $[a]_m = 0$. That means that the kernel of ψ consists of only the zero element of $\mathbb{Z}/m\mathbb{Z}$, namely $[0]_m$. Hence ψ is one-to-one.

To show that ψ is an isomorphism, we only need to show that ψ is onto. For that, we have two possible arguments. One uses the Chinese RemainderTheorem; the other reproves the Chinese Remainder Theorem.

Here is the argument that uses the Chinese Remainder Theorem:

Let $([b]_r, [c]_s)$ be an arbitrary element of $\mathbb{Z}/r\mathbb{Z} \times \mathbb{Z}/s\mathbb{Z}$. To show that $([b]_r, [c]_s) = ([a]_r, [a]_s) = \psi(a)$ for some integer a mod m, we must find an integer a so that

$$a \equiv b \quad (\text{mod } r),$$
$$a \equiv c \quad (\text{mod } s).$$

But since r and s are coprime, an integer a solving this pair of simultaneous congruences can always be found. Thus ψ is onto.

Conversely, if we can show that ψ is onto without using the Chinese Remainder Theorem, then for r, s coprime, the pair of congruences

$$x \equiv b \quad (\text{mod } r),$$
$$x \equiv c \quad (\text{mod } s),$$

has a solution for every b, c, and so the Chinese Remainder Theorem holds for sets of two congruences to coprime moduli.

Why is ψ onto? The argument that doesn't use the Chinese Remainder Theorem is a counting argument. We know that ψ is a one-to-one function from a set with m elements, namely, $\mathbb{Z}/m\mathbb{Z}$, to another set with m elements, namely, $\mathbb{Z}/r\mathbb{Z} \times \mathbb{Z}/s\mathbb{Z}$. A one-to-one function from a set R of m elements to another set S of m elements must be onto, because if ψ is one-to-one, then $\psi(R)$ must have the same number of elements as R does. Thus $\psi(R)$ is an m-element subset of the m-element set S. Hence $\psi(R) = S$. $\square$

For $m = rs$ with $(r, s) = 1$, the isomorphism $\psi : \mathbb{Z}/m\mathbb{Z} \to \mathbb{Z}/r\mathbb{Z} \times \mathbb{Z}/s\mathbb{Z}$ relates to the alternate method of solving the pair of congruences

$$x \equiv b \pmod{r}$$
$$x \equiv c \pmod{s}$$

presented in Section 12B. For if e_1 satisfies

$$x \equiv 1 \pmod{r}$$
$$x \equiv 0 \pmod{s},$$

then (dropping the bracket notation), $\psi(e_1) = (1, 0)$. Similarly, if e_2 satisfies

$$x \equiv 0 \pmod{r}$$
$$x \equiv 1 \pmod{s},$$

then $\psi(e_2) = (0, 1)$. To solve

$$x \equiv b \pmod{r}$$
$$x \equiv c \pmod{s}$$

means to find the unique a modulo m so that $\psi(a) = (a, a) = (b, c)$ in $\mathbb{Z}/r\mathbb{Z} \times \mathbb{Z}/s\mathbb{Z}$. But

$$\begin{aligned}
\psi(be_1 + ce_2) &= \psi(be_1) + \psi(ce_2) \\
&= \psi(b)\psi(e_1) + \psi(c)\psi(e_2) \\
&= (b, b)(1, 0) + (c, c)(0, 1) \\
&= (b, 0) + (0, c) = (b, c).
\end{aligned}$$

Since ψ is one-to-one, $be_1 + ce_2$ is the unique a modulo m that maps to (b, c), hence is the unique solution modulo m of the original congruences

$$x \equiv b \pmod{r}$$
$$x \equiv c \pmod{s}.$$

As we observed with the example of $\mathbb{Z}/6\mathbb{Z}$ above, units of $\mathbb{Z}/6\mathbb{Z}$ correspond to the units of $\mathbb{Z}/2\mathbb{Z} \times \mathbb{Z}/3\mathbb{Z}$. This is always the case. We have

Proposition 8. *If* $m = rs$ *with* r *and* s *coprime, and*

$$\psi : \mathbb{Z}/m\mathbb{Z} \longrightarrow \mathbb{Z}/r\mathbb{Z} \times \mathbb{Z}/s\mathbb{Z}$$

is the isomorphism of the theorem, then ψ *restricts to an isomorphism of groups from* U_m *to* $U_r \times U_s$.

This is a special case of Proposition 14, Section 11F. Here is a direct proof.

Proof. Observe that

$$U_r \times U_s = \{([b],[c]) \in \mathbb{Z}/r\mathbb{Z} \times \mathbb{Z}/s\mathbb{Z} | (b,r) = 1 \text{ and } (c,s) = 1\}.$$

If a is a unit of $\mathbb{Z}/m\mathbb{Z}$, then $(a,m) = 1$, hence $(a,r) = 1$ and $(a,s) = 1$. So $\psi([a]) = ([a],[a])$, and ψ maps the unit $[a]$ of U_m to the pair $([a],[a])$ in $U_r \times U_s$. Thus ψ defines a function ψ_u from U_m to $U_r \times U_s$.

Since ψ is a ring homomorphism, ψ_u is a group homomorphism.

ψ_u is one-to-one because ψ is one-to-one.

To show that ψ_u is onto, suppose $([b],[c])$ is in $U_r \times U_s$. Then there is some $[a]$ in U_m so that $\psi_u([a]) = ([a],[a]) = ([b],[c])$. Hence $[a] = [b]$ in $\mathbb{Z}/r\mathbb{Z}$ and $[a] = [c]$ in $\mathbb{Z}/s\mathbb{Z}$. But then a is coprime to r, and a is coprime to s, and so a is coprime to $rs = m$. Thus a is a unit modulo m, hence $[a]$ is in U_m. This shows that ψ_u maps U_m onto $U_r \times U_s$. □

This theorem yields a formula for Euler's phi function:

Corollary 9. *If* $m = rs$, r *and* s *coprime, then* $\phi(m) = \phi(r)\phi(s)$.

Proof. $\phi(m)$ is the number of units of $\mathbb{Z}/m\mathbb{Z}$, and $\phi(r)\phi(s)$ is the number of pairs $([b]_r,[c]_s)$ where $[b]_r$ is a unit of $\mathbb{Z}/r\mathbb{Z}$ and $[c]_s$ is a unit of $\mathbb{Z}/s\mathbb{Z}$. Since $\psi_u : U_m \to U_r \times U_s$ is an isomorphism, hence a bijection, the result follows from Proposition 8. □

Corollary 10. *Let* $m = p_1^{e_1} p_2^{e_2} \cdots p_g^{e_g}$ *be a product of prime powers. Then*

$$\mathbb{Z}/m\mathbb{Z} \cong \prod_{i=1}^{g} \mathbb{Z}/p_i^{e_i}\mathbb{Z},$$

$$U_m \cong \prod_{i=1}^{g} U_{p_i^{e_i}}$$

and

$$\phi(m) = \prod_{i=1}^{g} \phi(p_i^{e_i}).$$

The proof of this is a routine induction from the previous corollary.

Exercises.

47. Write down addition and multiplication tables for $\mathbb{Z}/2\mathbb{Z} \times \mathbb{Z}/2\mathbb{Z}$.

48. Write down the elements of $\mathbb{Z}/12\mathbb{Z}$ and of $\mathbb{Z}/3\mathbb{Z} \times \mathbb{Z}/4\mathbb{Z}$, and identify which elements correspond under the map ψ from $\mathbb{Z}/12\mathbb{Z}$ to $\mathbb{Z}/3\mathbb{Z} \times \mathbb{Z}/4\mathbb{Z}$.

49. Write down the elements of $\mathbb{Z}/10\mathbb{Z}$ and of $\mathbb{Z}/2\mathbb{Z} \times \mathbb{Z}/5\mathbb{Z}$, and identify which elements correspond under the map ψ from $\mathbb{Z}/10\mathbb{Z}$ to $\mathbb{Z}/2\mathbb{Z} \times \mathbb{Z}/5\mathbb{Z}$.

50. Show that if R, S are non-zero commutative rings then $R \times S$ always has zero divisors, and hence is never an integral domain or a field.

51. Extend the proof of Theorem 7 to show that if $m = q_1 q_2 \cdots q_g$ is a factorization of m into pairwise coprime factors, then $\mathbb{Z}/m\mathbb{Z}$ is isomorphic to $\mathbb{Z}/q_1\mathbb{Z} \times \mathbb{Z}/q_2\mathbb{Z} \times \ldots \times \mathbb{Z}/q_g\mathbb{Z}$.

52. Examine the map $\psi : \mathbb{Z}/24\mathbb{Z} \to \mathbb{Z}/6\mathbb{Z} \times \mathbb{Z}/4\mathbb{Z}$ given by $\psi([a]_{24}) = ([a]_6, [a]_4)$. What is the kernel of ψ? That is, which elements of $\mathbb{Z}/24\mathbb{Z}$ get mapped by ψ to the zero element of $\mathbb{Z}/6\mathbb{Z} \times \mathbb{Z}/4\mathbb{Z}$?

Which elements of $\mathbb{Z}/6\mathbb{Z} \times \mathbb{Z}/4\mathbb{Z}$ are in the image of ψ?

53. Suppose

$$x \equiv a \quad (\mathrm{mod}\ 8)$$
$$x \equiv b \quad (\mathrm{mod}\ 12)$$

has a solution $x = x_0$. How many solutions to this system of congruences are there modulo 24?

54. Find a number x_0 whose order modulo p is $(p-1)/2$ and whose order modulo q is $(q-1)/2$, where
  (i) $p = 7, q = 11$;
  (ii) $p = 11, q = 19$.
  (See Section 7B, E3).

The next exercises relate to determining the number of numbers $a < m$ for which m is an a-pseudoprime–see Section 11D.
  Recall that
$$U_m(r) = \{[a]_m \text{ in } U_m | a^r \equiv 1 \quad (\mathrm{mod}\ m)\},$$
the group of r-th roots of unity in $\mathbb{Z}/m\mathbb{Z}$. Then m is an a-pseudoprime if $a^{m-1} \equiv 1$ $(\mathrm{mod}\ m)$, which is the case if $[a]$ is in $U_m(m-1)$.

55. Show that if $a^t \equiv 1$ $(\mathrm{mod}\ p)$, where p is prime, then $a^{(t,p-1)} \equiv 1$ $(\mathrm{mod}\ p)$.

56. Let $m = pq$ with p, q distinct primes.
  (i) Show that $(m-1, p-1) = (m-1, q-1) = (p-1, q-1)$.
  (ii) Show that for a any integer, $a^r \equiv 1$ $(\mathrm{mod}\ m)$ if and only if $a^r \equiv 1$ $(\mathrm{mod}\ p)$ and $a^r \equiv 1$ $(\mathrm{mod}\ q)$.
  (iii) Show that $U_m(m-1) = U_m(d)$ where $d = (p-1, q-1)$.

57. Suppose p and q are primes with $p \equiv 3 \pmod 4$ and $(\frac{p-1}{2}, \frac{q-1}{2}) = 1$. Let $m = pq$, Use the last exercise to show that $U_m(m-1) = U_m(2)$. Show that $U_m(2)$ has four elements.

58. Use the last exercise to show that if $m = pq$ with p and q twin primes (that means: $q = p+2$), then $U_m(m-1)$ has four elements.

59. Show that if $m = pq$ with $p \equiv 3 \pmod 4$ and $q = 17$, then $U_m(m-1)$ has four elements.

60. Show that if $m = 2501 = 61 \cdot 41$, then $U_m(m-1) = U_m(20)$.

61. Show that $U_{65}(64) = U_{65}(4)$ and has order 16.

Part IV
Polynomials

Chapter 13
Polynomials

Beginning with this chapter we turn attention to polynomials with coefficients in a field. In broad outline the theory follows that for integers: we prove the analogue of the Fundamental Theorem of Arithmetic (Chapter 4), study irreducible polynomials (the analogue of primes), and develop the concepts of congruences and congruence classes, and analogues of Fermat's theorem and the Chinese remainder theorem. When the theory for polynomials is combined wih the theory for integers, what comes out in Chapters 23 and 24 is the theory of finite fields.

A. Polynomials and Functions

A polynomial with coefficients in a commutative ring R is an expression of the form

$$p(x) = a_n x^n + a_{n-1} x^{n-1} + \ldots + a_1 x + a_0,$$

where the coefficients $a_n, a_{n-1}, \ldots, a_0$ are elements of R, a commutative ring, x is a symbol, called an *indeterminate*, and n is some integer ≥ 0. The symbols $x^2, \ldots, x^n$ are powers of the indeterminate x: that is, $x^2 = x \cdot x, x^3 = x \cdot x \cdot x$, etc. By convention, $x^0 = 1$.

Some examples (with $R = \mathbb{R}$, the real numbers):

$$p(x) = x^2 - 3x + 2,$$
$$p(x) = -\frac{1}{3}x^3 + x,$$
$$p(x) = \pi \text{ (here } a_0 = \pi, \text{ and } 0 = a_1 = a_2 = \ldots),$$
$$p(x) = 0 \text{ (here all the coefficients are 0).}$$

We wrote $p(x)$ starting with the constant term on the right and writing decreasing powers of x from left to right. We could just as well reverse the order, and write

$$p(x) = a_0 + a_1 x + a_2 x^2 + \ldots + a_n x^n.$$

L.N. Childs, *A Concrete Introduction to Higher Algebra*, Undergraduate Texts in Mathematics, © Springer Science+Business Media LLC 2009

The order in which the terms are written doesn't matter (as long as we include the powers of x as placeholders–see Section C, below).

The notation $p(x)$ is suggestive of functional notation.

In calculus, a function such as $f(x) = x^3 + \sin x$ is presented by giving a description of what the value of the function f is on a "typical" or "indeterminate" real number x. Thus to find the value of the function f at the number 2, we simply replace x by 2, to get the real number $f(2) = 2^3 + \sin 2$.

So also with a polynomial: any polynomial $p(x)$ with coefficients in the ring R defines a function from R to R by sending r in R to $p(r)$, the element of R obtained by replacing the "indeterminate" element x in the expression $p(x)$ by the element r of R. Thus if $p(x) = 3x^2 - 2x + 5$ in $\mathbb{Q}[x]$, then $p(-3) = 3(-3)^2 - 2(-3) + 5 = 38$. If $p(x) = [3]x + [4]x^3$ in $\mathbb{Z}/6\mathbb{Z}$, then replacing x by $[2]$ in $\mathbb{Z}/6\mathbb{Z}$ yields

$$p([2]) = [3][2] + [4][2]^3 = [6 + 32] = [2]$$

in $\mathbb{Z}/6\mathbb{Z}$.

However, a polynomial with coefficients in a commutative ring R should not be thought of as a function described by its value at an indeterminate element of R, but rather as just a formal expression involving the symbol x and its powers.

The reason for making this distinction between polynomials and functions has to do with when two polynomials are equal, compared with when two functions are equal.

Two *polynomials*

$$p(x) = a_0 + a_1 x + \ldots + a_n x^n$$

and

$$q(x) = b_0 + b_1 x + \ldots + b_m x^m,$$

are equal if and only if the coefficients of each power of x are equal:

$$a_0 = b_0, a_1 = b_1, \ldots, a_n = b_n, \ldots, a_m = b_m.$$

In particular, if $n < m$, then

$$b_{n+1} = b_{n+2} = \ldots = b_m = 0.$$

Thus as polynomials with coefficients in $\mathbb{Z}/2\mathbb{Z}$, the polynomial

$$p(x) = a_3 x^3 + a_2 x^2 + a_1 x + a_0$$

is equal to

$$q(x) = x^3 + 1$$

if and only if $a_0 = a_3 = 1$ and $a_1 = a_2 = 0$.

On the other hand, two *functions* $f(x)$ and $g(x)$ defined on the set R are equal if and only if for all a in R, the numbers $f(a)$ and $g(a)$ are equal.

Any polynomial with coefficients in the commutative ring R defines a function on R, as we've seen. Thus two polynomials that are equal as polynomials are equal as functions. However, it is possible for two polynomials with coefficients in a ring R to be different as polynomials but be equal as functions. For example: in $R = \mathbb{Z}/2\mathbb{Z}$, let $p(x) = x + 1$, $q(x) = x^3 + 1$. Then $p(0) = 1 = q(0)$, and $p(1) = 0 = q(1)$. Thus as functions on $\mathbb{Z}/2\mathbb{Z}$, $p(x) = q(x)$. However, as polynomials, $p(x)$ and $q(x)$ are obviously different, since, for example, the coefficient of x^3 in $p(x)$ is 0 and in $q(x)$ is 1.

We will prove in the next chapter that if R is an infinite field, such as the real numbers, then two polynomials which are equal as functions on R must be equal as polynomials. The example above with $R = \mathbb{Z}/2\mathbb{Z}$ illustrates that the two notions of equality need not be the same if R is a finite field.

Exercises.

1. Using Fermat's theorem, for each prime number p find two different polynomials with coefficients in $\mathbb{Z}/p\mathbb{Z}$ which agree as functions on $\mathbb{Z}/p\mathbb{Z}$.

2. Find a polynomial $q(x)$ with coefficients in $\mathbb{Z}/6\mathbb{Z}$ such that $q(x)$ is equal to $p(x) = [3]x + [4]x^3$ as functions on $\mathbb{Z}/6\mathbb{Z}$ but $q(x)$ and $p(x)$ are not equal as polynomials.

B. The Commutative Ring $R[x]$

The set of all polynomials with coefficients in the commutative ring R is denoted by $R[x]$.

Earlier we observed that when p is prime, $\mathbb{Z}/p\mathbb{Z}$ is a field with p elements, and introduced the notation $\mathbb{F}_p$ for that field. When considering polynomials with coefficients in $\mathbb{Z}/p\mathbb{Z}$, we will generally use the notation $\mathbb{F}_p$ instead of $\mathbb{Z}/p\mathbb{Z}$ and write the set of polynomials as $\mathbb{F}_p[x]$, rather than $(\mathbb{Z}/p\mathbb{Z})[x]$.

The polynomial $p(x) = a_0 + a_1x + \ldots + a_nx^n$ has *degree n* if x^n is the highest power of x appearing in $p(x)$ with its coefficient a_n not zero. Then the coefficient a_n of x^n is called *the leading coefficient of* $p(x)$. The polynomial with $a_0 = a_1 = \ldots = 0$ is called the zero polynomial and is denoted by 0. By convention, the zero polynomial has degree $-\infty$. Every other polynomial $p(x)$ has a degree ≥ 0. The degree of a polynomial $p(x)$ is denoted by $\deg p(x)$. The ring R can be thought of as a subset of $R[x]$ by viewing an element a of R as a polynomial of degree 0 (if $a \neq 0$ or $-\infty$ (if $a = 0$).

Polynomials may be added and multiplied. The operations are defined just as for functions. If

$$p(x) = a_0 + a_1x + \ldots + a_nx^n$$

and

$$q(x) = b_0 + b_1x + \ldots + b_mx^m$$

then

$$p(x) + q(x) = (a_0 + a_1x + \ldots + a_nx^n) + (b_0 + b_1x + \ldots + b_mx^m);$$

if, say, $m > n$, we can collect terms and get

$$p(x) + q(x) = (a_0 + b_0) + (a_1 + b_1)x + \ldots + (a_n + b_n)x^n + b_{n+1}x^{n+1} + \ldots + b_m x^m.$$
$$(13.1)$$

Similarly, using the distributive law and collecting the coefficients of each power of x, multiplication of $p(x)$ and $q(x)$ is

$$
\begin{aligned}
p(x) \cdot q(x) \\
&= (a_0 + a_1 x + \ldots + a_n x^n)(b_0 + b_1 x + \ldots + b_m x^m) \\
&= a_0 b_0 + (a_0 b_1 + a_1 b_0)x + \ldots + [\sum_{i+j=k} a_i b_j]x^k + \ldots + a_n b_m x^{n+m}
\end{aligned}
$$
$$(13.2)$$

Thus if $a_n b_m \neq 0$ then the leading coefficient of $p(x)q(x)$ is the product of the leading coefficients of $p(x)$ and $q(x)$.

With these definitions of addition and multiplication, with 0 the zero polynomial, and with 1 the polynomial with $a_0 = 1$ and all other coefficients $= 0$, it is easy to see that $R[x]$ is a commutative ring.

Proposition 1. *Let R be a commutative ring. For every non-zero polynomials p and q in $R[x]$, if the leading coefficient of p is a non-zero divisor in R, then*

$$\deg(pq) = \deg(p) + \deg(q).$$

The formula holds for all non-zero polynomials p, q in $R[x]$ if R has no zero divisors.

Proof. Suppose $p(x), q(x)$ in $R[x]$ have degrees n and m, respectively. Let a_n be the leading coefficient of $p(x)$ and b_m the leading coefficient of $q(x)$. If a_n is not a zero divisor in R, then $a_n b_m \neq 0$, so is the leading coefficient of $p(x)q(x)$. Thus $p(x)q(x)$ has degree $n + m = \deg(p(x)) + \deg(q(x))$.   □

The convention that the zero polynomial has degree $-\infty$ together with the reasonable assumption that $-\infty + m = -\infty$ for m any integer or $m = -\infty$, allows the formula $\deg(fg) = \deg(f) + \deg(g)$ to extend to the case where one or both of f and g is the zero polynomial.

In the rest of the book, we will usually consider only polynomials with coefficients in a field F, rather than in a general commutative ring R. However, occasionally it is convenient to allow polynomials with coefficients in a commutative ring which is not a field. One example is $R = \mathbb{Z}/n\mathbb{Z}$, congruence classes of integers modulo n. Of course $\mathbb{Z}/n\mathbb{Z}$ is a field only when n is prime. Another example is $S = R[y]$, polynomials in the indeterminate y with coefficients in the the commutative ring R. Then $S[x]$ will be polynomials in x with coefficients in $R[y]$, that is, polynomials in two variables with coefficients in R. Then $S[x]$ is usually denoted by $R[y, x]$ or $R[x, y]$. An expression such as $x^2 + 3xy + y^2 - 2$ is a polynomial in $R[x, y]$. In a similar way we can define polynomials in three variables over R as polynomials with coefficients in the ring of polynomials in two variables with coefficients in R, etc.

Exercises.

3. Show that if R has no zero divisors, then $R[x]$ also has no zero divisors.

4. Show that if R has zero divisors, there exist polynomials f,g in $R[x]$ so that $\deg(fg) < \deg(f) + \deg(g)$.

5. Let F be a field. Using Proposition 1, show that for $p(x)$ in $F[x]$, there is some $q(x)$ with $p(x)q(x) = 1$, iff $p(x)$ has degree 0. Thus the units of $F[x]$ are precisely the polynomials of degree 0.

6. Let $R = \mathbb{Z}/4\mathbb{Z} = \{0,1,2,3\}$.
   (i) Show that $1 + 2x$ is a unit of $R[x]$.
   (ii) Show that every unit of $R[x]$ has the form $1 + 2f(x)$ for some $f(x)$ in $R[x]$.
   (iii) Find all of the zero divisors of $R[x]$.
   (iv) Find elements of $R[x]$ which are neither units nor zero divisors.

C. Detaching the Coefficients

We have defined polynomials in terms of an indeterminate, or formal symbol x, but it is possible, and sometimes more convenient, to define a polynomial strictly by its coefficients, without using x. The relevant information about the polynomial is its coefficients. Thus we can associate to

$$p(x) = 3x^4 + 2x^3 - 5x - 1,$$

the 5-tuple $(3,2,0,-5,-1)$, where the middle 0 is the coefficient of x^2. Here we must agree on the order in which the coefficients appear, so as not to think of that 5-tuple as representing the polynomial

$$q(x) = 3 + 2x - 5x^3 - x^4.$$

So our convention is that when describing polynomials by tuples of numbers, the tuples will always describe a polynomial written with decreasing powers of x from left to right.

With that convention, we can define a polynomial with coefficients in R as a sequence

$$(a_n, a_{n-1}, \ldots, a_1, a_0)$$

of elements of R. Two sequences

$$(a_n, a_{n-1}, \ldots, a_1, a_0)$$

and

$$(b_m, b_{m-1}, \ldots, b_1, b_0)$$

are equal if (say) $m \leq n$ and

$$a_0 = b_0, a_1 = b_1, \ldots, a_m = b_m,$$

and $a_{m+1}, a_{m+2}, \ldots, a_n$ are all zero. Thus two sequences are equal if one sequence is the rightmost part of the other, and all other entries in the longer sequence are zero.

Doing arithmetic operations on polynomials is easier when the polynomials are written as sequences. To see this, consider the way we work with natural numbers.

Suppose we want to multiply $7 \cdot 10 + 9$ and $3 \cdot 10 + 6$. We don't compute

$$
\begin{aligned}
(7 \cdot 10 + 9)(3 \cdot 10 + 6) &= (7 \cdot 3) \cdot 10^2 + ((7 \cdot 6) + (9 \cdot 3)) \cdot 10) + 9 \cdot 6 \\
&= (2 \cdot 10 + 1) \cdot 10^2 + (4 \cdot 10 + 2 + 2 \cdot 10 + 7) \cdot 10 + 5 \cdot 10 + 4 \\
&= (2 \cdot 10^3 + 10^2) + (6 \cdot 10^2 + 9 \cdot 10) + (5 \cdot 10 + 4) \\
&= 2 \cdot 10^3 + 7 \cdot 10^2 + (10 + 4) \cdot 10 + 4 \\
&= 2 \cdot 10^3 + 7 \cdot 10^2 + 10^2 + 4 \cdot 10 + 4 \\
&= 2 \cdot 10^3 + 8 \cdot 10^2 + 4 \cdot 10 + 4;
\end{aligned}
$$

instead, we write $7 \cdot 10 + 9$ as 79, $3 \cdot 10 + 6$ as 36 and multiply 79 and 36 using the well-known multiplication algorithm.

Similarly for polynomials.

Addition of two sequences is "componentwise", as follows:

$$
\begin{aligned}
(a_n, a_{n-1}, \ldots, a_1, a_0) &+ (b_m, b_{m-1}, \ldots, b_1, b_0) \\
&= (a_n + b_n, a_{n-1} + b_{n-1}, \ldots, a_1 + b_1, a_0 + b_0),
\end{aligned}
$$

where we have assumed here that $n > m$ and $b_{m+1} = \ldots = b_n = 0$.

For example, to add $4x^2 + 2x - 3$ and $x^3 - x + 8$, we add $(0, 4, 2, -3) + (1, 0, -1, 8) = (1, 4, 1, 5)$:

$$
\begin{array}{rrrr}
0 & 4 & 2 & -3 \\
+\quad 1 & 0 & -1 & 8 \\
\hline
= \quad 1 & 4 & 1 & 5
\end{array}
$$

Addition of polynomials is similar to addition of numbers, except that with polynomials there is no carrying.

Multiplication is not componentwise multiplication. The product of two polynomials $(a_n, a_{n-1}, \ldots, a_1, a_0)$ and $(b_m, b_m - 1, \ldots, b_1, b_0)$ is the tuple whose entries are the coefficients (in the appropriate order) of the polynomial $p(x) \cdot q(x)$ in formula (2) above. To illustrate, consider multiplying the two polynomials

$$p(x) = x^4 + 3x^3 - x^2 - 4x - 6 = (1, 3, -1, -4, -6)$$

and

$$q(x) = 3x^4 + x^2 + 5 = (3, 0, 1, 0, 5).$$

We set up the multiplication of five-tuples just like multiplication of decimal integers, and perform the algorithm in the same way, except with no carrying:

$$
\begin{array}{rrrrrr}
 & 1 & 3 & -1 & -4 & -6 \\
\cdot & 3 & 0 & 1 & 0 & 5 \\
\hline
 & 5 & 15 & -5 & -20 & -30 \\
 & 0 & 0 & 0 & 0 & 0 \\
 1 & 3 & -1 & -4 & 6 \\
 0 & 0 & 0 & 0 & 0 \\
3 & 9 & -3 & -12 & -18 \\
\hline
3 \quad 9 & -2 & -9 & -14 & 11 & -11 & -20 & -30
\end{array}
$$

Thus,

$$p(x)q(x) = 3x^8 + 9x^7 - 2x^6 - 9x^5 - 14x^4 + 11x^3 - 11x^2 - 20x - 30.$$

(In the array, the top two lines are the coefficients of the two polynomials, the next five lines are the coefficients of the polynomial obtained by multiplying $p(x)$ by 5, $0 \cdot x$, x^2, $0 \cdot x^3$ and $3x^4$, respectively, and the bottom line sums the coefficients of each power of x in the previous five lines.)

Just as with numbers, multiplying polynomials as above is efficient because the powers of x are never written down. Only the coefficients are in play. (Once you try a few examples, you should find multiplication of polynomials easier than multiplication of numbers. In particular, when the polynomials have entries in $\mathbb{F}_2 = \{0,1\}$, multiplication is extremely easy!)

If you have had some linear algebra, the identification of a polynomial with its sequence of coefficients is what we do when we write down the coordinates of a vector with respect to a basis. In fact, here the scalars are from R, the vector space is $R[x]$ (assuming R is a field), the basis is an infinite one: $\ldots, x^n, x^{n-1}, \ldots, x, 1$, and the tuple $(a_n, a_{n-1} \ldots, a_1, a_0)$ includes all the non-zero coordinates of the polynomial

$$p(x) = a_n x^n + a_{n-1} x^{n_1} + \cdots + a_1 + a_0$$

with respect to the basis $\ldots x^n, x^{n-1}, \ldots, x, 1$.

Exercises.

7. Detach the coefficients and multiply in $\mathbb{F}_2[x]$:
   (i) $(x^3 + x + 1)(x^4 + x^2 + 1)$;
   (ii) $(x^2 + x + 1)^2$;
   (iii) $(x^2 + x)(x^2 + x + 1)$;
   (iv) $(x^2 + x)(x^3 + x^2 + 1)(x^3 + x + 1)$.

8. Detach the coefficients and multiply in $\mathbb{F}_3[x]$:

(i) $(x^2 + 2x + 2)(2x^2 + 1)$;

(ii) $(x^3 + 1)(2x^2 + 2)$.

9. Detach the coefficients and multiply $(x^4 + x^3 + x^2 + x + 1)(x - 1)$ in $\mathbb{Q}[x]$. Then generalize the result to the case where the left factor has degree n.

D. Homomorphisms

Recall from Section 7D that a ring homomorphism from a ring S to a ring T is a function or "map" $\eta : S \rightarrow T$, so that for any s, s' in S, $\eta(s + s') = \eta(s) + \eta(s')$, $\eta(s \cdot s') = \eta(s) \cdot \eta(s')$, and $\eta(1) = 1$. There are several important ring homomorphisms whose domain is $R[x]$, the ring of polynomials with coefficients in a commutative ring R.

Example 1. Let $Funct(R,R)$ be the ring of functions from R to R, where addition and multiplication of functions are the usual operations as described above, and the function 1 is the constant function $1(s) = 1$ for all s in R.

Let $\varphi : R[x] \rightarrow Func(R,R)$ be the map given as follows: if $p(x)$ is a polynomial, $\varphi(p(x))$ is $p(x)$ thought of as a function on R. Then φ is a homomorphism, because addition and multiplication of polynomials as defined above, coincides with addition and multiplication of polynomial functions. This homomorphism φ, which tells us to view a polynomial with coefficients in R as a function on R, will often be applied implicitly, that is, without specific mention, as in statements like "think of $p(x)$ as a function on R." No confusion should arise. But it is useful to be explicit that polynomials and functions are different, and the identification of a polynomial as a polynomial function in fact defines a homomorphism that we have denoted by φ here. As we noted earlier (see Exercise 1), the homomorphism φ from $R[x]$ to $Func(R,R)$ need not be one-to-one. In fact φ is never one-to-one if R is a finite ring. This can be seen by counting: if R has n elements, then $Func(R,R)$ has n^n elements (why?), while $R[x]$ is an infinite set (why?).

Example 2. Related to φ is a collection of functions $\varphi_a : R[x] \rightarrow R$, one for each a in R: ϕ_a is "evaluation at a." For any $p(x)$ in $R[x]$, $\varphi_a(p(x))$ is defined to be the element $p(a)$ of R obtained by thinking of $p(x)$ as a function on R and evaluating the function at a. Then φ_a is a homomorphism; in fact it is the composite of the homomorphism φ of Example 1 and the map from $Func(R,R)$ to R given by taking a function $f(x)$ and evaluating it at a to get $f(a)$. The latter is a homomorphism because of the way we define addition and multiplication in $Func(R,R)$.

Example 3. Let $\psi : R \rightarrow S$ be a homomorphism. Then we get an induced homomorphism, which we'll also call ψ, from $R[x]$ to $S[x]$, defined by

$$\psi(a_n x^n + \ldots + a_1 x + a_0) = \psi(a_n)x^n + \ldots + \psi(a_1)x + \psi(a_0),$$

that is, replace the coefficients of a polynomial $p(x)$ by their images under the function ψ. This new function ψ is also a homomorphism, as is easily checked.

One particularly useful example of such a function ψ is the "reduce mod m" function that takes a polynomial $p(x)$ with coefficients in $\mathbb{Z}$ and yields the polynomial $\psi(p(x))$ with coefficients in $\mathbb{Z}/m\mathbb{Z}$ whose coefficients are the congruence classes mod m of the coefficients of $p(x)$. For example, if $p(x) = 5x^4 + 2x^3 - 7x + 3$, $m = 2$, and ψ is the homomorphism from $\mathbb{Z}$ to $\mathbb{F}_2 = \mathbb{Z}/2\mathbb{Z}$ given by taking a number n to its congruence class mod 2, then $\psi(p(x)) = [5]_2 x^4 + [2]_2 x^3 - [7]_2 x + [3]_2$, or if we, as usual, identify $\mathbb{F}_2$ as $\{0, 1\}$, then

$$\psi(p(x)) = x^4 + 0x^3 + x + 1 = x^4 + x + 1.$$

Another example is the homomorphism $\psi : \mathbb{Z}[x] \to \mathbb{Q}[x]$ given by taking a polynomial with coefficients in the ring of integers $\mathbb{Z}$ and thinking of it as having coefficients in the field of rational numbers $\mathbb{Q}$. Similar examples arise from thinking of $\mathbb{Z}$ (or $\mathbb{Q}$) as a subring of $\mathbb{R}$, the real numbers, or $\mathbb{C}$, the complex numbers.

Exercises.

10. Explain why if R is a finite ring, then $\varphi : R[x] \to Func(R, R)$ is not one-to-one.

11. For which, if any primes p, do $x^6 + 2x^2 + x$ and $x^9 + 8x^3 + x$ agree as functions on $\mathbb{Z}/p\mathbb{Z}$?

Chapter 14
Unique Factorization

In this chapter we show that every polynomial of degree >1 with coefficients in a field factors uniquely (in a sense to be defined) into a product of irreducible polynomials. To reach this result, we follow the same development as for natural numbers: the division theorem, Euclid's Algorithm and Bezout's Identity.

A. Division Theorem

Let $p(x) = a_d x^d + \ldots + a_1 x + a_0$ be a polynomial with coefficients in a field F. Recall that if $a_d \neq 0$, then d is the degree of $p(x)$, and a_d is called the leading coefficient of $p(x)$. If $p(x)$ has degree <0, then $p(x) = a_0$, so can be considered as an element of the field F, or a constant, or a scalar.

If d is the degree of $p(x)$ and the leading coefficient $a_d = 1$, then $p(x)$ is *monic*.

The main theme of this chapter is that the entire sequence of arguments in Chapters 3 and 4 which led to the Fundamental Theorem of Arithmetic is valid for polynomials with coefficients in any field. The fact that we can associate to a non-zero polynomial an integer ≥ 0, its degree, enables us to use induction to prove facts about polynomials, like the Division Theorem, Euclid's Algorithm and uniqueness of factorization, just as we did for numbers.

You may wish to review the definitions and results for $\mathbb{Z}$ in Chapters 3 and 4 at this point so you can anticipate what will happen in this chapter.

We will often let f, g, p, q, r, etc., denote polynomials, omitting the "(x)" in "$f(x)$."

Just as with the theory for natural numbers, the first step is:

Theorem 1 (Division Theorem for Polynomials). *Let R be a commutative ring. Let f, g be two polynomials in $R[x]$ with $f \neq 0$, and suppose that the leading coefficient of f is a unit of R. Then there are polynomials q (the quotient) and r (the remainder), with $\deg r < \deg f$, such that $g = fq + r$. If also $g = fq_1 + r_1$, then $q = q_1$ and $r = r_1$ (i.e., the quotient and the remainder are unique).*

L.N. Childs, *A Concrete Introduction to Higher Algebra*, Undergraduate Texts in Mathematics, © Springer Science+Business Media LLC 2009

Proof. We fix $f \neq 0$ and prove that for any g, there exists some q and r satisfying the statement of the theorem, using complete induction on the degree of g.

If $\deg g < \deg f$, then set $q = 0, r = g$: then obviously $g = fq + r$ with $\deg r < \deg f$.

Suppose $\deg g \geq \deg f$. Let $f = f(x) = a_d x^d + \ldots + a_0$ have degree d and that a_d is a unit of R. Let $g = g(x) = b_{d+s} x^{d+s} + \ldots + b_0$ have degree $d + s$, where $s \geq 0$. Let $g_1 = g - (b_{d+s}/a_d) x^s f$. Then $\deg g_1 < \deg g$, since the coefficient of x^{d+s} in g_1 is zero. By induction, $g_1 = fq_1 + r$ for some polynomials q_1 and r, with $\deg r < \deg f$. But then

$$g = g_1 + b_{d+s} a_d^{-1} x^s f$$
$$= fq_1 + r + b_{d+s} a_d^{-1} x^s f$$
$$= f(q_1 + b_{d+s} a_d^{-1} x^s) + r$$

proving the existence of a quotient and remainder for f and g. By induction, the existence of q and r is proven.

For uniqueness, suppose $g = fq + r = fq_1 + r_1$, with $\deg r < \deg f$ and $\deg r_1 < \deg f$. Then

$$f(q - q_1) = r_1 - r.$$

If $q - q_1 \neq 0$, let $s \geq 0$ be the degree of $q - q_1$. Since the leading coefficient of f is a unit, $f(q - q_1)$ has degree $\deg(f) + s$ by Proposition 1 of Chapter 13, while $r_1 - r$ has degree $< \deg(f)$, which is impossible. Thus $q - q_1 = 0$ and $r_1 - r = 0$. □

The argument which obtains g_1 from g in the first part of the proof is the first step in the familiar process of long division of polynomials, a process which computes q and r.

Corollary 2. *If F is a field, then the Division Theorem holds for all $f \neq 0$ and g in $F[x]$.*

A polynomial f *divides* a polynomial g if $g = fq$ for some polynomial q. For example, in $\mathbb{Q}[x]$, $x^2 - 1$ divides $x^4 - 1$ because $(x^2 - 1)(x^2 - 1) = x^4 - 1$. Similarly, $x^2 + x + 1$ divides $x^6 - 1$ (verify this), while $x - 1$ does not divide $x^3 - 2$, as follows immediately from the following useful criterion:

Theorem 3 (Remainder Theorem). *If $f(x)$ is a polynomial with coefficients in a field F, and a is in F, then $f(a)$ is the remainder when dividing $f(x)$ by $x - a$.*

Proof. Write $f(x) = (x - a)q(x) + r(x)$, by the Division Theorem. Then $\deg r(x) < \deg(x - a)$, so $r(x)$ is a constant, call it r, in F. That is,

$$f(x) = (x - a)q(x) + r.$$

Evaluating both sides at $x = a$, we have

$$f(a) = (a - a)q(a) + r = r.$$

Hence $f(a)$ is the remainder when $f(x)$ is divided by $x - a$. □

The Remainder Theorem yields as a special case:

Proposition 4 (Root Theorem). *If $f(x)$ is a polynomial with coefficients in a field F, and a is in F, then $f(a) = 0$ if and only if $x - a$ divides $f(x)$.*

The Root Theorem has the following useful consequence.

Theorem 5 (D'Alembert's Theorem). *A nonzero polynomial $f(x)$ of degree n in $F[x]$, F a field, has at most n distinct roots in F.*

Proof. Induction on n, the degree of f.

If $\deg f = 0$, then f is a nonzero constant polynomial, so has no roots in F.

Now suppose f is a polynomial of degree $n > 0$, and suppose it has r distinct roots $a_1, \ldots, a_r$ in F. We must show $r \leq n$. We have $f(a_r) = 0$, so by the Root Theorem, $f(x) = (x - a_r)g(x)$, where $g(x)$ has degree $n - 1$. Now for each i, $1 \leq i \leq r - 1$, $f(a_i) = (a_i - a_r)g(a_i)$ in F, so since $f(a_i) = 0$, $a_i \neq a_r$ and F has no zero divisors, we must have $g(a_i) = 0$. Hence $g(x)$ has roots $a_1, \ldots, a_{r-1}$. But $\deg g = n - 1$. By induction, $r - 1 \leq n - 1 = \deg g$. Hence $r \leq n = \deg f$. $\qquad\square$

Corollary 6. *If F is a field with infinitely many elements and $f(x)$ and $g(x)$ are two polynomials with coefficients in F, then $f(x)$ and $g(x)$ are equal as polynomials with coefficients in F if and only if $f(x) = g(x)$ as functions on F. That is, if F is an infinite field, then the homomorphism from $F[x]$ to $Func(F,F)$ given by viewing a polynomial as a function, is one-to-one.*

This result implies that no confusion can arise if over the real numbers, we think of polynomials as real valued functions, or view x as an "indeterminate real number". On the other hand, the assumption that F have infinitely many elements is necessary, as we observed in the last chapter.

Proof of Corollary. If $f(x) = g(x)$ as polynomials, then for any element a of F, $f(a) = g(a)$, as we observed in the last chapter. That is, $f(x)$ and $g(x)$ are equal as functions on F.

Conversely, suppose $f(x)$ and $g(x)$ are two polynomials, and $f(a) = g(a)$ for all a in F. Then $h(x) = f(x) - g(x)$ is a polynomial in $F[x]$ with the property that $h(a) = 0$ for every a in F. If $h(x)$ has degree n for some finite number n and F has infinitely many elements, then $h(x)$ has more than n roots in F. So $h(x)$ must be the zero polynomial, by D'Alembert's Theorem. Hence $f(x) = g(x)$ as polynomials, completing the proof. $\qquad\square$

D'Alembert's Theorem will enable us to prove the Primitive Root Theorem in Section 19A.

Exercises.

1. Find the quotient and remainder when the first polynomial is divided by the second (in $\mathbb{Q}[x]$):

(i) $x^3 - 7x - 1; x - 2$;
(ii) $x^4 - 2x^2 - 1; x^2 + 3x - 1$;
(iii) $2x^3 - 3x^2 + 1; x$;
(iv) $x^2 + x + 1; 2$;
(v) $3x^2 - x - 1; x^3 - 2$.

2. Does the Division Theorem always work for polynomials f and g in $\mathbb{Z}[x]$ if the leading coefficient of f is not a unit of $\mathbb{Z}$ but divides the leading coefficient of g? Explain.

3. Find an example of an f whose leading coefficient is a zero divisor, such that f divides some polynomial g but the quotient and remainder are not unique. (I found such an example with $R = \mathbb{Z}/8\mathbb{Z}$)

4. Without using long division of polynomials, find the remainder $\bigl($in $\mathbb{Q}[x]\bigr)$ when:
   (i) $x^3 - 2x + 4$ is divided by $x - 2$;
   (ii) $x^4 - 7x^2 + 3$ is divided by $x + 1$;
   (iii) $x^{40} - 8x^{12} + 3$ is divided by $x^4 - 1$.

5. (i) Does $x - 3$ divide $x^4 + x^3 + x + 4$ in $\mathbb{Q}[x]$? in $\mathbb{Z}[x]$?
   (ii) Since both $x - 3$ and $x^4 + x^3 + x + 4$ have coefficients in $\mathbb{Z}$, we can look at their images in $(\mathbb{Z}/m\mathbb{Z})[x]$ for any m. Find all $m \geq 2$ for which the image of $x - 3$ divides the image of $x^4 + x^3 + x + 4$ in $(\mathbb{Z}/m\mathbb{Z})[x]$.

6. Find all m so that the image of $x^3 + 3$ divides the image of $x^5 + x^3 + x^2 - 9$ in $(\mathbb{Z}/m\mathbb{Z})[x]$.

7. In $\mathbb{Q}[x]$, when f is divided by $(x^2 - 3)(x + 1)$, the remainder is $x^2 + 2x + 5$. What is the remainder when f is divided by $x^2 - 3$?

8. For which values of k in $\mathbb{Q}$ does $x - k$ divide $x^3 - kx^2 - 2x + k + 3$?

9. Show that if a, b, and $a + b$ are distinct nonzero elements of a commutative ring R, and $ab = 0$, then $f(x) = x^2 - (a + b)x$ has four distinct roots in R, namely, $a, b, a + b$, and 0. Give an example with $R = \mathbb{Z}/6\mathbb{Z}$.

10. Find all roots in $R[x]$ of $f(x) = x^2 - 2x$ when:
   (i) $R = \mathbb{Z}/15\mathbb{Z}$;
   (ii) $R = \mathbb{Z}/30\mathbb{Z}$

11. For every $n > 2$, can you find a commutative ring R and a polynomial $f(x)$ of degree 2 with at least n roots in R? (Try choosing $R = \mathbb{Z}/m\mathbb{Z}$ with m a product of many distinct primes.)

12. Let $\mathbb{F}_p = \mathbb{Z}/p\mathbb{Z}$ be the field of p elements, where p is an odd prime. Label the elements of $\mathbb{F}_p$ as $0, 1, 2, \ldots, p - 1$ modulo p. Prove Wilson's Theorem:

$$(p - 1)! \equiv -1 \pmod{p}$$

as follows: observe that, by Fermat's Theorem, the polynomial $f(x) = x^{p-1} - 1$ in $\mathbb{F}_p[x]$ has $1, 2, \ldots, p - 1$ as roots. Apply the Root Theorem to get a complete factorization of $x^{p-1} - 1$ into linear factors, and then compare the constant term of the product of the factors with the constant term of $f(x)$.

13. Let $\mathbb{F}_p$ be the field of p elements with p an odd prime, as in the last exercise. Explain why the polynomial $x^2 - 1$ in $\mathbb{F}_p[x]$ has only the roots 1 and $p - 1$. Show then that every number a with $1 < a < p - 1$ has an inverse modulo p that is not congruent to a modulo p. Using that observation, prove Wilson's Theorem.

B. Greatest Common Divisors

With the Division Theorem in hand, we can obtain Euclid's Algorithm and Bezout's Identity for polynomials, just as we did for natural numbers in Chapter 3.

Let f, g be in $F[x]$ where F is a field. A polynomial p in $F[x]$ is a *greatest common divisor* (g.c.d.) of f and g if p divides f and p divides g, and any q in $F[x]$ that divides f and g has a degree that is not larger than the degree of p. That is, p is a common divisor of f and g of largest degree.

We can find a greatest common divisor of two polynomials by using the Division Theorem repeatedly, just as for numbers. The process, called Euclid's Algorithm for polynomials, goes back at least to Simon Stevin, 1585, and works as follows:

Euclid's Algorithm. Given two polynomials f, g in $F[x]$ with $f \neq 0$, divide f into g, then the remainder into f, then that remainder into the previous remainder, etc., or symbolically,

$$g = f q_1 + r_1$$
$$f = r_1 g_2 + r_2$$
$$r_1 = r_2 q_3 + r_3$$

$$\vdots$$

$$r_{n-2} = r_{n-1} q_n + r_n$$
$$r_{n-1} = r_n q_{n+1} + 0$$

Since $\deg r_1 < \deg f, \deg r_2 < \deg r_1$, etc., the sequence of divisions ends after at most $\deg f$ steps. Then, just as with natural numbers, we have

Theorem 7. *In Euclid's Algorithm for f and g, the last nonzero remainder r_n is a greatest common divisor of f and g.*

Note that we have carefully said "a greatest common divisor," rather than "the greatest common divisor." Two polynomials have as many greatest common divisors as there are non-zero elements of the field of coefficients, and Euclid's Algorithm may in fact produce more than one of them (see Exercise 16). The reason this phenomenon didn't show up when we were finding greatest common divisors for numbers was that we looked only at natural numbers, that is, positive integers. If we were to have defined "greatest" for integers as "greatest in absolute value," then the greatest common divisor of two integers wouldn't be unique either: for example, -8 and 6 have two greatest common divisors in the absolute sense, namely 2 and -2.

Similarly for polynomials: $x^2 - 1$ and $5x^2 + 10x + 5$ have many greatest common divisors in $\mathbb{Q}[x]$: $x + 1, 2x + 2, (x/17) + 1/17$, etc. But with integers, it is obvious (is it?) that all greatest common divisors divide each other, and the same is true for polynomials.

Two greatest common divisors e and d of f and g must differ by a scalar multiple: $d = ae$ *for some a in* F. For let d be the greatest common divisor of f and g obtained by Euclid's Algorithm, and suppose d has degree r. If e is a common divisor of f and g, then e divides d by Exercise 17. If e is a greatest common divisor of f and g, then $\deg e = \deg d$, from which it follows that $d = ae$ where a has degree 0, that is, a is a nonzero element of the field F. But then $e = a^{-1}d$, so e and d divide each other.

Definition. Two polynomials d and e, such that each is a nonzero scalar multiple of the other, are *associates*.

For example, $x^2 + 2$ and $-5x^2 - 10$ are associates in $\mathbb{Q}[x]$.

Any polynomial with coefficients in a field is an associate of exactly one monic polynomial, namely, the polynomial obtained by dividing each coefficient by the leading coefficient. Thus, in $\mathbb{Q}[x]$, $3x^2 + 2x + 5$ is an associate of $x^2 + \frac{2}{3}x + \frac{5}{3}$, a monic polynomial.

Monic polynomials play a role similar to natural numbers: just as there is a unique natural number that is a greatest common divisor (in the sense of greatest in absolute value) of two integers, similarly there is a unique monic polynomial that is a greatest common divisor of two polynomials with coefficients in a field.

Thus if we refer to *the* greatest common divisor of two polynomials, we will mean the unique greatest common divisor that is monic.

As with numbers, we denote the greatest common divisor of two polynomials f and g by (f,g).

Returning to Euclid's Algorithm, we showed in Chapter 3 that if d is the greatest common divisor of two numbers a and b, then there are integers r and s so that $d = ra + sb$. So also with polynomials:

Theorem 8 (Bezout's Identity). *Every greatest common divisor d of two polynomials f and g in* $F[x]$, F *a field, can be written as* $d = rf + sb$ *for some polynomials r and s in* $F[x]$.

The proof of this is virtually identical to the proof for numbers, and is left as an exercise.

Say f and g are *coprime*, or *relatively prime*, if any greatest common divisor of f and g is a constant. In that case, 1 is a greatest common divisor (since 1 is an associate of any nonzero constant polynomial). So we can write $1 = rf + sg$ for some polynomials r and s.

Recall that Euclid's Algorithm gave a computational procedure for not only finding the greatest common divisor d of numbers a and b, but also writing that greatest common divisor d as $ra + sb$. So also with polynomials.

For example, consider $x^5 + 1$ and $x^3 + 1$ in $\mathbb{F}_2[x]$. Euclid's Algorithm is

$$x^5 + 1 = (x^3 + 1)x^2 + (x^2 + 1),$$
$$x^3 + 1 = (x^2 + 1)x + (x + 1),$$
$$x^2 + 1 = (x + 1)(x + 1) + 0.$$

Hence $x + 1$ is the greatest common divisor of $x^5 + 1$ and $x^3 + 1$. Then

$$\begin{aligned}
x + 1 &= (x^3 + 1) + (x^2 + 1)x \\
&= (x^3 + 1) + ((x^5 + 1) + (x^3 + 1)x^2)x \\
&= (x^5 + 1)(x) + (x^3 + 1)(x^3 + 1).
\end{aligned}$$

Thus the greatest common divisor is written according to Bezout's Identity, by successively substituting for remainders in the equation for the last nonzero remainder in Euclid's Algorithm.

We can also adapt the extended Euclidean algorithm matrix method that we used for numbers in Chapter 3.

Exercises.

14. Using Euclid's Algorithm, find a greatest common divisor in $\mathbb{F}_3[x]$ of $x^2 + 1$ and $x^5 + 1$.

15. Using Euclid's Algorithm, find a greatest common divisor in $\mathbb{F}_3[x]$ of $x^2 - x + 4$ and $x^3 + 2x^2 + 3x + 2$.

16. In $\mathbb{F}_5[x]$, find a greatest common divisor of

$$3x^3 + 4x^2 + 3 \text{ and } 3x^3 + 4x^2 + 3x + 4$$

in two ways, first dividing the left polynomial into the right one, then dividing the right one into the left one. Verify that the two greatest common divisors are associates but not equal.

17. (i) Prove that in Euclid's Algorithm for f and g, the last nonzero remainder is a common divisor of f and g.

(ii) Prove that if e is any common divisor of f and g, then e divides the last nonzero remainder in Euclid's Algorithm for f and g. Hence the last nonzero remainder is a greatest common divisor of f and g.

18. Prove Bezout's Identity for polynomials.

19. In $\mathbb{F}_3[x]$, write, if possible, the polynomial 1 in the form $f(x)p(x) + g(x)q(x)$, where $p(x) = x^3 + 1$ and $q(x) = x^3 + x + 1$.

20. In $\mathbb{F}_3[x]$, write, if possible, the polynomial 1 in the form $f(x)p(x) + g(x)q(x)$, where $p(x) = x^3 + x^2 + x + 2$ and $q(x) = x^3 + 2x^2 + 2$.

21. In $\mathbb{Q}[x]$, find the greatest common divisor of $x^6 - 1$ and $x^4 - 1$. Write the greatest common divisor as in Bezout's Identity.

22. In $\mathbb{F}_2[x]$, find some $r(x), s(x)$ so that $r(x)f(x) + s(x)g(x) = 1$, where $f(x) = x^2 + x + 1$, and $g(x) = x^3$;

23. In $\mathbb{F}_2[x]$, find some $r(x), s(x)$ so that $r(x)f(x) + s(x)g(x) = 1$, where $f(x) = x^6 + x^5 + x^3 + x$ and $g(x) = x^8 + x^7 + x^6 + x^4 + x^3 + x + 1$;

24. In $\mathbb{F}_2[x]$, find some $r(x), s(x)$ so that $r(x)f(x) + s(x)g(x) = 1$, where $f(x) = x^{15} - 1$ and $g(x) = x^4 + x^2 + x$.

25. (i) Find the greatest common divisor in $\mathbb{F}_3[x]$ of $x^5 + 2x^3 + x^2 + x + 1$ and $x^4 + 2x^3 + x + 1$.

(ii) Find the least common multiple in $\mathbb{F}_3[x]$ of $x^5 + 2x^3 + x^2 + x + 1$ and $x^4 + 2x^3 + x + 1$.

26. Find the greatest common divisor $d(x)$ in $\mathbb{Q}[x]$ of $f(x)$ and $g(x)$ and find polynomials $r(x)$ and $s(x)$ with $f(x)r(x) + g(x)s(x) = d(x)$, where $f(x) = x^2 - 3x + 2$ and $g(x) = x^2 + x + 1$;

27. (i) For every $m, n > 0$, show that the greatest common divisor in $\mathbb{Q}[x]$ of $x^m - 1$ and $x^n - 1$ is $x^d - 1$ where $d = (m, n)$.

(ii) Find polynomials $r(x)$ and $s(x)$ with $f(x)(x^m - 1) + g(x)(x^n - 1) = x^d - 1$.

28. Show that in $\mathbb{R}[x]$, $x^4 + x^2 + r^2$ and $x^2 - x + r$ are coprime for all $r \neq 0, 1$.

29. Find the monic polynomial $k(x)$ of smallest degree in $\mathbb{Q}[x]$ so that $(x^3 - 1)k(x)$ is a multiple of $x^2 - 1$.

30. Show that for f, g, h in $F[x]$, F a field, if $(f, g) = 1$ and h divides f, then $(h, g) = 1$.

31. Show that for f, g, h in $F[x]$, F a field, if $(f, g) = 1$, then $(fh, g) = (h, g)$.

32. For f, g, h in $F[x]$, F a field, show that if f divides gh, then f divides $(f, g)(f, h)$.

33. Show that for f, g in $F[x]$, F a field, if $(f, g) = d$, then $d = rf + sg$ where r, s in $F[x]$ may be chosen so that $\deg r < \deg g$ and $\deg s < \deg f$.

C. Factorization into Irreducible Polynomials

Recall that a *unit* of a ring R is an element f for which there is some element g in R with $fg = 1$. If F is a field, then in the ring $F[x]$, the only units are non-zero constants, that is, polynomials of degree zero.

Definition. A polynomial p in $F[x]$ is *irreducible* if p is not a unit, and if $p = fg$, then f or g must be a unit, that is, a constant polynomial.

Irreducible polynomials are like prime numbers. In particular:

Proposition 9. *If p is irreducible, and f is a polynomial which is not divisible by p, then the greatest common divisor of p and f is 1.*

Proof. Suppose $d = (f, p)$. Since p is irreducible and d divides p, either d is a unit, that is, a non-zero constant, or d is an associate of p. In the latter case, p divides f. In the former case, p and f are coprime, and the greatest common divisor of f and p is 1, an associate of d. □

Here are some examples of irreducible polynomials:
$x + a$ is irreducible in $F[x]$ for F any field;
$x^2 + 1$ is irreducible in $\mathbb{R}[x]$, but not in $\mathbb{C}[x]$;
$x^3 - 2$ is irreducible in $\mathbb{Q}[x]$, but not in $\mathbb{R}[x]$;
$x^2 + 1$ is irreducible in $\mathbb{Z}/3\mathbb{Z}[x]$, but not in $\mathbb{Z}/5\mathbb{Z}[x]$; and
$x^2 + x + 1$ is irreducible in $\mathbb{Z}/2\mathbb{Z}[x]$.
We will study the question of which polynomials are irreducible in several later chapters.

Irreducible polynomials in $F[x]$, F a field, are the multiplicative building blocks of nonconstant polynomials, just as primes are the building blocks of natural numbers > 1:

Theorem 10. *Every polynomial of degree ≥ 1 in $F[x]$, F a field, is irreducible or factors into a product of irreducible polynomials.*

The proof is virtually identical to that for numbers, an induction argument on the degree of the polynomial, and is left as an exercise, below.

It is also easy to prove that factorization of a polynomial into a product of irreducible polynomials is unique. The key lemma in the proof, as with numbers, is the following consequence of Bezout's Identity:

Theorem 11. *Let p be an irreducible polynomial in $F[x]$, F a field. For every two polynomials f, g in $F[x]$, if p divides fg, then p divides f or p divides g.*

This result is also proved in the same way as for integers, so is left as an exercise.
Here is the theorem on uniqueness of factorization:

Theorem 12. *In $F[x]$, F a field, if*

$$f = p_1 p_2 \cdot \ldots \cdot p_s = q_1 q_2 \cdot \ldots \cdot q_t$$

are two factorizations of the polynomial f into a product of irreducible polynomials in $F[x]$, then $s = t$ and there is a one-to-one correspondence between the factors $p_1, p_2, \ldots, p_s$ and $q_1, q_2, \ldots, q_t$, where if p_i corresponds with q_j, then p_i and q_j are associates.

Every factorization of an associate of f will have factors that are associates of a factorization of f. For example, in $\mathbb{F}_5[x]$,

$$x^2 + x + 3 = (x + 2)(x - 1),$$

and
$$3x^2 + 3x + 4 = (2x + 3)(4x + 3):$$

the two polynomials are associates of each other, and the factors $x + 2$ and $x - 1$ are associates of $4x + 3$ and $2x + 3$, respectively.

Since any polynomial is an associate of a unique monic polynomial, and the product of monic polynomials is monic, we can rephrase the theorem on unique factorization to require that f and all p_i and q_j be monic polynomials. In that case, the theorem becomes:

Theorem 13. *In $F[x]$, F a field, if*

$$f = p_1 p_2 \cdot \ldots \cdot p_s = q_1 q_2 \cdot \ldots \cdot q_t$$

are two factorizations of the monic polynomial f into a product of monic irreducible polynomials in $F[x]$, then $s = t$ and the sets $\{p_1, p_2, \ldots, p_s\}$ and $\{q_1, q_2, \ldots, q_t\}$ are equal.

We have left the proofs in this section as exercises because the theorems and the proofs are so similar to those in Chapter 4.

Just as with integers, we can write the factorization of a polynomial f in $F[x]$ in exponential notation, as

$$f = p_1^{e_1} p_2^{e_2} \cdot \ldots \cdot p_g^{e_g}$$

where $p_1, p_2 \ldots, p_g$ are distinct irreducible polynomials. If any e_i is bigger than 1, we shall say that f has a *multiple factor*: thus $f(x) = (x^2 + 2)^3(x + 1)$ in $\mathbb{R}[x]$ has a multiple factor, while $f(x) = (x^2 + 2)(x + 1)$ does not. If $f(x)$ has a multiple linear factor, then $f(x)$ is said to have a *multiple root* in F. An example is $f(x) = (x + 2)^3(x^2 + 1)$, which has the multiple root -2.

Just as with numbers, if the factorizations of f and g into products of irreducible polynomials are given in exponential notation, then it is easy to write down the greatest common divisor of f and g, and the least common multiple of f and g. If it is not clear how to do this, refer back to Section 4B.

Exercises.

34. Prove Theorem 10 using induction on the degree of the polynomial.

35. Prove Theorem 11.

36. Prove Theorem 13 by induction on s, just as with numbers.

37. Show that in $\mathbb{F}_2[x]$, two polynomials are associates if and only if they are equal.

38. In $F[x]$, F any field, show that if p, q are irreducible and monic, and p divides q, then $p = q$.

39. Let $f(x) = x^2 + bx + 4$ in $\mathbb{R}[x]$. For each b in $\mathbb{R}$, factor $f(x)$ into a product of irreducible polynomials in $\mathbb{R}[x]$.

40. (i) Show that in $\mathbb{R}[x]$, no polynomial of odd degree >1 is irreducible.
    (ii) Show that if $f(x)$ in $\mathbb{R}[x]$ has a multiple factor, then its derivative $f'(x)$ is not relatively prime to $f(x)$.
    (iii) Suppose a in $\mathbb{R}$ is a root of $f(x)$ in $\mathbb{R}[x]$. Show that a is a multiple root of $f(x)$, iff $f'(a) = 0$, iff the graph of $y = f(x)$ is tangent to the x-axis at $x = a$.
    (iv) Show that if $f(x)$ in $\mathbb{R}[x]$ has no multiple roots, and has odd degree, then $f(x)$ must have an odd number of real roots.

41. Find the greatest common divisor in $\mathbb{Q}[x]$ of
$(x^2 + 3x + 6)^2(x + 1)^3(x - 3)^2$ and $(x^2 + 3x + 6)(x + 1)^4(x - 2)^2$;

42. Find the greatest common divisor in $\mathbb{Q}[x]$ of $(x^2 - 3x - 4)^3(x - 3)^2$ and $(x - 4)^3(x^2 - 3x - 4)^2$

43. Factor $x^5 - x$ into a product of irreducible polynomials in $\mathbb{F}_5[x]$. (Hint: Recall Fermat's theorem.)

44. For any prime p, show that in $\mathbb{F}_p[x]$, $x^p - x$ factors into

$$x(x - 1)(x - 2) \cdot \ldots \cdot (x - (p - 1)).$$

    One way to factor a small number into a product of primes is by trial division. For example, to factor 60060, we factor out obvious small factors, namely 2, 2, 3, and 5 to get $60060 = 2 \cdot 2 \cdot 3 \cdot 5 \cdot 1001$, and then find by trial division that $1001 = 7 \cdot 13 \cdot 19$, so that $60060 = 2 \cdot 2 \cdot 3 \cdot 5 \cdot 7 \cdot 13 \cdot 19$. Similarly, this strategy is feasible for polynomials of low degree in $\mathbb{F}_2[x]$.

45. Prove that if $f(x)$ in $F[x]$, F any field, has degree $n > 1$ and is not irreducible, then $f(x)$ has an irreducible factor of degree $< n/2$.

46. Find all irreducible polynomials in $\mathbb{F}_2[x]$ of degree ≤ 4. There are eight of them. (The Root Theorem will be helpful.)

47. Using trial division in $\mathbb{F}_2[x]$, factor into a product of irreducible polynomials:
(i) $x^6 + x^4 + x$;
(ii) $x^8 + x^7 + x^6 + x^4 + 1$;
(iii) $x^7 + x^6 + x^4 + 1$;

48. Using trial division in $\mathbb{F}_2[x]$, factor into a product of irreducible polynomials:
(i) $x^8 - x$;
(ii) $x^{10} - 1$;
(iii) $x^{15} - 1$.

49. Let e be any natural number. Show that in $\mathbb{F}_p$, p prime, there are at most e units whose orders divide e.

50. Let R be a commutative ring and suppose a, b are nonzero elements of R such that $a \cdot b = 0$ (i.e., a and b are zero divisors).
(i) Show that unique factorization in $R[x]$ is false by finding two different factorizations of $f(x) = x^2 - (a+b)x$ into irreducible polynomials in $R[x]$.
(ii) Show that in $R[x]$ there exists an irreducible polynomial p that divides a product fg of two polynomials but divides neither f nor g.
(iii) Give explicit examples of (i) and (ii) when $R = \mathbb{Z}/21\mathbb{Z}[x]$.

51. Show that if F is an infinite field and $p(x)$ is an irreducible polynomial of degree d in $F[x]$, then $F[x]$ has infinitely many irreducible polynomials of degree d. (Hint: Try $p(x-a)$.)

Chapter 15
The Fundamental Theorem of Algebra

In Chapter 14 we showed that every nonconstant polynomial in $F[x]$, F a field, factors uniquely (up to associates and the order of the factors) into the product of irreducible polynomials. Irreducible polynomials therefore relate to all polynomials in the same way that primes do to all natural numbers. Thus one naturally asks: Which polynomials are irreducible? and, How does one factor a given polynomial into a product of irreducible polynomials?

When looking for irreducible polynomials over a field, we can restrict our attention to monic polynomials. Every polynomial is an associate of a monic polynomial.

The question, which polynomials are irreducible, depends on the field F of coefficients.

For example, consider the polynomial $x^3 - 2$. This is a polynomial with coefficients in $\mathbb{Q}$, and $\mathbb{Q} \subset \mathbb{R} \subset \mathbb{C}$, so we can ask how $x^3 - 2$ factors in $\mathbb{Q}$, in $\mathbb{R}$, and in $\mathbb{C}$.

In $\mathbb{Q}[x], x^3 - 2$ is irreducible.

In $\mathbb{R}[x], x^3 - 2 = (x - 2^{1/3})(x^2 + 2^{1/3}x + 4^{1/3})$.

In $\mathbb{C}[x], x^3 - 2 = (x - 2^{1/3})(x - \omega 2^{1/3})(x - \omega^2 2^{1/3})$ where $\omega = e^{2\pi i/3} = -(1/2) + (i\sqrt{3}/2)$ is a complex root of $x^3 - 1$.

Thus the answer to the question, which polynomials are irreducible, clearly depends on the field of coefficients.

When the field F is the real numbers, an additional reason to study irreducible polynomials was the discovery of the fundamental theorem of calculus by Newton and Leibniz in the last third of the seventeenth century. The fundamental theorem of calculus meant that previously intractable problems of finding areas, volumes, arc lengths, centroids, etc., could be solved by finding antiderivatives of functions. So attention turned to the problem of finding antiderivatives of all kinds of functions.

Antiderivatives of polynomial functions were easy.

The next natural class of functions to consider were rational functions, functions of the form $f(x)/g(x)$, where $f(x)$ and $g(x)$ are polynomials with real coefficients. For these functions, the method of partial fractions showed the need to know which polynomials with real coefficients were irreducible.

L.N. Childs, *A Concrete Introduction to Higher Algebra*, Undergraduate Texts in Mathematics, © Springer Science+Business Media LLC 2009

In this chapter we will begin by looking at rational functions and their partial fraction decompositions, then examine the Fundamental Theorem of Algebra and its antecedents.

A. Rational Functions

In the same way that the field of rational numbers is formed from the ring of integers, the field of rational functions may be constructed from the set of polynomials with coefficients in a field. A *rational function* with coefficients in the field F is an expression of the form $f(x)/g(x)$ where $f(x)$ and $g(x)$ are in $F[x]$ and $g(x) \neq 0$. Two rational functions are equal,

$$\frac{f(x)}{g(x)} = \frac{h(x)}{k(x)}$$

if $k(x)f(x) = g(x)h(x)$ in $F[x]$. Call the set of rational functions with coefficients in F by $F(x)$ (as opposed to $F[x]$, which denotes the set of polynomials with coefficients in F).

Addition and multiplication of rational functions is defined by the usual rules for fractions (we drop "(x)"):

$$\frac{f}{g} + \frac{h}{k} = \frac{fk+hg}{gk}; \frac{f}{g}\frac{h}{k} = \frac{fh}{gk}.$$

It is very easy to verify that $F(x)$ is a field. A polynomial f may be viewed as a rational function by thinking of it as $\frac{f}{1}$.

The terminology "rational function" is somewhat misleading. The elements of $F(x)$ are not functions on the field F, but formal expressions in the same sense as polynomials are. One can evaluate a rational function $f(x)/g(x)$ at any element a of F at which $g(a) \neq 0$, but two rational functions may agree when evaluated at every element of F and yet be different elements of $F(x)$, such as x and $x^3/(x^2+x+1)$ in $\mathbb{F}_2(x)$; and there may exist rational functions in $F(x)$ which cannot be defined as functions on F at all, such as $(x^3 - x + 1)/(x^3 - x)$ in $\mathbb{F}_3(x)$, whose denominator gives zero when evaluated at each element of $\mathbb{F}_3$.

However, as with polynomials, it can be proved that if F is an infinite field, then two rational functions which have the same values when evaluated on infinitely many elements of F must be equal.

Exercises.

1. Prove this last assertion.

2. Show that every rational function in $F(x)$ can be written uniquely as $\frac{h(x)}{k(x)}$ where $h(x)$ and $k(x)$ are coprime and $k(x)$ is monic.

B. Partial Fractions

The method of partial fractions is a way of decomposing a rational function f/g into a sum of terms with denominators of degrees smaller than $\deg g$ when a factorization of g is known. In case f/g is a rational function with real coefficients, viewed as a real-valued function, then partial fractions becomes an important technique of integration. In this section we shall describe the general method.

We assume $f(x)$ and $g(x)$ are in $F[x]$, where F is an arbitrary field.

Given f/g, we first use the division theorem, if necessary, to write $f = gq + r$, with $\deg r < \deg g$. Then $f/g = q + r/g$. The basic problem, to write f/g as a sum of terms with "nice" denominators, remains for r/g. So we shall assume that we started out with f/g, where $\deg f < \deg g$.

Here is the general description of partial fractions.

Theorem 1. *Let $g = p_1^{e_1} p_2^{e_2} \ldots p_r^{e_r}$ be a factorization of g into a product of powers of coprime polynomials p_i, and suppose that $\deg f < \deg g$. Then there are unique polynomials $h_i, i = 1, \ldots, r$, with $\deg h_i < \deg p_i^{e_i}$, such that*

$$\frac{f}{g} = \frac{h_1}{p_1^{e_1}} + \frac{h_2}{p_2^{e_2}} + \ldots + \frac{h_r}{p_r^{e_r}}.$$

Proof. Induction on r, $r = 1$ being trivial.

In order to pass from $r - 1$ to r, and thus prove the theorem, we let $a = p_1^{e_1} p_2^{e_2} \ldots p_{r-1}^{e_{r-1}}, b = p_r^{e_r}$ and prove the following, which is the induction step.

Lemma 2. *Let $g = ab$ where a and b are coprime, and suppose $\deg f < \deg g$. Then there are unique polynomials r, s with $\deg r < \deg a$, $\deg s < \deg b$, so that*

$$\frac{f}{g} = \frac{r}{a} + \frac{s}{b}.$$

The theorem follows. For using the lemma, we may write

$$\frac{f}{g} = \frac{r}{p_1^{e_1} p_2^{e_2} \ldots p_{r-1}^{e_{r-1}}} + \frac{s}{p_r^{e_r}},$$

use induction to write

$$\frac{r}{p_1^{e_1} p_2^{e_2} \ldots p_{r-1}^{e_{r-1}}} = \frac{h_1}{p_1^{e_1}} + \frac{h_2}{p_2^{e_2}} + \ldots + \frac{h_{r-1}}{p_{r-1}^{e_{r-1}}}$$

and set $h_r = s$. $\qquad\square$

Proof of lemma. To prove the lemma, we use Bezout's Identity: since a and b are coprime, there are polynomials s, r such that

$$as + br = f,$$

where we can choose $\deg s < \deg b$, $\deg r < \deg a$ by Exercise 33 of Section 14B. Then divide by $g = ab$:

$$\frac{f}{g} = \frac{r}{a} + \frac{s}{b}.$$

To show r, s are unique is easy.   □

Example 1. We decompose

$$\frac{5x+2}{x^2-4}$$

into partial fractions. To do so, we factor $x^2 - 4 = (x+2)(x-2)$, then we know we can write

$$\frac{5x+2}{x^2-4} = \frac{5x+2}{(x+2)(x-2)} = \frac{a}{x+2} + \frac{b}{x-2}$$

where a, b are constants (polynomials of degree <0). We can find a and b by putting the right side over the common denominator $(x-2)(x+2)$,

$$\frac{a}{x+2} + \frac{b}{x-2} = \frac{a(x-2)+b(x+2)}{(x+2)(x-2)},$$

and then solving the equation arising from setting the numerators equal:

$$5x+2 = a(x-2) + b(x+2)$$

to get $a = 2, b = 3$.

Example 2. In $\mathbb{F}_2[x]$, $x^3 + x^2 + 1$ and $x + 1$ are coprime, and hence

$$\frac{x^2}{x^4+x^2+x+1} = \frac{x^2}{(x^3+x^2+1)(x+1)}$$

$$= \frac{a}{x+1} + \frac{b(x)}{x^3+x^2+1}$$

where $b(x)$ has degree at most 2 and a is a constant. So we solve

$$x^2 = a(x^3+x^2+1) + b(x)(x+1)$$

for a and $b(x)$: we find that

$$x^2 = (x^3+x^2+1) + (x^2+x+1)(x+1);$$

thus $a = 1, b(x) = x^2 + x + 1$ and

$$\frac{x^2}{x^4+x^2+x+1} = \frac{1}{x+1} + \frac{x^2+x+1}{x^3+x^2+1}.$$

Once we have a rational function written as a sum of terms of the form f/p^e, we can further decompose f/g by representing the numerator in base p.

To write f in base p for any polynomial p means to write $f = r_0 + r_1 p + r_2 p^2 + \ldots + r_k p^k$, where $\deg r_i < \deg p$ for all i. In case $p(x) = x$, writing the polynomial $f(x)$ in base x is the way we usually write $f(x)$. If we can write f in base p, then f/p^e decomposes as

$$\frac{f}{p^e} = \frac{r_0 + r_1 p + r_2 p^2 + + r_k p^k}{p^e}$$

$$= \frac{r_0}{p^e} + \frac{r_1}{p^{e-1}} + \ldots + \frac{r_k}{p^{e-k}}$$

with $\deg r_i < \deg p$ for all i. In case $p(x) = x - r$ has degree 1, then all of the r_i are constants.

We write f in base p just as with integers, as follows. Divide p into f,

$$f = pq_0 + r_0$$

with $\deg r_0 < \deg p$; then divide p into the quotients, successively:

$$q_0 = pq_1 + r_1 \text{ with } \deg r_1 < \deg p$$
$$q_1 = pq_2 + r_2 \text{ with } \deg r_2 < \deg p$$

$$\vdots$$

$$q_{k-1} = pq_k + r_k \text{ with } \deg r_k < \deg p.$$

Then for all $i \geq 0$, $\deg q_{i+1} = \deg q_i - \deg p$, so for some k, $\deg q_{k-1} < \deg p$. Then $q_k = 0$ and $q_{k-1} = r_k$. Successively substituting then gives

$$f = r_k p^k + r_{k-1} p^{k-1} + \ldots + r_1 p + r_0.$$

as can be seen by substituting each successive equation into the first equation. Since the quotient and remainder in the division algorithm are unique, the representation of f in base p is unique.

Representation in base p for polynomials is thus essentially the same as for integers.

The complete decomposition of f/g is then achieved by partial fractions followed by writing the numerator in base p and reducing to lowest terms.

Example 3. Let

$$\frac{f(x)}{g(x)} = \frac{3x^4 + 5}{(x^2 + 1)^2 x}$$

in $\mathbb{R}(x)$.

Following the method as described above, we know that there is a polynomial $a(x)$ of degree < 3 and a constant b so that

$$\frac{3x^4 + 5}{(x^2 + 1)^2 x} = \frac{a(x)}{(x^2 + 1)^2} + \frac{b}{x}.$$

Putting everything over the common denominator $(x^2+1)^2x$ and comparing numerators gives

$$xa(x)+b(x^4+2x^2+1) = 3x^4+5.$$

We find that

$$x(-2x^3-10x)+5(x^4+2x^2+1) = 3x^4+5,$$

and so $a(x) = -2x^3-10x$, $b=5$ and

$$\frac{3x^4+5}{(x^2+1)^2x} = \frac{-2x^3-10x}{(x^2+1)^2} + \frac{5}{x}.$$

To expand $-2x^3-10x$ in base x^2+1 we divide $-2x^3-10x$ by x^2+1:

$$-2x^3-10x = (x^2+1)(-2x)-8x,$$

then

$$\frac{-2x^3-10x}{(x^2+1)^2} = \frac{(x^2+1)(-2x)-8x}{(x^2+1)^2} = \frac{-2x}{x^2+1} + \frac{-8x}{(x^2+1)^2}$$

and so

$$\frac{3x^4+5}{(x^2+1)^2x} = \frac{-2x}{x^2+1} + \frac{-8x}{(x^2+1)^2} + \frac{5}{x}.$$

Knowing the degrees of the numerators, we can find the coefficients of the numerators by setting up a system of linear equations: by the general theory, we know that the decomposition should be

$$\frac{3x^4+5}{(x^2+1)^2x} = \frac{ax+b}{x^2+1} + \frac{cx+d}{(x^2+1)^2} + \frac{e}{x}$$

for some real numbers a,b,c,d,e. We put the right side over a common denominator, collect coefficients, equate them to the coefficients on the left side, and solve the resulting system of linear equations:

$$\frac{3x^4+5}{(x^2+1)^2x} = \frac{(a+e)x^4+bx^3+(a+c+2e)x^2+(b+c)x+e}{(x^2+1)^2x},$$

so

$$a+e=3$$
$$b=0$$
$$a+c+2e=0$$
$$b+d=0$$
$$e=5$$

with solution $a=-2, b=d=0, c=-8, e=5$.

Exercises.

3. Write x^3 in base $x + 1$.

4. Write $(x^2 + 3x + 1)^4$
(i) in base $x + 2$;
(ii) in base $x^2 + x + 1$.

5. Decompose into partial fractions:
(i) $\frac{t+1}{(t-1)(t+2)}$
(ii) $\frac{1}{(t+1)(t^2+2)}$
(iii) $\frac{x^2+4}{(x+1)^2(x-2)(x+3)}$.

6. What is the analogue of partial fractions in $\mathbb{Z}$? Illustrate it with $17/180$.

7. Let x_0 a real number, and $f(x)$ be a polynomial in $\mathbb{R}[x]$ of degree n. The Taylor expansion of $f(x)$ about $x = x_0$ is

$$T_f(x) = f(x_0) + f'(x_0)(x - x_0) + \frac{f''(x_0)}{2!}(x - x_0)^2 + \ldots + \frac{f^{(n)}(x_0)}{n!}(x - x_0)^n$$

since the m-th derivative $f^{(m)}(x) = 0$ for $m > n$. Show that $T_f(x) = f(x)$, hence $T_f(x)$ is the expansion of f in base $p(x) = (x - x_0)$.

C. Irreducible Polynomials over $\mathbb{R}$

The theorem on partial fractions shows that any rational function $f(x)$ is a sum of rational functions $r(x)/p(x)^e$, where $p(x)$ is irreducible and $\deg r(x) < \deg p(x)$. In particular, this is true when $f(x), g(x)$ are polynomials with real coefficients. The antiderivative of a sum of functions is the sum of the antiderivatives, so partial fractions reduces the problem of finding the antiderivative of a complicated rational function to finding the antiderivatives of $\frac{r(x)}{p(x)^e}$ where $p(x)$ is an irreducible polynomial in $\mathbb{R}[x]$. The discoverers and early students of calculus in the seventeenth and eighteenth centuries knew partial fractions. Thus it was natural for them to want to know, which polynomials in $\mathbb{R}[x]$ are irreducible?

For F any field, if r is in F then the polynomial $f(x) = x - r$ is irreducible over any field, so, in particular, if F is the field of real numbers.

With only a bit more effort, we can find out which monic polynomials of degree 2 with real coefficients are irreducible.

Let $f(x) = x^2 + bx + c$. Then $f(x)$ is irreducible iff $f(x)$ has no real roots, by the Root Theorem. This occurs iff the graph of $y = f(x)$ doesn't cross the x-axis. To see what this means algebraically, we complete the square:

$$y = f(x) = x^2 + bx + c$$
$$= x^2 + bx + b^2/4 + c - b^2/4$$
$$= (x + b/2)^2 - (b^2 - 4c)/4.$$

Thus when $x = -b/2$, y takes on its minimum value, $y = -(b^2 - 4c)/4$. Of course, for x a large positive or a large negative real number, y is positive. So $f(x)$ crosses the x-axis iff $b^2 - 4c \geq 0$. To sum up:

Proposition 3. *If $f(x) = x^2 + bx + c$ is a polynomial of degree 2 in $\mathbb{R}[x]$, then $f(x)$ is irreducible iff $b^2 - 4c < 0$.*

What about polynomials of degree > 2?

If we think about the graph of $y = f(x)$ where $f(x)$ is a polynomial of odd degree, then it becomes clear that every polynomial of odd degree has a real root, and therefore by the Root Theorem is not irreducible. If $f(x) = x^n + a_{n-1}x^{n-1} + \cdots + a_1 x + a_0$ with n odd, then for x large and positive, $f(x) > 0$, while for x large and negative, $f(x) < 0$. Since $f(x)$ is a continuous function of x, the Intermediate Value Theorem implies that there is some x for which $f(x) = 0$, that is, $f(x)$ has a root.

Thus if there is an irreducible polynomial $f(x)$ of degree > 2 in $\mathbb{R}[x]$, then the degree of $f(x)$ must be even.

It turns out that

There are no irreducible polynomials in $\mathbb{R}[x]$ of degree > 2.

Several of the greatest mathematicians of the eighteenth century tried to prove this statement, notably Euler and Lagrange, but not until Gauss, in 1801, using complex numbers, was there a reasonably satisfactory proof of the result, which came to be known as the Fundamental Theorem of Algebra.

Integrating. Assuming that the only irreducible polynomials in $\mathbb{R}[x]$ have degree < 2, then by partial fractions, we know that any rational function $f(x)/g(x)$ may be written as a polynomial $q(x)$, plus a sum of terms of the form

$$\frac{a}{(x - d)^r} \tag{15.1}$$

and

$$\frac{px + q}{(x^2 + bx + c)^s}, \tag{15.2}$$

where $x^2 + bx + c$ is irreducible in $\mathbb{R}[x]$. Thus to find the indefinite integral (or antiderivative) of a rational function $f(x)/g(x)$, that is, to find a function $H(x)$ so that $H'(x) = f(x)/g(x)$, it suffices to find the integrals of expressions of the forms (15.1) and (15.2).

By completing the square and setting $e^2 = (4c - b^2)/4$, we can write the expression (15.2) as

$$\frac{px + q}{((x + b/2)^2 + e^2)^s}$$

which, after the change of variables $y = x + b/2$, becomes

$$\frac{my}{(y^2 + e^2)^s} + \frac{n}{(y^2 + e^2)^s}. \tag{15.3}$$

Thus assuming that the factorization of $g(x)$ into irreducible polynomials of degrees 1 and 2 can be found, the integral

$$\int \frac{f(x)}{g(x)} dx$$

reduces to a sum of integrals of the forms

$$\int p(x) dx$$

where $p(x)$ is a polynomial;

$$\int \frac{a}{(x - d)^r} dx \tag{15.4}$$

$$\int \frac{bx}{(x^2 + e^2)^s} dx \tag{15.5}$$

and

$$\int \frac{c}{(x^2 + e^2)^s} dx, \tag{15.6}$$

The integral of a polynomial is very easy.

If $r \neq 1$ the integral (15.4) equals

$$\frac{a}{(1 - r)(x - d)^{r-1}} + C;$$

if $r = 1$ it equals $a \log |x - d| + C$.

If we substitute $x^2 = u$ in integral (15.5), it becomes

$$\int \frac{bx}{(x^2 + e^2)^s} dx = \frac{b}{2} \int \frac{du}{(u + e^2)^s},$$

which is of the form (15.4).

The remaining integral, integral (15.6), is somewhat more difficult. In calculus textbooks an integral of type (15.6) is usually done by setting $x = e \tan(t)$ to transform it into

$$\int \frac{ce \sec^2(t) dt}{(e^2 \sec^2(t))^s} = \frac{c}{e^{2s-1}} \int \cos^{2s-2}(t) dt,$$

which is then done by a recurrence formula, derived by using integration by parts.

D. The Complex Numbers

Before proceeding further, we pause to review the complex numbers.

A complex number is an expression of the form $\alpha = a + bi$, where a and b are real numbers and $i = \sqrt{-1}$. If $\alpha = a + bi$, a is called the real part of α, b the imaginary part of α, and $\alpha = a + bi$ is the *normal form* of α. The set of all complex numbers is denoted by $\mathbb{C}$. The set $\mathbb{C}$ is a field, with addition:

$$(a + bi) + (c + di) = (a + c) + (b + d)i$$

and multiplication as forced by the distributive law:

$$(a + bi)(c + di) = ac + (ad + bc)i + bdi^2.$$

Using the property that $i^2 = -1$, this becomes

$$= (ac - bd) + (ad + bc)i.$$

Also,

$$0 = 0 + 0i \text{ is the zero element; and}$$

$$1 = 1 + 0i \text{ is the identity.}$$

If $\alpha = a + bi$ is in $\mathbb{C}$, then its complex conjugate, denoted $\overline{\alpha}$, is $\overline{\alpha} = a - bi$, and we have

$$\alpha\overline{\alpha} = (a + bi)(a - bi) = a^2 + b^2,$$

which $= 0$ if $a = b = 0$ and otherwise is always a positive real number. So if $\alpha = a + bi \neq 0$, then α has an inverse whose normal form is

$$\frac{a}{a^2 + b^2} - \frac{b}{a^2 + b^2}i.$$

So $\mathbb{C}$ is a field.

We may visualize $\mathbb{C}$ geometrically as the set of vectors (= directed line segments) in the plane, with $\alpha = a + bi$ corresponding to the vector from the origin to the point with coordinates (a, b). The horizonal axis is called the real axis, because vectors on the horizontal axis correspond to real numbers, complex numbers with imaginary part equal to zero. Similarly, the vertical axis is called the imaginary axis. The complex conjugate $\overline{\alpha}$ of α is the reflection of the vector α across the real axis.

The real number $|\alpha| = \sqrt{\alpha\overline{\alpha}} = \sqrt{a^2 + b^2}$ is the length of the vector α.

A convenient way to represent elements of $\mathbb{C}$ is in terms of polar coordinates. If $\alpha = a + bi$ is a complex number, $|\alpha| = r$ is the distance from the origin to the point (a, b) (i.e., $r = \sqrt{a^2 + b^2}$) and θ is the angle (measured counterclockwise) from the positive real axis to the vector α, then $a = r\cos\theta, b = r\sin\theta$, so $\alpha = a + bi = r\cos\theta + ir\sin\theta$. The angle θ is called the argument of α, and is sometimes denoted arg α. Multiplication of complex numbers when described in polar coordinates works rather neatly, thanks to some trigonometric formulas:

$$(r\cos\theta + ir\sin\theta)(s\cos\phi + is\sin\phi)$$
$$= rs((\cos\theta\cos\phi - \sin\theta\sin\phi) + i(\cos\theta\sin\phi + \sin\theta\cos\phi)) \qquad (15.7)$$
$$= rs(\cos(\theta + \phi) + i\sin(\theta + \phi)).$$

That is, when multiplying two complex numbers, lengths multiply and arguments add.

If you have had some acquaintance with infinite series, then you probably know the Taylor series for the exponential function $e^x = \exp(x)$:

$$e^x = 1 + x + \frac{x^2}{2!} + \frac{x^3}{3!} + \cdots$$

as well as for $\sin x$ and $\cos x$:

$$\sin x = x - \frac{x^3}{3!} + \frac{x^5}{5!} - \frac{x^7}{7!} + \cdots$$

and

$$\cos x = 1 - \frac{x^2}{2!} + \frac{x^4}{4!} - \frac{x^6}{6!} + \cdots$$

all three of which converge for all real numbers x. Then $\cos x + i\sin x$ has a Taylor series

$$1 + ix - \frac{x^2}{2!} - \frac{ix^3}{3!} + \frac{x^4}{4!} + \frac{ix^5}{5!} - \frac{x^6}{6!} + \cdots$$

which would be the same as the Taylor series for the complex function e^{ix} if we knew what e^{ix} was. So we define e^{ix} for x real by setting $e^{ix} = \cos x + i\sin x$: that is, we define e^{ix} by replacing x by ix in the Taylor series for e^x. Then an arbitrary complex number α can be written in *polar form* as

$$\alpha = r(\cos\theta + i\sin\theta) = re^{i\theta}.$$

If β is another complex number, $\beta = se^{i\phi}$, then the multiplication of formula (15.7) above becomes

$$\alpha\beta = re^{i\theta}se^{i\phi} = rse^{i(\theta + \phi)},$$

which is exactly what one would expect from the laws of exponents.

Thus, we have two ways to represent complex numbers α. There is normal form:

$$\alpha = a + bi,$$

and polar form:

$$\alpha = re^{i\theta}.$$

Normal form is convenient for addition. Polar form is convenient for multiplication.

An interesting special case of multiplication is de Moivre's formula

$$(\cos\theta + i\sin\theta)^n = (e^{i\theta})^n = e^{i(n\theta)} = \cos n\theta + i\sin n\theta.$$

This formula is useful for understanding the roots of unity in $\mathbb{C}$. If we set $\zeta = e^{2\pi i/n}$, then $\zeta^n = e^{2\pi i} = \cos 2\pi + i\sin 2\pi = 1$. So ζ and all of its powers are roots of the polynomial $x^n - 1$, hence are called n-th roots of unity. The n-th roots of unity may be drawn as follows: take the circle of radius one with center at the origin (the "unit circle") and starting with the point $(1, 0)$, mark along the circumference n equally spaced points, each separated from the next by an arc of the circle of length $2\pi/n$. The vectors from the origin to these points are the n-th roots of unity in the complex plane.

For example, the cube roots of unity in $\mathbb{C}$ are 1 and

$$e^{\frac{2\pi i}{3}} = \cos\frac{2\pi}{3} + i\sin\frac{2\pi}{3} = -\frac{1}{2} + \frac{\sqrt{3}}{2}i$$

$$e^{\frac{4\pi i}{3}} = \cos\frac{4\pi}{3} + i\sin\frac{4\pi}{3} = -\frac{1}{2} - \frac{\sqrt{3}}{2}i.$$

The fourth roots of unity are $1, e^{\pi i/2} = i, e^{\pi i} = -1$ and $e^{3\pi i/2} = -i$.

We note that roots of unity occur in fields and rings other than subrings of $\mathbb{C}$, and can look much different than roots of unity in $\mathbb{C}$. as we observed in Section 11B. For example, in $\mathbb{Z}/13\mathbb{Z}$, the congruence class of 5 is a fourth root of unity since (denoting $[\ ]_{13}$ by $[\ ]$):

$$[5]^2 = [25] = [-1], [5]^4 = [1],$$

and the congruence class of 3 is a cube root of unity:

$$[3]^3 = [27] = [1]$$

in $\mathbb{Z}/13\mathbb{Z}$.

Exercises.

8. Solve $\alpha x = \beta$ in $\mathbb{C}$, where:
   (i) $\alpha = 3 + 2i$ and $\beta = 1 - i$;
   (ii) $\alpha = 1 - i$ and $\beta = 3 + 2i$;
   (iii) $\alpha = 3 - 2i$ and $\beta = 3 + 2i$.

9. Find the inverses in normal form of
   (i) $1 + i$,
   (ii) $1 + \sqrt{2}i$,
   (iii) $1 + 6i$.

10. Write in polar form:
   (i) $(1 + i)/2$,

(ii) $4 - 4i$,

(iii) $8i$,

(iv) -1.

11. Let b, c be real numbers with $b^2 - 4c < 0$.

(i) Write down the two roots α and β of $x^2 + bx + c$. Show that α and β are complex conjugates of each other.

(ii) Find a formula for the length of α in terms of the coefficients b and c.

12. Use de Moivre's Theorem for $n = 3$ to write $\cos(3x)$ as a polynomial in $\cos(x)$.

13. Find all of the roots in $\mathbb{C}$ of:

(i) $x^3 - 1$;

(ii) $x^8 - 1$;

(iii) $x^{12} - 1$;

(iv) $x^3 + 1$.

14. Show that complex conjugation is a one-to-one ring homomorphism from $\mathbb{C}$ onto $\mathbb{C}$.

15. (i) What is the complex conjugate of $re^{i\theta}$?

(ii) What is the inverse of $re^{i\theta}$?

16. Show that $\overline{\alpha} = \alpha$ if and only if α is in $\mathbb{R}$.

17. Show that if $f(x)$ is a polynomial with real coefficients and $\alpha = r + is$ in $\mathbb{C}$ is a root of $f(x)$, then so is $\overline{\alpha} = r - is$, the complex conjugate of z.

18. Given that $f(x) = x^4 - 4x^3 + 3x^2 + 14x + 26$ has the root $3 + 2i$, factor f into a product of irreducible polynomials in $\mathbb{R}[x]$.

19. Find all roots in $\mathbb{C}$ of:

(i) $x^3 - 2$;

(ii) $x^4 + 2$;

(iii) $x^5 - 2i$.

20. If

$$f(x) = x^n + \alpha_1 x^{n-1} + \alpha_2 x^{n-2} + \ldots + \alpha_{n-1} x + \alpha_n$$

is a polynomial with coefficients $\alpha_1, \ldots, \alpha_n$ in $\mathbb{C}$, let

$$\overline{f}(x) = x^n + \overline{\alpha_1} x^{n-1} + \overline{\alpha_2} x^{n-2} + \ldots + \overline{\alpha_{n-1}} x + \overline{\alpha_n}.$$

(i) Show that $g(x) = f(x)\overline{f}(x)$ is a polynomial with real coefficients.

(ii) Show that if γ is a root of $g(x)$, then either γ or $\overline{\gamma}$ is a root of $f(x)$.

21. Let $f(x) = x^2 - (3 - 2i)x + (5 - i)$. Find a polynomial $g(x)$ with real coefficients such that every root of $f(x)$ is a root of $g(x)$.

E. Root Formulas

Quadratic equations. The problem of finding roots of polynomials has its origins in work of the ancient Babylonians (before 1500 B.C.), who began the study of finding roots of quadratic and cubic polynomials, or equivalently solving quadratic and cubic equations. "Roots" to all mathematicians before the sixteenth century A.D. meant positive real roots, and negative coefficients were not permitted, so even the study of quadratic equations was complicated by the need to look at several different cases:

$$ax^2 + bx = c,$$
$$ax^2 + c = bx,$$

and

$$ax^2 = bx + c,$$

where a, b, c are all positive integers.

By the time of Euclid (300 B.C.) mathematicians knew how to complete the square to solve these equations. For example, given

$$ax^2 + bx = c,$$

add $\frac{b^2}{4a}$ to both sides to give

$$a(x + \frac{b}{2a})^2 = c + \frac{b^2}{4a},$$

from which x can be found as soon as one finds the square root of a and of $c + b^2/4a = (b^2 + 4ac)/4a$. The question of imaginary numbers never really arose in this context, because a, b, and c were all chosen to be nonnegative, and all square roots involved positive numbers. The fact that the square roots were often not natural numbers was avoided by thinking of the problem geometrically, as the problem of constructing a line segment whose length was x. Since square roots of positive numbers can be constructed by straightedge and compass techniques, and the problem of solving the equation was reduced to that of constructing square roots of positive quantities, the problem was solved.

Long before the Greeks, the Babylonians knew how to find square roots of positive quantities, and also knew how to solve problems such as the following:

Given the area a and the perimeter 2q of a rectangle, find the lengths of the sides.

If we denote the sides by x and y, then we need to solve

$$x + y = q$$
$$xy = a.$$

This pair of equations is called the *Babylonian normal equations*. Here is how the Babylonians solved them. Introduce a new unknown, z, so that

$$x = \frac{q}{2} + z, y = \frac{q}{2} - z.$$

Then $x + y = q$, clearly. Substituting into the second equation, we obtain

$$\left(\frac{q}{2} + z\right)\left(\frac{q}{2} - z\right) = a,$$

or

$$z^2 = \frac{q^2}{4} - a.$$

Take the square root of the right side to find

$$z = \sqrt{\frac{q^2}{4} - a} = \frac{\sqrt{q^2 - 4a}}{2},$$

then substitute to find x and y.

We can use the Babylonian normal equations to obtain the quadratic formula, as follows:

Suppose $f = x^2 + bx + c$ has roots r and s. Then

$$x^2 + bx + c = (x - r)(x - s) = x^2 - (r + s)x + rs.$$

Thus the roots r and s satisfy the Babylonian normal equations

$$r + s = -b$$
$$rs = c$$

The Babylonian method then finds r and s:

$$r = -\frac{b}{2} + z, s = -\frac{b}{2} - z$$

with

$$z = \frac{\sqrt{b^2 - 4c}}{2}.$$

Cubic equations. For many centuries, from the time of the Babylonians to the sixteenth century A.D., mathematicians sought a method to find a root of the cubic polynomial. The problem had particular import because of its close relationship with the classical geometrical problem of trisecting an angle by ruler and compass. For example, to trisect the angle of 60 degrees is equivalent to constructing the cosine of 20 degrees, and since $\cos 3\theta = 4\cos^3 \theta - 3\cos\theta$, $\cos 20°$ is a root of the cubic polynomial $4x^3 - 3x = 1/2$.

Unfortunately, the geometrical methods of the ancient Greeks were doomed to be unsuccessful, for it turns out to be impossible to find the root of a general cubic by straightedge and compass methods.

Some ancient Greek mathematicians, and Arabic mathematicians of the ninth and tenth centuries, including Omar Khayyam (the poet), discovered that a root of a cubic could be obtained geometrically as the intersection of two conics. For example, the equation $x^3 + px = q$ (with $p, q > 0$) can be rewritten as $x^3 + a^2 x = a^2 b$; then the positive real solution x of the equation is then the x that solves the two equations $x^2 = ya, y^2 = x(b - x)$, equations of a parabola and a circle, respectively. (Of course, a parabola cannot be constructed with straightedge and compass.) [Berggren (1986), p. 126ff.]

But it wasn't until the assimilation of the methods of algebra, introduced into western Europe in the middle ages via Latin translations of Al-Khwarismi's The Compendious Book on Calculation by Al-jabr and Al-muqabala, written in Baghdad around A.D. 825, that a general method was discovered for solving the cubic, by the Italian mathematician del Ferro sometime prior to his death in 1526, and later by Tartaglia in 1535. The method was first published by Cardano in his Ars Magna in 1545.

The solution of the cubic equation

$$x^3 + ax = b \qquad (a, b > 0),\tag{15.8}$$

by del Ferro was perhaps the first major new discovery in mathematics since 212 B.C., the time of the death of Archimedes.

The solution of (15.8) was as follows:

Set $x = u + v$, and substitute into (15.8), to get

$$u^3 + v^3 + 3uv(u + v) + a(u + v) = b.$$

This can be solved if we set

$$u^3 + v^3 = b \text{ and } 3uv = -a.$$

Cubing the second equation yields

$$u^3 + v^3 = b \qquad u^3 v^3 = -(a^3/27).\tag{15.9}$$

Equations (15.9) are Babylon normal equations for u^3 and v^3. The solutions of (15.9) are

$$u^3 = \frac{b}{2} + \frac{1}{2}\sqrt{b^2 + \frac{4a^3}{27}}$$

and

$$v^3 = \frac{b}{2} - \frac{1}{2}\sqrt{b^2 + \frac{4a^3}{27}}.$$

Taking the real cube roots of u^3 and v^3 to find u and v, we obtain $x = u + v$, a positive solution to (15.8). Notice that for $a, b > 0$, (15.8) has a unique real root, and that root is positive, since the function

$$f(x) = x^3 + ax - b$$

has $f(0) = -b < 0$ and $f'(x) = 3x^2 + a > 0$ for all x.

Example 4. (from Cardano's Ars Magna.) Consider

$$x^3 + 6x = 20.$$

Set $x = u + v$, to get

$$u^3 + v^3 + 3uv(u + v) + 6(u + v) = 20.$$

Set $u^3 + v^3 = 20, 3uv = -6$, and solve for u^3 and v^3 to get

$$u^3 = 10 + \sqrt{108}, v^3 = 10 - \sqrt{108}.$$

To get the obvious solution $x = 2$, we note that

$$10 + \sqrt{108} = 10 + 6\sqrt{3} = (1 + \sqrt{3})^3,$$

so $u = 1 + \sqrt{3}$, and similarly, $v = 1 - \sqrt{3}$, so $x = u + v = 2$.

Example 5. Consider the equation

$$x^3 + 3x = 14. \tag{15.10}$$

Setting $x = u + v$, as above, we obtain

$$u^3 + v^3 = 14, uv = -1,$$

then

$$u^3 = 7 + \sqrt{50}$$
$$v^3 = 7 - \sqrt{50}.$$

If we choose for u and v the real cube roots of u^3 and v^3, then uv is a positive real number whose cube is $u^3 v^3 = -1$, hence $uv = -1$. Then

$$x = u + v = (7 + \sqrt{50})^{1/3} + (7 - \sqrt{50})^{1/3}$$

is the desired solution to (15.10). (See Exercise 23, below.)

Del Ferro solved one particular case of the cubic, namely the case (15.8) above, $x^3 + ax = b$. The sixteenth century Italians, not recognizing negative numbers, had to consider three different cases of the cubic:

$$x^3 + ax = b,$$

$$x^3 + b = ax,$$

and

$$x^3 = ax + b,$$

where in each case, a and b are >0. (The case $x^3 + ax + b = 0, a, b > 0$, did not arise because that equation has a unique real solution which is negative, so, from their point of view, had no solution of interest.) Each of the three cases involved slightly different methods of solution. Cardano, in the Ars Magna (1545) was the first to publish solutions to all three cases.

The general cubic equation has the form

$$t^3 + at^2 + bt + c = 0.$$

Cardano showed how to reduce this general equation to one of the three cases above: eliminate the t^2 term by making an appropriate substitution, namely, $x = t + a/3$; then $f(t)$ is transformed into a polynomial of the form

$$p(x) = x^3 + qx + r$$

for some q and r. Which case this represents depends on the signs of q and r.

The "casus irreducibilis". One of the cases solved by Cardano was particularly mysterious, namely the case where the cubic polynomial has three real roots. For those polynomials, the solution of Cardano involves imaginary numbers.

We illustrate this situation with the equation

$$x^3 = 7x + 6.$$

We set $x = u + v$ in this case, to get

$$u^3 + 3u^2v + 3uv^2 + v^3 = 7u + 7v + 6,$$

which is solved if we can solve

$$u^3 + v^3 = 6, 3uv = 7.$$

We set $u^3 = 3 + z, v^3 = 3 - z$, then clearly $u^3 + v^3 = 6$, while

$$343/27 = u^3v^3 = 9 - z^2,$$

hence

$$z^2 = 9 - 343/27 = -100/27,$$

so

$$z = \pm \frac{10\sqrt{-3}}{9}.$$

Thus

$$u^3 = 3 + \frac{10\sqrt{-3}}{9}, v^3 = 3 - \frac{10\sqrt{-3}}{9}.$$

If we observe that

$$u^3 = [\frac{9+\sqrt{-3}}{6}]^3, v^3 = [\frac{9+\sqrt{-3}}{6}]^3,$$

then we can set

$$u = \frac{9+\sqrt{-3}}{6}, v = \frac{9-\sqrt{-3}}{6}$$

and

$$x = u + v = (9+\sqrt{-3})/6 + (9-\sqrt{-3})/6] = 3.$$

Now let

$$\omega = \frac{-1+\sqrt{-3}}{2},$$

a cube root of unity in $\mathbb{C}$. Then u^3 is also the cube of

$$(\frac{9+\sqrt{-3}}{6})\omega = \frac{-3+2\sqrt{-3}}{3}$$

and of

$$(\frac{9+\sqrt{-3}}{6})\omega^2 = \frac{-3-5\sqrt{-3}}{6}.$$

So we can let

$$x = \frac{-3+2\sqrt{-3}}{3} + \frac{-3-2\sqrt{-3}}{3} = -2.$$

or

$$x = \frac{-3+5\sqrt{-3}}{6} + \frac{-3-5\sqrt{-3}}{6} = -1.$$

Thus $x = -1, 2$ and 3 are the solutions of the equation

$$x^3 - 7x - 6 = (x-3)(x+1)(x+2).$$

In this example, Cardano's method finds the three real roots of the polynomial $x^3 - 7x - 6$, but expresses all three as the sums of complex numbers.

Cardano's solution of the cubic in this case is the first situation in the history of mathematics in which complex numbers appeared in an essential way in the solution of a "real" problem.

We can show that for an equation such as $x^3 = 7x + 6$, Cardano's method must always express the real roots as differences of non-real complex numbers. To do so, we first obtain a criterion for a cubic to have three real roots.

Proposition 4. *Let* $f(x) = x^3 - px + q$ *have distinct complex roots. Then* $f(x)$ *has three real roots if and only if* $27q^2 < 4p^3$.

Proof. The derivative $f'(x) = 3x^2 - p$. If $p < 0$, then $27q^2 > 4p^3$, and also $f'(x) > 0$ for all x, so $f(x)$ has exactly one real root. If $p > 0$, write $p = s^2$ with $s > 0$. Then $f'(x) = 0$ for $x = s/\sqrt{3}$ and $x = -s/\sqrt{3}$, and f has three real roots iff $f(-s/\sqrt{3}) > 0$ and $f(s/\sqrt{3}) < 0$. Now

$$f\left(\frac{-s}{\sqrt{3}}\right) = \frac{-s^3}{3\sqrt{3}} + s^2 \frac{s}{\sqrt{3}} + q = \frac{2s^3}{3\sqrt{3}} + q > 0$$

iff

$$q > \frac{-2s^3}{3\sqrt{3}},$$

and similarly, $f(\frac{s}{\sqrt{3}}) < 0$ iff

$$q < \frac{2s^3}{3\sqrt{3}}.$$

Thus f has three real roots iff

$$|q| < |\frac{2s^3}{3\sqrt{3}}|,$$

which is equivalent to

$$q^2 < \frac{4s^6}{27} = \frac{4p^3}{27},$$

the desired inequality.   □

Proposition 5. *Let $f(x) = x^3 - px + q$ have distinct complex roots. Then $f(x)$ has three real roots iff Cardano's method gives roots that are sums of non-real complex numbers.*

Proof. In Cardano's method we set $x = u + v$, then set

$$u^3 + v^3 = -q$$
$$3uv = p;$$

we introduce z so that

$$u^3 = -\frac{q}{2} + z, v^3 = -\frac{q}{2} - z$$

from which it follows that

$$z^2 = \frac{q^2}{4} - \frac{p^3}{27}.$$

Cardano's method will yield u and v non-real complex numbers if and only if $z^2 < 0$. As shown above, $f(x)$ has three real roots if and only if

$$q^2 < \frac{4p^3}{27},$$

exactly the condition that $z^2 < 0$.   □

Vieta's method. Later in the 16th century, Vieta (1593) discovered a way to use trigonometry to find the three real roots of the polynomial

$$f(x) = x^3 - px - q$$

when $27q^2 < 4p^3$. The method uses the trigonometric identity

(∗) $4\cos^3\theta = 3\cos\theta + \cos 3\theta.$

Multiply (*) by $2m^3$ for any real number m to get

$$8m^3 \cos^3 \theta = 6m^3 \cos \theta + 2m^3 \cos 3\theta.$$

If we set $x = 2m\cos\theta$, then this equation becomes

$$x^3 - 3m^2 x - 2m^3 \cos 3\theta = 0.$$

If we can solve the equations

$$p = 3m^2, \quad q = 2m^3 \cos 3\theta$$

for m and 3θ in terms of p and q, then $f(x)$ will have the solutions $x = 2m\cos\theta$, $x = 2m\cos(\theta + 2\pi/3)$ and $x = 2m\cos(\theta - 2\pi/3)$. Now the equation $q = 2m^3 \cos 3\theta$ is solvable iff $-1 \le \frac{q}{2m^3} \le 1$, or equivalently,

$$q^2 \le 4m^6.$$

When $p = 3m^2$, this becomes

$$q^2 \le 4\frac{p^3}{27}.$$

But $f(x)$ has three real roots iff $27q^2 < 4p^3$. Thus whenever $f(x)$ has three real roots we can solve for 3θ and find the three roots of $f(x)$ by Vieta's trigonometric method. [see Hartshorne, http://math.berkeley.edu/robin/Viete/construction.html]

Quartic equations. Once having learned how to solve a cubic, it was only a short time before Cardano's student Ferrari (born 1522) discovered how to solve a quartic, sometime before 1541. Here is how it is done.

Given the polynomial equation

$$y^4 + ay^3 + by^2 + cy + d = 0,$$

we first make the substitution $y = z - a/4$ to get a new equation in z in which the coefficient of z^3 is 0. Thus we reduce to an equation of the form

$$z^4 + pz^2 + qz + r = 0.$$

Now isolate the term z^4 and put the other terms on the right side, then add to both sides $t^2 z^2 + t^4/4$, to get

$$z^4 + t^2 z^2 + t^4/4 = t^2 z^2 + t^4/4 - pz^2 - qz - r.$$

The left side is a perfect square, namely $(z^2 + t^2/2)^2$, and we can solve the equation easily if we can choose t so that the right side is also a perfect square. We write the right side as

$$(t^2 - p)z^2 - qz + (t^4/4 - r) = \alpha z^2 + \beta z + \gamma.$$

The right hand side is a perfect square if $\beta^2 - 4\alpha\gamma = 0$: for if we complete the square:

$$\alpha z^2 + \beta z + \gamma = \alpha(z^2 + \beta z/a) + \gamma$$
$$= \alpha(z^2 + \beta z/\alpha + \beta^2/4\alpha^2) + \gamma - \beta^2/4\alpha$$
$$= \alpha\left(z + \frac{\beta}{2\alpha}\right)^2 + \left(\frac{4\alpha\gamma - \beta^2}{4\alpha}\right),$$

we will obtain a perfect square if we can find some t so that $4\alpha\gamma - \beta^2 = 0$.
    Now the condition

$$4\alpha\gamma - \beta^2 = 0$$

becomes

$$4(t^2 - p)(-r + t^4/4) - (-q)^2 = 0$$

or

$$t^6 - pt^4 - 4rt^2 + (4pr - q^2) = 0.$$

Setting $t^2 = x$ yields

$$x^3 - px^2 - 4rx + (4pr - q^2) = 0,$$

a cubic that we already know how to solve!

    It is interesting that there is no formula to find the roots of a polynomial of degree ≥ 5; that is, there is a polynomial of degree 5 whose roots cannot be described by taking the coefficients and manipulating them by the usual algebraic operations together with the operation of taking n-th roots (forming radicals) in the way we did for polynomials of degree 2 or 3. This famous theorem is due to N.H. Abel (1802-1829), the Norwegian mathematician for whom the term "abelian" ("abelian group") was named.

Exercises.

22. A man walks 6 miles in time t hours. If he walked 6 miles in time $t - 2$ hours, the rate would be 2 miles per hour more. Put this problem into Babylonian normal form, then find t and the rate $r = 6/t$.

23. In Example 3, show that $(7 + \sqrt{50})^{1/3} + (7 - \sqrt{50})^{1/3} = 2$ by showing that

$$7 + \sqrt{50} = 7 + 5\sqrt{2} = (1 + \sqrt{2})^3.$$

24. Find a solution of $x^3 + 3x = 5$.

25. (i) Verify that $x^3 - 49x + 120$ has three real roots.
    (ii) Find the three roots by Cardano's method.
    (iii) Find the three roots by Vieta's method.

26. Find a solution of $x^3 + 6x + 8 - 6x^2$.

27. The equation $y^4 + 5y = 6$ has the obvious solution $y = 1$. Use Ferrari's method to show that $y = 1$ is a solution.

F. The Fundamental Theorem

The Fundamental Theorem of Algebra was finally given a proof by Gauss in 1801.

By the time of Gauss it was natural to express the theorem as a result over the complex numbers, namely:

Theorem 6 (Fundamental Theorem of Algebra). *Every polynomial $p(x)$ in $\mathbb{C}[x]$ of degree ≥ 1 has a root in $\mathbb{C}$.*

Of course this implies that the only irreducible polynomials in $\mathbb{C}[x]$ are of degree one.

From the complex version of the fundamental theorem we can easily obtain Euler's conjectured real version:

Corollary 7. *No polynomial $f(x)$ in $\mathbb{R}[x]$ of degree >2 is irreducible in $\mathbb{R}[x]$.*

We prove Euler's version from Gauss's version.

Proof of Corollary. Let $f(x)$ in $\mathbb{R}[x]$ have degree >2. We will show that $f(x)$ is not irreducible. We can assume that $f(x)$ has no real roots, by the Root Theorem.

Suppose α is a nonreal complex root of $f(x)$. Let

$$p(x) = (x - \alpha)(x - \overline{\alpha}),$$

where, if $\alpha = a + bi$, then $\overline{\alpha} = a - bi$ is the complex conjugate of α. Then

$$p(x) = x^2 - 2\alpha x + (a^2 + b^2)$$

is in $\mathbb{R}[x]$ (and is irreducible in $\mathbb{R}[x]$ since its two roots are nonreal). Now divide $f(x)$ by $p(x)$ in $\mathbb{R}[x]$,

$$f(x) = p(x)q(x) + r(x), \tag{15.11}$$

with deg $r(x) \leq 1$. Let $r(x) = r + sx$. Evaluate equation (15.11) at $x = \alpha$. We get $r(\alpha) = 0$, since α is a root of both $f(x)$ and $p(x)$. But then $r + s\alpha = 0$, and so unless $r = s = 0$, we conclude that α is real, a contradiction. Thus $p(x)$ divides $f(x)$, and since deg $p(x) = 2 < $ deg $f(x)$, $f(x)$ is not irreducible. $\qquad\square$

Proof of the Fundamental Theorem of Algebra. The rest of this section is devoted to one of the half-dozen or more distinctly different proofs of the Fundamental Theorem of Algebra. The proof we present is essentially a proof of Argand, 1814.

It involves a minimal acquaintance with functions of two (real) variables. and may be omitted without loss of continuity.

In the proof we shall assume that $p(z)$ is monic, that is, has leading coefficient $= 1$.

Before beginning the proof we describe some facts we need which go into the proof. Let $z = x + iy$. We think of a complex number as represented in the real (x, y)-plane by corresponding $z = x + iy$ with (x, y).

A polynomial $p(z)$ in $\mathbb{C}[z]$ may be written as

$$p(z) = p(x + iy) = p_1(x, y) + ip_2(x, y),$$

where $p_1(x, y)$ and $p_2(x, y)$ are real polynomials in the real variables x, y. Then $|p(z)|$ may be written as

$$|p(z)| = \sqrt{p_1(x, y)^2 + p_2(x, y)^2}.$$

Since $p_1(x, y)$ and $p_2(x, y)$ are real polynomials in x, y, $p_1(x, y)^2 + p_2(x, y)^2$ is a nonnegative real-valued continuous function of x and y, and since the positive square root $\sqrt{t}$ of t is a continuous function of t for $t > 0$, $|p(z)| = \sqrt{p_1(x, y)^2 + p_2(x, y)^2}$ is continuous as a function of x and y.

A basic fact from calculus is that a function continuous on a closed disk

$$D = \{(x, y) | x^2 + y^2 \leq R\}$$

in the x-y-plane has a minimum value in D.

Before beginning the proof, we prove a result that yields an upper bound on the size of the roots of a polynomial.

Proposition 8. Let $f(z) = z^n + a_{n-1}z^{n-1} + \ldots + a_1z + a_0$ in $\mathbb{C}[x]$. For every $M \geq 0$, if

$$|z| \geq M + 1 + |a_{n-1}| + \ldots + |a_1| + |a_0|,$$

then $|f(z)| > M$.

Proof. From the triangle inequality:

$$|a + b| \leq |a| + |b|$$

we obtain

$$|a| = |a + b - b| \leq |a + b| + |b|$$

so

$$|a + b| \geq |a| - |b|. \qquad (*)$$

We prove that if $|z| \geq 1$, then

$$|f(z)| \geq |z| - (|a_{n-1}| + \ldots + |a_1| + |a_0|)$$

by induction on $n = \deg f(z)$.

If $\deg f = 1$, then $f(z) = z + a_0$, so $|f(z)| \geq |z| - |a_0|$ by $(*)$.

If $\deg f = n > 1$, let $f(z) = zf_1(z) + a_0$ with

$$f_1(z) = z^{n-1} + a_{n-1}z^{n-2} + \ldots + a_2z + a_1.$$

We assume by induction that if $|z| \geq 1$, then

$$|f_1(z)| \geq |z| - (|a_{n-1}| + \ldots + |a_1|).$$

Then

$$\begin{aligned}
|f(z)| &= |zf_1(z) + a_0| \\
&\geq |z||f_1(z)| - |a_0| \text{ by } (*) \\
&\geq |f_1(z)| - |a_0| \\
&\geq |z| - (|a_{n-1}| + \ldots + |a_1| + |a_0|).
\end{aligned}$$

If we assume that

$$|z| \geq M + (1 + |a_{n-1}| + \ldots + |a_1| + |a_0|),$$

then

$$|f(z)| \geq M + 1.$$

$\square$

Corollary 9. *If r is a root of $f(z)$, then*

$$|r| < 1 + |a_{n-1}| + \ldots + |a_1| + |a_0|.$$

See Section 26B for other bounds on the roots of a polynomial.

Our proof that $p(z)$ in $\mathbb{C}[z]$ has a root in $\mathbb{C}$ has two parts.

(I) There is a point z_0 in the complex plane such that $|p(z_0)| \leq |p(z)|$ for all z in $\mathbb{C}$ (not just in some disk).

(II) If z_0 is the point found in (1), where $|p(z_0)|$ is a minimum, then $p(z_0) = 0$.

Part I. For the first part of the proof, let

$$p(z) = z^n + a_{n-1}z^{n-1} + \ldots + a_1z + a_0$$

in $\mathbb{C}[x]$. Using Proposition 8, choose $M = 1 + |a_0|$. Then there is some $R \geq 1$ so that for $|z| > R$, $|p(z)| \geq M$. Let $D = \{z : |z| < R\}$. From calculus it is known that there is some z_0 in D such that $|p(z_0)| \leq |p(z)|$ for all z in D. Now, by the way that we have chosen D, $|p(z_0)| \leq |p(z)|$ for all z. For if z is not in D, $|z| > R$, so $|p(z)| \geq 1 + |a_0| > |a_0| = |p(0)|$. Since 0 is in D, $|p(0)| \geq |p(z_0)|$. Thus $|p(z_0)| < |p(z)|$ for all z, in D or not. That completes the first part of the proof.

Part II. Let z_0 be the point found in Part I such that $|p(z_0)| \leq |p(z)|$ for all z. We are going to make two changes of variables to put z_0 at the origin and make our polynomial easy to work with.

First make a change of variables $w = z - z_0$. Then $p(z) = p(w + z_0) = q_1(w)$ is a polynomial in w and $|q_1(0)| = |p(z_0)| < |p(z)| = |q_1(w)|$ for all w: thus $|q_1(w)|$ has its minimum at $w = 0$.

We want to show that $q_1(0) = p(z_0) = 0$. If that is the case, we are done. So for the rest of the proof we assume that $q_1(0) = a_0 \neq 0$; from that assumption we shall reach a contradiction.

Assuming $a_0 \neq 0$, let $q_2(w) = (1/a_0)q_1(w)$. Then $|q_2(w)|$ has a minimum at $w = 0$ iff $q_1(w)$ does. Now $q_2(w)$ has the form

$$q_2(w) = 1 + bw^m + b_1 w^{m+1} + \ldots + b_k w^{m+k}$$

for some $m \geq 1$, where $b \neq 0$ in $\mathbb{C}$ and $m + k = n = $ the degree of $q_2(w) = $ the degree of $p(z)$.

Let r be an m-th root of $-1/b$ in $\mathbb{C}$. Then $r^m b = -1$. Let $w = ru$, and set $q(u) = q_2(ru) = q_2(w)$. Then $|q(u)|$ has a minimum at $u = 0$ iff $|q_2(w)|$ has a minimum at $w = 0$. Now $q(u)$ has the form

$$q(u) = 1 + b(ru)^m + b_1(ru)^{m+1} + \ldots + b_k(ru)^{m+k}$$
$$= 1 - u^m + u^{m+1}Q(u) \text{ (since } r^m b = -1),$$

where

$$Q(u) = c_1 + c_2 u + \ldots + c_k u^{k-1}$$

is in $\mathbb{C}[u]$, with $c_j = b_j r^{m+j}$ for each j, $1 \leq j \leq k$.

Note that $q(0) = 1$, so 1 is the minimum value of $|q(u)|$.

Let t be a real number > 0. Setting $u = t$,

$$|Q(t)| = |c_1 + c_2 t + \ldots + c_k t^{k-1}|$$
$$\leq |c_1| + |c_2|t + \ldots + |c_k|t^{k-1} = Q_0(t)$$

by the triangle inequality. This last polynomial $Q_0(t)$ is a polynomial with real coefficients, and is > 0 when t is real and ≥ 0.

As $t \to 0$, $tQ_0(t) \to 0$. Choose t with $0 < t < 1$ so that $tQ_0(t) < 1$.

We show that for this choice of t, setting $u = t$ gives $|q(t)| < 1 = |q(0)|$, contradicting the assumption that $|q(u)|$ had its minimum at $u = 0$. Here is why $|q(t)| < 1$:

$$|q(t)| = |1 - t^m + t^{m+1}Q(t)|$$
$$\leq |1 - t^m| + |t^{m+1}Q(t)| \text{ (by the triangle inequality)}$$
$$= (1 - t^m) + t^m t|Q(t)| \text{ (since } 0 < t < 1)$$
$$\leq (1 - t^m) + t^m(tQ_0(t)).$$

Since t is chosen so that $tQ_0(t) < 1$, this last number is

$$< (1 - t^m) + t^m = 1 = |q(0)|.$$

Since $t \neq 0$, $|q(u)|$ does not have its minimum at $u = 0$. We have reached a contradiction, and the proof is complete.

The above proof is given in articles by C. Fefferman (1967) and F. Terkelson (1976), and is essentially in Chrystal (1904, Chapter XII).

While we know which polynomials in $\mathbb{R}[x]$ or $\mathbb{C}[x]$ are irreducible, it is substantially harder to see how to factor or to find the roots of a polynomial we know is not irreducible. In general, the roots must be obtained by approximation. Such problems form an important part of the subject of numerical analysis.

Exercises.

28. Give an example where $p(z) = z^n + a_{n-1}z^{n-1} + \ldots + a_2z^2 + a_1z + a_0$ and $|z| > |a_{n-1}| + \ldots + |a_1| + |a_0|$, but $p(z) = 0$.

29. Where in the proof of the fundamental theorem did we write a polynomial in a new base?

30. Let $p(z) = (1+i)z^3 + (2-i)z^2 + 4z + 2i$. Write $p(z) = p_1(x,y) + ip_2(x,y)$ and determine $|p(z)|$, as in the proof.

31. Let $f(z) = z^3 + iz^2 + 8z + 3$. Find some $R > 0$ so that for all z with $|z| > R$, $|f(z)| > 20$.

G. The Derivative and Multiple Roots

If you have seen any calculus at all, you have learned to find the derivative of a polynomial function. If not, it doesn't matter, because finding the derivative of a polynomial is really an algebraic process that can be performed without limits, slopes of tangent lines, or other interpretations that arise in calculus.

We are interested in finding derivatives of polynomials because the derivative of a polynomial, together with Euclid's algorithm, helps determine if a polynomial has a multiple factor.

Let F be a field (not necessarily the real numbers). The *differentiation operator*

$$D : F[x] \to F[x]$$

is defined by the following properties:
(1) $D(x^n) = nx^{n-1}$ for all $n \geq 0$. In particular, $D(1) = D(x^0) = 0$.
(2) D is a linear transformation. Thus for all $f(x), g(x)$ in $F[x]$ and a in F,
    (a) $D(af(x)) = aD(f(x))$ and
    (b) $D(f(x) + g(x)) = D(f(x)) + D(g(x))$.

Since every polynomial $f(x)$ of degree n,

$$f(x) = a_n x^n + a_{n-1} x^{n-1} + \ldots a_1 x + a_0,$$

is a unique linear combination of $1, x, x^2, \ldots, x^n$, it follows that D is uniquely defined on $F[x]$ by the two rules (1) and (2). In the language of linear algebra, $\{1, x, \ldots, x^n\}$ is a basis of the F-vector space of polynomials of degree $<n$, and so the linear transformation D is uniquely determined by rule (1).

For $f(x)$ in $F[x]$, let $f'(x) = D(f(x))$ and call $f'(x)$ the *derivative* of $f(x)$.
For example, if $f(x) = x^3 + 6x^2 - 5x - 10$, rule (2) gives:

$$D(x^3 + 6x^2 - 5x - 10) = D(x^3) + 6D(x^2) + (-5)D(x) + (-10)D(1);$$

then applying rule (1) we get

$$= 3x^2 + 12x - 5.$$

There is a subtlety in rule (1), $D(x^n) = nx^{n-1}$. The "n" in x^n is a natural number: x^n means n copies of x multiplied together. However, the coefficient "n" in nx^{n-1} means the image in F of the natural number n under the map from $\mathbb{Z}$ to F given by $n \mapsto 1 + 1 + \ldots + 1$ (n summands) in F. Recall that if R is a commutative ring with unity (denoted by 1), then R has characteristic zero if

$$n \cdot 1 = 1 + 1 + \ldots + 1 (n \text{ summands }) \neq 0 \text{ for every } n > 0,$$

and has characteristic p if

$$1 + 1 + \ldots + 1 \ (p \text{ summands}) = 0.$$

If R has characteristic p, then $nx^{n-1} = 0$ if p divides n. For example, if $F = \mathbb{F}_3$, then

$$D(x^6) = 6x^5 = 0,$$

since the coefficient 6 really means $[6]_3 = 0$, the zero element of $\mathbb{F}_3$.
From rules (1) and (2) follow the Product Rule:

$$D(f(x)g(x)) = f'(x)g(x) + f(x)g'(x)$$

and the Power Rule:

$$D(f(x)^e) = ef(x)^{e-1} f'(x).$$

Their derivations are left as exercises.
Here is our main reason for introducing the derivative of a polynomial:

Theorem 10. *Let F be a field, $f(x)$ be in $F[x]$.*
   (a) If $f(x)$ has a multiple factor, then $f(x)$ and $f'(x)$ are not coprime.
   (b) If the field F has characteristic zero or is a finite field of characteristic $p \neq 0$, and $f(x)$ and $f'(x)$ are not coprime, then $f(x)$ has a multiple factor.

Proof. By unique factorization, $f(x)$ and $f'(x)$ are not coprime if and only if there is an irreducible polynomial $q(x)$ that is a common divisor of $f(x)$ and $f'(x)$.

To prove (a), suppose $f(x)$ has a multiple factor: $f(x) = q(x)^e h(x)$ with $e > 1$. Then using the product and power rules,

$$f'(x) = eq(x)^{e-1}q'(x)h(x) + q(x)^e h'(x)$$
$$= q(x)^{e-1}[eq'(x)h(x) + q(x)h'(x)],$$

so that if $e > 1$, then $q(x)$ is a common factor of $f(x)$ and $f'(x)$, and $f(x)$ and $f'(x)$ are not coprime.

For (b), we need:

Lemma 11. *If $q(x)$ is a polynomial of degree ≥ 1, and $q(x)$ divides $q'(x)$, then $q'(x) = 0$.*

Proof. Since $\deg q'(x) < \deg q(x)$, the Division Theorem gives

$$q'(x) = 0 \cdot q(x) + q'(x).$$

Thus in order for $q(x)$ to divide $q'(x)$, the remainder $q'(x) = 0$.   □

Now suppose $q(x)$ is an irreducible common divisor of $f(x)$ and $f'(x)$. Then $f(x) = q(x)h(x)$ for some polynomial $h(x)$. So by the product rule,

$$f'(x) = q'(x)h(x) + q(x)h'(x).$$

Now $q(x)$ divides $f'(x)$, so $q(x)$ must divide $q'(x)h(x)$. Since $q(x)$ is irreducible, $q(x)$ must therefore divide $q'(x)$ or $h(x)$.

If $q(x)$ divides $h(x)$, then $h(x) = q(x)k(x)$ for some $k(x)$, and so $f(x) = q(x)h(x) = q(x)q(x)k(x)$, and $q(x)$ is a multiple factor of $f(x)$, proving the theorem.

We show that if $q(x)$ is irreducible and F is as in (b), then the case, $q(x)$ divides $q'(x)$, cannot occur.

If $q(x)$ divides $q'(x)$, then $q'(x) = 0$, by Lemma 11.

Suppose the polynomial $q(x)$ has degree $n \geq 1$,

$$q(x) = a_n x^n + a_{n-1}x^{n-1} + \ldots + a_1 x + a_0,$$

where $n \neq 0$. Then

$$q'(x) = na_n x^{n-1} + (n-1)a_{n-1}x^{n-2} + \ldots + a_1.$$

If F has characteristic zero, $n \cdot a_n$ is not equal to 0 in F, and so $na_n x^{n-1}$ is not zero in $F[x]$. Hence $q'(x)$ is a nonzero polynomial of degree $< n$. Therefore $q(x)$ cannot divide $q'(x)$ by Lemma 11.

If F has characteristic p for some prime p, and $q(x)$ divides $q'(x)$, then $q'(x) = 0$, so $ra_r = 0$ in F for $r = 0, 1, \ldots, n$. This implies that the only non-zero coefficients of $q(x)$ are the a_r where p, the characteristic of F, divides r. Thus

$$q(x) = a_{mp}x^{mp} + a_{(m-1)p}x^{(m-1)p} + \ldots + a_{2p}x^{2p} + a_p x^p + a_0$$

for some m.

To see that such a polynomial $q(x)$ cannot be irreducible, we observe that since F is a finite field, each element of F is a pth power (since the Frobenius map $a \mapsto a^p$ is a one-to-one ring homomorphism from F to F, hence maps onto F–see Section 12E). So in particular, each non-zero coefficient a_{kp} of $q(x)$ can be written as $a_{kp} = c_k^p$ for some c_k in F. Hence

$$q(x) = c_m^p x^{mp} + c_{m-1}^p x^{(m-1)p} + \ldots + c_2^p x^{2p} + c_1^p x^p + c_0^p.$$

Since $(a+b)^p = a^p + b^p$ for all a, b in F, we can write this as

$$q(x) = (c_m x^m + c_{m-1}x^{m-1} + \ldots + c_2 x^2 + c_1 x + c_0)^p.$$

But then $q(x)$ is not irreducible. Thus if $q(x)$ is an irreducible polynomial in $F[x]$ where F is a finite field, then $q(x)$ does not divide $q'(x)$. That completes the proof of the theorem.   □

Example 6. Let $f(x) = x^3 + x^2 - 8x - 12$ in $\mathbb{Q}[x]$. Then $f'(x) = 3x^2 + 2x - 8$; doing Euclid's algorithm on f and f' gives:

$$f(x) = f'(x)q(x) + r(x)$$

where $q(x) = \frac{1}{3}x + \frac{1}{9}$ and

$$r(x) = -\frac{50}{9}x - \frac{100}{9} = \frac{-50}{9}(x+2).$$

Then $r(x)$ divides $f'(x)$, in fact,

$$3x^2 + 2x - 8 = (x+2)(3x-4).$$

So $x + 2$ is a greatest common divisor of f and f'. Hence $r(x)$ is a multiple factor of $f(x)$. If we divide $f(x)$ by $(x+2)^2$, we find that the quotient is $x - 3$, so $f(x)$ factors as

$$f(x) = (x+2)^2(x-3).$$

Example 7. In $\mathbb{F}_3[x]$, $x + 1$ is a common factor of $f(x) = x^5 + 2x^4 + 2x^2 + x + 1$ and $f'(x) = 2x^4 + 2x^3 + x + 1$. We find that $f(x) = (x+1)^2(x^3 + 2x + 1)$.

Using the derivative helps to simplify a polynomial $f(x)$ over a field of characteristic zero (like $\mathbb{Q}$ or $\mathbb{R}$) when we wish to understand the roots of $f(x)$. For suppose $f(x)$ is a polynomial of the form:

$$f(x) = p_1(x)^{e_1} p_2(x)^{e_2} \ldots p_g(x)^{e_g},$$

a product of powers of distinct irreducible polynomials. Then for each i, the highest power of $p_i(x)$ that divides $f'(x)$ is $p_i^{e_i - 1}$, and so the greatest common divisor of

$f(x)$ and $f'(x)$ is

$$(f(x), f'(x)) = p_1(x)^{e_1-1} p_2(x)^{e_2-1} \ldots p_g(x)^{e_g-1}.$$

Hence

$$\frac{f(x)}{(f(x), f'(x))} = p_1(x) p_2(x) \ldots p_g(x),$$

the *squarefree part of* $f(x)$. It is obvious that $f(x)$ and $\frac{f(x)}{(f(x), f'(x))}$ have the same irreducible factors in $F[x]$, except that in $\frac{f(x)}{(f(x), f'(x))}$ all the irreducible factors have multiplicity 1. In particular, if we are interested in just finding the distinct roots of $f(x)$, then the squarefree part of $f(x)$ has the same roots and, if not equal to $f(x)$, will have lower degree than $f(x)$. For some root-finding algorithms over the real numbers, such as Sturm's Theorem, knowing that a polynomial has no multiple roots is necessary for locating the roots.

Exercises.

32. Prove the product rule: $D(f \cdot g) = fD(g) + gD(f)$, as follows: First show it in case both $f(x)$ and $g(x)$ are monomials, that is, when $f(x) = ax^n, g(x) = bx^m$ for some a, b in F, and some natural numbers m, n. Then using that D is a linear transformation, reduce $D(f \cdot g)$ for any two polynomials f and g to the case of a product of monomials.

33. From the product rule, prove the power rule: for every natural number $e \geq 1$,

$$D(f(x)^e) = ef(x)^{e-1} f'(x).$$

34. Let $f(x) = x^4 - 3x^3 + x^2 + 3x - 2$ in $\mathbb{Q}[x]$. Find the greatest common divisor of $f(x)$ and $f'(x)$. Find the squarefree part of $f(x)$.

35. Let $f(x) = x^4 - 2x^3 + 3x^2 - 2x + 1$ in $\mathbb{Q}[x]$. Show that a greatest common divisor of $f(x)$ and $f'(x)$ is $x^2 - x + 1$. Then factor $f(x)$.

36. If $f(x)$ is in $\mathbb{Q}[x]$, then $f(x)$ can also be thought of as having coefficients in the larger field $\mathbb{C}$. Show that $f(x)$ has a multiple factor in $\mathbb{Q}[x]$ if and only if $f(x)$ has a multiple factor in $\mathbb{C}[x]$.

37. Let $f(x) = x^4 + x^2 + x + 1$ in $\mathbb{F}_2[x]$. Show that $f(x)$ has no multiple factor.

38. Let $f(x) = x^4 + x^2 + 1$ in $\mathbb{F}_2[x]$. Show that $f'(x) = 0$. Factor $f(x)$.

39. In $\mathbb{F}_2[x]$ test the following polynomials for multiple factors:
  (i) $x^5 + x^4 + x^2 + x$;
  (ii) $x^7 + x^6 + x^5 + x^3 + 1$.

40. In $\mathbb{F}_2[x]$ test the following polynomials for multiple factors:
(i) $x^7 + x^4 + x^2 + x + 1$;
(ii) $x^7 + x^6 + x^5 + x^4 + x^3 + x^2 + x + 1$;
(iii) $x^7 + x^6 + x^4 + x^2 + x$.

41. In $\mathbb{Q}[x]$ test $f(x) = x^4 + 5x^3 + 9x^2 + 7x + 2$ for a multiple factor. Find the square-free part of $f(x)$.

42. Suppose $f(x)$ is in $F[x]$, F a field, and the characteristic of F does not divide the degree of $f(x)$. Show that if $f(x)$ and $f'(x)$ are not coprime, $f(x)$ must have a multiple factor.

43. Factor $x^{15} + 3x^{10} + 2x^5 + 4$ in $\mathbb{F}_5[x]$.

44. Factor $x^5 + x^4 + x^3 + x^2 + x + 1$ in $\mathbb{F}_2[x]$.

45. Factor $x^4 + 2x^3 + 2x^2 + x + 4$ in $\mathbb{F}_5[x]$.

46. Prove that $f^n = g^n + h^n$ has no solutions in $\mathbb{R}[x]$ with $n > 2$, f, g, h each of degree at least 1, and f, g, and h pairwise coprime. (This is the analogue for polynomials of Fermat's Last Theorem for natural numbers. The problem can be done by looking at the possible degrees of f, g, and h; it may help to start out by taking the derivative of both sides of the equation.)

Chapter 16
Polynomials in $\mathbb{Q}[x]$

In this chapter we begin considering the question of how to factor polynomials with coefficients in $\mathbb{Q}$, the field of rational numbers.

Here the situation is much different from the situation over $\mathbb{R}$ or $\mathbb{C}$. Over $\mathbb{Q}$ there are many irreducible polynomials of every degree, and determining which polynomials are irreducible is difficult, compared to the real or complex case. On the other hand, finding roots (and therefore irreducible factors of degree 1) of a polynomial in $\mathbb{Q}[x]$ is easy, and we will eventually give two different explicit procedures for determining the complete factorization of any polynomial with rational coefficients in a finite number of steps.

The starting point for all the results on $\mathbb{Q}[x]$ is the fact that factoring in $\mathbb{Q}[x]$ is "the same" as factoring in $\mathbb{Z}[x]$. The first part of this chapter is devoted to showing that fact.

A. Gauss's Lemma

Recall that if $f(x)$ and $g(x)$ are two polynomials with coefficients in a field F, and there is some non-zero element c of F so that $f(x) = cg(x)$, then $f(x)$ and $g(x)$ are *associates*, or $g(x)$ is an associate of $f(x)$. Polynomials that are associates have essentially the same factorizations into products of irreducible polynomials.

If $f(x) = a_n x^n + \ldots + a_1 x + a_0$ is a polynomial with rational coefficients (i.e., $a_n, \ldots, a_1, a_0$ are in $\mathbb{Q}$), then we can multiply $f(x)$ by the least common multiple of the denominators of the coefficients, call it s, to get a polynomial $g(x) = sf(x)$ with integer coefficients that is an associate of $f(x)$ in $\mathbb{Q}[x]$. If we have a factorization of $f(x)$, multiplying one of the factors of $f(x)$ by s will then give a factorization of $g(x)$. Hence the factorizations of $g(x)$ and $f(x)$ into products of irreducible polynomials in $\mathbb{Q}[x]$ will be the same, up to associates. So in studying factorization of a polynomial in $\mathbb{Q}[x]$, we can always assume that the polynomial has integer coefficients.

We can ask for more.

L.N. Childs, *A Concrete Introduction to Higher Algebra*, Undergraduate Texts in Mathematics, © Springer Science+Business Media LLC 2009

Definition. A polynomial $f(x)$ with rational coefficients is *primitive* if $f(x)$ has integer coefficients and the greatest common divisor of those coefficients is 1.

Every polynomial $f(x)$ with integer coefficients is an associate of a primitive polynomial: simply divide $f(x)$ by the greatest common divisor of its coefficients. The resulting polynomial is an associate of $f(x)$ and is primitive. Hence *any polynomial in $\mathbb{Q}[x]$ is an associate of a primitive polynomial*. For example, a primitive associate of $8x^3 + (\frac{10}{3})x + \frac{6}{5}$ is

$$60x^3 + 25x + 9 = \left(\frac{15}{2}\right)\left(8x^3 + \left(\frac{10}{3}\right)x + \frac{6}{5}\right).$$

A convenient way to characterize primitive polynomials is by looking at them modulo p.

Let p be a prime number. Given a polynomial $f(x)$ with integer coefficients, we can obtain a polynomial with coefficients in $\mathbb{Z}/p\mathbb{Z}$ by replacing the coefficients of $f(x)$ by the congruence classes of those coefficients modulo p. Let us denote by γ_p the map that does this. Thus if

$$a_n x^n + \ldots + a_1 x + a_0$$

has integer coefficients, then

$$\gamma_p(a_n x^n + \ldots + a_1 x + a_0) = [a_n]x^n + \ldots + [a_1]x + [a_0].$$

Then $f(x)$ in $\mathbb{Z}[x]$ is primitive if and only if no prime number divides all the coefficients of $f(x)$, if and only if for every prime p, $\gamma_p(f(x)) \neq 0$. For example,

$$\gamma_2(60x^3 + 25x + 9) = x + [1]_2.$$

Notice that the map $\gamma_p : \mathbb{Z}[x] \longrightarrow (\mathbb{Z}/p\mathbb{Z})[x]$ is a homomorphism: in particular,

$$\gamma_p(f(x)) \cdot \gamma_p(g(x)) = \gamma_p(f(x)g(x))$$

for all polynomials $f(x), g(x)$ in $\mathbb{Z}[x]$.

Using the map γ_p we can easily prove:

Proposition 1. *The product of two primitive polynomials is again a primitive polynomial.*

Proof. Suppose $f(x)$ and $g(x)$ are primitive. Then for every prime p, $\gamma_p(f(x)) \neq 0$, and $\gamma_p(g(x)) \neq 0$ in $(\mathbb{Z}/p\mathbb{Z})[x]$. But $(\mathbb{Z}/p\mathbb{Z})[x]$ has no zero divisors since $\mathbb{Z}/p\mathbb{Z}$ is a field. So $\gamma_p(f(x)g(x)) = \gamma_p(f(x)) \cdot \gamma_p(g(x)) \neq 0$. Since this is true for every prime p, therefore $f(x)g(x)$ is primitive.   □

We also need:

Lemma 2. *If $g(x)$ is primitive, $f(x)$ is in $\mathbb{Z}[x]$, and $f(x) = ag(x)$ for some rational number a, then a is in $\mathbb{Z}$. If $f(x)$ is also primitive, then $a = 1$ or -1.*

Proof. Write $a = r/s$ with r, s coprime integers. Then

$$sf(x) = rg(x).$$

Since r and s are coprime, s must divide all the coefficients of $g(x)$. But $g(x)$ is primitive, so $s = 1$ or -1. If also $f(x)$ is primitive, then by the same argument, r must be 1 or -1. Hence $a = 1$ or -1. □

The main result of this section is the very useful

Theorem 3 (Gauss's Lemma). *Let $f(x)$ be a polynomial with integer coefficients. Suppose $f(x) = a(x)b(x)$ with $a(x)$ and $b(x)$ in $\mathbb{Q}[x]$. Then there are polynomials $a_1(x)$ and $b_1(x)$ in $\mathbb{Z}[x]$, associates of $a(x)$ and $b(x)$, respectively, so that $f(x) = a_1(x)b_1(x)$.*

Gauss's Lemma means that if we wish to find a factorization of a polynomial with integer coefficients, we need only look for factors that have integer coefficients.

Example 1. Consider the polynomial

$$x^4 - 3x^2 + x + 5.$$

Suppose we seek a factorization into the product of two polynomials of degree 2:

$$x^4 - 3x^2 + x + 5 = (x^2 + ax + b)(x^2 + cx + d).$$

If there is such a factorization in $\mathbb{Q}[x]$, there is one in which the coefficients a, b, c, d are integers, according to Gauss's Lemma. We multiply out the right side and equate coefficients of the various powers of x. Comparing coefficients of x^3, we see that $c = -a$; then comparing coefficients of x^2, x and 1 yields

$$-3 = b + d - a^2,$$
$$1 = ad - ab = a(d - b),$$
$$5 = bd.$$

Since a, b and d are integers, the second equation yields $d - b = 1$ or -1, so b and d differ by 1; while the third equation yields $b = \pm 1, d = \pm 5$ or $b = \pm 5, d = \pm 1$. So there is no factorization of the desired form.

If we had been unable to assume that a, b and d were integers, we would have had infinitely many possibilities for a, b and d, and showing that the equations had no solution for every possible choice of a, b and d would have been more difficult.

Proof of Gauss's Lemma. Let $f(x)$ be in $\mathbb{Z}[x]$, and suppose $f(x) = a(x)b(x)$, where $a(x)$ and $b(x)$ are in $\mathbb{Q}[x]$. Let $a_1(x)$ and $b_1(x)$ be primitive polynomials in $\mathbb{Z}[x]$ that are associates of $a(x)$ and $b(x)$, respectively, so that

$$a(x) = ca_1(x), \quad b(x) = db_1(x)$$

with c, d some rational numbers. Then

$$f(x) = cda_1(x)b_1(x).$$

Now $a_1(x)b_1(x)$ is primitive, by Proposition 1, and so cd is in $\mathbb{Z}$, by Lemma 2. Hence

$$f(x) = (cda_1(x))b_1(x),$$

a factorization in $\mathbb{Z}[x]$ where $cda_1(x)$ and $b_1(x)$ are associates of $a(x)$ and $b(x)$, respectively. That completes the proof.   □

Corollary 4. *If $f(x)$ is in $\mathbb{Z}[x]$ and $f(x) = g(x)h(x)$ in $\mathbb{Q}[x]$ with $g(x)$ primitive, then $h(x)$ is in $\mathbb{Z}[x]$.*

Proof. By Gauss's Lemma, $f(x) = cg(x) \cdot dh(x)$ for some c, d in $\mathbb{Q}$ with $cd = 1$ and $cg(x), dh(x)$ in $\mathbb{Z}[x]$. But since $cg(x)$ is in $\mathbb{Z}[x]$ and $g(x)$ is primitive, c must be in $\mathbb{Z}$ by Lemma 2. Hence $h(x) = cdh(x) = c(dh(x))$ is in $\mathbb{Z}[x]$.   □

Applications of Gauss's Lemma. Here is a well-known criterion, due to Descartes (1637), for finding roots of a polynomial with integer coefficients:

Theorem 5 (Descartes' Rational Root Theorem). *Let*

$$f(x) = a_n x^n + a_{n-1} x^{n-1} + \ldots + a_1 x + a_0$$

be in $\mathbb{Z}[x]$. Suppose r/s is a rational root of $f(x)$ where r, s are in $\mathbb{Z}$ with $(r, s) = 1$. Then s divides a_n and r divides a_0.

Proof. Since r/s is a root of $f(x)$, we can write $f(x) = (sx - r)g(x)$ for some polynomial

$$g(x) = b_{n-1} x^{n-1} + \ldots + b_1 x + b_0$$

in $\mathbb{Q}[x]$. By Corollary 4, since $sx - r$ is primitive, $g(x)$ is in $\mathbb{Z}[x]$, and clearly $b_{n-1}s = a_n, b_0 r = a_0$.   □

Here is an alternate proof:

Proof. Suppose

$$0 = f(\frac{r}{s}) = a_n(\frac{r}{s})^n + a_{n-1}\frac{r}{s} + \cdots + a_1\frac{r}{s} + a_0.$$

Multiply through by s^n, to get

$$0 = s^n f(\frac{r}{s}) = a_n r^n + a_{n-1} r^{n-1} s + \cdots + a_1(rs^{n-1}) + a_0 s^n.$$

Then s must divide $a_n r^n$, and since r and s are coprime, s must therefore divide a_n. Similarly, r must divide $a_0 s^n$; since r and s are coprime, r must divide a_0.   □

Since a_n and a_0 each have only a finite number of divisors, the theorem implies that finding the roots of a polynomial with integer coefficients is reduced to testing a finite collection of rational numbers r/s depending on a_n and a_0 (see Exercise 12).

Example 2. The only possible roots of the polynomial

$$x^4 + 8x^3 + 15x^2 - 6x - 9$$

are $x = 1, -1, 3, -3, 9$, and -9, the six divisors of 9.

Example 3. Consider finding a root of the polynomial

$$p(x) = x^5 - 141x^4 + 142x^3 - 281x^2 + 176x - 5040$$

Since $5040 = 2^4 \cdot 3^2 \cdot 5 \cdot 7$, it has 120 (positive or negative) divisors, and hence, using Descartes' criterion, there are 120 candidates for a possible root of $p(x)$. It would be helpful to find a way to reduce the number of possibilities. One way is to get a bound B on the size of the roots of $p(x)$ that is smaller than $B = 5040$. We will consider that question in Chapter 26. (It turns out that 140 is a root of $p(x)$.)

Here is an example that illustrates a way to improve the effectiveness of Descartes' criterion.

Example 4. Consider finding roots of

$$f(x) = x^4 + 5x^3 - 9x^2 - 14x + 24.$$

Since 24 has 16 divisors we would need to check as many as 16 numbers as possible roots of $f(x)$. But notice that $f(1) = 7$, which has only four divisors. If b is a root of $f(x)$, then since $f(x)$ is monic, b must be in $\mathbb{Z}$, and so $x - b$ is primitive. Therefore $x - b$ divides $f(x)$ in $\mathbb{Z}[x]$, and so

$$1 - b \text{ divides } f(1) = 7.$$

Thus the possible roots b of $f(x)$ must satisfy

$$1 - b = 1, \text{ or}$$
$$1 - b = -1, \text{ or}$$
$$1 - b = 7, \text{ or}$$
$$1 - b = -7.$$

That is, $b = 0, 2, -6$ or 8. Testing those four possibilities, we find that $f(-6) = 0$.

Before leaving this section we note that, if we wish, we can restrict our factorizations of polynomials in $\mathbb{Q}[x]$ to *monic* polynomials with integer coefficients. For suppose given a polynomial

$$f(x) = a_n x^n + \ldots + a_1 x + a_0$$

with integer coefficients. If we set $x = y/a_n$, we get

$$g(y) = a_n \frac{y^n}{a_n^n} + a_{n-1} \frac{y^{n-1}}{a_n^{n-1}} + \ldots + a_1 \frac{y}{a_n} + a_0.$$

Multiplying $g(y)$ by a_n^{n-1} yields

$$h(y) = y^n + a_{n-1} y^{n-1} + a_{n-2} a_n y^{n-2} + \ldots + a_1 a_n^{n-2} y + a_0 a_n^{n-1},$$

a monic polynomial with integer coefficients. Any factorization of $f(x)$ would correspond to a factorization of $g(y)$, and hence of $h(y)$, and conversely. In particular, if $y = r$ is a root of $h(y)$, then $x = r/a_n$ is a root of $f(x)$.

Exercises.

1. Find a primitive polynomial which is an associate of
   (i) $f(x) = (4/3)x^4 + 6x^3 + (2/9)x^2 + (9/2)x + 18$
   (ii) $f(x) = \frac{36}{35}x^3 + \frac{24}{5}x^2 + \frac{180}{7}x + 12$.

2. Try to find a factorization of $x^4 - 3x^2 + 9$ into a product of two polynomials of degree 2 in $\mathbb{Z}[x]$.

3. (i) Show: if $f(x)$ in $\mathbb{Z}[x]$ is monic, it is primitive.
   (ii) Show that if $f(x)$ in $\mathbb{Z}[x]$ is monic and factors as $f(x) = g(x)h(x)$ where $g(x)$ and $h(x)$ are in $\mathbb{Q}[x]$, then $g(x)$ and $h(x)$ are associates of monic polynomials $g_1(x)$ and $h_1(x)$ in $\mathbb{Z}[x]$ such that $f(x) = g_1(x)h_1(x)$.

4. Show the converse of Proposition 1: if $f(x)$ and $g(x)$ are in $\mathbb{Z}[x]$ and $f(x)g(x)$ is primitive, then $f(x)$ and $g(x)$ are both primitive.

5. Show that if $f(x)$ and $g(x)$ are primitive in $\mathbb{Z}[x]$ and $f(x)$ divides $g(x)$ in $\mathbb{Q}[x]$, then the quotient is primitive in $\mathbb{Z}[x]$. (Use the previous exercise.)

6. (i) Show that if $f(x)$ is monic in $\mathbb{Z}[x]$ and $f(x) = g(x)h(x)$ in $\mathbb{Q}[x]$ where $g(x)$ and $h(x)$ are monic, then $g(x)$ and $h(x)$ are in $\mathbb{Z}[x]$.
   (ii) Show that if $f(x)$ is monic in $\mathbb{Z}[x]$ and f factors into a product of monic irreducible polynomials in $\mathbb{Q}[x]$, then all of the monic irreducible factors of f are in $\mathbb{Z}[x]$.

7. Show that if the greatest common divisor of $a_n, \ldots, a_1, a_0$ is d, then the greatest common divisor of $a_n/d, \ldots, a_1/d, a_0/d$ is 1.

8. If f and g are monic polynomials in $\mathbb{Z}[x]$, does their (monic) greatest common divisor in $\mathbb{Q}[x]$ necessarily have coefficients in $\mathbb{Z}$?

9. Find all rational roots of:
  (i) $x^3 - x + 1$;
  (ii) $x^3 + x^2 + x + 1$;
  (iii) $x^3 - x^2 - 3x + 6$;
  (iv) $x^4 + 7x^3 + 11x^2 + 6x + 5$;
  (v) $x^4 - x^3 + 5x^2 + x - 6$.

10. Find all rational roots of:
  (i) $6x^3 + x^2 - 5x - 2$;
  (ii) $2x^2 - 3x - 4$;
  (iii) $3x^3 + 7x^2 - 7x - 3$.

11. (i) For the polynomial $f(x) = 6x^3 + x^2 - 5x - 2$, find the roots of the monic polynomial $h(y) = 6^2 f(\frac{y}{6})$, then find the roots of $f(x)$.
  (ii) Repeat with $f(x) = 2x^2 - 3x - 4$;
  (iii) Repeat with $f(x) = 3x^3 + 7x^2 - 7x - 3$.

12. Observe that if $f(x) = x^4 + 15x^3 + 72x^2 + 137x + 174$, then $f(-7) = -1$. What are the possible roots of $f(x)$?

13. Let $d(n)$ be the number of positive divisors of $n \geq 1$ (including 1 and n).
  (i) Show that $d(a)d(b) = d(ab)$ if a and b are coprime.
  (ii) Find $d(p^m)$ for p prime. Find $d(n)$ for any n.
  (iii) Show that if $f(x)$ is a monic polynomial in $\mathbb{Z}[x]$ and $f(0) = n$, then there are $2d(n)$ potential roots of $f(x)$ according to Theorem 5.
  (iv) Find the comparable number if $f(x)$ is not monic, but has leading coefficient a_0 and $f(0) = a_n$.

14. Find a polynomial $f(x)$ whose constant term $f(0)$ has at least 12 divisors, but such that $f(a) = 1$ for some $a \neq 0$.

B. Testing for Irreducibility

How do we decide if a polynomial $f(x)$ in $\mathbb{Q}[x]$ is irreducible? Suppose $f(x)$ has degree $d > 1$.

One way is to reduce modulo m for some number $m > 1$.

Let $\gamma_m : \mathbb{Z}[x] \to (\mathbb{Z}/m\mathbb{Z})[x]$ be the homomorphism which replaces each coefficient of $f(x)$ by its congruence class modulo m. Then if the leading coefficient of $f(x)$ is a unit modulo m, then $\gamma_m(f(x))$ will also have degree d.

Suppose $f(x) = a(x)b(x)$ where $a(x)$ and $b(x)$ have degrees r and s, respectively, where $r + s = d$, and the leading coefficient of $f(x)$ is a unit modulo m. Then the leading coefficients of $a(x)$ and $b(x)$ must also be units modulo m, and so

$$\gamma_m(f(x)) = \gamma_m(a(x)b(x)) = \gamma_m(a(x))\gamma_m(b(x)).$$

is a factorization of $f(x)$ modulo m into polynomials of the same degrees r and s. This observation implies.

Proposition 6. *Let $f(x)$ be in $\mathbb{Z}[x]$. If for some number $m \geq 2$, the leading coefficient of $f(x)$ is a unit modulo m and $\gamma_m(f(x))$ is irreducible for some m, then $f(x)$ is irreducible.*

In particular, this last result is true when f is monic.

Example 5. Let $f(x) = x^5 - 4x^4 + 2x^3 + x^2 + 18x + 3$. Then $\gamma_2(f(x)) = x^5 + x^2 + 1$, which is easily shown to be irreducible in $\mathbb{F}_2[x]$. So $f(x)$ is irreducible in $\mathbb{Q}[x]$.

Example 6. Let $f(x) = x^5 + 4x^4 + 2x^3 + 3x^2 - x + 5$. Then $\gamma_2(f(x)) = x^5 + x^2 + x + 1$, which is reducible since 1 is a root in $\mathbb{F}_2$; however, $\gamma_3(f(x)) = x^5 + x^4 + 2x^3 + 2x + 2$ is irreducible in $\mathbb{F}_3[x]$ (see Exercise 19), and so $f(x)$ is irreducible in $\mathbb{Q}[x]$.

This last example shows that a polynomial in $\mathbb{Q}[x]$ can be irreducible but factor modulo m for some modulus m. In fact, there are monic irreducible polynomials in $\mathbb{Z}[x]$ that factor modulo p for every prime p–see Section C, below.

Example 7. Let $f(x) = 3x^4 + 6x^3 + 12x^2 + 13x + 31$. Modulo 3, $f(x) = x + 1$, a polynomial of degree <4, so reducing $f(x)$ modulo 3 is not helpful. But $\gamma_2(f(x)) = x^4 + x + 1$, an irreducible polynomial of degree 4 in $\mathbb{F}_2[x]$. So $f(x)$ is irreducible in $\mathbb{Q}[x]$.

Example 8. Let $f(x) = (2x + 1)(x^2 - x + 1) = 2x^3 - x^2 + x + 1$, obviously not irreducible in $\mathbb{Q}[x]$. But $\gamma_2(f(x)) = x^2 + x + 1$ is irreducible in $\mathbb{Z}/2\mathbb{Z}[x]$. This shows that the condition on the leading coefficient of $f(x)$ is necessary for Proposition 6 to be valid.

Proposition 6 implies that if for some degree d and some m there is an irreducible polynomial $h(x)$ of degree d with coefficients in $\mathbb{Z}/m\mathbb{Z}$, then there are infinitely many primitive polynomials of degree d with coefficients in $\mathbb{Z}$ that are irreducible in $\mathbb{Q}[x]$, namely, all polynomials of degree d in $\mathbb{Z}[x]$ that reduce modulo m to $h(x)$.

It is known (see Chapter 27, Proposition 8) that there are irreducible polynomials of every degree >0 in $\mathbb{Z}/p\mathbb{Z}[x]$ for every prime p, and hence there are infinitely many irreducible polynomials of every degree >0 in $\mathbb{Q}[x]$. But we can also show that fact directly, using:

Theorem 7 (Eisenstein's Irreducibility Criterion). *Suppose $f(x) = a_n x^n + \ldots + a_1 x + a_0$ is in $\mathbb{Z}[x]$ and there exists a prime p such that p does not divide a_n, p does divide $a_{n-1}, a_{n-2}, \ldots, a_1, a_0$, but p^2 does not divide a_0. Then $f(x)$ is irreducible in $\mathbb{Q}[x]$.*

Proof. Given the hypotheses on $f(x)$, if $\gamma_p : \mathbb{Z}[x] \to \mathbb{Z}/p\mathbb{Z}[x]$ is the "reduce the coefficients mod p" map, then $\gamma_p(f(x)) = [a_n]x^n$, where $[a_n] \neq [0]$ in $\mathbb{Z}/p\mathbb{Z}$. Assume $n \geq 2$.

Suppose $f(x) = g(x)h(x)$, where $\deg g(x) = r \geq 1$, $\deg h(x) = s \geq 1$, and $r + s = n$. Then $\gamma_p(f(x)) = \gamma_p(g(x))\gamma_p(h(x))$ in $\mathbb{Z}/p\mathbb{Z}[x]$. By unique factorization in $\mathbb{Z}/p\mathbb{Z}[x]$, we must have $\gamma_p(g(x)) = [b]x^r$ and $\gamma_p(h(x)) = [c]x^s$, where $bc \equiv a_n \pmod{p}$. Hence $\gamma_p(g(0)) = \gamma_p(h(0)) = 0$; that is, $g(0) \equiv 0 \pmod{p}$ and $h(0) \equiv 0 \pmod{p}$. Then p divides $g(0)$ and p divides $h(0)$. But then p^2 divides $g(0)h(0) = f(0) = a_0$, contradicting the hypothesis on a_0. Thus $f(x)$ must be irreducible. $\square$

Example 9. It is easy to construct examples where Eisenstein's criterion applies. The simplest are radical polynomials

$$x^n - b$$

where b has a prime factor p such that p^2 does not divide b, such as

$$x^7 - 12$$

or

$$x^4 - 45.$$

Example 10. Let p be an odd prime, and let

$$\Phi(x) = x^{p1} + x^{p-2} + \ldots + x + 1 = \frac{x^p - 1}{x - 1}.$$

Then setting $x = y + 1$,

$$q(y) = \phi(x+1) = \frac{(y+1)^p - 1}{y}$$

$$= y^{p-1} + \binom{p}{1}y^{p-2} + \binom{p}{2}y^{p-3} + \ldots + \binom{p}{p-2}y + \binom{p}{p-1}.$$

Then $q(y)$ is irreducible by Eisenstein's criterion, since p divides $\binom{p}{k}$ for $1 \le k \le p-1$, and $\binom{p}{p-1} = p$ is not divisible by p^2. Hence $\Phi(x)$ is irreducible.

We observe that $\Phi(x)$ has as roots all of the p-th roots of unity in the complex numbers other than 1. Since $\Phi(x)$ is irreducible, $\Phi(x)$ is the polynomial of smallest degree whose roots are the p-th roots of unity.

Chebyshev polynomials. For each $n \ge 1$, the Chebyshev polynomial of the first kind $T_n(x)$ is defined by $T_n(\cos\theta) = \cos(n\theta)$. For example,

$$T_1(x) = x \qquad \text{since } \cos(\theta) = \cos(\theta);$$
$$T_2(x) = 2x^2 - 1 \qquad \text{since } 2\cos^2(\theta) - 1 = \cos(2\theta);$$
$$T_3(x) = 4x^3 - 3x \qquad \text{since } 4\cos^3(\theta) - 3\cos(\theta) = \cos(3\theta);$$

etc. The Chebyshev polynomials are polynomials in $\mathbb{Z}[x]$.

Vieta used the formula $T_3(\cos(\theta)) = \cos(3\theta)$ to help solve cubic polynomials with three real roots (see Section 15E).

For p prime, we can show that $T_p(x) = xQ(x)$ where $Q(x)$ is irreducible in $\mathbb{Q}[x]$, using Eisenstein's criterion with the prime p.

To do so, we first obtain a formula for $T_n(x)$ by using DeMoivre's formula for θ:

$$(\cos(\theta) + i\sin(\theta))^n = \cos(n\theta) + i\sin(n\theta)$$

and for $-\theta$:

$$(\cos(-\theta) + i\sin(-\theta))^n = \cos(-n\theta) + i\sin(-n\theta),$$

which becomes

$$(\cos(\theta) - i\sin(\theta))^n = \cos(n\theta) - i\sin(n\theta).$$

Adding the DeMoivre equations for θ and $-\theta$ gives

$$2\cos(n\theta) = (\cos(\theta) + i\sin(\theta))^n + (\cos(\theta) - i\sin(\theta))^n.$$

If we set $x = \cos(\theta)$ and

$$i\sin(\theta) = \sqrt{\cos^2(\theta) - 1} = \sqrt{x^2 - 1}$$

for $-1 \le x \le 1$, we have

$$T_n(x) = \frac{(x + \sqrt{x^2 - 1})^n + (x - \sqrt{x^2 - 1})^n}{2}.$$

This formula for $T_n(x)$ yields

Theorem 8. *Assume n is odd. Then*
  (i) $T_n(0) = 0$;
  (ii) The leading coefficient of $T_n(x)$ is 2^{n-1};
  (iii) The coefficient of x in $T_n(x)$ is $(-1)^{\frac{n-1}{2}} n$;
  (iv) If p is an odd prime, then $T_p(x) \equiv x^p \pmod{p}$

Proof. For n odd, expanding the expression for $T_n(x)$ using the Binomial Theorem gives

$$T_n(x) = \frac{(x + \sqrt{x^2 - 1})^n + (x - \sqrt{x^2 - 1})^n}{2}$$

$$= \frac{1}{2}\left(\sum_{k=0}^{n} \binom{n}{k} x^{n-k} (\sqrt{x^2 - 1})^k + \sum_{k=0}^{n} \binom{n}{k} x^{n-k} (-1)^k (\sqrt{x^2 - 1})^k \right).$$

The terms for k odd cancel, and the k even terms are equal, so we have

$$T_n(x) = \sum_{l=0}^{(n-1)/2} \binom{n}{2l} x^{n-2l} (x^2 - 1)^l.$$

Since $n - 2l$ is odd for all l, $T_n(0) = 0$, proving (i).

The coefficient of x^n is then

$$\sum_{l=0}^{\frac{n-1}{2}} \binom{n}{2l}.$$

Since n is odd,

$$2\sum_{l=0}^{\frac{n-1}{2}} \binom{n}{2l} = \sum_{l=0}^{\frac{n-1}{2}} \binom{n}{2l} + \sum_{l=0}^{\frac{n-1}{2}} \binom{n}{n-2l}$$

$$= \sum_{k=0}^{n} \binom{n}{k} = (1+1)^n = 2^n.$$

So the coefficient of x^n in $T_n(x)$ is 2^{n-1}, proving (ii).

In the expression for $T_n(x)$, the coefficient of x only arises for $l = \frac{n-1}{2}$, with coefficient

$$\binom{n}{n-1}(-1)^{\frac{n-1}{2}} = (-1)^{\frac{n-1}{2}}n,$$

proving (iii).

Finally, if $n = p$, an odd prime, then $\binom{p}{2l} \equiv 0 \pmod{p}$ for all $l > 0$, so

$$T_p(x) = \sum_{l=0}^{(n-1)/2} \binom{p}{2l} x^{p-2l}(x^2-1)^l$$

$$\equiv x^p \pmod{p},$$

proving (iv). $\square$

Now we get our claimed result:

Theorem 9. *For each odd prime p, $T_p(x) = xQ_p(x)$ where $Q_p(x)$ is irreducible.*

Proof. $T_p(x) = xQ_p(x)$ by the Root Theorem, since $T_p(0) = 0$. Since $T_p(x) \equiv x^p$ (mod p), it is clear that $Q_p(x) \equiv x^{p-1}$ (mod p), and so the prime p divides every coefficient of $Q_p(x)$ except the leading coefficient. But the constant term of $Q_p(x)$ is $(-1)^{(p-1)/2}p$ by part (iii) of the Theorem. So $Q_p(x)$ satisfies the conditions of Eisenstein's Irreducibility Criterion. $\square$

In Chapters 17 and 26 we shall describe two different techniques for systematically factoring a polynomial $f(x)$ in $\mathbb{Q}[x]$. One uses an analogue of the Chinese remainder theorem for polynomials. The other involves the idea of factoring $f(x)$ modulo m and then seeing if any such factorization lifts to a possible factorization in $\mathbb{Q}[x]$.

Exercises.

15. Prove that for any $f(x)$ in $\mathbb{Z}[x]$ and any c in $\mathbb{Z}$, evaluating $f(x)$ at $x = c$ and then reducing modulo p, is the same as evaluating $\gamma_p(f(x))$ at $x = [c]$ in $\mathbb{Z}/p\mathbb{Z}$. Where was this fact implicitly used in the proof of Eisenstein's criterion?

16. Give an example of a monic polynomial $f(x)$ in $\mathbb{Z}[x]$ that is irreducible in $\mathbb{Q}[x]$ but factors modulo 2 and modulo 3.

17. Adapt Eisenstein's criterion to prove that $2x^4 - 8x^2 + 3$ is irreducible in $\mathbb{Q}[x]$.

18. Show that $f(x) = x^5 + x^2 + 1$ is irreducible in $\mathbb{F}_2[x]$ as follows:
(i) Show that if $f(x)$ is not irreducible, then it has an irreducible factor of degree 1 or degree 2.
(ii) Show that $f(x)$ has no factors of degree 1 (check roots).
(iii) Show that $f(x)$ is not divisible by $x^2 + x + 1$, hence has no irreducible factors of degree 2.

19. Show that $f(x) = x^5 + x^4 + 2x^3 + 2x + 2$ is irreducible in $\mathbb{F}_3[x]$ as follows:
(i) Show that $f(x)$ has no roots in $\mathbb{F}_3$.
(ii) Suppose $f(x) = (ax^2 + bx + c)(dx^3 + ex^2 + fx + g)$, then you can assume $a = d = 1$ (why); multiply the right side together, collect coefficients of $x^4, x^3, x^2, x, 1$ together and equate them to the coefficients of $f(x)$ to get five equations in the five unknowns b, c, e, f, g. Show that the system of five equations has no solution in $\mathbb{F}_3$.

20. (i) Suppose $f(x)$, monic in $\mathbb{Z}[x]$, factors modulo 3 into the product of two irreducible polynomials of degree 2, and factors modulo 2 into the product of an irreducible polynomial of degree 3 and a polynomial of degree 1. Show that $f(x)$ is irreducible in $\mathbb{Q}[x]$.
(ii) Illustrate part (i) with $f(x) = x^4 + 5x^3 + 3x^2 + 2x + 5$.

C. Polynomials that Factor Modulo Every Prime

In the last section, it was claimed that there are polynomials in $\mathbb{Z}[x]$ that factor modulo p for every prime p but are irreducible in $\mathbb{Q}[x]$. In this section we describe a class of such examples. Their existence will be of interest in Section 27B.

Our examples are polynomials of the form $x^4 + ax^2 + b^2$ for integers a and b.

We first show that these polynomials factor modulo any prime. Then we will find conditions on a and b so that the polynomials are irreducible in $\mathbb{Q}[x]$.

Proposition 10. *For all integers a, b and all primes p, the polynomial $f(x) = x^4 + ax^2 + b^2$ factors modulo p.*

Proof. First suppose $p = 2$. Then $f(x)$ is congruent modulo 2 to one of the following polynomials:
$$x^4 + x^2 + 1 = (x^2 + x + 1)^2$$
$$x^4 + 1 = (x^2 + 1)^2$$
$$x^4 + x^2 = x^2(x^2 + 1) = (x^2 + x)^2$$
$$x^4.$$

Each of these is reducible modulo 2.

Now suppose p is an odd prime. Choose s so that $a \equiv 2s \pmod{p}$. Then

$$f(x) \equiv x^4 + 2sx^2 + b^2 \pmod{p},$$

and we may write this polynomial in three different ways:

$$\begin{aligned} x^4 + 2sx^2 + b^2 &= (x^2 + s)^2 - (s^2 - b^2) \\ &= (x^2 + b)^2 - (2b - 2s)x^2 \\ &= (x^2 - b)^2 - (-2b - 2s)x^2 \end{aligned}$$

Then $f(x)$ will be the difference of two squares modulo p, and will therefore factor modulo p, if one of $s^2 - b^2, 2b - 2s$, or $-2b - 2s$ is a square modulo p.

From Example 24 of Section 11G, we know that in the group U_p of units modulo p, the product of two non-squares is a square. Thus, suppose $2b - 2s$ and $-2b - 2s$ are nonsquares modulo p. Then their product $(2s)^2 - (2b)^2 = 4(s^2 - b^2)$ is a square. Since 4 is a square, therefore $s^2 - b^2$ is a square modulo p. □

Now we prove:

Proposition 11. *For a, b in $\mathbb{Z}$,*

$$f(x) = x^4 + ax^2 + b^2$$

factors in $\mathbb{Q}[x]$ if and only if at least one of $a^2 - 4b^2$, $2b - a$ and $-2b - a$ is a square in $\mathbb{Z}$.

Proof. If $a^2 - 4b = c^2$ is a square, then $f(x)$ factors as

$$f(x) = (x^2 - \tfrac{a}{2})^2 - (\tfrac{c}{2})^2 = (x^2 - (\tfrac{a}{2} - \tfrac{c}{2}))(x^2 - (\tfrac{a}{2} + \tfrac{c}{2})).$$

If $2b - a = r^2$ is a square, then $f(x)$ factors as

$$f(x) = (x^2 + rx + b)(x^2 - rx + b).$$

If $-2b - a = r^2$ is a square, then $f(x)$ factors as

$$f(x) = (x^2 + rx - b)(x^2 - rx - b).$$

Conversely, we will show that if $f(x)$ factors, then one of $a^2 - 4b$, $2b - a$ and $-2b - a$ is a square in $\mathbb{Z}$.

If $f(x)$ factors, then it has a factor of degree 1 or degree 2. If $f(x)$ has a factor of degree 1, then $f(x)$ has a root, m.

If $m = 0$ then $b = 0$ and $a^2 - 4b$ is a square.

If $m \neq 0$ and $f(m) = 0$, then $f(-m) = 0$. Thus $x - m$ and $x + m$ divide $f(x)$, hence $x^2 - m^2$ divides $f(x)$. Thus if $f(x)$ has a factor of degree 1 other than x, then $f(x)$ has a factor of degree 2.

Suppose, then, that $f(x)$ has a factor of degree 2. Then $f(x)$ factors in $\mathbb{Z}[x]$ as

$$f(x) = (x^2 + rx + s)(x^2 + tx + u)$$

for some integers r, s, t, u. Equating coefficients of x^3 yields $t = -r$. Then equating coefficients of $1, x, x^2$ yields the equations:

$$us = b^2$$
$$r(u - s) = 0$$
$$s + u = r^2 + a.$$

If $r = 0$, then $a^2 - 4b^2 = (s + u)^2 - 4su = (s - u)^2$, a square.
If $r \neq 0$, then $u = s$, hence $s^2 = b^2$. Hence if $u = s = b$, then $2b - a = r^2$, a square, while if $u = s = -b$, then $-2b - a = r^2$, a square.

Thus any factorization of $f(x)$ implies that one of $a^2 - 4b$, $2b - a$ and $-2b - a$ is a square. $\qquad\square$

Corollary 12. *Suppose $2b > a > 0$. Then $f(x) = x^4 + ax^2 + b^2$ is irreducible in $\mathbb{Q}[x]$ if and only if $2b - a$ is not a square in $\mathbb{Z}$.*

Proof. If $2b > a > 0$, then $a^2 - 4b^2 < 0$ and $-2b - a < 0$, so cannot be squares in $\mathbb{Z}$. The corollary then follows immediately from the last theorem. $\qquad\square$

Letting $2b - a = c$ we can rephrase the last corollary as:

Corollary 13. *Suppose $0 < c < 2b$. Then $f(x) = x^4 + (2b - c)x^2 + b^2$ factors in $\mathbb{Q}[x]$ if and only if c is a square in $\mathbb{Z}$.*

This makes finding examples especially easy. For example, if we let $b = 20$, then

$$f(x) = x^4 + (40 - c)x^2 + 400$$

is irreducible in $\mathbb{Q}[x]$ for all c with $0 < c < 40$ except for $c = 1, 4, 9, 16, 25$ and 36.

For an interesting article on irreducibility and factorization modulo m of biquadratic polynomials such as these, see Driver, Leonard and Williams [Monthly 112 (2005), 876-890].

Exercises.

21. Show that $x^4 - x^2 + 1$ is irreducible in $\mathbb{Q}[x]$.

22. Show that $x^4 + 1$ is irreducible in $\mathbb{Q}[x]$.

23. Show that $x^4 + 2x^2 + 4$ is irreducible in $\mathbb{Q}[x]$.

24. Show that if s, b are numbers such that $0 < b < s$ and $s^2 - b^2$ is not a square in $\mathbb{Z}$, then $x^4 + 2sx^2 + b^2$ is irreducible in $\mathbb{Q}[x]$.

25. Show that if s, b are numbers such that $0 < s < b$ and $2b - 2s$ is not a square in $\mathbb{Z}$, then $x^4 + 2sx^2 + b^2$ is irreducible in $\mathbb{Q}[x]$.

26. Can Eisenstein's irreducibility criterion be applied to any of the examples of Proposition 10?

27. Show that $x^4 - 15x^2 + 1$ factors into two coprime polynomials of degree 2 modulo every odd prime, and also factors modulo 16. What about modulo 32?

Chapter 17
Congruences and the Chinese Remainder Theorem

We develop the idea of congruence for polynomials with coefficients in a field. The properties of congruence for polynomials are very similar to those for congruence for integers, as presented in Chapter 5.

A. Congruence Modulo a Polynomial

Here is the basic definition:

Definition. Let F be a field, m a polynomial with coefficients in F. For f and g in $F[x]$, say that f is congruent to g modulo m, written

$$f \equiv g \pmod{m}$$

if p divides $f - g$, or equivalently, if $f = g + hm$ for some polynomial h in $F[x]$.

Basic properties. Congruence modulo m in $F[x]$ has identical properties to congruence modulo m in the integers, where m is any natural number. In particular, the arithmetic properties of congruence hold:

Proposition 1. *For $f, f_1, f_2, g, g_1, g_2, k$ in $F[x]$,*
If $f \equiv g \pmod{m}$, then $kf \equiv kg \pmod{m}$;
If $f_1 \equiv g_1 \pmod{m}$ and $f_2 \equiv g_2 \pmod{m}$, then $f_1 + f_2 \equiv g_1 + g_2 \pmod{m}$, and $f_1 f_2 \equiv g_1 g_2 \pmod{m}$;
If $f \equiv g \pmod{m}$ then $f^n \equiv g^n \pmod{m}$ for every number $n \geq 0$.

Proof. All of the proofs are straightforward and essentially identical to those for integers. We illustrate this by proving the multiplication property.
    If $f_1 \equiv g_1 \pmod{m}$, there is a polynomial h_1 so that $f_1 = g_1 + mh_1$. Similarly, if $f_2 \equiv g_2 \pmod{m}$, there is a polynomial h_2 so that $f_2 = g_2 + mh_2$. Then

$$f_1 f_2 = (g_1 + mh_1)(g_2 + mh_2)$$
$$= g_1 g_2 + m(h_1 g_2 + g_1 h_2 + mh_1 h_2).$$

L.N. Childs, *A Concrete Introduction to Higher Algebra*, Undergraduate Texts in Mathematics, © Springer Science+Business Media LLC 2009

Hence
$$f_1 f_2 \equiv g_1 g_2 \quad (\text{mod } m).$$

□

As with congruence in $\mathbb{Z}$, congruence modulo a polynomial m is an equivalence relation: for all f, g, h in $F[x]$,

(R) (reflexive) $f \equiv f \pmod{m}$,

(S) (symmetric) If $f \equiv g \pmod{m}$ then $g \equiv f \pmod{m}$,

(T) (transitive) If $f \equiv g \pmod{m}$ and $g \equiv h \pmod{m}$, then $f \equiv h \pmod{m}$.

These properties make congruence modulo m much like equality.

As with congruence in $\mathbb{Z}$, the only property of equality that requires some care is cancellation. The cancellation property of equality:

if $ra = rb$ and $r \neq 0$, then $a = b$

becomes, for congruence in $\mathbb{Z}$:

if $ra \equiv rb \pmod{m}$ and r and m are coprime, then $a \equiv b \pmod{m}$.

The same result is true for congruence with polynomials:

Proposition 2. *Let f, g, h, m be polynomials with coefficients in a field, and assume $m \neq 0$. If*
$$hf \equiv hg \quad (\text{mod } m)$$

and h and m are coprime, then

$$f \equiv g \quad (\text{mod } m)$$

Proof. The hypothesis is equivalent to

$$m \text{ divides } h(f - g).$$

Since $(m, h) = 1$, it follows from Bezout's Identity (Section 14B) that m divides $f - g$, and hence
$$f \equiv g \quad (\text{mod } m).$$

□

Least degree residues. The first proposition in Chapter 5 showed that every number is congruent modulo m to a number in the set $\{0, 1, \ldots, m - 1\}$. Here is the corresponding result for polynomials:

Proposition 3. *Let m be a polynomial of degree ≥ 0. If f is any polynomial in $F[x]$, then f is congruent modulo m to a unique polynomial of degree $< \deg(m)$, called the residue of least degree modulo m.*

Proof. Apply the Division Theorem: write $f = mq + r$ with $\deg(r) < \deg(m)$. Then $f \equiv r \pmod{m}$.

The residue of least degree r is unique by the uniqueness of the remainder in the Division Theorem.   □

Proposition 4. *Two polynomials a and b are congruent modulo m if and only if their residues of least degree modulo m are equal.*

Proof. Using the Division Theorem, let $a = mq + r$, $b = ms + t$, where r, t both have degree less than the degree of m.

If $r = t$, then $a \equiv b \pmod{m}$ by transitivity and symmetry.

Conversely, if $a \equiv b \pmod{m}$, then $r \equiv t \pmod{m}$, again by symmetry and transitivity of congruence. But if $r \equiv t \pmod{m}$, then m divides $r - t$. The polynomial $r - t$ has degree less than the degree of m, so if m divides $r - t$, then $r - t = 0$. Hence $r = t$.   □

Proposition 5. *For any field F, any element r of F and any $f(x)$ in $F[x]$,*

$$f(x) \equiv f(r) \pmod{x - r}.$$

Proof. Apply the Remainder Theorem: if we divide $f(x$ by $x - r$, the remainder is $f(r)$:
$$f(x) = (x - r)q(x) + f(r)$$
where $q(x)$ is the quotient. Hence $f(x) \equiv f(r) \pmod{x - r}$.   □

Example 1. Let $m(x) = x^2 + x + 1$ in $\mathbb{F}_3[x]$. When we divide $f(x) = x^5 + 2x^3 + x^2 + 1$ by $m(x)$, we find that

$$f(x) = m(x)(x^2 + 2x + 2) + (x + 1).$$

So $x + 1$ is the residue of least degree of $f(x)$ modulo $m(x)$.

An alternative way to see Proposition 5 is to observe that $x - r \equiv 0 \pmod{x - r}$, so that
$$x \equiv r \pmod{x - r}.$$
Then
$$x^k \equiv r^k \pmod{x - r}$$
for all $k > 1$. Thus we may replace x^k by r^k everywhere in $f(x)$ and get an element of F that is congruent to $f(x)$ modulo $x - r$: if

$$f(x) = a_n x^n + a_{n-1} x^{n-1} + \ldots + a_1 x + a_0,$$

then
$$f(x) \equiv a_n r^n + a_{n-1} r^{n-1} + \ldots + a_1 r + a_0 \pmod{x - r}$$

and the right side of this last congruence is $f(r)$. This method is analogous to "casting out 9's" or "casting out 11's" with numbers: think of a number such as 365 as a polynomial evaluated at 10:

$$365 = 3 \cdot 10^2 + 6 \cdot 10 + 5,$$

then since $10 \equiv -1 \pmod{11}$, we can replace 10^r by $(-1)^r$ modulo 11 to get

$$365 \equiv (-1)^2 3 + (-1)6 + 5 \equiv 3 - 6 + 5 \equiv 2 \pmod{11}.$$

This idea extends well to moduli of degree > 1.

Example 2. In $\mathbb{F}_2[x]$, let $m(x) = x^3 + x + 1$. Then every polynomial $f(x)$ in $\mathbb{F}_2[x]$ is congruent modulo $m(x)$ to a polynomial of degree < 2 by Proposition 3. We find the least degree residues of x^n for every $n \geq 3$. Since

$$x^3 + x + 1 \equiv 0 \pmod{m(x)},$$

and $-1 = 1$ in $\mathbb{F}_2$, we have

$$\begin{aligned}
x^3 &\equiv -x - 1 \equiv x + 1 \pmod{m(x)}, \\
x^4 &\equiv x(x+1) \equiv x^2 + x \pmod{m(x)}, \\
x^5 &\equiv x(x^2 + x) \equiv x^3 + x^2 \equiv x^2 + x + 1 \pmod{m(x)}, \\
x^6 &\equiv (x+1)^2 \equiv x^2 + 1 \pmod{m(x)}, \\
x^7 &\equiv x(x^2 + 1) \equiv x^3 + x \equiv 1 \pmod{m(x)}.
\end{aligned}$$

Thus for any n, if $n = 7q + r$, then

$$x^n \equiv x^{7q} x^r \equiv x^r \pmod{m(x)}.$$

Example 3. Let $m(x) = x^2 + x + 1$ in $\mathbb{F}_3[x]$. For any n we find the residue of least degree in $\mathbb{F}_3[x]$ modulo $m(x)$ of x^n. To do so, we note that

$$x^3 - 1 = (x - 1)(x^2 + x + 1),$$

so

$$x^3 \equiv 1 \pmod{m(x)}.$$

Hence for any n, if $n = 3q + r$ with $r < 3$, then

$$x^n = x^{3q} x^r \equiv x^r \pmod{m(x)}.$$

Then, to find the least degree residue of $f(x) = x^5 + 2x^3 + x^2 + 1$, we substitute for the powers of x:

$$\begin{aligned}
x^5 + 2x^3 + x^2 + 1 &\equiv x^2 + 2 + x^2 + 1 \\
&\equiv 2x^2 \\
&\equiv 2(-x - 1) = x + 1 \pmod{m(x)}.
\end{aligned}$$

Linear congruences. We can solve linear congruences modulo a polynomial in the same way we do for numbers. To solve

$$au \equiv b \pmod{m},$$

for some polynomial u in $F[x]$, we find the greatest common divisor of a and m, call it d. If d doesn't divide b, then there will be no solution of the congruence, because if u were a solution, then

$$au + mv = b$$

for some polynomial v. But then the greatest common divisor d of a and m would divide $au + mv = b$.

On the other hand, if d does divide b, say $b = df$, then by Bezout's Identity, we can find polynomials s, t so that

$$as + mt = d.$$

Multiplying both sides by f gives

$$asf + mtf = df = b$$

and so $u = sf$ is a solution of the congruence

$$au \equiv b \pmod{m}.$$

Note that the theory is identical to that for congruence in $\mathbb{Z}$.

As with integers, if we can find one solution u_1 of

$$au \equiv b \pmod{m}$$

then the general solution is found by finding all solutions of the corresponding homogeneous congruence

$$au \equiv 0 \pmod{m};$$

if u_0 is a solution of the homogeneous congruence, then $u = u_0 + u_1$ is a solution of the original non-homogeneous congruence $au \equiv b \pmod{m}$, and conversely, every solution of the non-homogeneous congruence has that form for some solution u_0 of the homogeneous congruence.

In particular, the congruence

$$au \equiv 1 \pmod{m}$$

has a solution if and only if a and m are coprime.

Example 4. In $\mathbb{F}_3[x]$, let $m(x) = x^3 + x + 2$ and $f(x) = x^2 + 2x$. We try to find $z(x)$ so that

$$f(x)z(x) \equiv 1 \pmod{m(x)}.$$

That is the same as finding $z(x), w(x)$ so that

$$f(x)z(x) + m(x)w(x) = 1$$

To solve this, we do Euclid's algorithm with $m(x)$ and $f(x)$:

$$x^3 + x + 2 = (x^2 + 2x)(x+1) + (2x+2);$$
$$x^2 + 2x = (2x+2)(2x+2) + 2.$$

Solving for $1 = -2$ in terms of $f(x)$ and $m(x)$:

$$\begin{aligned}
1 &= (2x+2)(2x+2) - (x^2+2x)\\
&= 2(x^2+2x) + (2x+2)((x^3+x+2) - (x^2+2x)(x+1))\\
&= (x^3+x+2)(2x+2) + (x^2+2x)((2x+2)(2x+2)+2)\\
&= (x^3+x+2)(2x+2) + (x^2+2x)(x^2+2x).
\end{aligned}$$

So $z(x) = x^2 + 2x$.

The Extended Euclidean Algorithm (row operations) of Chapter 3 works well here also.

Complete set of representatives. Just as with numbers (mod m), a *complete set of representatives modulo m* in $F[x]$ is defined to be a set of polynomials with the property that every polynomial in $F[x]$ is congruent modulo m to exactly one polynomial in the set.

If the field F is infinite (such as $\mathbb{Q}$ or $\mathbb{R}$), then for any modulus m of degree >0, a complete set of representatives will be an infinite set. But for F finite (such as $\mathbb{F}_p$ for p prime), any complete set of representatives will be a finite set. Here is an example.

Example 5. Continuing Example 2, we find a complete set of representatives modulo $m(x) = x^3 + x + 1$ in $\mathbb{F}_2[x]$.

One complete set of representatives consists of the eight polynomials of degree <2 in $\mathbb{F}_2[x]$,

$$0, 1, x, x+1, x^2, x^2+1, x^2+x, x^2+x+1.$$

In Example 2 we found the residues of least degree of the powers of x, and found that

$$\begin{aligned}
1 &\equiv x^7 \pmod{m(x)}\\
x &\equiv x \pmod{m(x)}\\
x+1 &\equiv x^3 \pmod{m(x)}\\
x^2 &\equiv x^2 \pmod{m(x)}\\
x^2+1 &\equiv x^6 \pmod{m(x)}\\
x^2+x &\equiv x^4 \pmod{m(x)}\\
x^2+x+1 &\equiv x^5 \pmod{m(x)}.
\end{aligned}$$

So instead of using the residues of least degree as a complete set of representatives, we could use

$$\{0, 1, x, x^2, x^3, x^4, x^5, x^6\}$$

with $x^7 \equiv 1 \pmod{m}$. This latter set is convenient for multiplying. For example,

$$(x^2 + x)(x^2 + x + 1) \equiv x^4 \cdot x^5 \equiv x^9 \equiv x^7 \cdot x^2 \equiv x^2,$$

or

$$(x^2 + x)(x^2 + 1) \equiv x^4 \cdot x^6 \equiv x^{10} \equiv x^3 \equiv x + 1.$$

The polynomial x is a primitive root modulo $x^3 + x + 1$ in $\mathbb{F}_2[x]$.

Example 6. We find a complete set of representatives modulo $m(x) = x^2 + x + 1$ in $\mathbb{F}_3[x]$, and find the product of two representatives as another representative.

By the Division Theorem, a complete set of representatives modulo $m(x)$ consists of polynomials of degree < 1, that is, polynomials of the form $ax + b$, for a, b running through the elements of $\mathbb{F}_3 = \{0, 1, -1\}$. There are nine such polynomials:

$$0, 1, -1, x, x+1, x-1, -x, -x+1, -x-1.$$

To multiply these polynomials is easy when one of the polynomials has degree < 0:

$$0 \cdot (ax + b) = 0; \qquad 1 \cdot (ax + b) = ax + b; \qquad -1 \cdot (ax + b) = -ax - b.$$

If both polynomials have degree 1, their product will have degree 2, and we want to replace the polynomial of degree 2 by the polynomial of degree < 1 to which it is congruent.

A convenient way to do this is to notice that since $m(x) = x^2 + x + 1$, we have $x^2 \equiv -x - 1 \pmod{m(x)}$, which we can use to substitute. So, for example,

$$x(x - 1) = x^2 - x \equiv (-x - 1) - x$$
$$\equiv -2x - 1 \equiv x - 1 \pmod{m(x)}.$$

Here is the table of multiplications for polynomials of degree 1 modulo $m(x) = x^2 + x + 1$:

$\cdot$	x	$x+1$	$x-1$	$-x$	$-x-1$	$-x+1$
x	$-x-1$	-1	$x-1$	$x+1$	1	$-x+1$
$x+1$	-1	x	$-x+1$	1	$-x$	$x-1$
$x-1$	$x-1$	$-x+1$	0	$-x+1$	$x-1$	0
$-x$	$x+1$	1	$-x+1$	$x-1$	-1	$x-1$
$-x-1$	1	$-x$	$x-1$	-1	x	$-x+1$
$-x+1$	$-x+1$	$x-1$	0	$x-1$	$-x+1$	0

Note that the entries below the diagonal are the reflection of those above (since multiplication is commutative).

In this example, multiplication modulo $x^2 + x + 1$ in $\mathbb{F}_3[x]$, certain polynomials $(x, x+1, -x, -x-1, 1, -1)$ were units modulo $m(x)$, and the others $(x-1, -x+1)$ were zero divisors. (Note that $(x-1)^2 = x^2 + x + 1$.)

If $f(x)$ is a unit modulo $m(x)$, then we can always solve a congruence of the form

$$f(x)z(x) \equiv h(x) \quad (\text{mod } m(x))$$

for $z(x)$: simply multiply both sides of the congruence by the inverse (modulo $m(x)$) of $f(x)$. For example, in $\mathbb{F}_3[x]$, to solve

$$z(x)x \equiv x+1 \quad (\text{mod } x^2+x+1)$$

in $\mathbb{F}_3[x]$, we find the inverse of x, namely $-x-1$, from the table, and multiply both sides of the congruence by $-x-1$, to get

$$z(x) \equiv (x+1)(-x-1) \equiv -x \quad (\text{mod } x^2+x+1).$$

On the other hand, if $f(x)$ is not a unit, we may not be able to solve (*) at all: for example,

$$(x-1)z(x) \equiv x \quad (\text{mod } x^2+x+1)$$

has no solution, as can be seen either by looking at the table we constructed above or by setting up the congruence with an unknown $z(x) = ax+b$, and trying to find a and b.

The general result about units and zero divisors modulo $m(x)$ is:

Proposition 6. *Let F be a field and let $m(x)$ in $F[x]$ have degree >0. For $f(x)$ in $F[x]$,*
  a) $f(x) \equiv 0 \ (\text{mod } m(x))$ iff $m(x)$ divides $f(x)$;
  b) $f(x)$ is a unit modulo $m(x)$ iff the greatest common divisor $(m(x), f(x)) = 1$;
  c) $f(x)$ is a zero divisor modulo $m(x)$ iff $f(x)$ is not divisible by $m(x)$ and $(m(x), f(x))$ has degree ≥ 1.

The proof is just the same as for integers modulo m and is left as an exercise.

Exercises.

1. Find the residue of least degree modulo $m(x)$ of $f(x)$ in $F[x]$, where $F = \mathbb{F}_3$, $m(x) = x^2+x$, and
  (i) $f(x) = x^3+2$;
  (ii) $f(x) = x^4+2x^2+1$;
  (iii) $f(x) = x^r$ for $r = 6,7,8$, etc.

2. Identify the units and the zero divisors in $\mathbb{F}_3[x]$ modulo $m(x) = x^2+x$.

3. Find the residue of least degree modulo $m(x)$ of $f(x)$ in $F[x]$, where $F = \mathbb{F}_5$, $m(x) = x^4-2$, and
  (i) $f(x) = x^5$;
  (ii) $f(x) = x^4+2x^2+1$;
  (iii) $f(x) = x^r$ for $r = 4,5,6$, etc.

4. Write down the multiplication table for the polynomials of degree <1 in $\mathbb{F}_2[x]$ modulo x^2+x+1.

5. Write down the multiplication table for the polynomials of degree <2 in $\mathbb{F}_2[x]$ modulo x^3+1.

6. Write down the multiplication table for the polynomials of degree <2 in $\mathbb{F}_2[x]$ modulo x^3+x^2+1.

7. In $\mathbb{F}_3[x]$, find a complete set of representatives modulo x^2-1. For each element of the set, decide whether it is a unit, a zero divisor, or zero modulo x^2-1.

8. In $\mathbb{F}_5[x]$, decide whether or not there is a complete set of representatives consisting of 0 and powers of x modulo
    (i) x^2+3x+4;
    (ii) x^2+3x+3;
    (iii) x^2+x+4.

9. Show that there is no complete set of representatives consisting of 0 and powers of x modulo $x^4+x^3+x^2+x+1$ in $\mathbb{F}_2[x]$.

10. Find the residue of least degree in $\mathbb{F}_2[x]$ modulo x^4+x+1 of:
    (i) x^5;
    (ii) x^9;
    (iii) x^{13}.

11. In $\mathbb{F}_2[x]$ modulo x^4+x+1,
    (i) Find the inverse of x^3+x;
    (ii) Find the inverse of x^2+x+1;
    (iii) Find the inverse of x^2+x.

12. Let $m(x)=x^3+4x+1$ in $\mathbb{F}_5[x]$. Find polynomials $f(x),g(x)$ in $\mathbb{F}_5[x]$ so that

$$(x-3)f(x) \equiv (x-3)g(x) \pmod{m(x)}$$

but

$$f(x) \not\equiv g(x) \pmod{m(x)}.$$

13. Let p be a prime, $m(x)$ a polynomial in $\mathbb{F}_p[x]$ of degree d. How many elements does a complete set of representatives for $\mathbb{F}_p[x]$ modulo $m(x)$ have?

14. Solve, if possible:
    (i) $(x^3+x+1)f(x) \equiv 1 \pmod{x^4+x+1}$ in $\mathbb{F}_2[x]$;
    (ii) $(2x+1)f(x) \equiv x^3 \pmod{x^2+1}$ in $\mathbb{F}_3[x]$;
    (iii) $x^9 f(x) \equiv 1 \pmod{x^2+2}$ in $\mathbb{Q}[x]$.

15. Prove Proposition 6.

16. (i) Let m be a number >1. Euler's theorem says that for certain integers a and some number $\phi(m)$,

$$a^{\phi(m)} \equiv 1 \quad (\text{mod } m).$$

For which integers a does Euler's theorem apply? Describe the number $\phi(m)$.

(ii) Let p be a prime number, and $m(x)$ be in $\mathbb{F}_p[x]$. The analogue of Euler's theorem is that for certain polynomials $a(x)$ in $\mathbb{F}_p[x]$, and some number $\phi_p(m)$,

$$a(x)^{\phi_p(m)} \equiv 1 \quad (\text{mod } m(x)).$$

For which polynomials $a(x)$ does this theorem apply? What is the number $\phi_p(m)$?

(iii) Let $p = 3$, $m(x) = x^2 + x + 1$ and $a(x) = x + 1$. Find $\phi_p(m)$ and verify that

$$a(x)^{\phi_p(m)} \equiv 1 \quad (\text{mod } m(x)).$$

17. Let $p(x)$ be a polynomial in $F[x]$ of degree 0, F a field. Show that every two polynomials $f(x)$ and $g(x)$ in $F[x]$ are congruent modulo $p(x)$.

18. Let $f(x), g(x)$ be monic polynomials with integer coefficients. Show that $f(x)$ and $g(x)$ agree as functions on $\mathbb{F}_p$, p a prime, iff $f(x) \equiv g(x) \ (\text{mod } x^p - x)$.

B. The Chinese Remainder Theorem

Since nearly all of the properties of congruence for integers are also valid for polynomials, it should be hardly surprising that the Chinese Remainder Theorem is also valid for polynomials. After we present the theorem for polynomials, we show how the Chinese Remainder Theorem relates to interpolation. Interpolation, in turn, yields the theorem, dating from 1793, that every polynomial with integer coefficients can be factored in a finite number of steps.

Here is the general theorem for congruences modulo a collection of pairwise coprime polynomials.

Theorem 7 (The Chinese Remainder Theorem). *Let F be a field. Let $a_1(x),\dots,$ $a_d(x)$ be arbitrary polynomials, and $m_1(x),\dots,m_d(x)$ be pairwise coprime polynomials in $F[x]$. Then there exists a polynomial $f(x)$ in $F[x]$ such that*

$$f(x) \equiv a_1(x) \quad (\text{mod } m_1(x)),$$

$$\vdots \tag{17.1}$$

$$f(x) \equiv a_d(x) \quad (\text{mod } m_d(x)).$$

If $f_1(x)$ and $f_2(x)$ are two solutions, then

$$f_1(x) \equiv f_2(x) \quad (\text{mod } m_1(x) \cdot \dots \cdot m_d(x)).$$

This theorem may be proved in any of the ways we proved the Chinese Remainder Theorem for integers. Here is one proof.

Proof. We first find $h_i(x)$ so that

$$h_i(x) \equiv 0 \pmod{m_j(x)} \text{ for } j \neq i$$
$$h_i(x) \equiv 1 \pmod{m_i(x)}.$$

To find $h_i(x)$, observe that since $m_i(x)$ is coprime to $m_j(x)$ for every $j \neq i$, $m_i(x)$ is coprime to the product

$$l_i(x) = m_1(x)m_2(x) \cdot \ldots \cdot m_{i-1}(x)m_{i+1}(x) \cdot \ldots \cdot m_{d-1}(x)m_d(x).$$

Thus we can solve the equation

$$1 = g_i(x)m_i(x) + k_i(x)l_i(x)$$

for $k_i(x)$ and $g_i(x)$ by Bezout's Lemma. Then $k_i(x)l_i(x) = h_i(x)$ satisfies

$$h_i(x) \equiv 1 \pmod{m_i(x)}),$$
$$h_i(x) \equiv 0 \pmod{m_j(x)}) \text{ for all } j \neq i$$

Once we have the polynomials $h_1(x), h_2(x), \ldots, h_d(x)$, we can solve (17.1) by setting $f(x) = f_0(x)$ where

$$f_0(x) = a_1(x)h_1(x) + a_2(x)h_2(x) + \ldots + a_d(x)h_d(x).$$

As with numbers, if $f_0(x)$ is one solution of (17.1) then any solution $f(x)$ of (17.1) will satisfy

$$f(x) \equiv f_0(x) \pmod{m(x)},$$

where $m(x) = m_1(x) \cdot \ldots \cdot m_d(x)$, since the $m_i(x)$ are pairwise coprime. In particular, there is a unique solution of (17.1) whose degree is less than the degree of $m(x)$. $\square$

As noted in Section 12B, this method of solving (17.1) is particularly useful if there is a need to solve several systems (17.1) with the same moduli $m_1, \ldots, m_d$ but with different choices of $a_1, \ldots, a_d$.

The Chinese remainder theorem has a different interpretation if we apply the Remainder Theorem (Chapter 14) in the following form:

Proposition 8 (Remainder Theorem). *Let $g(x)$ be a polynomial with coefficients in a field F. Then $g(x) \equiv b \pmod{x-a}$ if and only if $g(a) = b$.*

If we choose $d + 1$ distinct elements $n_0, n_1, \ldots, n_d$ of F and any elements $s_0, s_1, \ldots, s_d$ of F, the Chinese Remainder Theorem implies that there exists a polynomial $q(x)$ such that

$$q(x) \equiv s_i \pmod{x - n_i}$$

for $i = 0, \ldots, d$. Applying the Remainder Theorem yields immediately.

Corollary 9 (Interpolation Theorem). *If $n_0, \ldots, n_d$ are distinct elements of F and $s_0, \ldots, s_d$ are arbitrary elements of F, there exists a unique polynomial $q(x)$ in $F[x]$ of degree $\leq d$ such that $q(r_i) = s_i$ for each $i = 0, \ldots, d$.*

We shall show how to write down such a $q(x)$ explicitly in the next section.

Exercises.

19. Solve for $f(x)$ in $\mathbb{F}_2[x]$:

(i)

$$f(x) \equiv x \quad (\text{mod } x^2 + x),$$
$$f(x) \equiv 1 \quad (\text{mod } x^2 + x + 1);$$

(ii)

$$f(x) \equiv x^2 \quad (\text{mod } x^4 + x^2 + x),$$
$$f(x) \equiv x \quad (\text{mod } x^4 + x^3 + 1);$$

(iii)

$$f(x) \equiv x \quad (\text{mod } x^2 + 1),$$
$$f(x) \equiv 1 \quad (\text{mod } x).$$

20. Find a polynomial $q(x)$ of degree <2 in $\mathbb{Q}[x]$ so that $q(1) = 3$ and $q(-1) = 2$.

21. Find a polynomial $q(x)$ of minimal degree in $\mathbb{F}_3[x]$ so that

$$q(x) \equiv x \quad (\text{mod } x^2 + 2)$$
$$q(x) \equiv 1 \quad (\text{mod } x)$$
$$q(x) \equiv x + 1 \quad (\text{mod } x^2 + 2x + 2)$$

22. (Euler's Function for Polynomials). For $m(x)$ in $\mathbb{F}_p[x]$, let $\phi_p(m)$ denote the number of polynomials $f(x)$ in $\mathbb{F}_p[x]$ so that the degree of $\deg f(x) < \deg m(x)$ and $f(x)$ is coprime to $m(x)$.
  (i) Find $\phi_p(m)$ if $m(x)$ is irreducible.
  (ii) Find $\phi_p(m)$ if $m(x) = q(x)^e$ is the power of an irreducible polynomial.
  (iii) Suppose $m(x)$ and $n(x)$ are coprime. Use the Chinese Remainder Theorem for polynomials to show that $\phi_p(mn) = \phi_p(m)\phi_p(n)$.

23. Prove the Chinese Remainder Theorem for polynomials by one of the other proofs given for the Chinese Remainder Theorem for integers in Chapter 12.

C. The Method of Lagrange Interpolation

Corollary 9 of the last section showed that there is a unique polynomial $f(x)$ with real coefficients of degree $\leq d$ whose graph $y = f(x)$ passes through any $d + 1$ specified points with distinct x-coordinates. Finding a polynomial whose graph passes through a given set of points is called interpolation. In this section we show how to use the proof of the Chinese Remainder Theorem to construct such a polynomial. The method is called Lagrange interpolation.

Choose $d+1$ distinct points $n_0,\ldots,n_d$. For each i between 0 and d, we want to find a polynomial $h_i(x)$ with the property that

$$h_i(n_i) = 1$$
$$h_i(n_j) = 0$$

for $j \neq i$. Here is how to construct $h_i(x)$: Let

$$g(x) = (x-n_0)(x-n_1)\cdot\ldots\cdot(x-n_d).$$

Let

$$g_i(x) = \frac{g(x)}{(x-n_i)}.$$

Then $g_i(n_j) = 0$ for $j \neq i$. We set

$$h_i(x) = \frac{g_i(x)}{g'(n_i)} = \frac{g(x)}{(x-n_i)g'(n_i)}$$

where $g'(x)$ is the derivative of $g(x)$. Then for all $j \neq i$, $h_i(n_j) = 0$. As for $h_i(n_i)$, note that by the product rule,

$$g'(x) = (x-n_1)(x-n_2)\cdots(x-n_d)$$
$$+ (x-n_0)(x-n_2)\cdots(x-n_d)$$
$$+\ldots$$
$$+ (x-n_0)(x-n_1)\cdots(x-n_{i-1})(x-n_{i+1})\cdots(x-n_d)$$
$$+\ldots$$
$$+ (x-n_0)(x-n_1)\cdots(x-n_{d-1}),$$

and so

$$g'(n_i) = (n_i-n_0)(n_i-n_1)\cdots(n_i-n_{i-1})(n_i-n_{i+1})\cdots(n_i-n_d) = g_i(n_i).$$

Thus $h_i(n_i) = 1$.

The polynomials $h_i(x)$ all have degree d.

Example 7. For $d = 2$, let n_0, n_1 and n_2 be three distinct real numbers. Then

$$h_0(x) = \frac{(x-n_1)(x-n_2)}{(n_0-n_1)(n_0-n_2)}$$

so $h_0(n_0) = 1, h_0(n_1) = h_0(n_2) = 0$;

$$h_1(x) = \frac{(x-n_0)(x-n_2)}{(n_1-n_0)(n_1-n_2)}$$

so $h_1(n_1) = 1, h_1(n_0) = h_1(n_2) = 0$;

$$h_2(x) = \frac{(x-n_0)(x-n_1)}{(n_2-n_0)(n_2-n_1)}$$

so $h_2(n_2) = 1, h_2(n_0) = h_2(n_1) = 0$.

Once the polynomials $h_0(x), \ldots, h_d(x)$ have been found, then for any vector $\mathbf{s} = (s_0, s_1, \ldots, s_d)$ of numbers, we can find a polynomial $a_{\mathbf{s}}(x)$ of degree $\leq d$ with $a_{\mathbf{s}}(n_0) = s_0, a_{\mathbf{s}}(n_1) = s_1, \ldots, a_{\mathbf{s}}(n_d) = s_d$, by simply setting

$$a_{\mathbf{s}}(x) = s_0 h_0(x) + s_1 h_1(x) + \ldots + s_d h_d(x).$$

Then $a_{\mathbf{s}}(x)$ is a polynomial of degree $\leq d$ that interpolates (whose graph passes through) the points $(n_0, s_0), \ldots (n_d, s_d)$. We call such a polynomial a Lagrange interpolator.

Example 8. Let $n_0 = 1, n_1 = 3, n_2 = 7$, then

$$h_0(x) = \frac{(x-3)(x-7)}{(1-3)(1-7)} = \frac{1}{12}(x-3)(x-7) = \frac{1}{12}(x^2 - 10x + 21)$$

$$h_1(x) = \frac{(x-1)(x-7)}{(3-1)(3-7)} = -\frac{1}{8}(x-1)(x-7) = \frac{-1}{8}(x^2 - 8x + 7)$$

$$h_2(x) = \frac{(x-1)(x-3)}{(7-1)(7-3)} = \frac{1}{24}(x-1)(x-3) = \frac{1}{24}(x^2 - 4x + 3).$$

If $s_0 = 5, s_1 = 17, s_2 = -11$, we let $\mathbf{s} = (5, 17, -11)$. To find $a_{\mathbf{s}}(x)$ of degree ≤ 2 so that

$$a_{\mathbf{s}}(1) = 5, a_{\mathbf{s}}(3) = 17, a_{\mathbf{s}}(7) = -11,$$

we set

$$\begin{aligned}
a_{\mathbf{s}}(x) &= 5h_0(x) + 17h_1(x) - 11h_2(x) \\
&= \frac{5}{12}(x^2 - 10x + 21) - \frac{17}{8}(x^2 - 8x + 7) - \frac{11}{24}(x^2 - 4x + 3) \\
&= -\frac{13}{6}x^2 + \frac{88}{6}x - \frac{45}{6}.
\end{aligned}$$

Then $a_{\mathbf{s}}(x)$ satisfies $a_{\mathbf{s}}(1) = 5, a_{\mathbf{s}}(3) = 17, a_{\mathbf{s}}(7) = -11$.

The polynomial $a_{\mathbf{s}}(x)$ is the unique polynomial of degree $\leq d$ such that $a_{\mathbf{s}}(n_0) = s_0, a_{\mathbf{s}}(n_1) = s_1, a_{\mathbf{s}}(n_2) = s_2, \ldots, a_{\mathbf{s}}(n_d) = s_d$. This follows by the Chinese Remainder Theorem, or we can argue as follows:

Suppose two polynomials $a(x)$ and $b(x)$ of degrees $\leq d$ both have the values $s_0, s_1, \ldots, s_d$ at $n_0, n_1, \ldots, n_d$. Then $c(x) = a(x) - b(x)$ is a polynomial of degree $\leq d$ with the property that $n_0, n_1, \ldots, n_d$ are roots of $c(x)$. By D'Alembert's Theorem (Section 14A), a polynomial of degree $\leq d$ cannot have $d+1$ roots in a field unless the polynomial is the zero polynomial. Thus $c(x) = 0$, and $a(x) = b(x)$.

Exercises.

24. Find a polynomial $f(x)$ of degree ≤ 2 so that $f(1) = 1, f(3) = 2, f(7) = -1$.

25. Find a polynomial $f(x)$ of degree ≤ 2 so that $f(-1) = 1, f(0) = 0, f(1) = 2$.

D. Factoring Polynomials in $\mathbb{Z}[x]$ is a Finite Process

In Chapter 16 we showed that when factoring a polynomial with integer coefficients, we can always assume that the factors have integer coefficients. In this section we use that information to describe a procedure for factoring any polynomial in $\mathbb{Z}[x]$ in a finite number of steps. The method is traditionally attributed to Kronecker, around 1883, but is apparently due originally to Schubert, 1793. Schubert's method obviously predates the computer era, and in recent years other methods have been found that are much faster. But the fact that there is a finite algorithm, however slow, for completely factoring a polynomial in $\mathbb{Z}[x]$ is of at least theoretical interest.

Shubert's method generalizes the idea that there can be only a finite set of possible roots of a polynomial with integer coefficients. Recall from Chapter 16 that if $f(x)$ in $\mathbb{Z}[x]$ has leading coefficient a_d and constant term a_0, then any root of $f(x)$ must be of the form $x = r/s$ where s is a divisor of a_d and r is a divisor of a_0. Since a_d and a_0 both have only a finite number of divisors, there are only a finite number of fractions r/s where the denominator divides a_d and the numerator divides a_0. By checking the possibilities, one either finds a root or proves that none exists.

Finding roots is equivalent to finding factors of $f(x)$ of degree 1, by the Root Theorem.

To generalize this idea, Schubert's method uses interpolation to find possible factors of a polynomial.

Let $p(x)$ in $\mathbb{Z}[x]$ be a polynomial that we wish to factor. Since $p(x)$ has integer coefficients, then for every integer r, $p(r)$ is an integer.

If $p(x)$ has degree m and is not irreducible it has a factor of degree $\leq m/2$. So let $d = m/2$ if m is even, $d = (m-1)/2$ if m is odd.

Let $n_0, \ldots, n_d$ be distinct integers, and let $p(n_0) = r_0, \ldots, p(n_d) = r_d$. Then $r_0, \ldots, r_d$ are integers. For each vector $\mathbf{s} = (s_0, \ldots, s_d)$ of integers so that s_i divides r_i for $i = 0, \ldots, d$, we find the unique polynomial $a_\mathbf{s}(x)$ in $\mathbb{Q}[x]$ of degree $\leq d$ with $a_i(n_i) = s_i$ for all i. Since each r_i has only a finite number of (positive or negative) divisors s_i, there is only a finite number of possible vectors $\mathbf{s} = (s_0, \ldots, s_d)$ and hence only a finite number of polynomials $a_\mathbf{s}(x)$, one for each of the possible vectors $\mathbf{s}$.

It turns out that any divisor $a(x)$ of $p(x)$ in $\mathbb{Z}[x]$ of degree $\leq d$ must be an $a_\mathbf{s}(x)$ for some $\mathbf{s}$. For suppose $a(x)b(x) = p(x)$ for some $b(x)$ in $\mathbb{Z}[x]$. Then for each n_i, $a(n_i)b(n_i) = p(n_i)$ in $\mathbb{Z}$, so $a(n_i)$ divides $p(n_i) = r_i$. Thus the vector $\mathbf{s} = (a(n_0), \ldots, a(n_d))$ is a vector of divisors of $(r_0, r_1, \ldots, r_d)$. Lagrange interpolation gives a unique polynomial $a_\mathbf{s}(x)$ of degree $\leq d$ with $a_\mathbf{s}(n_0) = a(n_0), a_\mathbf{s}(n_1) = a(n_1), \ldots, a_\mathbf{s}(n_d) = a(n_d)$. Thus, since $a(x)$ and $a_\mathbf{s}(x)$ both have degree $\leq d$ and have the same value on $d+1$ elements of $\mathbb{Q}$, they must be equal. We conclude that any divisor $a(x)$ of $p(x)$ of degree $\leq d$ must be among the finite number of polynomials $a_s(x)$ constructed from the finite number of vectors s of divisors of $(r_0, r_1, \ldots, r_d)$.

Call $a_s(x)$ a Lagrange interpolator for $f(x)$. (The Lagrange interpolators also depend on the set of abscissas $n_0, n_1, \ldots, n_d$.)

Having found all possible Lagrange interpolators $a_s(x)$ for $p(x)$, to determine whether $p(x)$ is irreducible or not in $\mathbb{Z}[x]$, we divide $p(x)$ by each of the polynomials $a_s(x)$ to see whether $a_s(x)$ is a divisor of $p(x)$. If some $a_s(x)$ of degree ≥ 1 and $\leq \deg(p(x))/2$ divides $p(x)$, then an explicit factorization of $p(x)$ has been found. Otherwise, $p(x)$ must be irreducible.

From this procedure, we can obtain easily, by induction:

Theorem 10. *The complete factorization of any polynomial in $\mathbb{Z}[x]$ can be achieved in a finite number of steps.*

Example 9. Here is an illustration of how the factoring method works. Let $p(x) = x^4 + x + 1$ in $\mathbb{Z}[x]$. If $p(x)$ factors it must have a factor of degree ≤ 2. (Of course reducing mod 2 we already know it is irreducible!) Now $p(-1) = 1$, $p(0) = 1$, $p(1) = 3$. Thus for each $s = (s_{-1}, s_0, s_1)$ dividing $(1, 1, 3)$, the corresponding Lagrange interpolator $a_s(x)$ is

$$a_s(x) = \frac{s_{-1}}{2}x(x-1) - s_0(x-1)(x+1) + \frac{s_1}{2}x(x+1).$$

Collecting the coefficients of the powers of x gives

$$a_s(x) = \left(\frac{s_1}{2} + \frac{s_{-1}}{2} - s_0\right)x^2 + \left(\frac{s_1}{2} - \frac{s_{-1}}{2}\right)x + s_0.$$

The following table gives all possible vectors $s = (s_{-1}, s_0, s_1)$, and the corresponding polynomials $a_s(k)$:

$s = (s_1, s_0, s_1)$	$a_s(x)$
$(1, 1, 3)$	$x^2 + x + 1$
$(1, 1, 1)$	1
$(1, 1, -3)$	$-2x^2 - 2x + 1$
$(1, 1, -1)$	$-x^2 - x + 1$
$(-1, 1, 3)$	$2x + 1$
$(-1, 1, 1)$	$-x^2 + x + 1$
$(-1, 1, -3)$	$-3x^2 - x + 1$
$(-1, 1, -1)$	$-2x^2 + 1$
$(1, -1, 3)$	$3x^2 + x - 1$
$(1, -1, 1)$	$2x^2 - 1$
$(1, -1, -3)$	$-2x - 1$
$(1, -1, -1)$	$x^2 - x - 1$
$(-1, -1, 3)$	$2x^2 + 2x - 1$
$(-1, -1, 1)$	$x^2 + x - 1$
$(-1, -1, -3)$	$-x^2 - x - 1$
$(-1, -1, -1)$	-1

If $x^4 + x + 1$ factors we can assume all factors have integer coefficients and are primitive. Since $x^4 + x + 1$ is monic, the leading coefficients of any such factors must be ± 1. Now eight of the $a_s(x)$ are primitive but do not have leading coefficient ± 1, so cannot be factors. Two more $a_s(x)$ are uninteresting since they have degree 0. So only six of the $a_s(x)$ must be checked as possible factors of $x^4 + x + 1$:

$$x^2 + x + 1; \quad -x^2 - x - 1;$$
$$x^2 + x - 1; \quad -x^2 - x + 1;$$
$$x^2 - x - 1; \quad -x^2 + x + 1.$$

Since the three on the right are associates of the three on the left, we need only check the three on the left. Three divisions show that none are factors.

Notice that the number of possible factors $a_s(x)$ of $p(x)$ depends on d ($\leq \frac{1}{2} \deg(p)$) but, more significantly, also depends on the number of divisors of $p(n_i)$. The number of possible factors in the example we just did was kept small by the fact that $p(1) = p(0) = 1$, which has only two factors in $\mathbb{Z}$. In general one is not so fortunate, and the number of $a_s(x)$ can become unpleasantly large (see Exercise 30, below).

In recent years a more efficient method of factoring has been developed based on factoring modulo M for an appropriate number M. See Chapter 26.

Exercises.

26. Let $f(x) = x^4 - 7x^2 + 1$. Then $f(-1) = f(1) = -5$, $f(0) = 1$. Show that the Lagrange interpolator $a_s(x)$ divides $f(x)$ for $s = (5, 1, -1)$ and $s = (-1, 1, 5)$.

27. Let $f(x) = x^6 + 2x^5 - x^4 + 2x^2 - x + 1$. Then $f(-2) = -5, f(-1) = 2$, $f(0) = 1, f(1) = 4$. Find the Lagrange interpolator $a_s(x)$ for $s = (-5, 1, 1, 1)$ and verify that $a_s(x)b(x) = f(x)$ for some $b(x)$. To which s does $b(x)$ correspond?

28. Why is it that of the sixteen polynomials $a_s(x)$ arising as Lagrange interpolators for $x^4 + x + 1$, eight are associates of the other eight?

29. For every polynomial $f(x)$ in $\mathbb{Z}[x]$ and every $n_0, \ldots, n_d$, show that $a_s(x) = 1$ and $a_s(x) = -1$ always arise as Lagrange interpolators. Under what circumstances can a constant polynomial $a_s(x) = c$ with $c \neq 1$ or -1 arise as a Lagrange interpolator?

30. If $f(x)$ has degree $2d$, and $n_1, \ldots, n_d$ are distinct integers such that $f(n_i) = p_i$ is prime for each i, $1 \leq i \leq d$, how many polynomials $a_s(x)$ arise as possible divisors of $f(x)$ using Lagrange interpolation? Do they all pair off as associates?

31. Construct a proof of Theorem 10 by complete induction on the degree of the polynomial to be factored.

Chapter 18
Fast Polynomial Multiplication

The development of computers since 1940 has led to new advancements in very old mathematics. In Section 7B we looked at Karatsuba multiplication, a faster way of multiplying numbers. In Section 9F we described a more efficient way to raise a number to a power modulo m. and in Section 9G we presented Montgomery multiplication, a way of avoiding long division modulo m.

In this chapter we look at multiplication of polynomials.

The efficiency of multiplication. Given

$$f(x) = a_0 + a_1 x + a_2 x^2 + \ldots + a_d x^d,$$

and

$$g(x) = b_0 + b_1 x + b_2 x^2 + \ldots + b_d x^d,$$

two polynomials of degree d. The standard way to multiply

$$f(x)g(x) = (a_0 + a_1 x + a_2 x^2 + \ldots + a_d x^d) \cdot (b_0 + b_1 x + b_2 x^2 + \ldots + b_d x^d)$$

is to multiply each $a_i x^i$ by each $b_j x^j$, and then collect the coefficients of each power of x, to get

$$f(x)g(x) = c_0 + c_1 x + c_2 x^2 + \ldots + c_{2d} x^{2d},$$

where

$$c_0 = a_0 b_0,$$
$$c_1 = a_0 b_1 + a_1 b_0,$$
$$c_2 = a_0 b_2 + a_1 b_1 + a_2 b_0,$$

$$\vdots$$

$$c_d = a_0 b_d + a_1 b_{d-1} + \ldots + a_{d-1} b_1 + a_d b_0,$$
$$c_{d+1} = a_1 b_d + a_2 b_{d-1} + \ldots + a_{d-1} b_2 + a_d b_1,$$

$$\vdots$$

$$c_{2d} = a_d b_d.$$

L.N. Childs, *A Concrete Introduction to Higher Algebra*, Undergraduate Texts in Mathematics, © Springer Science+Business Media LLC 2009

A measure of efficiency of this method is to count the number of coefficient multiplications required. If f and g both have degree d, then both have $d + 1$ coefficients, and each coefficient of $f(x)$ is multiplied by every coefficient of $g(x)$. So multiplying two polynomials of degree d in the standard way requires $(d + 1)^2$ number multiplications.

We used digit multiplication as a measure of efficiency for multiplication of integers, and by that measure found that Karatsuba multiplication was significantly more efficient than ordinary multiplication. Presumably we could use Karatsuba for polynomials as well. But in this section we present another method that for large degree polynomials is more efficient than either method.

Multiplying using the Chinese Remainder Theorem. In Section 12B we suggested a way to multiply numbers using the Chinese Remainder Theorem as follows.

To multiply a and b, find a set $q_1, q_2, \ldots, q_s$ of pairwise coprime moduli whose product is $> ab$. Then

Step I. find $a \bmod q_i$ and $b \bmod q_i$ for $i = 1, \ldots, s$;

Step II. Multiply $(a \bmod q_i)$ and $(b \bmod q_i)$ to get $(ab \bmod q_i)$ for $i = 1, \ldots, s$;

Step III. Solve the Chinese Remainder Theorem problem: given $ab \bmod q_i$ for $i = 1, \ldots, s$, find $ab \bmod q_1 q_2 \cdots q_s$.

This same strategy works for polynomials in $\mathbb{C}[x]$.

To multiply $f(x)$ and $g(x)$, pick $2d + 1$ different complex numbers $a_1, \ldots, a_{2d+1}$. Then,

Step I. For $i = 1, \ldots, 2d + 1$, find $f(x)$ and $g(x) \bmod (x - a_i)$. By the Remainder Theorem, $f(x) \equiv f(a_i) \pmod{x - a_i}$, so finding $f(x) \bmod (x - a_i)$ is the same as evaluating $f(x)$ at a_i, and also for $g(x)$.

Step II. Multiply $f(a_i)$ and $g(a_i)$ for $i = 1, \ldots, 2d + 1$.

Step III. Find a polynomial $h(x)$ such that for $i = 1, \ldots, 2d + 1$,

$$h(x) \equiv f(x)g(x) \pmod{x - a_i}.$$

This is the same as finding a polynomial $h(x)$ of degree $\leq 2d$ so that

$$h(a_i) = f(a_i)g(a_i)$$

for $i = 1, \ldots, 2d + 1$. This is an interpolation problem.

To sum up, using the Chinese Remainder Theorem to multiply two polynomials $f(x)$ and $g(x)$ of degree d with coefficients in the complex numbers $\mathbb{C}$, we have three steps: I. Evaluate, II. Multiply, and III. Interpolate.

This seems more complicated than ordinary multiplication. But note that in Step II, we do $2d + 1$ number multiplications, rather than the $(d+1)^2$ number multiplications when we multiply $f(x)$ and $g(x)$ by the standard method.

Thus if there is an efficient way to do Step I, evaluation, and Step III, interpolation, then it is possible that that this strategy will be more efficient than the standard method.

Before looking at the efficiency of I and III, we need to be certain that the $h(x)$ which comes out in III is really $f(x)g(x)$. This follows from Corollary 9 of Section 17B:

Theorem 1. *Let $a_1, \ldots, a_e$ be distinct complex numbers and $h_1, h_2, \ldots, h_e$ be any complex numbers. Then there exists a unique polynomial $h(x)$ of degree $\leq e$ so that $h(a_i) = h_i$ for all i, $i = 1, \ldots, e$.*

Thus if we find a polynomial $h(x)$ of degree $\leq 2d$ so that $h(a_i) = f(a_i)g(a_i)$ for $i = 1, 2, \ldots, 2d + 1$, then since $f(x)g(x)$ is a polynomial of degree $\leq 2d$ with the same values at $a_1, \ldots, a_{2d+1}$ as $h(x)$, then the polynomials $h(x)$ and $f(x)g(x)$ must be equal. That is, the three-step strategy for finding $f(x)g(x)$ really will work.

Example 1. Let $f(x) = x + 1$, $g(x) = x - 2$. We find $f(x)g(x)$ by the three-step strategy. Choose $a_1 = 0, a_2 = 3, a_3 = -1$. Then Steps I and II are quick:

$f(0) = 1$, $g(0) = -2$, so $f(0)g(0) = -2$;
$f(3) = 4$, $g(3) = 1$ so $f(3)g(3) = 4$;
$f(-1) = 0$, $g(-1) = -3$ so $f(-1)g(-1) = 0$;

For Step III we need to interpolate a polynomial $h(x)$ of degree ≤ 2 with

$$h(0) = -2, \quad h(3) = 4, \text{ and } h(-1) = 0.$$

We can accomplish this the same way we finished the process of multiplying numbers–we use Lagrange interpolators. We find

$$h_0(x) = \frac{(x-3)(x+1)}{-3},$$

so that $h_0(0) = 1, h_0(3) = h_0(-1) = 0$;

$$h_3(x) = \frac{x(x+1)}{12},$$

so that $h_3(0) = 0 = h_3(-1), h_3(3) = 1$;

$$h_{-1}(x) = \frac{x(x-3)}{4},$$

so that $h_{-1}(0) = 0 = h_{-1}(3)$ and $h_{-1}(-1) = 1$. Then

$$h(x) = -2h_0(x) + 4h_3(x) + 0h_{-1}(x) = x^2 - x - 2$$

satisfies $h(0) = -2, h(3) = 4$ and $h(-1) = 0$, so must be $f(x)g(x)$.

Returning to the general situation, let $f(x)$ and $g(x)$ each have degree $\leq d$. The key to making the three-step strategy work efficiently is to choose a good set of points $a_1, \ldots, a_{2d+1}$.

Roots of unity. We recall a definition from Section 15D.

Definition. For an integer $f > 0$, an f-th root of unity in a field F is an element ω of F so that $\omega^f = 1$.

Example 2. 1 and -1 are 2nd roots of unity in $\mathbb{R}$.

If p is a prime number, then any unit $[a]$ of $\mathbb{F}_p$ is a $p-1$-st root of unity, since $[a]^{p-1} = [1]$ (by Fermat's theorem).

By the Fundamental Theorem of Algebra, the polynomial $x^f - 1$ has f roots in $\mathbb{C}$. Every root of $x^f - 1$ is an f-th root of unity in $\mathbb{C}$.

If a is a root of unity in F, the order of a is the smallest exponent $e > 0$ so that $a^e = 1$. If the order of a is e, then a is called a *primitive* e-th root of unity.

Example 3. In $\mathbb{Z}$, -1 is a primitive 2nd root of unity.

In $\mathbb{C}$, i is a primitive 4-th root of unity.

In $\mathbb{F}_7$, $[2]$ is a primitive 3rd root of unity.

In $\mathbb{C}$, $\omega = \cos(2\pi/m) + i\sin(2\pi/m)$ is a primitive m-th root of unity in $\mathbb{C}$.

In any ring, if ω is a primitive m-th root of unity, and $(r, m) = 1$, then ω^r is also a primitive m-th root of unity.

Proposition 2. *In $\mathbb{C}$, for any n, there exists a primitive n-th root of unity. In $\mathbb{F}_p$, for any divisor d of $p - 1$, there exists a primitive d-th root of unity.*

Proof. The statement about $\mathbb{C}$ follows from the fact that

$$\omega = \cos(2\pi/n) + i\sin(2\pi/n)$$

is a primitive nth root of unity. The statement about $\mathbb{F}_p$ follows from the fact that $\mathbb{F}_p$ has a primitive element, that is, an element α of order exactly $p - 1$; if $de = p - 1$, then α^e has order d.   □

Evaluation and the Fast Fourier Transform. Let $f(x)$ and $g(x)$ each have degree $\leq d$. Let $2^{r-1} < 2d + 1 \leq 2^r$, and let ω be a primitive 2^rth root of unity in $\mathbb{C}$. In the three-step strategy, we will evaluate $f(x)$ and $g(x)$ at the powers of ω. Here is a crucial fact that makes the strategy efficient:

Theorem 3. *Let ω be a primitive 2^rth root of unity in the field F, and let $f(x)$ be a polynomial in $F[x]$ of degree $d < 2^{r-1}$. Then evaluating $f(x)$ at $1, \omega, \omega^2, \ldots, \omega^{2^r-1}$ requires at most $2^r(r - 1)$ multiplications of elements of F.*

The proof of Theorem 3 describes the algorithm known as the "Fast Fourier Transform." Evaluating the polynomial $f(x)$ at $1, \omega, \omega^2, \ldots, \omega^{2^r-1}$ is the same as applying a discrete Fourier transform to $f(x)$ (we'll give a matrix version in the proof of Theorem 6, below) and the proof shows how to obtain the evaluation quickly.

Proof. We proceed by induction. First, let $r = 2$. Then $2^{r-1} = 2$ and $f(x)$ has degree $d = 1$.

Write $f(x) = a_0 + a_1 x$, with a_0 and a_1 in F, and assume ω is a primitive 4th root of unity. To compute

$$f(1) = a_0 + a_1,$$
$$f(\omega) = a_0 + a_1 \omega,$$
$$f(\omega^2) = a_0 + a_1 \omega^2, \text{ and}$$
$$f(\omega^3) = a_0 + a_1 \omega^3,$$

requires at most four multiplications (in fact, three multiplications: $a_1 \omega, a_1 \omega^2$, and $a_1 \omega^3$).

Now consider $r = 3$, then ω is now a primitive 8th root of unity, and $f(x)$ has degree $\leq 3 < 4 = 2^{3-1}$:

$$f(x) = a_0 + a_1 x + a_2 x^2 + a_3 x^3.$$

To evaluate $f(x)$ at the eight powers $1, \omega, \omega^2, \dots, \omega^7$, we let $y = x^2$ and write

$$f(x) = (a_0 + a_2 x^2) + x(a_1 + a_3 x^2)$$
$$= (a_0 + a_2 y) + x(a_1 + a_3 y),$$
$$= g_0(y) + x g_1(y),$$

where $g_0(y) = a_0 + a_2 y$ and $g_1(y) = a_1 + a_3 y$.

Evaluating $f(x)$ at $1, \omega, \omega^2, \dots, \omega^7$ is the same as evaluating $g_0(y) = g_0(x^2)$ and $g_1(y) = g_1(x^2)$ at $y = 1, \omega^2, \omega^4$, and ω^6, and then multiplying $g_1(x^2)$ by x for $x = 1, \omega, \omega^2, \dots, \omega^7$.

To evaluate $g_0(y) = g_0(x^2)$ at $y = 1, \omega^2, \omega^4$, and ω^6 requires at most four multiplications in F, by the case $r = 2$.

To evaluate $g_1(y) = g_1(x^2)$ at $y = 1, \omega^2, \omega^4$, and ω^6 requires at most four multiplications in F, also by the case $r = 2$.

To multiply $g_1(x^2)$, once evaluated, by x for $x = 1, \omega, \omega^2, \dots, \omega^7$ requires at most 8 multiplications in F (actually at most 7, since multiplying a number by 1 is the same as not multiplying at all).

Thus to evaluate $f(x)$ of degree $3 < 2^2$ at the $2^3 = 8$ powers of a primitive 8th root of unity takes at most $4 + 4 + 8 = 16 = 2^3 \cdot 2$ multiplications in F.

The case for general $r > 2$ is just like the case for $r = 3$.

Suppose by induction that to evaluate a polynomial $g(y)$ of degree $\leq 2^{r-1}$ at all of the powers of a primitive 2^rth root of unity requires at most $M_{r-1} = 2^r(r-1)$ multiplications in F.

Let $f(x)$ be a polynomial of degree $\leq 2^r$. We wish to evaluate it at all of the powers $1, \omega, \omega^2, \dots, \omega^{2^{r+1}-1}$ of a primitive 2^{r+1} th root of unity. As in the case of a polynomial of degree 3, we write $f(x)$ as the sum of its even powers of x, which is a polynomial $g_0(x^2)$, plus the sum of its odd powers of x, which can be written as x times a polynomial $g_1(x^2)$. That is,

$$f(x) = g_0(x^2) + xg_1(x^2).$$

Set $y = x^2$.

To evaluate $g_0(x^2)$ at $x = 1, \omega, \omega^2, \ldots, \omega^{2^{r+1}-1}$ is the same as to evaluate $g_0(y)$ at $y = 1, \omega^2, \omega^4, \ldots, \omega^{2(2^r-1)}$. But ω^2 is a primitive 2^rth root of unity, and we are evaluating $g_0(y)$, a polynomial of degree $\leq 2^{r-1}$, at the powers of ω^2. By the induction hypothesis, this requires at most M_{r-1} multiplications in the field F.

Similarly, at most M_{r-1} multiplications in F are needed to evaluate $g_1(y)$ at $y = 1, \omega^2, \ldots, \omega^{2(2^r-1)}$.

Finally, we need at most 2^{r+1} multiplications in F to multiply $g_1(y) = g_1(x^2)$ by x for $x = 1, \omega, \omega^2, \ldots, \omega^{2^{r+1}-1}$.

Thus the total number of multiplications needed in the field F is at most M_r, where

$$\begin{aligned} M_r &= M_{r-1} + M_{r-1} + 2^{r+1} \\ &= 2^r(r-1) + 2^r(r-1) + 2^{r+1} \quad = 2^{r+1}r \end{aligned}$$

completing the proof by induction.   □

Let us return to our three-step strategy for finding the product $f(x)g(x)$ of two polynomials of degree d.

For Step I, we suppose $2^{r-2} \leq d < 2^{r-1}$, and we evaluate $f(x)$ and $g(x)$ at all the powers of a primitive 2^rth root of unity ω. By the last theorem this takes at most $2 \cdot 2^r(r-1)$ multiplications in F. Now observe that $2^{r-1} \leq 2d < 2^r \leq 4d$, so $r - 1 \leq \log_2 2d$, and we have

$$2 \cdot 2^r(r-1) \leq 8d \log_2 2d.$$

That is,

Corollary 4. *Step I of our strategy for finding the product of two polynomials of degree d requires at most* $8d \log_2 2d$ *multiplications in F.*

This number is much less than $(d+1)^2$ for large d.

We previously observed that Step II of the polynomial multiplication strategy takes only $2d + 1$ multiplications in F, which is much less than $(d+1)^2$ for large d.

Interpolation. We are left with Step III, interpolation.

Before dealing with interpolation, we note a most useful fact about roots of unity.

Proposition 5. *Let* ζ *be an e-th root of unity. Then*

$$1 + \zeta + \zeta^2 + \ldots + \zeta^{e-1} = 0 \text{ if } \zeta \neq 1,$$

and

$$1 + \zeta + \zeta^2 + \ldots + \zeta^{e-1} = e \text{ if } \zeta = 1.$$

Proof. The second equality is obvious. For the first, notice that ζ is a root of the polynomial

$$x^e - 1 = (x - 1)(1 + x + \ldots + x^{e-1})$$

but is not a root of $x - 1$ if $\zeta \neq 1$. Hence ζ must be a root of the polynomial $1 + x + x^2 + \ldots + x^{e-1}$. That is,

$$1 + \zeta + \zeta^2 + \ldots + \zeta^{e-1} = 0.$$

$\square$

Here is the main result about interpolation.

Theorem 6. *Let $e = 2^r$. Let ω be a primitive e-th root of unity in F. Suppose $h(x)$ is a polynomial with coefficients in F of degree $<e$ and let*

$$h(1) = c_0, h(\omega) = c_1, h(\omega^2) = c_2, \ldots, h(\omega^{e-1}) = c_{e-1}.$$

Let $c(x)$ be the polynomial

$$c(x) = c_0 + c_1 x + c_2 x^2 + \ldots + c_{e-1} x^{e-1}.$$

Then we can find the coefficients of $h(x)$ by evaluating $c(x)$, namely:

$$h(x) = \frac{c(1)}{e} + \frac{c(\omega^{e-1})}{e} x + \ldots + \frac{c(\omega^{e-(e-1)})}{e} x^{e-1}.$$

In short, we can *interpolate* a polynomial $h(x)$ such that $h(1) = c_0, h(\omega) = c_1, \ldots, h(\omega^{e-1}) = c_{e_1}$ for given $c_0, c_1, \ldots, c_{e_1}$, by *evaluating* the polynomial $c(x)$ at $1, \omega^{e-1}, \omega^{e-2}, \ldots, \omega$.

Thus, interpolation is as efficient as evaluation.

Proof. It is convenient to use vector and matrix notation. If

$$h(x) = h_0 + h_1 x + h_2 x^2 + \ldots + h_{e-1} x^{e-1},$$

then

$$c_i = h(\omega^i) = h_0 + h_1 \omega^i + h_2 \omega^{2i} + \ldots + h_{e-1} \omega^{(e-1)i}$$

for all i. Thus the row vector

$$\mathbf{C} = (c_0, c_1, \ldots, c_{e-1}) = (h(1), h(\omega), h(\omega^2), \ldots, h(\omega^{e-1}))$$

can be written as $\mathbf{C} = \mathbf{HF}$ where

$$\mathbf{H} = (h_0, h_1, h_2, \ldots, h_{e-1})$$

is the row vector of coefficients of h and $\mathbf{F}$ is the $e \times e$ matrix

$$\mathbf{F} = \begin{pmatrix} 1 & 1 & 1 & \cdots & 1 \\ 1 & \omega & \omega^2 & & \omega^{e-1} \\ 1 & \omega^2 & \omega^4 & & \omega^{2(e-1)} \\ \vdots & & & & \vdots \\ 1 & \omega^k & \omega^{2k} & \cdots & \omega^{k(e-1)} \\ \vdots & & & & \vdots \\ 1 & \omega^{e-1} & \omega^{2(e-1)} & & \omega^{(e-1)(e-1)} \end{pmatrix}$$

called the discrete Fourier transform. (Note that since ω is an eth root of unity, $\omega^{(e-1)(e-1)} = \omega$.)

The inverse of the matrix $\mathbf{F}$ turns out to be the matrix $\frac{1}{e}\hat{\mathbf{F}}$, where the entries of $\hat{\mathbf{F}}$ are the inverses of the entries of $\mathbf{F}$.

The matrix $\hat{\mathbf{F}}$ is called the inverse discrete Fourier transform of $\mathbf{F}$.

To see that $\frac{1}{e}\hat{\mathbf{F}}$ is the inverse of $\mathbf{F}$, notice that the ith row of $\hat{\mathbf{F}}$ is

$$(1, \omega^{-i}, \omega^{-2i}, \ldots, \omega^{-(e-1)i})$$

and the jth column of $\mathbf{F}$ is the transpose of the row vector

$$(1, \omega^{j}, \omega^{2j}, \ldots, \omega^{(e-1)j}).$$

Multiplying the ith row of $\hat{\mathbf{F}}$ with the jth column of $\mathbf{F}$ gives

$$\begin{aligned} q_{ij} &= 1 + \omega^{-i}\omega^{j} + \omega^{-2i}\omega^{2j} + \ldots + \omega^{-(e-1)i}\omega^{(e-1)j} \\ &= 1 + \omega^{j-i} + \omega^{2j-2i} + \ldots + \omega^{(e-1)j-(e-1)i} \\ &= 1 + \omega^{(j-i)} + \omega^{2(j-i)} + \ldots + \omega^{(e-1)(j-i)}. \end{aligned}$$

If we let $\zeta = \omega^{j-i}$, then ζ is an eth root of unity, and so by Proposition 5,

$$q_{ij} = 1 + \zeta + \zeta^2 + \ldots + \zeta^{e-1} = 0$$

for $i \neq j$, while $q_{ii} = e$. Thus $\hat{\mathbf{F}}\mathbf{F} = e\mathbf{I}$ (where $\mathbf{I}$ is the $e \times e$ identity matrix), and so the inverse of $\mathbf{F}$ is $\frac{1}{e}\hat{\mathbf{F}}$.

Now since $\mathbf{C} = \mathbf{HF}$, we have $\mathbf{C}\hat{\mathbf{F}} = e\mathbf{H}$. But this says that:

$$c(1) = eh_0,$$
$$c(\omega^{e-1}) = eh_1,$$
$$c(\omega^{e-2}) = eh_2,$$
$$\vdots$$
$$c(\omega) = eh_{e-1},$$

and so the coefficients of h are obtained by evaluating $c(x)$ at powers of ω.   □

Because we use Theorem 3 in Steps I and III of the algorithm, the entire three-step procedure will be called "FFT polynomial multiplication".

Counting multiplications. We summarize how many multiplications in the field F the three-step FFT multiplication needs:

Theorem 7. *Let $f(x)$ and $g(x)$ have degree $d < 2^{r-1}$. The FFT polynomial multiplication algorithm to find $f(x)g(x)$ requires at most $2^r(3r - 1)$ multiplications in the field F.*

Proof. We count the number of multiplications needed for each step.
   Step I. *Evaluate $f(x)$ and $g(x)$ at the powers of ω, a primitive 2^rth root of unity.* Evaluating $f(x)$ takes $2^r(r - 1)$ multiplications, as does also evaluating $g(x)$.

   Step II. *Multiply $f(\omega^i)g(\omega^i) = h(\omega^i)$ for $i = 0, 1, 2, \ldots, 2^r - 1$.* This consists of 2^r multiplications.

   Step III. *Interpolate $h(x)$ from the $h(\omega^i)$, for $i = 0, 1, \ldots, 2^r - 1$.* This is the same as to evaluate

$$c(x) = h(1) + h(\omega)x + h(\omega^2)x^2 + \ldots + h(\omega^{2^r-1})x^{2^r-1}$$

at $x = 1, \omega^{e-1}, \omega^{e-2}, \ldots, \omega$ where $e = 2^r$. By Exercise 1 below, this takes $2^r r$ multiplications.

Thus the entire FFT strategy involves at most

$$2^r(r - 1) + 2^r(r - 1) + 2^r + 2^r r = 2^r(3r - 1)$$

multiplications in the field F.   □

Suppose $f(x)$ and $g(x)$ have degree $63 = 2^6 - 1$. Then $r = 7$, and the FFT algorithm requires $2^7(20) = 2560$ multiplications. By comparison, the usual algorithm requires $64^2 = 2^{12} = 4096$ multiplications. For $r \geq 7$,

$$2^r(3r - 1) < (d + 1)^2 = (2^{r+1})^2 = 4^{r-1}$$

so for polynomials of $2^{r-1} - 1$ digits, the FFT algorithm is more efficient than the usual algorithm for $r \geq 7$.
   More generally, if $2^{r-2} < d < 2^{r-1}$, then $2^r < 4d$ and

$$2^r(3r - 1) < 2^r 3r < 4d(3\log_2 4d).$$

For $d \geq 103$, $4d(3\log_2 4d)$ is smaller than $(d + 1)^2$, and as d increases, the ratio $4d(3\log_2 4d)/(d + 1)^2$ goes to zero.
   We note that all of this works over $\mathbb{C}$; but if we start with f, g in $\mathbb{Z}[x]$ and choose M to be a prime so large that all coefficients of $f(x)g(x)$ are in absolute value less

than $M/2$ and so that $\mathbb{Z}/M\mathbb{Z}$ has a primitive 2^rth root of unity for $2^r > d$, then we can do all this in $(\mathbb{Z}/M\mathbb{Z})[x]$ and the result will translate immediately to $\mathbb{Z}[x]$.

The process of going from the vector $\mathbf{H}$ of coefficients of the polynomial $h(x)$ to the vector $\mathbf{C} = (h(1), h(\omega), \ldots, h(\omega^{e-1}))$ by multiplying by $\mathbf{F}$, the discrete Fourier transform, can be done by the Fast Fourier Transform method of Theorem 3. That method was published in 1965 by J.W. Cooley of IBM and J.W. Tukey of Princeton University [see Cochran (1967)]. The fast Fourier transform has been called "the most valuable numerical algorithm in our lifetime." See Cipra (May, 1993).

Exercises.

1. Let ω be a primitive 2^r-th root of unity. Show by induction that to evaluate a polynomial of degree $< 2^r$ at $x = 1, \omega, \omega^2, \ldots, \omega^{2^r - 1}$ by the method of Theorem 3 requires at most $2^r \cdot r$ multiplications in the field F.

2. Suppose $f(x)$ and $g(x)$ in $\mathbb{C}[x]$ have degree $\leq d$, every coefficient of $f(x)$ is in absolute value $< B$ and every coefficient of $g(x)$ is in absolute value $< C$. Show that every coefficient of $f(x)g(x)$ is in absolute value $\leq (d+1)BC$.

3. Verify that $d_0 = 103$ is the smallest d_0 such that $4d(3\log_2 4d)$ is smaller than $(d+1)^2$ for all $d > d_0$.

4. Let $F = \mathbb{F}_5$. Then 2 is a primitive 4th root of unity in F. Let $h(x) = 3 + 2x + x^3$ in $F[x]$. Using the method of Theorem 3:
   (i) Evaluate $h(x)$ at $1, 2, 2^2, 2^3$ in F.
   (ii) Let $c(x) = h(1) + h(2)x + h(2^2)x^2 + h(2^3)x^3$. Evaluate $c(x)$ at $x = 1, 2, 2^2, 2^3$.

5. Repeat the last exercise with $h(x) = 4 + 4x + 2x^2 + x^3$.

6. In $\mathbb{F}_{17}$, 2 is a primitive 8th root of unity. Using the method of Theorem 3, evaluate

$$f(x) = 7x^3 + 8x^2 + 3x + 5$$

at the eight powers of 2 in $\mathbb{F}_{17}$. Verify that the method requires at most 16 multiplications in $\mathbb{F}_{17}$.

7. (i) Verify that 5 is a primitive 4th root of unity in $\mathbb{F}_{13}$.
   (ii) Let $\mathbf{F}$ be the 4×4 matrix whose (i, j)th entry is 5^{ij} in $\mathbb{F}_{13}$ for $i, j = 0, 1, 2, 3$. Compute $\hat{\mathbf{F}}$ and verify that $\mathbf{F}\hat{\mathbf{F}} = \mathbf{I}$.

8. If we wish to multiply two polynomials f and g in $\mathbb{Z}[x]$, we could work in $\mathbb{Z}/p\mathbb{Z}$ where p is a prime so large that all coefficients of $f \cdot g$ are $\leq p/2$, and so that $\mathbb{Z}/p\mathbb{Z}$ has a primitive 2^rth root of unity for large r. For example, if the Fermat number $F(r) = 2^{2^r} + 1$ were prime, then $\mathbb{Z}/F(r)\mathbb{Z}$ would have a primitive 2^{2^r}th root of unity by the Primitive Root Theorem. (Of course no Fermat numbers are known to be prime for $r > 5$: see Section 10B.) Almost as good would be prime numbers of the form $p = a \cdot 2^b + 1$, where b is large, such as $3 \cdot 2^6 + 1 = 193$: $\mathbb{Z}/193\mathbb{Z}$ has a primitive 64th root of unity.

Can you find a prime number of the form $a \cdot 2^b + 1$ with $b > 16$? For how large an exponent b can you find a prime number of that form?

9. Use Dirichlet's Theorem on primes in an arithmetic progression to show that for every r, there is a prime p so that $\mathbb{F}_p$ contains a primitive 2^r-th root of unity.

10. For polynomials $f(x), g(x)$ of degree $d = 2^{r-1} - 1$, check that multiplying $f(x)$ and $g(x)$ by the Karatsuba method (Section 7B) requires 3^{r-1} multiplications in the field F.

11. What is the smallest r so that for polynomials $f(x), g(x)$ of degree $d = 2^{r-1} - 1$ the FFT method uses fewer multiplications than the Karatsuba method?

Part V
Primitive Roots

Chapter 19
Cyclic Groups and Cryptography

In this chapter we prove the Primitive Root Theorem, which says that if p is prime, there is a unit b of order $p - 1$ modulo p. This is equivalent to the statement that the group U_p of units of $\mathbb{Z}/p\mathbb{Z}$ is a cyclic group. Then we determine all numbers m for which U_m is a cyclic group, and conclude with a look at discrete logarithm cryptography.

A. The Exponent of an Abelian Group

Recall (from Section 11C) that for G a finite group, the *order* of G is defined to be the number of elements of G.

Let G be a finite abelian group of order g, with operation multiplication and with identity element e. Then for every a in G, $a^g = e$, by Lagrange's theorem. The *order* of a is the smallest number $d > 0$ so that $a^d = e$. Then, just as with units modulo m, we have:

Proposition 1. *In a finite abelian group G,*
  (i) Every a in G has an order.
  (ii) If d is the order of a, and m is any number with $a^m = e$, then d divides m.
  (iii) The order of a divides g, the order of G;
  (iv) If d is the order of a, then a^r has order $d/(r,d)$, where (r,d) is the greatest common divisor of r and d.

The proofs are the same as for units modulo m in Chapter 9, and are left as exercises.
  Since the order of every element of G divides g, the set of numbers that are orders of elements of G is a finite set of numbers.

Definition. The *exponent* λ of a finite abelian group G is the number that is maximal among all orders of elements of G.

L.N. Childs, *A Concrete Introduction to Higher Algebra*, Undergraduate Texts in Mathematics, © Springer Science+Business Media LLC 2009

To rephrase the definition of λ:

The exponent λ is the order of some element of G; and for every element b of G, if h is the order of b, then $h \leq \lambda$.

Example 1. Let $G = U_{15}$, the group of units of $\mathbb{Z}/15\mathbb{Z}$. Then

$$G = \{[1], [2], [4], [7], [8], [11], [13], [14]\}$$

has order 8, so for every $[a]$ in U_{15}, $[a]^8 = [1]$. But no element of U_{15} has order 8. The orders of the elements of G are as follows:

element	order
[1]	1
[2]	4
[4]	2
[7]	4
[8] = [−7]	4
[11] = [−4]	2
[13] =[−2]	4
[14] =[−1]	2

Therefore, the exponent of G is 4.

The main theorem of this section is:

Theorem 2. *Let λ be the exponent of a finite abelian group G. Then the order of every element b of G divides λ.*

To prove this theorem we need one more fact about orders, beyond facts (i)-(iv) above:

Proposition 3. *Let a, b be elements of a finite abelian group. If a has order r, and b has order s, and $(r, s) = 1$, then ab has order rs.*

Proof of Proposition 3. First note that $(ab)^{rs} = a^{rs}b^{rs} = e$, so the order of ab is $\leq rs$.

Now, let $d > 0$ so that $(ab)^d = e$. Then

$$e = (ab)^{dr} = a^{dr}b^{dr} = e^d b^{dr} = b^{dr}$$

since $a^r = e$. Since the order of b is s, therefore, by (ii) of Proposition 1 , s divides dr. But $(r, s) = 1$, so s divides d. Similarly,

$$e = (ab)^{ds} = a^{ds}b^{ds} = a^{ds}e^d = a^{ds}$$

since $b^s = e$. But r is the order of a, and so r divides ds. Since $(r, s) = 1$, it follows that r divides d. Hence d is a common multiple of r and s, hence is a multiple of $[r, s] = rs$.

Since $(ab)^{rs} = e$, and every $d > 0$ with $(ab)^d = e$ is a multiple of rs, therefore the order of ab is rs.   □

Now for the proof of Theorem 2:

Proof. Let b be an element of G, and let m be the order of b. We know that $m \leq \lambda$. We must show that m divides λ. By definition, λ is the order of some element a of G.

If m does not divide λ, there must be some prime number p so that a higher power of p divides m than divides λ. We'll use this assumption to find an element of G whose order is greater than λ, contradicting the definition of λ.

Suppose p^r is the highest power of p that divides m, and p^s is the highest power of p which divides λ, where $r > s$.

Since b in G has order m, then

$$d = b^{m/p^r}$$

has order p^r by property (iv) of Proposition 1. Since a in G has order λ, then

$$c = a^{p^s}$$

has order λ / p^s, again by property (iv).

But p^r and λ / p^s are coprime, since p^s is the highest power of p dividing λ. So by Proposition 3, the element cd of G has order $(\lambda / p^s)p^r = \lambda p^{r-s}$, which is larger than λ. This violates the assumption that λ is the exponent of G.

Hence the order m of b must divide λ. We have therefore shown that the order of every element of G divides the exponent λ, and the proof is complete.   □

Corollary 4. *If λ is the exponent of a finite abelian group G, then $a^\lambda = e$ for every a in G.*

This follows immediately from property (ii) of Proposition 1 and from Theorem 2. This result yields

Theorem 5 (The Primitive Root Theorem). *For every prime p there exists a primitive root modulo p.*

Proof. We know that $\mathbb{Z}/p\mathbb{Z}$ is a field, and the group of units U_p of $\mathbb{Z}/p\mathbb{Z}$ has order $p - 1$. By D'Alembert's Theorem (Chapter 14), a polynomial of degree d cannot have more than d roots in a field. If the exponent of U_p is λ, then λ divides $p - 1$ by Proposition 1, (iii). Also, by the last corollary, every element of U_p is a root of the polynomial $x^\lambda - 1$. Since U_p has $p - 1$ elements, therefore $\lambda \geq p - 1$. Therefore $\lambda = p - 1$. Hence there is an element $[b]_p$ of U_p of order $p - 1$, and b is a primitive root modulo p.   □

Proposition 3 can help in finding a primitive root modulo p.

For example, modulo 13, it is easy to see that $3^3 = 27 \equiv 1 \pmod{13}$, and $5^2 = 25 \equiv -1 \pmod{13}$, hence 3 has order 3, and 5 has order 4 modulo 13. Hence $3 \cdot 5 = 15 \equiv 2$ has order 12 modulo 13.

The exercises have some other examples.

Theorem 5 generalizes easily:

Theorem 6. *Every finite subgroup of the multiplicative group of non-zero elements of a field is cyclic.*

Proof. Let U be a finite subgroup of the units of a field F. Suppose U has n elements and exponent λ. Then $\lambda \leq n$ by Proposition 1, (iii), and $a^\lambda = 1$ for all a in U. Thus the polynomial $x^\lambda - 1$ has n roots in F. By D'Alembert's Theorem, $n \leq \lambda$. Thus $n = \lambda$, so U has an element of order n, hence is cyclic. $\square$

Corollary 7. *The multiplicative group of units of a finite field is cyclic.*

This generalizes the Primitive Root Theorem from $\mathbb{Z}/p\mathbb{Z}$ to every finite field. We'll see many examples of new finite fields later in the book.

Exercises.

1. Find the exponent of:
   (i) U_8;
   (ii) U_9;
   (iii) U_{10}.

2. Find the exponent λ of G and verify that $a^\lambda = 1$ for all a in G, where $G =$
   (i) U_{14};
   (ii) U_{16};
   (iii) U_{20}.

3. Prove (i) of Proposition 1, that if G is a finite group of order n, then every a in G has an order that is $\leq n$.

4. Let a be an element of of order d in a finite abelian group. Show (ii) of Proposition 1, that if m is any number with $a^m = e$, then d divides m.

5. Prove (iii) of Proposition 1, that for every finite abelian group G and every element a of G, the order of a divides g, the order of G.

6. Let G be a finite abelian group. Prove (iv) of Proposition 1, that if a is an element of G and d is the order of a, then a^r has order $d/(r,d)$, where (r,d) is the greatest common divisor of r and d.

7. What is the exponent of the group (under addition) $\mathbb{Z}/m\mathbb{Z}$?

8. Recall that $[r,s] = rs$ if and only if $(r,s) = 1$. Suppose you wish to generalize Proposition 3 to:
   If a has order r and b has order s, then ab has order $[r,s]$.
Show that this proposed generalization is false.

9. In $\mathbb{Z}/31\mathbb{Z}$,
  (i) find an element of order 5;
  (ii) find an element of order 10;
  (iii) show that 5 has order 3 modulo 31;
  (iv) then, using Proposition 3, find a primitive root modulo 31.

10. Modulo 61, show that
  (i) $11^2 \equiv -1$;
  (ii) $3^5 \equiv -1$;
  (iii) $13^3 \equiv 1$.
  Then find a primitive root modulo 61.

11. It is true that 5 has order 25 modulo 401, and 30 has order 16 modulo 401. Without computing anything, show that 6 has order $400 = 25 \cdot 16$ modulo 401.

12. Observe that $33^2 = 1089 \equiv -1 \pmod{109}$. It is a fact that 3 has order 27 modulo 109. Show that 11 is a primitive root modulo 109 with no further computations.

13. Let $m = p_1 p_2 \cdots p_r$ be a product of distinct primes, and let U_m be the group of units of $\mathbb{Z}/m\mathbb{Z}$. Let e be the least common multiple of $p_1 - 1, p_2 - 1, \ldots, p_r - 1$.
  (i) Show that the order of every element a of U_m divides e.
  (ii) Explain how to use the Chinese Remainder Theorem to find an element of U_m of order e.

B. Finite Cyclic Groups

We recall the definition of cyclic group from Section 11B.
  Suppose G is a group with operation $*$ and identity e, and let a be an element of G of order d. If G is finite, then the subset of G consisting of all non-negative powers of a,

$$\langle a \rangle = \{e, a, a^2, a^3, \ldots, a^r, \ldots,\}$$

is the cyclic subgroup of G generated by a.
  The word "cyclic" comes from the idea that if a has order d, then $a^d = e$, so

$$\langle a \rangle = \{e, a, a^2, a^3, \ldots, a^{d-1}\},$$

and the powers of a repeatedly cycle through the elements of $\langle a \rangle$ as follows:

$$
\begin{array}{ccc}
e & a & \cdots & a^{d-1} \\
a^d & a^{d+1} & \cdots & a^{2d-1} \\
a^{2d} & a^{2d+1} & & a^{3d-1} \\
\vdots & & & \vdots
\end{array}
$$

where the elements in each column are all equal to each other.

If a group G is generated by a, then a is called a *generator* of G. The same group may have several generators, as we'll see.

Some examples of cyclic subgroups:

Example 2. In U_{14}, the group of units of $\mathbb{Z}/14\mathbb{Z}$, let's look at the cyclic subgroups generated by each of the elements of U_{14} (we denote $[a]_{14}$ by a):

$$\langle 1 \rangle = \{1\}$$
$$\langle 3 \rangle = \{3, 9, 13, 11, 5, 1\}$$
$$\langle 5 \rangle = \{5, 11, 13, 9, 3, 1\}$$
$$\langle 9 \rangle = \{9, 11, 1\}$$
$$\langle 11 \rangle = \{11, 9, 1\}$$
$$\langle 13 \rangle = \{13, 1\}.$$

Thus U_{14} has four cyclic subgroups, $\langle 1 \rangle, \langle 13 \rangle, \langle 9 \rangle = \langle 11 \rangle$, and $\langle 3 \rangle = \langle 5 \rangle = U_{14}$. In particular, U_{14} is a cyclic group and $[3]$ and $[5]$ are generators of U_{14}.

Definition. A finite group G of order n is *cyclic* if there is an element a of G so that the cyclic subgroup of G generated by a is all of G.

It follows from Theorem 2 that a finite group G of order n is cyclic if and only if there is an element a of G of order n, if and only if the exponent of G is equal to the order of G.

Many of the groups we have seen thus far in the book are cyclic groups. For example:

For every m, $\mathbb{Z}/m\mathbb{Z}$ with the operation $+$ and identity element 0 is a cyclic group, generated by $[1]_m$, the congruence class of the number 1.

If p is prime and b is a primitive root modulo p, then $\langle [b]_p \rangle = U_p$, so U_p is a cyclic group.

U_8 is not a cyclic group, since $U_8 = \{[1], [3], [-3], [-1]\}$, and all four elements have order 1 or 2.

Here is a useful characterization of a cyclic group. Assume G has operation $*$ and identity e, and let a^r denote $a * a * \ldots * a$ (r factors).

Proposition 8. *A finite abelian group G is cyclic if and only if for every integer $r > 0$, the equation $x^r = e$ has at most r solutions in G.*

Here is a proof that if G is not cyclic, then there is some $r > 0$ so that the equation $x^r = e$ has more than r solutions in G:

Proof. We proved in Theorem 2, Section A above, that if λ is the exponent of G, then $g^\lambda = e$ for every element g of G. Suppose that G is not cyclic. Then G has order n and has no element of order n. So the exponent $\lambda < n$. Since every element g of G satisfies $g^\lambda = e$, it follows that $x^\lambda = e$ has more than λ solutions in G. $\qquad\square$

The converse follows from the more precise result:

Proposition 9. *If $G = \langle b \rangle$ is a cyclic group of order n, then for every $r > 0$, the equation $x^r = e$ has exactly $d = (n, rd)$ solutions.*

Proof. For $1 \leq s \leq n$, $(b^s)^r = e$ if and only if n divides sr, if and only if n/d divides sr/d Since $(\frac{n}{d}, \frac{r}{d}) = 1$, $(b^s)^r = e$ if and only if s is a multiple of n/d. Thus the solutions of $x^r = e$ are $x = b^{\frac{n}{d}k}$ for $1 \leq k \leq d$. $\qquad\square$

As an immediate corollary, we have

Corollary 10. *If G is a cyclic group, then every subgroup of G is cyclic.*

Proof. For every $r > 0$, the equation $x^r = e$ has at most r solutions in G, and hence has at most r solutions in each subgroup of G. Hence every subgroup of G must be cyclic. $\qquad\square$

We can determine all of the subgroups of a finite cyclic group G by the following result:

Proposition 11. *Let G be a cyclic group of order n and let b be a generator of G. For every divisor d of n, the cyclic subgroup $\langle b^d \rangle$ generated by b^d is a subgroup of G of order n/d. Let H be a subgroup of G. If d is the least number ≥ 1 so that b^d is in H, then $H = \langle b^d \rangle$.*

Proof. Let $n = dq$. Then

$$\langle b^d \rangle = \{b^d, b^{2d}, \ldots, b^{qd} = b^n = e\}$$

is a cyclic subgroup of G of order $n/d = q$. Now let H be a subgroup of G, and let d be the least positive exponent so that b^d is in H. Since G is cyclic, every element of H is a power of b. Let b^s be in H. We show d divides s. For by the Division Theorem, $s = dq + r$ for some numbers q and r with $0 \leq r < d$. Then $b^r = b^s * (b^{dq})^{-1}$ is in H. If $r > 0$, then b^d would not be the least power of b in H. So $r = 0$ and d divides s. Hence $H = \langle b^d \rangle$. $\qquad\square$

An easy consequence of this proposition is that if G is cyclic of order n, then there is a one-to-one correspondence between subgroups of G and divisors of n. If b is a generator of G, then corresponding to the divisor d of n is the subgroup of order d generated by $b^{n/d}$.

Example 3. The group of units modulo 19 is a cyclic group of order 18, generated by 2 mod 19. The divisors of 18 are 1, 2, 3, 6, 9 and 18. The cyclic subgroups of orders 18, 9, 6, 3, 2, and 1, respectively, are generated by $2, 2^2 = 4, 2^3 = 8, 2^6 \equiv 7, 2^9 \equiv -1$ and $2^{18} \equiv 1 \pmod{19}$.

Exercises.

14. Show that a finite abelian group G is cyclic if and only if the exponent of $G =$ the order of G.

15. Show that the group U_{10} of units of $\mathbb{Z}/10\mathbb{Z}$ is cyclic. Find a generator of U_{10}.

16. Show that the group U_{25} of units of $\mathbb{Z}/25\mathbb{Z}$ is cyclic. Find a generator of U_{25}.

17. Show that the group U_{49} of units of $\mathbb{Z}/49\mathbb{Z}$ is cyclic. Find a generator of U_{49}.

18. Find the order and the exponent of U_{15}.

19. Find the order and the exponent of U_{28}.

20. Find the order and the exponent of U_{24}. Find all the cyclic subgroups of U_{24}.

21. Find all possible generators of the cyclic group $\mathbb{Z}/n\mathbb{Z}$ under addition.

22. If p is prime, then U_p is a cyclic group because there is a primitive root modulo p. Show that U_p has $\phi(p-1)$ primitive roots.

23. Show that $\mathbb{F}_9 = \{a+bi | a,b$ in $\mathbb{Z}/3\mathbb{Z}, i^2 = -1\}$ is not a cyclic group under addition.

24. Let $M_2(\mathbb{Z}/2\mathbb{Z})$ denote the ring of 2×2 matrices with entries in $\mathbb{Z}/2\mathbb{Z}$. Is $M_2(\mathbb{Z}/2\mathbb{Z})$ a cyclic group under addition? Is the group of invertible elements of $M_2(\mathbb{Z}/2\mathbb{Z})$ cyclic?

C. Primitive Roots Modulo p^e

Here is the main result of this section

Theorem 12. *For every odd prime p and every $e > 1$, there is a primitive root modulo p^e.*

This means: there is some number b so that every number coprime to p is congruent modulo p^e to a power of b.

In other terminology,

Corollary 13. *For every odd prime p and every exponent $e > 0$, U_{p^e} is a cyclic group.*

Example 4. We see that 2 is a primitive root modulo 9: 2 has order 6 (mod 9), and

$$U_9 = \{[1],[2],[4],[5],[7],[8]\} = \{[1],[2],[2^2],[2^5],[2^4],[2^3]\}.$$

For the proof of the theorem, we need a consequence of the Binomial Theorem.

Lemma 14. *Let p be an odd prime and r be coprime to p. Then for each $k \geq 0$,*
$(1+pr)^{p^k} = 1 + p^{k+1}s$ *for some number s coprime to p.*

Proof of Lemma. The lemma is obvious if $k = 0$. For $k = 1$ we have

$$(1+pr)^p = 1 + \binom{p}{1}pr + \binom{p}{2}p^2r^2 + \cdots$$

by the Binomial Theorem, where, since $p \geq 3$, all of the terms not written out are
multiples of p^3. Since $\binom{p}{1} = p$, and $\binom{p}{2} = \frac{p(p-1)}{2}$, a multiple of p, we have $(1 + pr)^p \equiv 1 + p^2r \pmod{p^3}$, and so

$$(1+pr)^p = 1 + p^2r + p^3t = 1 + p^2(r+pt)$$

for some integer t. Then $r + pt$ is coprime to p because r is. Thus the lemma is true
for $k = 1$.

Suppose for some $k \geq 1$, we have $(1 + pr)^{p^k} = 1 + p^{k+1}s$ for some s coprime to
p. Then

$$(1+pr)^{p^{k+1}} = (1+p^{k+1}s)^p$$

$$= 1 + \binom{p}{1}p^{k+1}s + \binom{p}{2}(p^{k+1}s)^2 \pmod{p^{k+3}}$$

since $(p^{k+1})^m$ is a multiple of p^{k+3} for $m \geq 3$. Now $\binom{p}{2}(p^{k+1}s)^2$ is a multiple of
$p^{2(k+1)+1}$ and $2(k+1)+1 \geq k+3$, so

$$(1+pr)^{p^{k+1}} \equiv 1 + \binom{p}{1}p^{k+1}s = 1 + p^{k+2}s \pmod{p^{k+3}}.$$

Thus

$$(1+pr)^{p^{k+1}} = 1 + p^{k+2}t$$

where $t \equiv s \pmod{p}$, and hence t is coprime to p.

The lemma is therefore proved by induction.   □

Proof of Theorem 12. Let b be a primitive root modulo p. Let d be the order of b
modulo p^e. The group of units U_{p^e} has order $p^{e-1}(p-1)$, so d divides $p^{e-1}(p-1)$.
Now $b^d \equiv 1 \pmod{p^e}$, so $b^d \equiv 1 \pmod{p}$. Since b is a primitive root modulo p,
the order of b modulo p is $p-1$, and so $p-1$ must divide d. Thus the order modulo
p^e of b must be $p^l(p-1)$ for some l, $0 \leq l \leq e-1$.

We first do the case $e = 2$.

Let b be a primitive root modulo p. If b has order $p(p-1)$, then b is a primitive
root modulo p^2. Otherwise, b has order $p-1$ modulo p^2. In that case, consider
$b+p$, which also has order $p-1$ modulo p. We compute $(b+p)^{p-1}$ by the Binomial
Theorem and look at the result modulo p^2:

$$(b+p)^{p-1} \equiv b^{p-1} + (p-1)b^{p-2}p$$

$$\equiv 1 - pb^{p-2} \pmod{p^2}.$$

Thus $(b+p)^{p-1} \not\equiv 1 \pmod{p^2}$, and hence must have order $p(p-1)$ modulo p^2. So if b is not a primitive root modulo p^2, then $b+p$ is a primitive root modulo p^2. In either case, $\mathbb{Z}/p^2\mathbb{Z}$ has a primitive root.

Now assume b is a primitive root modulo p^2, that is, the order of b modulo p^2 is $p(p-1)$. We show that b is then a primitive root modulo p^e for all $e > 2$.

We know that b has order $(p-1)$ modulo p, so $b^{p-1} = 1 + pr$ for some number r. Since b has order $p(p-1)$ modulo p^2, $b^{p-1} \not\equiv 1 \pmod{p^2}$, and so r is coprime to p. We showed that b has order $(p-1)p^l$ modulo p^e, for some $1 \le l < e$. We claim $l = e - 1$. For by Lemma 14, since $(r, p) = 1$,

$$(b^{(p-1)})^{p^{e-2}} = (1 + rp)^{p^{e-2}} = 1 + sp^{e-1} \not\equiv 1 \pmod{p^e},$$

and therefore the order of b cannot be less than $(p-1)p^{e-1}$. Hence b is a primitive root modulo p^r for all $r > 2$. □

Exercises.

25. Find a primitive root modulo 27.

26. Find a primitive root modulo 625.

27. Find a primitive root modulo 49.

28. Let p be an odd prime and $e > 1$. Show that if b is a primitive root $(\bmod\ p^e)$, then b is a primitive root $(\bmod\ p)$.

29. Let p be an odd prime, and a be a number $< p$. Show that among the p numbers b with $0 < b < p^2$ and $b \equiv a \pmod{p}$, exactly one satisfies $b^{p-1} \equiv 1 \pmod{p^2}$.

D. The Exponent of U_m

In this section we find all moduli m for which there is a primitive root modulo m. There is a primitive root modulo m if and only if the group of units U_m modulo m is cyclic.

Recall that the exponent of a finite abelian group G is the largest number λ so that λ is the order of some element of G. If G has order n, then G is cyclic if and only if there is an element a of G of order n, if and only if the order n of G is also the exponent of G. Thus given a modulus m, to decide if U_m, the group of units of $\mathbb{Z}/m\mathbb{Z}$, is cyclic, we can compare the exponent $\lambda(m)$ of U_m with $\phi(m)$, the order of U_m. Then U_m is cyclic if and only if $\lambda(m) = \phi(m)$.

To compute $\lambda(m)$, we have

Proposition 15. *Let $m = rs$ with r, s coprime. Then $\lambda(m)$ is the least common multiple of $\lambda(r)$ and $\lambda(s)$.*

Compare this result to that for $\phi(m)$: if $m = rs$ with r, s coprime, then $\phi(m) = \phi(r)\phi(s)$.

To prove the proposition, we need

Lemma 16. *Let b be a unit modulo m. If $m = rs$ with r, s coprime, then the order of b modulo m is the least common multiple of the order of b modulo r and the order of b modulo s.*

Proof of lemma. Suppose the integer b has order e (mod r) and has order f (mod s). If $d = [e, f]$ is the least common multiple of e and f, then e and f divide d, so

$$b^d \equiv 1 \quad (\text{mod } r)$$

and

$$b^d \equiv 1 \quad (\text{mod } s),$$

hence

$$b^d \equiv 1 \quad (\text{mod } m).$$

On the other hand, if $b^g \equiv 1$ (mod m) for some g, then

$$b^g \equiv 1 \quad (\text{mod } r)$$

so e divides g, and

$$b^g \equiv 1 \quad (\text{mod } s)$$

so f divides g. Hence g is a common multiple of e and f, hence is divisible by $d = [e, f]$. Thus d is the minimal exponent $g > 0$ so that $b^g \equiv 1$ (mod m), so d is the order of b modulo m. $\square$

Proof of Proposition 15. Let b and c be integers so that b has order $\lambda(r)$ (mod r), and c has order $\lambda(s)$ (mod s). By the Chinese Remainder Theorem we can find an integer a so that

$$a \equiv b \quad (\text{mod } r)$$
$$a \equiv c \quad (\text{mod } s).$$

Then a is a unit modulo m and the order of a modulo m is $[\lambda(r), \lambda(s)]$. So $[\lambda(r), \lambda(s)]$ divides the exponent $\lambda(m)$ of U_m by Theorem 2 of Section A.

Conversely, let a have order $\lambda(m)$ modulo m. If a has order e modulo r, then e divides $\lambda(r)$. If a has order f modulo s, then f divides $\lambda(s)$. By Lemma 16, the order of a modulo $m = rs$ is the least common multiple of e and f. So $\lambda(m) = [e, f]$. Since e divides $\lambda(r)$ and f divides $\lambda(s)$, $[e, f] = \lambda(m)$ divides $[\lambda(r), \lambda(s)]$. Hence $\lambda(m) = [\lambda(r), \lambda(s)]$. $\square$

An easy induction argument on the number of prime power factors of m describes the exponent of m as the least common multiple of the exponents of the prime power factors of m:

Corollary 17. *Let* $m = 2^r p_1^{e_1} p_2^{e_2} \cdot \ldots \cdot p_g^{e_g}$. *Then*

$$\lambda(m) = [\lambda(2^r), \phi(p_1^{e_1}), \phi(p_2^{e_2}), \ldots, \phi(p_g^{e_g})],$$

To complete the results needed to compute $\lambda(m)$, we need to deal with the prime 2.

Proposition 18. *For* $r \geq 3$, $\lambda(2^r) = 2^{r-2}$

Proof. First observe that 5 has order 2 modulo $8 = 2^3$, and has order $4 = 2^2$ modulo $16 = 2^4$, because $5^2 \equiv 9 \pmod{16}$ while

$$5^{2^2} = 625 = 1 + 39 \cdot 2^4 \pmod{16}.$$

In general, for $k \geq 4$, if

$$5^{2^{k-2}} = 1 + 2^k r$$

with r odd, then

$$\begin{aligned} 5^{2^{k+1}} &= (1 + 2^k r)^2 \\ &= 1 + 2(2^k r) + (2^k r)^2 \\ &= 1 + 2^{k+1}(r + 2^{k-1} r^2) \\ &= 1 + 2^{k+1} s \end{aligned}$$

with s an odd number. Hence, by induction, 5 has order $2^{r-2} \mod 2^r$ for all $r \geq 4$.

Since the order of U_{2^r} is $\phi(2^r) = 2^{r-1}$, the exponent $\lambda(2^r)$ of U_{2^r} must be either 2^{r-2} or 2^{r-1}. If $\lambda(2^r) = 2^{r-1}$, then U_{2^r} would be a cyclic group. But notice that the polynomial $x^2 - 1$ has four roots in U_{2^r}, namely, 1, -1, $1 + 2^{r-1}$ and $-1 + 2^{r-1}$, and if $r \geq 3$, these are all different modulo 2^r. Thus by Proposition 8, U_{2^r} cannot be cyclic, and so $\lambda(2^r) = 2^{r-2}$ for $r \geq 3$. □

To decide for which m there is a primitive root modulo m, the following result is helpful:

Corollary 19. *For all* $m > 2$, $\lambda(m)$ *is an even number.*

Proof. The result is true for $m = 2^r$ when $r \geq 2$.

If m is divisible by an odd prime p, write $m = p^e q$ with $(p, q) = 1$. Then $\lambda(m) = [\lambda(p^e), \lambda(q)]$ is a multiple of $\lambda(p^e) = \phi(p^e) = p^e(p-1)$, an even number. So $\lambda(m)$ is even. □

Theorem 20. *The group* U_m *of units of* $\mathbb{Z}/m\mathbb{Z}$ *is cyclic if and only if* $m = 2$ *or* 4, *or* $m = p^e$ *or* $2p^e$ *for some odd prime* p.

Proof. U_m is a cyclic group iff $\lambda(m) = \phi(m)$. We know that $\lambda(m) = \phi(m)$ if $m = 2, 4$ or p^e for p an odd prime. Since for p an odd prime,

$$\lambda(2p^e) = [\lambda(2), \lambda(p^e)] = [1, \lambda(p^e)] = \lambda(p^e),$$

$$\phi(2p^e) = \phi(2)\phi(p^e) = \phi(p^e)$$

and $\lambda(p^e) = \phi(p^e)$, we have $\lambda(2p^e) = \phi(2p^e)$.

Now we consider all other possible values of m.

We showed above that for $r \geq 3$, $\lambda(2^r) < \phi(2^r)$.

If m is divisible by an odd prime p, let $m = qp^e$ with $(p,q) = 1$. If $q > 2$, then $\lambda(q)$ is divisible by 2, and $\lambda(p^e)$ is even, so the greatest common divisor of $\lambda(q)$ and $\lambda(p^e)$ is a multiple of 2. Hence

$$\lambda(m) = [\lambda(q), \lambda(p^e)] = \frac{\lambda(q)\lambda(p^e)}{(\lambda(q), \lambda(p^e))} < \lambda(q)\lambda(p^e) \leq \phi(q)\phi(p^e) = \phi(m).$$

Thus in all cases except when $m = 2$ or 4 or $m = qp^e$ with p an odd prime and $q \leq 2$, $\lambda(m) < \phi(m)$ and U_m is not cyclic. That completes the proof. $\qquad \square$

Corollary 21. *There is a primitive root modulo m iff $m = 4, p^e$ or $2p^e$ for p an odd prime.*

For b is a primitive root modulo m iff the group of units U_m is cyclic with generator $[b]_m$.

Exercises.

30. Let $m = rs$ with $r, s > 2$ and r and s coprime.

(i) Show that there are at least four roots of the polynomial $x^2 - 1$ modulo m. [Hint: use the Chinese Remainder Theorem.]

(ii) Show that U_m is not cyclic.

31. (i) Show that if $G = \langle b \rangle$ is a finite cyclic group (written multiplicatively) of even order $2m$, then the product of all the elements of G is b^m.

(ii) Use part a) with $G = U_p$ to derive Wilson's Theorem:
if p is prime, then $(p-1)! \equiv -1 \pmod{p}$.

32. Let e be some exponent > 0 so that $a^e \equiv 1 \pmod{m}$ for all numbers a coprime to m. Show that if $\lambda(m)$ is the exponent of the group U_m of units modulo m, then $\lambda(m)$ divides e.

33. (i) Show that for every r and s, the exponent of $U_r \times U_s$ is the least common multiple of $\lambda(r)$ and $\lambda(s)$.

(ii) Suppose $m = rs$ and r and s are coprime. Then by Section 12D,

$$U_m \cong U_r \times U_s.$$

Use this fact and part (i) to prove Proposition 15, that

$$\lambda(m) = [\lambda(r), \lambda(s)].$$

E. Pseudoprimes

Let m be an odd number. Recall from Section 10B that m is an a-pseudoprime if $a^{m-1} \equiv 1 \pmod{m}$, if $[a]^{m-1} = [1]$ in $\mathbb{Z}/m\mathbb{Z}$.

In this section we ask: for how many numbers a with $1 \leq a \leq m$ is m an a-pseudoprime.

If a is not a unit modulo m, then $a^{m-1} \not\equiv 1 \pmod{m}$, so we need only look for numbers a that are units modulo m.

Let U_m be the group of units modulo m, and let $U_m(e)$ be the subgroup of U_m consisting of units $[a]$ with $[a]^e = [1]$, the group of e-th roots of unity in $\mathbb{Z}/m\mathbb{Z}$.

The answer to our question is the size of $U_m(m-1)$.

Proposition 22. *For every m and every $e > 0$,*

$$U_m(e) = U_m(d)$$

where $d = (e, \phi(m))$, the greatest common divisor of e and the order $\phi(m)$ of U_m.

Proof. If $d = (e, \phi(m))$, then there are integers r, s so that

$$d = re + s\phi(m).$$

For $[a]$ in U_m,

$$[a]^d = [a^{er+\phi(m)s}] = [a]^{er}[a]^{\phi(m)s} = [a]^{er}.$$

Thus if $[a]$ is in $U_m(e)$, then $[a]$ is in $U_m(d)$.

Conversely, since d divides e, $U_m(d) \subseteq U_m(e)$. $\square$

Proposition 23. *Let $m = rs$ with $(r, s) = 1$. Then for every $e \geq 1$,*

$$U_m(e) \cong U_r(e) \times U_s(e).$$

Proof. We know that the homomorphism θ from U_m to $U_r \times U_s$ defined by $\theta([a]_m) = ([a]_r, [a]_s)$ defines an isomorphism

$$U_m \cong U_r \times U_s.$$

If $a^e \equiv 1 \pmod{m}$, then $a^e \equiv 1 \pmod{r}$ and also $\pmod{s}$, so θ maps $U_m(e)$ into $U_r(e) \times U_s(e)$. To show that θ is onto $U_r(e) \times U_s(e)$, let b, c satisfy $[b]_r^e = [1]_r$ and $[c]_s^e = [1]_s$. We find an integer a satisfying

$$a \equiv b \pmod{r}$$
$$a \equiv c \pmod{s}.$$

Then $a^e \equiv 1 \pmod{r}$ and $a^e \equiv 1 \pmod{s}$. Since r and s are coprime, then $a^e \equiv 1 \pmod{m}$. Then $\theta([a]_m) = ([b]_r, [c]_s)$ in $U_r(e) \times U_s(e)$. Hence θ yields an isomorphism

$$U_m(e) \cong U_r(e) \times U_s(e).$$ $\square$

Corollary 24. *If* $m = q_1 q_2 \cdots q_g$ *with* $q_i = p_i^{e_i}$ *with* $p_1, \ldots, p_g$ *distinct primes, then*

$$U_m(e) = U_{q_1}(gcd(e, \phi(q_1))) \times \ldots \times U_{q_g}(gcd(e, \phi(q_g))).$$

We also have:

Proposition 25. *If* U_q *is cyclic and* d *divides* $\phi(q)$*, then* $U_q(d)$ *has* d *elements.*

Proof. Since U_q is cyclic, if $U_q = \langle [b] \rangle$, then $U_q(d) = \langle [b^{\phi(q)/d}] \rangle$ is cyclic of order d. □

Using these results, for every m we can count the number of elements in $U_m(m-1)$. (Recall that (r, s) is the greatest common divisor of r and s.)

Proposition 26. *Let* m *be odd,* $m = q_1 q_2 \cdots q_g$ *with* $q_i = p_i^{e_i}$ *with* $p_1, \ldots, p_g$ *distinct primes. Then*

$$|U_m(m-1)| = (m-1, \phi(q_1)) \cdots (m-1, \phi(q_g)).$$

Proof. For each i, since U_{q_i} is a cyclic group, $U_{q_i}(m-1)$ has $(m-1, \phi(q_i))$ elements, by Proposition 9. The result then follows from Corollary 24. □

Example 5. Let $m = 225 = 5^3 \cdot 3^2$. Then

$$U_{225}(224) = U_{25}(224) \times U_9(224)$$

has $(224, 20) \cdot (224, 6) = 4 \cdot 2 = 8$ elements. Hence

$$\frac{|U_{225}(224)|}{\phi(225)} = \frac{8}{120} = \frac{1}{15}.$$

This means that 225 is an a-pseudoprime for 1/15 of the units modulo 225.

Example 6. Let $m = 187 = 11 \cdot 17$. Then

$$U_{187}(186) = U_{11}(186) \times U_{17}(186)$$

has $2 \cdot 2 = 4$ elements. Thus

$$\frac{|U_{187}(186)|}{\phi(187)} = \frac{4}{160} = \frac{1}{40},$$

and so 187 is an a-pseudoprime for only 1/40 of the units modulo 187.

Example 7. Let $m = 561 = 3 \cdot 11 \cdot 17$. Then

$$U_{561}(560) = U_3(560) \times U_{11}(560) \times U_{17}(560)$$

has

$$(560, 2) \cdot (560, 10) \cdot (560, 16) = 2 \cdot 10 \cdot 16 = 320 = \phi(561)$$

elements. Thus 561 is an a-pseudoprime for every a coprime to 561, hence is a Carmichael number (as we knew from before).

Exercises.

34. Find $|U_m(m-1)|$ and $|U_m(m-1)|/\phi(m)$ for:
  (i) $m = 91$
  (ii) $m = 105$
  (iii) $m = 1001$.

35. Find m so that $|U_m(m-1)|/\phi(m) < 1/100$.

36. For the 2-pseudoprime $m = 1194649 = 1093^2$ of Section 10B, find $|U_m (m-1)|/\phi(m)$.

37. Show that for p an odd prime, $|U_{p^e}(p^e - 1)| = p - 1$.

F. Discrete Logarithms

Let G be a cyclic group of order n with generator g. Every element b of G may be written as

$$b = g^r$$

for some integer r with $0 \le r < n$. The number r is called the *logarithm* of b to the base g, written

$$r = \log_g(b)$$

or just $r = \log(b)$ if the generator g is understood.

The logarithm relative to a generator of a finite cyclic group is called a *discrete logarithm*, to contrast it with the functions $\ln(x)$ or $\log_{10}(x)$, which are continuous functions of x for $x > 0$. The discrete logarithm is an integer defined modulo the order n of g, since

$$g^r = g^{r+kn}$$

for every k.

The classical logarithm was invented by Napier and Briggs in the early part of the 17th century for computational purposes. The key property of the base 10 logarithm of Briggs (1624) was that $\log(ab) = \log(a) + \log(b)$. This property enables users to transform multiplications of many-digit numbers into addition by the use of logarithm tables. To multiply a and b, they would look up $\log(a)$ and $\log(b)$, add the logarithms, and then find the number c with

$$\log(c) = \log(a) + \log(b).$$

Logarithms "revolutionized the art of numerical computation" [Edwards (1979), p. 154].

In a similar way, if $b = g^r, a = g^s$, then $ab = g^{r+s}$, so

$$\log_g(ab) \equiv \log_g(a) + \log_g(b) \pmod{n}.$$

Thus with a table of discrete logarithms for a finite cyclic group, multiplying in the group can be transformed into addition modulo n, the order of the group.

Example 8. Let $G = U_{11}$. Then U_{11} is generated by $[2]_{11}$, which we will abbreviate as 2 in this example. Here is a logarithm table for U_{11}:

$a = 2^b$	$b = \log_2(a)$
1	10
2	1
3	8
4	2
5	4
6	9
7	7
8	3
9	6
10	5

Thus if we want to multiply $5 \cdot 8$, we have

$$\log_2(5) = 4,$$
$$\log_2(8) = 3,$$
$$4 + 3 \equiv 7 \pmod{10}, \text{ and}$$
$$2^7 \equiv 7.$$

Thus
$$5 \cdot 8 \equiv 2^4 \cdot 2^3 = 2^7 \equiv 7 \pmod{11}.$$

The discrete logarithm is helpful when evaluating a polynomial with coefficients modulo 11.

Example 9. Let
$$p(x) = x^6 + 3x^5 + 8x^3 + 7x^2 + 5,$$

and suppose we want to find $p(6)$ modulo 11. Then, modulo 11 (and recalling that the exponents of 2 are modulo 10),

$$
\begin{aligned}
p(6) &= 6^6 + 3 \cdot 6^5 + 8 \cdot 6^3 + 7 \cdot 6^2 + 5 \\
&\equiv (2^9)^6 + 2^8 \cdot (2^9)^5 + 2^3 \cdot (2^9)^3 + 2^7 \cdot (2^9)^2 + 5 \\
&\equiv 2^{54} + 2^{53} + 2^{30} + 2^{25} + 5 \\
&\equiv 2^4 + 2^3 + 1 + 2^5 + 5 \\
&\equiv 5 + 8 + 1 + 10 + 5 \\
&\equiv 7 \pmod{11}.
\end{aligned}
$$

Observe that we did no multiplications, rather just additions, reduction of exponents modulo 10 and reduction of numbers modulo 11.

We'll see in Chapter 25 that a discrete logarithm table is very helpful for evaluating polynomials in a field of 16 elements.

Suppose we let $G = U_p$ for p a large prime number (e.g. of 100 digits), let g be a generator of G, that is, a primitive root modulo p, and we wish to find $g^m = a$. for some m. That is computationally easy—we just write m in base 2 and use the squaring modulo p method of section 9F. But now suppose we are given the generator g and the number a. How do we find $m = \log_g(a)$? That is much harder. The problem is an example of

The Discrete Logarithm Problem. Let G be a finite cyclic group with generator g. Given an element a in G, then $a = g^x$ for some number x. Find $x = \log_g(a)$.

The problem of finding efficient ways to produce $\log_g(a)$ has been an intense area of research ever since Diffie and Hellman (1976) introduced public key cryptography with a scheme whose security depended on the difficulty of the discrete logarithm problem. We present their scheme in the next section.

Exercises.

38. In Example 9,
(i) compute $p(7)$ modulo 11;
(ii) compute $p(3)$ modulo 11.

39. Set up a discrete logarithm table for integers modulo 13, using the primitive root 2.

40. Let $f(x) = x^8 + 7x^6 + 3x^5 + 4x^4 + x^2 + 9x + 1$. Use the log table of the last exercise to find
(i) $f(3)$ modulo 13;
(ii) $f(9)$ modulo 13;
(iii) $f(10)$ modulo 13.

G. Discrete Logarithm Cryptography

We reintroduce Alice, Bob and Eve.

Alice and Bob wish to communicate privately. Eve can see everything sent between Alice and Bob. Alice and Bob don't want Eve to be able to read messages from Alice to Bob, so need to encrypt the messages so that Eve cannot read them.

In their celebrated paper, W. Diffie and M. Hellman (1976) introduced the idea of a public key cryptosystem. They presented a method by which Alice and Bob can obtain a common private key, and in order for Eve to know the key, Eve would apparently have to solve the discrete logarithm problem.

A common private key was a fundamental part of all cryptosystems prior to Diffie and Hellman's paper. For example, during World War II, the German Enigma machines were used to encrypt and decrypt messages between the German Admiralty and the North Atlantic submarine fleet, and both sender and receiver had to share a common private key in order to configure the machines to encrypt and decrypt compatibly. But how to communicate a common private key by radio between Berlin and the North Atlantic while the British are listening?

Diffie and Hellman's scheme solves the problem of communicating a private key through a public channel. The scheme works as follows.

Alice and Bob agree on a finite cyclic group G of large order, and a generator g of G. The shared key will be an element of G.

Alice chooses a random number a between 0 and the order of G, computes $g^a = A$ in G and sends the resulting group element A to Bob, while keeping the exponent a secret.

Bob chooses a random number b between 0 and the order of G, computes $g^b = B$ in G and sends the resulting group element B to Allice, while keeping the exponent b secret.

Alice and Bob's shared private key is then $K = g^{ab}$ in G.

Alice can compute K by computing $B^a = (g^b)^a$, which she can do because she knows a and she received B from Bob. Bob can compute K by computing $A^b = (g^a)^b$, which he can do because he knows b and he received A from Alice.

Assume that Eve can eavesdrop on the transmissions between Alice and Bob. Then Eve knows G, g, A and B. She wants to learn the key K. Since $K = A^b = B^a = g^{ab}$, she can determine K if she can learn a or b or ab. We can state Eve's problem as follows:

Diffie-Hellman problem: Given a group G, an element g in G, and elements A and B in $\langle g \rangle$ where $A = g^a$ for some unknown number a and $B = g^b$ for some unknown number b, determine $K = g^{ab}$ in G.

We can restate the problem.

Definition. Given a cyclic group G and a generator g of G, define the function

$$DH_g : G \times G \to G$$

by $DH_g(g^a, g^b) = g^{ab}$.

The Diffie-Hellman problem is then to compute $K = DH_g(A, B)$ for every pair (A, B) of elements of $G = \langle g \rangle$.

In the Diffie-Hellman cryptographic scheme, the eavesdropper Eve will know G, g, A and B, so to find the secret key K and crack the code, Eve needs to compute $DH_g(A, B)$.

Evidently, one way to determine DH on a pair of elements of G is to solve the discrete logarithm problem: find $a = \log_g(A)$ and $b = \log_g(B)$; then

$$DH_g(A, B) = g^{ab} = g^{\log_g(A)\log_g(B)}.$$

Since $DH(A, B) = B^{\log_g(A)} = A^{\log_g(B)}$, it suffices only to know either $\log_g(A)$ or $\log_g(B)$.

It is an open question whether it is possible to compute $DH(A, B)$ directly, without knowing the value of $\log_g(A)$ or $\log_g(B)$. It is widely thought that the two problems, the Diffie-Hellman problem and the discrete logarithm problem are of the same order of difficulty.

A nice feature of the Diffie-Hellman scheme is that it can be used with any cyclic group G of large order for which there is an efficient procedure for multiplying elements in G. Some examples:

- $G = U_p$, the group of units modulo a large prime p. The group U_p is cyclic of order $p - 1$.
- G is the group of non-zero elements of any finite field E. For examples, see Chapter 23.
- G is a cyclic subgroup of large order inside the group of points of an elliptic curve over a finite field. Cryptosystems based on elliptic curves have been widely studied during the past 20 years, but are a bit beyond the scope of this book. See Koblitz (1994) for a description.

DSA. The Digital Signature Standard of the National Institute of Standards and Technology, U. S. Department of Commerce, was issued in 1994 and is applicable for the implementation of public-key based signature systems for use by Federal agencies. The Digital Signature Algorithm (DSA) is based on the discrete logarithm.

Suppose Alice regularly sends messages to Bob. To apply a signature to those messages, Alice makes public a large prime p, and a number g of prime order q modulo p, where q is a large prime divisor of $p - 1$. As in the Diffie-Hellman cryptosystem, the group G used is the cyclic group $\langle g \rangle$ of order q generated by g.

Alice chooses a random private key a and makes public $A = g^a$, her public key.

For each message m Alice sends to Bob, Alice picks a random number $k \bmod q$. To sign the message m, she computes $H(m)$, where H is a public hash function that creates a "digest" of the message m in $\mathbb{Z}/q\mathbb{Z}$. There is a federal standard for such hash functions, the Secure Hash Algorithm. Then she computes her signature (r, s) where

- $r = g^k \bmod q$
- $s = k^{-1}(H(m) + ar) \bmod q$.

She sends Bob the triple (m, r, s) consisting of the message m and the signature (r, s). The signature depends on Alice's private key a, on the message-specific information $H(m)$, and on the random number k generated specifically for the message m. Since k is random, $r = g^k$ will be a random element of G.

Bob receives (m', r', s') and wants to be sure that that the message m' is the same as the message m that Alice sent, and that (r', s') is Alice's signature (r, s). He knows the modulus p, the group generator g, the order q of g, and Alice's public key A. To verify Alice's signature he computes $H(m')$, computes w, the inverse of s' modulo q, and then computes

$$v = g^{H(m')w} A^{r'w}.$$

If $v \equiv r \pmod{q}$, then Bob decides to accept as authentic the message from Alice.

To see why, we first observe that if $(m', r', s') = (m, r, s)$, then

$$v = g^{H(m)w} A^{rw}$$
$$= g^{H(m)w} g^{arw}$$
$$= g^{H(m)w + arw}$$
$$= g^{(H(m) + ar)w}.$$

Since $s \equiv k^{-1}(H(m) + ar) \pmod{q}$ and $w \equiv s^{-1} \pmod{q}$, we have

$$k \equiv (H(m) + ar)w \pmod{q}$$

and so $v = g^k \equiv r \pmod{q}$.

Suppose Eve wished to impersonate Alice. Eve would construct a message m, pick a random k, compute $r = g^k$, $H(m)$ and k^{-1}, and try to determine the correct s. But $s = k^{-1}(H(m) + ar)$ depends on Alice's secret key a. In order for Eve to succeed in misleading Bob, Eve would need to find s so that $v = r$. But if Eve found such an s, then

$$g^k = r = v = g^{H(m)w} A^{rw}$$

with $w \equiv s^{-1} \pmod{q}$, so

$$g^{ks} = g^{H(m)} A^r;$$

letting z be the inverse of r modulo q, this becomes

$$g^{(ks - H(m))z} = A.$$

But $g^a = A$, so this equation yields Alice's secret key a:

$$(ks - H(m))z \equiv a \pmod{q}.$$

Thus for Eve, finding s is equivalent in difficulty to solving the discrete logarithm problem for Alice's secret key:

Given g and $A = g^a$, find a.

In the implementation of DSA (as specified in 1994), the prime q (and hence the order of g) should satisfy $2^{159} < q < 2^{160}$, and the prime p should lie between 2^{L-1} and 2^L where $512 \le L \le 1024$ and L is a multiple of 64.

For the full National Institute of Standards and Technology description of DSA, see www.itl.nist.gov/fipspubs/fip186.htm.

Exercises.

41. Let $G = \langle 3 \rangle \subset U_{83}$. Then 3 has order 41 modulo 83. Suppose m is a message with $H(m) = 24$. Alice's secret exponent is $a = 23$, and she publishes $A = 49$. For the message m she selects the random number $k = 17$. Compute Alice's signature for the message m.

42. With $G = \langle 3 \rangle \subset U_{83}$, suppose you receive $(H(m), r, s) = (33, 20, 27)$ from Alice and know that $A = 49$. Verify Alice's signature.

H. Pseudorandom Numbers

A sequence of (uniformly distributed) random numbers, chosen from the interval between 1 and some fixed integer M, is a sequence of numbers $x_0, x_1, \ldots$ such that for each i, the value of x_i has a $1/M$ chance of being any given number between 1 and M, independent of what values $x_0, x_1, \ldots, x_{i-1}$ took on.

For example, M could be 6 and x_i could be the number of dots showing on the ith toss of a fair die. Or M could be equal to 13 and x_i could be the value of the ith card drawn from a well-shuffled deck of 52 standard playing cards (with J = 11, Q = 12, K = 13), where before drawing the next card the last card drawn is shuffled back into the deck. Or M could be 38 and (with 0 = 37, 00 = 38) x_i could be the result of the ith spin of a roulette wheel.

Random numbers are useful in many contexts related to computing. Here are some examples.

(1) In order to provide individualized web-based quizzes for calculus students, web designers can provide variations of given problems by randomly changing parameters in the problem. For example, suppose the problem is to find the derivative of a function of the form $\sin(ax^2 + bx + c)$. The parameters a, b and c for each problem could be randomly chosen triples of integers from the set $\{-4, -3, -2, -1, 1, 2, 3, 4, 5\}$. There are 729 possible triples. If we number them, then when a student selects that problem, a random number between 1 and 729 would determine the parameters for that student.

The random selection of parameters would insure that the student cannot know in advance exactly which problem will be asked.

Similar uses of random numbers create unpredictability and variety in many computer games.

(2) Suppose f is a positive real function, such as

$$f(x) = \frac{1}{1 + x^2 + x^3}$$

or

$$f(x) = e^{-x^2},$$

and we wish to determine

$$f(x) = \int_0^1 f(x)dx,$$

the area under the curve $y = f(x)$ between $x = 0$ and $x = 1$. For many functions f, such as these two examples, there is no formula for the antiderivative, and so the integral cannot be explicitly determined and must be approximated. If f happens to have a number of jump discontinuities, or is not differentiable, even some of the standard calculus methods, such as Simpson's rule, or expansion into an infinite series (Newton's favorite method), don't work well.

Suppose $0 \leq f(x) \leq 1$ for $0 \leq x \leq 1$. We can use random numbers to estimate the value of $\int_0^1 f(x)dx$ as follows: generate N pairs of random numbers,

$$(z_1, w_1), (z_2, w_2), \ldots, (z_n, w_n),$$

with $0 \leq z_i \leq M, 0 \leq w_i \leq M$ and M is large. Let $x_i = z_i/M, y_i = w_i/M$. Then x_i and y_i lie between 0 and 1, so (x_i, y_i) are the coordinates of a point in the square with corners $(0, 0), (0, 1), (1, 0), (1, 1)$.

Some of those N points (x_i, y_i) will lie on or beneath the curve $y = f(x)$, namely, those for which $y_i \leq f(x_i)$; the others will lie above the curve. We can estimate the area A under the curve $y = f(x)$ by

$$A = \int_0^1 f(x)dx \approx \#\{(x_i, y_i)|y_i \leq f(x_i)\}/N.$$

If we choose the N points randomly, rather than systematically, then for any function $f(x)$, there is a high probability that the resulting estimate for the area will not differ greatly from the true area. By contrast, any particular systematic choice of the N points may be quite good for estimating the area for most functions, but can be extremely poor for estimating the area under certain functions. It is desirable to have a method which we can confidently apply to any function. The random point method is such a method.

(3) As we shall see in Section 20C, random numbers are useful in testing a large number to see whether it is composite or "probably prime."

In such applications, computations are efficiently done if sequences of random numbers can be generated as needed by the computer.

But a computer is a deterministic machine. It cannot compute random numbers, for there is no chance built into its computations.

To circumvent this problem, and to be able to replicate computations, a strategy adopted by computer scientists is to have the computer generate sequences of numbers that look random, even though in fact they are not.

Such numbers are called *pseudorandom numbers*.

In this section we will look at a classic strategy for generating sequences of pseudorandom numbers.

Lehmer's multiplicative congruential method. The first widely used and understood method for computing pseudorandom numbers is the multiplicative congruential method developed by D.H. Lehmer in 1949.

Lehmer's method constructs sequences of pseudorandom numbers by the simple process of computing successive powers of some number A mod m.

To illustrate the construction for an unreasonably small value of M, let $M = 41$, and $A = 23$. Then the powers of A mod M, starting from $A^0 = 1$, are

$$1, 23, 37, 31, 16, 40, 18, 4, 10, 25, 1, 23, \ldots .$$

This is a sequence of numbers ranging between 1 and 40. The numbers don't appear to have much of a pattern to them except that they repeat with a period of 10. This is because 23 has order 10 modulo 41.

In implementing Lehmer's method, one objective is that if the numbers are natural numbers $< M$, then as many as possible of the numbers $< M$ should be part of the sequence. This was clearly not the case with $M = 41$, $A = 23$. If we choose $A = 13$, rather than $A = 23$, the powers of A mod M are

$$13, 5, 24, 25, 38, 2, 26, 10, 7, 9, 35, 4,$$
$$11, 20, 14, 18, 29, 8, 22, 40, 28, 36, 17, 16,$$
$$3, 39, 15, 31, 34, 32, 6, 37, 30, 21, 27, 23,$$
$$12, 33, 19, 1, 13, 5, 24, 25, 38, 2, \ldots,$$

which includes all numbers between 1 and 40 and appears not to have much of a pattern to it. Since we obtain all 40 numbers mod 41, we see that 13 is a primitive root modulo 41. Choosing A to be a primitive root modulo M gives the longest possible period before the sequence cycles.

The general method of Lehmer is the following:

Pick a number M, the modulus.

Pick a starting number X_0.

Pick a multiplier A.

Generate the sequence as follows:

X_1 is the remainder on dividing AX_0 by M; that is, X_1 is AX_0 mod M.

X_2 is the remainder on dividing AX_1 by M; that is, X_2 is AX_1 mod M.

$$\vdots$$

X_{n+1} is the remainder on dividing AX_n by M; X_{n+1} is AX_n mod M ....

Thus each of $X_0, X_1, X_2, \ldots$ is a number between 0 and M, and each X_n satisfies

$$X_{n+1} = AX_n \bmod M.$$

In the example above, $X_0 = 1$, $M = 41$ and $A = 13$. Notice that $X_n \equiv A^n X_0 \pmod{M}$ for all $n > 0$. Thus for every X_0, the sequence always starts repeating itself eventually:

Proposition 27. *For every A, M, X_0, the sequence $X_0, X_1, \ldots, X_n, \ldots$ begins repeating after at most M terms.*

Proof. There are only M remainders modulo M, so if we consider the sequence of numbers $X_0, X_1, \ldots, X_M$, two of them must be equal. Suppose $X_r = X_s$ for some $s > r$. Then

$$X_{r+1} \equiv AX_r \equiv AX_s = X_{s+1} \pmod{M}.$$

Since X_{r+1} and X_{s+1} are both between 0 and $M - 1$, they must be equal. The same argument shows that $X_{r+k} = X_{s+k}$ for all $k > 0$. $\square$

We can be more precise about the period of the sequence:

Proposition 28. *In the multiplicative congruential method, if A and X_0 are both coprime to M, then the length of the sequence*

$$\{X_0, X_1, \ldots, X_k, \ldots\}$$

before repetition is equal to the order of A modulo M.

Proof. This follows because $X_n \equiv A^n X_0 \pmod{M}$. If $X_n = A^n X_0 \equiv X_0 \pmod{M}$, then, since $(X_0, M) = 1$, we have $A^n \equiv 1 \pmod{M}$. The smallest such $n > 0$ is both the order of $A \pmod{M}$ and the smallest n so that the sequence begins repeating at X_n. $\square$

To implement Lehmer's method in practice to get a good sequence of pseudorandom numbers, one should use a large modulus M, and then a multiplier A so that the order of A modulo M is large.

From Sections 19A and C, we know that if M is prime, there are numbers A of order $M - 1$ modulo M; if $M = p^e$ with p an odd prime, there are numbers A of order $p^{e-1}(p - 1)$ modulo M, and if $M = 2^e$, there are numbers A of order 2^{e-2} modulo M.

A large modulus M commonly used in practice is $M = 2147483647 = 2^{31} - 1$. This modulus M is prime, so that there are elements of order $M - 1 \pmod{M}$. This M also has an advantage that taking remainders by M can be done very quickly on a computer.

(In fact, finding remainders is like "casting out 9's". With $M = 2^{31} - 1$, if A and X are $< 2^{31} - 1$, then $AX < (2^{31})^2$, so

$$W = AX = Q \cdot 2^{31} + R$$

for some $Q, R \leq 2^{31} - 1$. But since $2^{31} \equiv 1 \pmod{M}$,

$$AX \equiv Q + R \pmod{M}.$$

Then $Q + R < 2M$, and so either $Q + R$ or $Q + R - M$ is the remainder on dividing W by M. This means that finding the remainder on dividing AX by $2^{31} - 1$ does not require the division algorithm, but simply an addition and perhaps one subtraction.)

Park and Miller (1988) suggested three criteria for judging the suitability of a pseudorandom number generator. The sequence generated should:

- have a period as long as possible;
- behave as much like a random sequence of numbers from the interval between 1 and $M - 1$ as possible; and
- be able to be generated efficiently with 32-bit arithmetic.

Based on all three criteria, they recommended $M = 2^{31} - 1$ and $A = 48271$, a primitive root modulo M.

Knowing that a sequence $\{X_0, X_1, \ldots\}$ has maximal or near-maximal period is no guarantee of its suitability as a sequence of pseudorandom numbers. For a number of years, IBM computers used a pseudorandom number generator called RANDU, using the modulus $M = 2^{31}$ and the multiplier $A = 65539 = 2^{16} + 3$, a number of order 2^{29} modulo M. The RANDU generator has been called "truly horrible" [Knuth (1998), p. 188] for its lack of random behavior. If one plots points $p_i = (x_{3i+1}, x_{3i+2}, x_{3i+3})$ for $i = 1, 2, \ldots$ in the cube of side 1, where $x_j = X_j / 2^{31}$, it turns out that all the points lie in one of fifteen planes. For a beautiful demonstration, see http://etsuodt.tamu-commerce.edu/AcademicOrganizations/sigmaxi/hopftorii/testap6.html.

For a comprehensive study of the random behavior of these sequences of pseudorandom numbers from a statistical viewpoint, see Knuth (1998), Chapter 3.

For cryptographic use, Lehmer-type pseudorandom number generators are not considered acceptable. We will look at the newer Blum-Blum-Shub generator in Section 22C.

Exercises.

43. (i) Find a primitive root b modulo 29 and a primitive root c modulo 31.

(ii) Let $e_r = b^r \pmod{29}$ and $f_r = c^r \pmod{31}$, where $0 < e_r < 29, 0 < f_r < 31$, and let $x_r = e_r / 29$, and $y_r = f_r / 31$. Plot the points (x_r, y_r) for $r = 3$ to 13. Do they look like randomly chosen points in the unit square?

44. Show that if $a, b < 100$ and $ab = Q \cdot 100 + R$, then $ab \pmod{99}$ is either $Q + R$ or $Q + R - 99$. Illustrate with $a = 67, b = 83$.

45. Let $M = 957, A = 15$. Then $(A, M) = 3$. Without computing $A, A^2, A^3, \ldots$, can you identify the period of the sequence $A, A^2, A^3, \ldots$, as the order of some number to some appropriate modulus?

Chapter 20
Carmichael Numbers

This chapter returns to the question of deciding whether a given odd number m is prime. The a-pseudoprime test of Chapter 10B will not work on Carmichael numbers. We first describe an idea of Alford that shows that there are many Carmichael numbers. Then we develop the strong a-pseudoprime test and prove that every composite number m fails the strong a-pseudoprime test for at least half of the numbers $a < m$. Thus there are no composite numbers that are "strong Carmichael numbers".

A. Lots of Carmichael Numbers

Recall (from Section 10B) that a number m passes the a-pseudoprime test if

$$a^{m-1} \equiv 1 \pmod{m}.$$

Any prime number p passes the a-pseudoprime test for all a coprime to p, by Fermat's Theorem. Most composite numbers m fail some a-pseudoprime tests, as we observed in Section 10B. If m fails an a-pseudoprime test for some a coprime to m, then m fails a-pseudoprime tests for at least half of the numbers $a < m$, as we showed in Sections 11D and 19E.

If m is composite and passes the a-pseudoprime test, then m is called an a-pseudoprime.

A number m is Carmichael if m is an a-pseudoprime for all a coprime to m.

Carmichael numbers exist, as we observed in Chapter 10: the first three are 561, 1105, and 1729. The existence of Carmichael numbers means that trial a-pseudoprime testing can be unreliable as a primality test. A Carmichael number m passes a-pseudoprime tests for all a coprime to m but is not prime.

In this section we study odd Carmichael numbers and show that there are many of them.

Assume for the remainder of this section that m is odd.

We begin with a characterization of Carmichael numbers dating from 1899.

L.N. Childs, *A Concrete Introduction to Higher Algebra*, Undergraduate Texts in Mathematics, © Springer Science+Business Media LLC 2009

Theorem 1 (Korselt's Criterion). *A number m is Carmichael if and only if m is squarefree and for all primes p dividing m, p − 1 divides m − 1.*

Proof. Suppose m is squarefree and $p - 1$ divides $m - 1$ for all primes p dividing m. Let b be coprime to m. Then for all p dividing m, b is coprime to p, so $b^{p-1} \equiv 1$ (mod p) by Fermat's theorem. Since $p - 1$ divides $m - 1$, $b^{m-1} \equiv 1$ (mod p). Now since m is squarefree, m is the least common multiple of the primes which divide m. So if $b^{m-1} \equiv 1$ (mod p) for all p dividing m, then $b^{m-1} \equiv 1$ (mod m) So m is Carmichael.

Conversely, suppose m is Carmichael, and suppose p is any (odd) prime divisor of m. Let $m = p^e q$, where $(p, q) = 1$. Let b be a primitive root modulo p^e, and let a be a number such that

$$a \equiv b \pmod{p^e}$$
$$a \equiv 1 \pmod{q}$$

Then a is coprime to m. If m is Carmichael, then

$$a^{m-1} \equiv 1 \pmod{m}$$

so

$$a^{m-1} \equiv 1 \pmod{p^e}$$

But the order of a modulo p^e is $p^{e-1}(p - 1)$. So $p^{e-1}(p - 1)$ divides $m - 1$, and hence $p - 1$ divides $m - 1$. Also, if $e > 1$, then p divides $m - 1$. But since p divides m, p cannot divide $m - 1$. Thus $e = 1$.

Since this is true for all primes p dividing m, therefore m must be squarefree. □

Korselt's criterion is quite useful for identifying Carmichael numbers.

Example 1. By Korselt's criterion, 2821 is Carmichael: 2821 factors as $7 \cdot 13 \cdot 31$, and 2820 is divisible by 6, 12 and 30: in fact, $2820 = 12 \cdot 235 = 30 \cdot 94 = 6 \cdot 470$.

Here is a simple consequence of Korselt's criterion:

Corollary 2. *If m is Carmichael, then m must be a product of at least three primes.*

Proof. Note that m must be square-free, so, since m is composite, m must be a product of at least two distinct primes.

If $m = pq$ with $p < q$, primes, then $q - 1$ divides $m - 1$, so $q - 1$ divides $m - 1 - p(q - 1) = pq - 1 - pq + p = p - 1$, impossible since $p < q$.   □

Korselt's criterion has been used to find many Carmichael numbers.

An early use of Korselt's criterion by Chernick (1939) was a strategy for finding Carmichael numbers that are products of three primes.

Proposition 3. *Let $m = (6k + 1)(12k + 1)(18k + 1)$. Then m is Carmichael for all k for which all three factors of m are prime.*

Proof. Observe that

$$m = (6k+1)(12k+1)(18k+1) = 1296k^3 + 396k^2 + 36k + 1$$
$$= 36k(36k^2 + 11k + 1) + 1,$$

so $m - 1$ is divisible by $6k$, $12k$ and $18k$. □

Some examples where m is Carmichael are for $k = 1, 6, 35, 45, 51, 55, 56$ and 100. See Exercises 3, 4 and 5 below.

If it were known that there are infinitely many values of k for which all three factors of m in Proposition 3 are prime, then it would follow that there are infinitely many Carmichael numbers. However, while it is known by a famous theorem of Dirichlet that each factor is prime for infinitely many k, it is not known if there are infinitely many k for which all three factors are simultaneously prime. Thus it is apparently an open question whether or not there exist infinitely many Carmichael numbers that are products of exactly three primes. See Ribenboim (1980), p. 298.

In 1956, Erdos proposed a different way to use Korselt's criterion to produce Carmichael numbers.

Theorem 4 (Erdos' Criterion). *Given a number L, let $\mathscr{P}$ be the set of all primes p coprime to L so that $p - 1$ divides L. If C is a product of distinct primes from $\mathscr{P}$ so that $C \equiv 1 \pmod{L}$, then C is Carmichael.*

Proof. This follows quickly from Korselt's criterion. For if $C \equiv 1 \pmod{L}$, then L divides $C - 1$. Each prime p dividing C is in $\mathscr{P}$, so $p - 1$ divides L, hence $p - 1$ divides $C - 1$. □

Erdos' criterion works well if L has many divisors, so is divisible by $p - 1$ for many primes p. Here is a small example.

Example 2. Let $L = 2^3 \cdot 3 \cdot 5 = 120$. The primes $p > 5$ so that $p - 1$ divides 120 are 7, 11, 13, 31, 41, and 61. To find a Carmichael number by Erdos' criterion we must find some product C of some of these six primes so that C is congruent to 1 modulo 120.

A simple counting argument suggests that this should be possible. There are $\phi(120) = 32$ units modulo 120, and $2^6 - 1 = 63$ different products of one or more of the six primes. So we should expect perhaps two different products of the primes which are congruent to 1 modulo 120 and hence are Carmichael.

Since such a product C must be congruent to 1 $\pmod{10}$, we can see easily that if 7 is a factor of C, so is 13.

So we need to check all products involving three or more of the primes 11, 31, 41, and 61; and all products involving $91 = 7 \cdot 13$ and one or more of 11, 31, 41, and 61. This is a total of nineteen products. After some computation, we find that

$$11 \cdot 31 \cdot 41 \cdot 61 \equiv 1 \pmod{120}$$
$$11 \cdot 41 \cdot 91 \equiv 1 \pmod{120}$$
$$31 \cdot 61 \cdot 91 \equiv 1 \pmod{120}$$

We have found three Carmichael numbers.

In 1992, W. R. Alford strengthened Erdos' criterion using the following idea. Pick L so that $p - 1$ divides L for a large set

$$\mathscr{P} = \{p_1, p_2, \ldots, p_N\}$$

of primes. Suppose that we can find a subset, say

$$\mathscr{P}_0 = \{p_1, \ldots, p_s\}$$

where $s < N$, so that *every* unit modulo L can be represented by a product of distinct primes from $\mathscr{P}_0$. Then take any product q of zero or more distinct primes from the remaining $N - s$ primes

$$\mathscr{P}_1 = \{p_{s+1}, \ldots, p_N\}.$$

That product q is a unit modulo L. The inverse of that unit is represented by some product q' of primes from $\mathscr{P}_0$. But then $C = qq'$ is a product of distinct primes from $\mathscr{P}$, and $C \equiv 1 \pmod{L}$. So C is Carmichael, by Erdos' criterion.

An idea like could only be utilized effectively with a computer. To see if every unit mod L can be represented by a product of distinct primes from a given set $\mathscr{P}_0$ is not easy to do by hand, even if $L = 120$. And 120 is too small.

Example 3. Let $L = 2^6 \cdot 3^5 \cdot 5 \cdot 7 \cdot 11$. Then there are 77 primes $p > 11$ such that $p - 1$ divides L. Since $\phi(L) = 2^5 \cdot 3^4 \cdot 2 \cdot 4 \cdot 6 \cdot 10 = 1244160$, while $2^{21} = 2097152$, one would expect that if we take $\mathscr{P}_0$ to be the set of the first 21 primes that divide L, then distinct products of these primes would yield all units modulo L. If so, then the set $\mathscr{P}_1$ would contain 56 primes. Each product q of distinct primes in $\mathscr{P}_1$, when multiplied by a suitable product q' of primes in $\mathscr{P}_0$, would give a Carmichael number C. But there are 2^{56} possible products of distinct primes in $\mathscr{P}_1$. So this strategy should give us 2^{56} Carmichael numbers!

Alford in fact did better than that. He let $L = 2^6 \cdot 3^3 \cdot 5^2 \cdot 7^2 \cdot 11$, and found a set $\mathscr{P}$ of 155 primes $p \geq 13$ so that $p - 1$ divides L. Now $\phi(L) = 4,838,400$ and $2^{27} = 134,217,728$. So it seems very likely, and Alford in fact showed (not by hand!), that products of the first 27 primes of $\mathscr{P}$ give all possible units modulo L. That leaves 128 primes in $\mathscr{P}$. Thus Alford found 2^{128} Carmichael numbers, one for each product of distinct primes from the last 128 primes of $\mathscr{P}$.

In particular, he found a Carmichael number divisible by at least 128 primes.

Goaded by this discovery, Alford, Granville, and Pomerance (1994) proved that for every n sufficiently large, there are at least $n^{2/7}$ Carmichael numbers $< n$. In particular, the number of Carmichael numbers is infinite, thereby settling a question raised by Carmichael in 1912.

An exposition of the Alford, Granville, Pomerance result may be found in Granville (1992).

Exercises.

1. Using Korselt's criterion, show that 1105 is Carmichael.

2. Show that 2465 is Carmichael.

3. Show that $6k + 1, 12k + 1$, and $18k + 1$ cannot all be prime if k is congruent to 2, 3 or 4 (mod 5).

4. Show that if m is a Carmichael number of the form in Proposition 3, then m is either of the form

$$m = (30r + 1)(60r + 1)(90r + 1)$$

or of the form

$$m = (30r + 7)(60r + 13)(90r + 19).$$

5. (i) Show that if $m = (30r + 1)(60r + 1)(90r + 1)$ is a product of three primes, then r cannot be congruent to 1, 3, or 5 (mod 7).
   (ii) Find the analogous result for $m = (30r + 7)(60r + 13)(90r + 19)$.

6. (i) Let s be a number which is a sum

$$s = a_1 + a_2 + \ldots + a_g$$

of some of its proper divisors (where $a_1 < a_2 < \ldots < a_g$). Show that

$$n = (a_1 sk + 1)(a_2 sk + 1) \cdots (a_g sk + 1)$$

is Carmichael for every k for which each factor $a_j sk + 1$ is prime.
   (ii) Show that if $s = 2^e q$, where $e \geq 1$ and q is an odd number $< 2^{e+1}$, then s is a sum of some of its proper divisors.
   (iii) Suppose s is a number that is a sum

$$s = a_1 + a_2 + \ldots + a_g$$

of some of its proper divisors (as in (i)), and let $d_i = s/a_i$. Show that then we can write 1 as a sum of distinct reciprocals:

$$1 = \frac{1}{d_1} + \frac{1}{d_2} + \cdots + \frac{1}{d_g}$$

and, conversely, if 1 is a sum of distinct reciprocals, then the least common multiple of the d's is a sum of some of its proper divisors.
   (iv) In (iii) can you find such an s so that $d_1, \ldots, d_g$ are pairwise relatively prime?

7. (i) Show that the eight numbers, 1, 11, 31, 41, 61, 71, 91, and 101 form a subgroup G of the units of $\mathbb{Z}/120\mathbb{Z}$.
   (ii) Show that the homomorphism from $\mathbb{Z}/120\mathbb{Z}$ to $\mathbb{Z}/24\mathbb{Z}$ defined by $[a]_{120} \mapsto [a]_{24}$ induces an isomorphism of groups from G onto U_{24}.

8. Define an isomorphism of groups,

$$U_{24} \to U_8 \times U_3$$
$$\to \langle 3 \rangle \times \langle 5 \rangle \times \langle 2 \rangle$$
$$\to \mathbb{Z}_2 \times \mathbb{Z}_2 \times \mathbb{Z}_2.$$

by

$$(a \bmod 24) \mapsto (a \bmod 8, a \bmod 3)$$
$$\mapsto (3^r 5^s \bmod 8, 2^t \bmod 3)$$
$$\mapsto (r, s, t).$$

Then the composite of this map with the map of Exercise 7 (ii) yields an isomorphism from G to $\mathbb{Z}_2 \times \mathbb{Z}_2 \times \mathbb{Z}_2$.

(i) Show that $91 \mapsto 19 \bmod 24 \mapsto (3 \bmod 8, 1 \bmod 3) \mapsto (1, 0, 0)$. Find the images in $\mathbb{Z}_2 \times \mathbb{Z}_2 \times \mathbb{Z}_2$ of the other elements of G.

(ii) In $\mathbb{Z}_2 \times \mathbb{Z}_2 \times \mathbb{Z}_2$, write $(0, 0, 0)$ as a sum of distinct elements (excluding $(1, 1, 1)$ and $(0, 1, 1)$). For each such sum, identify the corresponding Carmichael number as in Example 2.

9. Let $L = 360$. In addition to the six primes $p > 5$ so that $p - 1$ divides 120, the set $\mathscr{P}$ of primes with $p - 1$ dividing 360 includes 19, 37, 73, and 181. Find some Carmichael numbers other than those found in Example 1.

10. Show that no unit in $\mathbb{Z}/120\mathbb{Z}$ that is congruent to 4 modulo 5 is a product of distinct primes $p > 5$ so that $p - 1$ divides 120.

11. Write a computer program to show that every unit modulo 360 is a product of one or more distinct primes from the set $\mathscr{P}$ of primes p so that $p - 1$ divides 360. Or, if you want to try this by hand, first try to show that each of the 24 units that are congruent to 1 modulo 5 is a product of distinct primes from $\mathscr{P}$.

12. Show that there are no even composite numbers m so that

$$a^{m-1} \equiv 1 \pmod{m}$$

for all a with $(a, m) = 1$.

B. Strong a-Pseudoprimes

Because of Carmichael numbers, the "trial a-pseudoprime test" is not completely effective as a probabilistic primality test. So we would like a new test that is stronger than the pseudoprime test, in the sense that every prime would pass the new test, but every composite number, no matter how cleverly chosen, would fail the new test most of the times the test is applied.

Here is such a test.

The Strong a-Pseudoprime Test. Suppose m is an odd number we wish to test for primeness. Write $m - 1 = 2^e q$ where q is odd and $e > 0$. Suppose a is a number coprime to m such that either

  $a^q = 1 \pmod{m}$

or

  there is some $k < e$ so that $a^{2^k q} \equiv -1 \pmod{m}$.

Then m passes the strong a-pseudoprime test.

Definition. A composite number m that passes the strong a-pseudoprime test is called a *strong a-pseudoprime*.

 The test is called the strong a-pseudoprime test because a number that passes the strong a-pseudoprime tests will certainly pass the a-pseudoprime test. For if either $a^q = 1 \pmod{m}$ or $a^{2^k q} \equiv -1 \pmod{m}$ for some $k < e$, then $a^{2^e q} \equiv 1 \pmod{m}$.

 We'll see shortly that any prime number m passes the strong a-pseudoprime test for all a, $1 \le a < m$.

 On the other hand, since a number m is a strong a-pseudoprime if it is an a-pseudoprime and also satisfies additional conditions, it is plausible that there are composite numbers m that are a-pseudoprimes but not strong a-pseudoprimes. We'll see that in fact this is so.

 Here is another way of looking at the strong a-pseudoprime test. Let $m - 1 = 2^e q$, with q odd, and consider the set of numbers mod m,

$$(a^q, a^{2q}, \ldots, a^{2^e q}),$$

that we obtain, starting from a^q modulo m, by successively squaring modulo m. We'll call this sequence of numbers the "strong a-pseudoprime sequence."

 Suppose $a^{2^e q} \equiv 1 \pmod{m}$, so that m passes the a-pseudoprime test. Suppose also that either:

  all numbers in the sequence are $\equiv 1 \pmod{m}$

or

  the rightmost number not $\equiv 1 \pmod{m}$ is $\equiv -1 \pmod{m}$.

Then m passes the strong a-pseudoprime test. The strong a-pseudoprime sequence in that case looks like

$$(1, 1, \ldots, 1) \text{ or } (\ldots, -1, 1, \ldots, 1).$$

 On the other hand, if m is an a-pseudoprime but the strong a-pseudoprime sequence for m looks like

$$(\ldots, b, 1, \ldots, 1)$$

with $b \not\equiv 1$ or $-1 \pmod{m}$, then m fails the strong a-pseudoprime test, so m must be composite. The number m cannot be prime because b is a square root of 1 modulo m, and so 1 has at least three square roots, $1, -1$ and $b \pmod{m}$ (See Proposition 5, below).

We illustrate the strong a-pseudoprime test with some examples.

Example 4. Let $m = 91$. We ignore the fact that 91 is easily factored. Since $m - 1 = 90 = 2 \cdot 45$, the strong a-pseudoprime sequence is short:

$$(a^{45}, a^{90}).$$

We try $a = 3$. Then $3^{45} \equiv 27$ while $3^{90} \equiv 1 \pmod{91}$. So while 91 is a 3-pseudoprime, the strong 3-pseudoprime sequence is $(27, 1)$, so 91 fails the strong 3-pseudoprime test, and 91 is not prime. (In particular, 27, 1 and -1 are three square roots of 1 modulo 91).

If we try $a = 2$, then $2^{45} \equiv 57$, while $2^{90} \equiv 64 \pmod{91}$: the strong 2-pseudoprime sequence is $(57, 64)$. So 91 is not a 2-pseudoprime, hence is not a strong 2-pseudoprime.

Example 5. Let $m = 97$, a prime. Then $m - 1 = 96 = 2^5 \cdot 3$, so modulo 97, the strong a-pseudoprime sequence is

$$(a^3, a^6, a^{12} a^{24}, a^{48}, a^{96}).$$

Once we find $a^3 \pmod{97}$, the rest of the sequence is determined by squaring modulo 97, so the sequence is easy to compute for any a. Here is a table of some examples:

For $a =$	the sequence is
2	$(8, 64, 22, -1, 1, 1)$
3	$(27, 50, -22, -1, 1, 1)$
5	$(28, 8, 64, 22, -1, 1)$
13	$(63, 89, 64, 22, -1, 1, 1)$
17	$(63, 89, 64, 22, -1, 1)$
61	$(1, 1, 1, 1, 1, 1)$

Thus 97 passes the strong a-pseudoprime test for all of these a's.

Here is the proof that any prime p passes the strong a-pseudoprime test for any number a not divisible by p.

Proposition 5. *If m fails the strong a-pseudoprime test for some a not divisible by m, then m is composite. If m is an a-pseudoprime but not a strong a-pseudoprime, then a yields a factorization of m.*

Proof. Write $m - 1 = 2^e q$, where q is odd. If $a^{2^e q} \not\equiv 1 \pmod{m}$. then m is composite by Fermat's theorem. Suppose $a^{2^e q} \equiv 1 \pmod{m}$ and suppose the sequence

$$(a^q, a^{2q}, \ldots, a^{2^e q})$$

is congruent modulo m to

$$(\ldots, b, 1, \ldots, 1)$$

where $b \equiv a^{2^k q}$ is the rightmost element of the sequence that is not congruent to 1 (mod m). If m fails the strong a-pseudoprime test, $b \not\equiv 1$ or -1 (mod m). Now b^2 is congruent modulo m to the element of the sequence immediately to the right of b, so $b^2 \equiv 1$ (mod m). Hence m divides $b^2 - 1 = (b-1)(b+1)$. Since $b \not\equiv 1$ or -1 (mod m), m does not divide $b-1$ or $b+1$. So m cannot be prime, and, in fact, $m = (m, b-1)(m, b+1)$ is a non-trivial factorization of m.   □

Here are some more examples.

Example 6. Let $m = 341$, a known 2-pseudoprime. The sequence for 341 is

$$(a^{85}, a^{170}, a^{340}).$$

We try $a = 2$. Then $2^{85} \equiv 32$ (mod 341), while $2^{170} \equiv 1$ (mod 341), so the sequence is

$$(32, 1, 1)$$

and 341 is not a strong 2-pseudoprime.

Example 7. Let $m = 561$, the smallest Carmichael number. Then the sequence is

$$(a^{35}, a^{70}, a^{140}, a^{280}, a^{560}).$$

Here is how the sequence looks for various numbers a:

If $a =$	the sequence becomes (mod 561)
2	(263, 166, 67, 1, 1)
5	(23, 529, 463, 67, 1)
7	(241, 298, 166, 67,1)
13	(208, 67, 1, 1, 1)
19	(76, 166, 67, 1, 1)
101	(−1,1,1,1,1)
103	(1, 1, 1, 1, 1)

For all but the last two numbers a, 561 fails the strong a-pseudoprime test. Since 561 fails the strong a-pseudoprime test for at least one a, 561 cannot be prime.

Finally, note that computing the strong a-pseudoprime sequence is essentially no more work than computing a^{m-1} (mod m), because an efficient way to compute a^{m-1} (mod m) is to write $m - 1 = 2^e q$, write q in base 2 and compute a^q (mod m) using the method in chapter 9F, then square the result e times   (mod m). The strong a-pseudoprime sequence is just the sequence of the last e partial results in that computation.

Exercises.

13. Show that 1729 is not prime because it is not a strong 2-pseudoprime.

14. Find some number $a \not\equiv 1$ or -1 (mod 91) so that 91 is a strong a-pseudoprime.

15. Find all numbers $a < 21$ for which 21 is:
   (i) an a-pseudoprime;
   (ii) a strong a-pseudoprime.

16. Find all numbers $a < 35$ for which 35 is:
   (i) an a-pseudoprime;
   (ii) a strong a-pseudoprime.

17. Find all numbers $a < 65$ for which 65 is:
   (i) an a-pseudoprime;
   (ii) a strong a-pseudoprime.

C. Strong Carmichael Numbers

In this section we show that there are no "strong" Carmichael numbers. In other words, every odd composite number m fails the strong a-pseudoprime test for some number a coprime to m.

First, some alternate terminology. If for some number a, the number m fails the strong a-pseudoprime test, then m is necessarily a composite number: in that case the number a is called a *witness to the compositeness of m*, or, for short, a *witness for m*. If m is composite but is a strong a-pseudoprime, the number a is called a *false witness for m*.

In this section we prove:

Theorem 6. *If m is an odd composite number, then at most half of the numbers $a < m$ are false witnesses for m.*

M. O. Rabin (1980) proved a stronger version of Theorem 6. He showed that for m odd and composite, at most one fourth of the numbers $a < m$ are false witnesses. His result leads to the following probabilistic primality test:

For m odd, pick N random numbers a with $1 < a < m - 1$ and do the strong a-pseudoprime test on m.

If m fails any test, then m is composite.

If m passes all N tests, then m is probably prime, in the sense that if m were composite, the chance that m passed all N tests for randomly chosen a is at most $1/4^N$.

Before beginning the proof of Theorem 6, we recall from Chapter 11 the subgroups of roots of unity of the multiplicative group $U = U_m$ of units of $\mathbb{Z}/m\mathbb{Z}$. Let

$$U(k) = \{[a] \text{ in } U \,|\, [a]^k = [1]\},$$

the group of k-th roots of unity in U, for any number k. Then $U(k)$ is a subgroup of U. A number m is Carmichael if and only if $a^{m-1} \equiv 1 \pmod{m}$ for all a coprime to m, hence if and only if $U(m-1) = U$.

Proof. We consider three cases:

1) If a number m is not a Carmichael number, then $U(m-1)$, the set of $[a]$ in U for which m is an a-pseudoprime, is a proper subgroup of U. As we showed in Section 11D, if $q \geq 2$ is the number of cosets of $U(m-1)$ in U, that is, the index of $U(m-1)$ in U, then at most $1/q$ of the units in U are not witnesses for m.

Write $m-1 = 2^e t$ for t odd, and assume that m is Carmichael. Then $U = U(m-1) = U(2^e t)$. Let s be the least number so that $U = U(2^s t)$. Then $s > 0$ since $[-1]^t = [-1]$ is not in $U(2^0 t)$. Then

$$U(2^{s-1}t) = \{[a] \in U \,|\, [a]^{2^{s-1}t} = 1\}$$

is a proper subgroup of U. So the index of $U(2^{s-1}t)$ in U is $q \geq 2$.

2) Suppose there is no $[b]$ so that $[b]^{2^{s-1}t} = -1$. Then every $[a]$ not in $U(2^{s-1}t)$ is a witness to the compositeness of m, and so at most $1/q$ of the units in U are false witnesses for m.

3) Suppose $[b]$ in U satisfies $[b]^{2^{s-1}t} = [-1]$. Then every element $[b']$ of $[b]U$ $(2^{s-1}t)$ satisfies $[b']^{2^{s-1}t} = [-1]$. Moreover, if $[b'']^{2^{s-1}t} = [-1]$, then $[a] = [b''][b]^{-1}$ is in $U(2^{s-1}t)$. Hence every false witness for m is in $U(2^{s-1}t)$ or in $[b]U(2^{s-1}t)$.

We construct two other cosets of $U(s^{s-1}t)$.

Since m is Carmichael, m is divisible by at least two odd primes. So write $m = fg$ with f, g coprime. Since $b^{2^{s-1}t} \equiv -1 \pmod{m}$, we have $b^{2^{s-1}t} \equiv -1 \pmod{f}$ and also mod g.

Let c satisfy $c \equiv b \pmod{f}, c \equiv 1 \pmod{g}$, and let d satisfy $d \equiv 1 \pmod{f}$, $d \equiv b \pmod{g}$. Then

$$c^{2^{s-1}t} \equiv -1 \pmod{f}, c^{2^{s-1}t} \equiv 1 \pmod{g},$$

while

$$d^{2^{s-1}t} \equiv 1 \pmod{f}, d^{2^{s-1}t} \equiv -1 \pmod{g}.$$

Thus $1, b^{2^{s-1}t}, c^{2^{s-1}t}$ and $d^{2^{s-1}t}$ are different square roots of 1 in U, and so

$$U(2^{s-1}t), [[b]^{2^{s-1}t}]U(2^{s-1}t), [[c]^{2^{s-1}t}]U(2^{s-1}t) \text{ and } [[d]^{2^{s-1}t}]U(2^{s-1}t)$$

are distinct cosets. All the false witnesses for m lie in the first two of these cosets. Thus there are at least as many witnesses for m as there are false witnesses. $\square$

Example 8. Consider $1729 = 13 \cdot 7 \cdot 19$, a Carmichael number because

$$1728 = 2^6 \cdot 3^3,$$

and $13 - 1 = 12 = 2^2 \cdot 3, 7 - 1 = 6 = 2 \cdot 3$, and $19 - 1 = 18 = 2 \cdot 3^2$ all divide 1728.
As an application of the Chinese Remainder Theorem, we have

$$U = U_{1729} \cong U_{13} \times U_7 \times U_{19}$$

so the order of U is $12 \cdot 6 \cdot 18 = 1296$ and the exponent of U is the least common multiple of 6, 12 and 18, $[6, 12, 18] = 36$. Hence if $U(k)$ denotes the kth roots of unity in U, then

$$U = U(36).$$

Thus

$$U = U(1728) = U(2^6 \cdot 9) = U(2^2 \cdot 9) = U(36),$$

while $U(18) \neq U$. In the notation of the proof of Theorem 6, $s = 2, t = 9$.

Now every number b coprime to 1729 satisfies $b^6 \equiv 1 \pmod 7$, and so $b^{18} \equiv 1$ $\pmod 7$. Hence no b can satisfy $b^{18} \equiv -1 \pmod{1729}$, and we are in case 2) of the proof of Theorem 6. Thus any coset $bU(18) \neq U(18)$ will consist of witnesses for 1729. To find such a coset, we just need to find a number b so that $b^{18} \not\equiv 1$ $\pmod{1729}$.

We note that

$$2^{18} \equiv 1 \quad \pmod{19},$$

$$2^{18} \equiv 2^{12+6} \equiv 2^6 \equiv -1 \quad \pmod{13}, \text{ and}$$

$$2^{18} \equiv (2^6)^3 \equiv 1 \quad \pmod 7.$$

So

$$2^{18} \not\equiv 1 \text{ or } -1 \quad \pmod{1729}.$$

In fact, $2^{18} \equiv 1065 \pmod{1729}$. So the coset $[2]U(18)$ consists of witnesses for the compositeness of 1729, because every $[c]$ in $[2]U(18)$ satisfies $c^{18} \equiv 1065$ $\pmod{1729}$ while $(c^{18})^2 \equiv 1 \pmod{1729}$.

Example 9. Consider $8911 = 7 \cdot 19 \cdot 67$, a Carmichael number because $8910 = 2 \cdot 3^4 \cdot 5 \cdot 11$, while

$$7 - 1 = 6 = 2 \cdot 3,$$

$$19 - 1 = 18 = 2 \cdot 3^2,$$

$$67 - 1 = 66 = 2 \cdot 3 \cdot 11,$$

all of which divide 8910.

The strong a-pseudoprime sequence for any $[a]$ in U_{8911} looks like (a^{4455}, a^{8910}). Since $(-1)^{4455} = -1$, we are in case 3) of the proof of Theorem 6.

Using the Chinese Remainder Theorem, we can find numbers c that satisfy

$$c \equiv \pm 1 \quad \pmod 7$$

$$c \equiv \pm 1 \quad \pmod{19}$$

$$c \equiv \pm 1 \quad \pmod{67}$$

for each of the eight choices of $+$ and $-$. For all $+$, $c = 1$; for all $-$, $c = -1$. For the other six choices, we get $c = \pm 267, \pm 6364$ and ± 2813. Each of those satisfies $c^{4455} \equiv c \pmod{8911}$ and $c^2 \equiv 1 \pmod{8911}$. So U_{8911} partitions into the following eight cosets:

$$U(4455) = \{[a] \in U_{8911} | [a]^{4455} = [1]\}$$
$$[-1]U(4455) = \{[a] \in U_{8911} | [a]^{4455} = [-1]\}$$
$$[267]U(4455) = \{[a] \in U_{8911} | [a]^{4455} = [267]\}$$
$$[-267]U(4455) = \{[a] \in U_{8911} | [a]^{4455} = [-267]\}$$
$$[6364]U(4455) = \{[a] \in U_{8911} | [a]^{4455} = [6364]\}$$
$$[-6364]U(4455) = \{[a] \in U_{8911} | [a]^{4455} = [-6364]\}$$
$$[2813]U(4455) = \{[a] \in U_{8911} | [a]^{4455} = [2813]\}$$
$$[-2813]U(4455) = \{[a] \in U_{8911} | [a]^{4455} = [-2813]\}$$

The elements of the last six cosets are all witnesses to the compositeness of 8911, because their strong pseudoprime sequences exhibit a non-trivial square root of 1 modulo 8911, namely, $\pm 267, \pm 6364,$ or ± 2813.

Trial strong a-pseudoprime testing is used in practice. An example that appeared in the mathematics research literature involves the eighth Fermat number, $F_8 = 2^{2^8} + 1$. Brent and Pollard (1981) found the smallest prime factor of F_8, namely $p_8 = 1,238,926,361,552,897$. The other factor, $q_8 = F_8/p_8$, is a 62-digit number. Brent and Pollard applied Rabin's test to q_8, and concluded: "the application of more than 100 trials of Rabin's probabilistic algorithm led us to suspect that the cofactor q_8 was prime". (H.C. Williams subsequently gave a nonprobabilistic proof that q_8 is in fact prime.)

The strong a-pseudoprime test has been used to test for primeness in several standard computer algebra systems, such as MAPLE and Mathematica [see Pinch (1993)].

Rabin's test is a probabilistic version of a primality test proposed by Miller in 1975. Miller's test assumes the validity of a certain unsolved conjecture in number theory known as the Generalized Riemann Hypothesis (GRH). As improved by Bach in 1990, Miller's test is the following:

Suppose the GRH is true. If m is a composite number, then m will fail the strong a-pseudoprime test for some $a < 4(\log m)^2$.

Thus if the GRH is true, then a d digit number m will be prime if it passes the strong a-pseudoprime test for all $a, 1 < a < 4(\log 10^d)^2 < (21.4)d^2$. If m has 100 digits, $(21.4)d^2 = 214,000$. So even if GRH is true, going from a conclusion that m is probably prime, using 20 or 100 a's, to a conclusion that m is certainly prime, using all the numbers $a < 4(\log m)^2$, would involve considerable additional computation, even for numbers of under 100 digits. See also the comments by Knuth (1998, p. 395).

A final note. Alford, Granville, and Pomerance, as a byproduct of their proof that there are infinitely many Carmichael numbers, also proved that given any fixed set B of numbers a, there are infinitely many Carmichael numbers which are strong

a-pseudoprimes for all a in the set B: "strong B-pseudoprimes." This means that if you fix the set B in advance, rather than choosing potential witnesses a at random, then the analogue of Rabin's theorem is false (see Granville (1992)).

Exercises.

18. Verify the theorem for the Carmichael number $m = 1105$.

19. Verify the theorem for the Carmichael number $m = 1729$.

20. Verify the theorem for the Carmichael number $m = 41041$.

D. RSA Codes and Carmichael Numbers

In this section we apply ideas of this chapter to make some observations about RSA codes and Carmichael numbers.

(i) Carmichael Numbers in RSA Codes. In Chapter 10A we examined the RSA cryptosystem, which encrypts a message word a by replacing a by a^e (mod m) where the modulus m is the product of two large prime numbers p and q.

To find a large prime p we could proceed as follows: first pick an interval of numbers of the desired size and sieve out all the composite numbers with small prime numbers as factors, as in section 6G. Then we use the a-pseudoprime test or the strong a-pseudoprime test to test the remaining unsieved numbers for primeness, as in section 10B.

Suppose, after sieving, we found a potential prime number q, and we used the a-pseudoprime test repeatedly, checking $a^{q-1} \equiv 1$ (mod q) for a collection of numbers a.

If q is prime, q will pass this test for any a.

If q is composite and Carmichael, q will also pass this test for any a not relatively prime to q.

If q is composite and not Carmichael, then the set of a (mod q) for which $a^{q-1} \equiv 1$ (mod q) is a proper subgroup of the group of units of $\mathbb{Z}/q\mathbb{Z}$, so the probability that q passes the a-pseudoprime test for a randomly selected number a is at most $1/2$. For such numbers q, repeated testing with randomly chosen numbers a will almost surely reveal that q is composite.

Carmichael numbers are very much rarer than primes. So if we had a number q which passed repeated a-pseudoprime tests, it would be reasonable for us to assume that q is prime, not Carmichael.

But suppose we were wrong? Suppose q were Carmichael?

Suppose p and q are coprime Carmichael numbers. Let $m = pq$ and set up an RSA code with modulus m. Believing that p and q are primes, we would assume that $\phi(m) = (p-1)(q-1)$. As in section 10A, we would pick an encoding exponent

e by choosing any number e coprime to $(p-1)(q-1)$. We would find the decoding exponent via Bezout's identity: since $(e,(p-1)(q-1)) = 1$, there is some d,k so that $ed - k(p-1)(q-1) = 1$. Then for any integer a,

$$a^{ed} = a^{1+k(p-1)(q-1)}.$$

If p,q are primes, we know by Euler's Theorem that

$$a^{1+k(p-1)(q-1)} \equiv a \pmod{m}.$$

But what happens if p and q are not primes, but are Carmichael numbers? Then the construction still works:

Proposition 7. *If p and q are primes or Carmichael numbers, then for any $a < m$, $a^{1+k(p-1)(q-1)} \equiv a \pmod{m}$.*

Proof. First note that since p and q are each assumed to be either prime or Carmichael, each is squarefree. Since p and q are coprime, it follows that m is squarefree. Hence

$$a^{1+k(p-1)(q-1)} \equiv a \pmod{m}$$

iff

$$a^{1+k(p-1)(q-1)} \equiv a \pmod{c}$$

where c is any prime divisor of m.

If c is a prime divisor of m, then c divides p or c divides q. Suppose c divides p. If $c = p$, then $c - 1 = p - 1$; if p is Carmichael, $c - 1$ divides $p - 1$ by Korselt's criterion.

Now by Fermat's theorem, for any a coprime to c,

$$a^{c-1} \equiv 1 \pmod{c},$$

so

$$a^{h(c-1)+1} \equiv a \pmod{c}$$

for every h, and in particular if $h = k(q-1)(p-1)/(c-1)$, an integer since $c - 1$ divides $p - 1$.

But this last congruence is also true if a is divisible by c, and so it is true for every a.

Thus for any prime c dividing m,

$$a^{(p-1)(q-1)+1} \equiv a \pmod{c}.$$

Since m is squarefree, it follows that

$$a^{(p-1)(q-1)+1} \equiv a \pmod{m},$$

for every $a < m$, as we wished to show. $\qquad\square$

Thus for setting up RSA codes, Carmichael numbers work as well as primes. Of course, if $m = pq$ is a product of Carmichael numbers then the prime factors of m will be much smaller than m, so will be easier to find than if p and q were prime. So the security of the RSA code will be less than it would be if p and q are primes.

(ii) Factoring Carmichael Numbers. A Carmichael number m is an a-pseudoprime for all a coprime to m, but as we saw in the last section, m is not a strong a-pseudoprime for most numbers a coprime to m. This fact leads to the following perhaps surprising fact:

Proposition 8. *An odd Carmichael number is easy to factor.*

Proof. Let m be an odd Carmichael number. Then for any number a relatively prime to $m, a^{m-1} \equiv 1 \pmod{m}$. By Rabin's theorem, m is a strong a-pseudoprime for at most 1/4 of all numbers $a < m$. So barring exceptionally bad luck, it should be easy to find a number a for which m is not a strong a-pseudoprime. Such a number a yields a factorization of m by Proposition 5.   □

Example 10. To illustrate with a small example, 561 is a 2-pseudoprime but not a strong 2-pseudoprime: $2^{140} \equiv 67 \pmod{561}$, while $2^{280} \equiv 67^2 \equiv 1 \pmod{561}$. Thus setting $b = 67$, we have

$$561 = (561, 66) \cdot (561, 68).$$

Here $(561, 66) = 33$, while $(561, 68) = 17$.

It is ironic that numbers whose compositeness is the most difficult to verify by pseudoprime tests are at the same time so easy to factor.

Exercises.

21. It is a fact that $2^{54} \equiv 1065 \pmod{1729}$, while $2^{108} \equiv 1 \pmod{1729}$. Factor 1729.

22. It is a fact that $b = 2^{73602} \equiv 262144 \pmod{294409}$, while $b^2 \equiv 1 \pmod{294409}$. Factor 294409.

23. Factor the Carmichael number 29341.

24. Factor the Carmichael number 252601.

25. Factor the Carmichael number 3215031751.

26. Suppose m is odd, m divides $b^2 - 1$ and m does not divide $b - 1$ or $b + 1$ Show that $m = (m, b+1)(m, b-1)$. (Since $m = (m, b^2 - 1)$, this follows from a result in Chapter 4 if $b + 1$ and $b - 1$ are coprime. What if they are not?)

27. We showed that Fermat numbers, numbers of the form $F_n = 2^{2^n} + 1$, are 2-pseudoprimes. Show that the factorization idea used for Carmichael numbers will not work for Fermat numbers, because a composite Fermat number is not only a 2-pseudoprime but a strong 2-pseudoprime.

28. We showed that Mersenne numbers, numbers of the form $M_p = 2^p - 1$ with p prime, are 2-pseudoprimes. Show that the factorization idea used for Carmichael numbers will not work for Mersenne numbers, because a composite Mersenne number is not only a 2-pseudoprime but a strong 2-pseudoprime.

(iii) Choosing the Primes for an RSA Modulus. The idea that a composite number m is easy to factor whenever it is an a-pseudoprime but not a strong a-pseudoprime is a consideration in choosing primes p and q for a modulus in an RSA code.

We want the modulus m to be hard to factor by anyone who does not already know the prime factors p and q of m. Thus we want m to be an a-pseudoprime for as few numbers a as possible.

If we use in an RSA code a modulus $m = pq$ where p and q are primes and $(p-1)/2$ and $(q-1)/2$ are coprime, then m will be an a-pseudoprime for only four numbers $a \pmod{m}$, and a strong a-pseudoprime for half of those, namely $a = 1$ and -1. Thus the chance of coming up with some number a which will lead to a factorization of m is $2/m$, a chance which is smaller than that of finding a number $< m$ that is not coprime to m.

In particular, if $(p-1)/2$ and $(q-1)/2$ are primes, then $m = pq$ will be safe from this kind of attack. Thus a prime p so that $(p-1)/2$ is prime is called a safeprime.

Exercises.

29. Factor $m - 2741311$ by finding some number a for which m is an a-pseudoprime but not a strong a-pseudoprime.

30. Verify that if p and q are safeprimes with $p, q, (p-1)/2$ and $(q-1)/2$ all distinct primes, then $m = pq$ is an a-pseudoprime for only four numbers $a \pmod{m}$, and a strong a-pseudoprime only for $a = 1$ and $-1 \pmod{m}$.

31. Let $m = pq$ where p, q are primes with $p > 2q$, and suppose p is a safeprime (that is, $\frac{p-1}{2}$ is prime). Show that m is an a-pseudoprime but not a strong a-pseudoprime for only two numbers $a < m$.

32. Let $p \equiv 3 \pmod 4$ and suppose p and $2p - 1 = q$ are primes (e.g. $p, q = 31, 61$). Let $m = pq$. Then $m - 1 = 2t$ for t odd
   (i) Show that $U_m(m-1)$ has order $(p-1)^2$, hence has index 2 in U_m.
   (ii) Show that $U_m(t)$ has order $(\frac{p-1}{2})^2$, hence has index 4 in U_m.
   (iii) Show that all false witnesses for m lie in $U_m(t)$ or in $[1]U_m(t)$.
   (iv) Show that one-fourth of the elements of U_m will yield a factorization of m via Proposition 5.

33. Find an analogous result for $m = pq$ where p and $q = 2p - 1$ are prime and $p \equiv 1 \pmod 4$.

(iv) A User of an RSA Code Can Factor the Modulus. When the RSA code was first publicized, one idea was that a user of RSA codes who received coded messages from a number of different sources could use the same modulus for all of the messages.

For example, suppose a stock broker receives orders from various clients around the world, by phone or wire. The broker would like to be certain that the orders it receives are authentic. The RSA code could be used as a kind of verified signature on the order. It would work as follows:

The client and broker have an RSA code with modulus m. The client is given a secret encoding exponent e, and the broker assigns for the client's account, and makes public as appropriate (e.g., for audits) the decoding exponent d. The client places an order by taking the order message a, encoding it by his secret exponent e to get $c \equiv a^e \pmod m$, and sends the encoded order c to the broker. The broker decodes it by replacing c by $c^d \pmod m$, which will yield the original order a.

Since only the client knows the exponent e, the broker will know that the order is authentic. The encoding of the order using e is like adding the client's signature to the order.

The broker might find some technical advantage in using the same modulus m for every client. For example, equipment that computes modulo the common modulus m could be used for all the broker's clients.

However, DeLaurentis (1984) pointed out that a client who knows the modulus m and both the encoding exponent e and the decoding exponent d would, with high probability, be able to factor m, and thereby be able to compromise the authenticity of all other clients' signatures.

The idea is the same as that noted above for Carmichael numbers.

Suppose m is the modulus, $m = pq$, p, q unknown primes, and e, d are the encoding and decoding exponents known to the client. Then $ed - 1$ is a multiple of $(p-1)(q-1) = \phi(m)$. Write $ed - 1 = 2^h r$, where r is odd. Since $\phi(m)$ is a product of two even numbers, $h \geq 2$.

Suppose $p - 1 = 2^u v$ and $q - 1 = 2^t w$, with v, w odd, and suppose $u \leq t$. Note that $u \geq 1$, and $h \geq u$.

For every a coprime to m, $a^{2^h r} \equiv 1 \pmod m$. We consider the sequence:

$$\{a^r, a^2 r, \ldots, a^{2^h r}\} \pmod m$$

just as in the strong a-pseudoprime test. Then the proof of Theorem 6 applies without change to show that at least half of the units modulo m yield a non-trivial square root of 1 in the sequence

$$\{a^r, a^{2r}, a^{2^2} r, \ldots, a^{2^h} r\} \pmod m$$

Thus at least half of all numbers a modulo m will yield a factorization of m.

Repeatedly choosing randomly numbers a will therefore yield with arbitrarily high probability a factorization of m.

Exercises.

34. For a relatively prime to 8509, $a^{374220} \equiv 1 \pmod{8509}$. Factor 8509.

35. Let $m = 77123, e = 79, d = 33919$ be the components of an RSA code. Factor m.

36. Let m be as in the last exercise, and suppose you are assigned the e and d of the last exercise. Suppose a business rival has the same m, and you know her encoding exponent $e = 133$. Crack her code by finding a decoding exponent d.

Chapter 21
Quadratic Reciprocity

In this chapter we describe a procedure for deciding efficiently whether or not a given number is a square modulo m. The main result, known as the law of quadratic reciprocity, was first proved by Gauss (1801) and is a cornerstone of number theory. The last section gives some applications of quadratic reciprocity to primality testing. In the next chapter we give a variety of other applications of the ideas in this chapter.

A curious aspect of the main result of this chapter is that while quadratic reciprocity makes it easy to determine whether or not the congruence $x^2 \equiv a \pmod{m}$ has a solution, the method provides very little help in determining what number is a solution if there is one. (See Section 22C for some implications of this fact.)

A. Reduction to the Odd Prime Case

If a and m are coprime, then the number a is a *square* modulo m, or, in number theory terminology, a is a *quadratic residue* modulo m, if the congruence $x^2 \equiv a \pmod{m}$ has a solution, and a is a *non-square* modulo m, or a is a *quadratic nonresidue* modulo m, if the congruence has no solution.

This section will show that to decide whether a number a coprime to m is a square modulo m for any m, it suffices to be able to decide whether a is a square modulo p when p is an odd prime.

First, we use the Chinese Remainder Theorem to reduce to the case that m is a prime power.

Theorem 1. *Let $m = p_1^{e_1} p_2^{e_2} \cdots p_r^{e_r}$. Then the number a is a quadratic residue mod m iff a is a square modulo each prime power divisor $p_i^{e_i}$ of m.*

Proof. If there is some number b so that $b^2 \equiv a \pmod{m}$, then $b^2 \equiv a \pmod{d}$ for every divisor d of m. Conversely, suppose we can find numbers $b_1, \ldots, b_r$ so that

$$b_1^2 \equiv a \quad (\text{mod } p_1^{e_1}),$$
$$b_2^2 \equiv a \quad (\text{mod } p_2^{e_2}),$$
$$\vdots$$
$$b_r^2 \equiv a \quad (\text{mod } p_r^{e_r}).$$

By the Chinese Remainder Theorem there is a unique number b modulo $m = p_1^{e_1} \cdots p_r^{e_r}$ such that

$$b \equiv b_1 \quad (\text{mod } p_1^{e_1}),$$
$$\vdots$$
$$b \equiv b_r \quad (\text{mod } p_r^{e_r}).$$

The number b then satisfies $b^2 \equiv a$ (mod $p_i^{e_i}$) for each i, and hence,

$$b^2 \equiv a \quad (\text{mod } m).$$

$\square$

Thus to decide whether a number a is a square modulo m, it suffices to decide if a is a square modulo the prime power divisors of m. To do that we must consider separately the case where the prime is odd, and the case where the prime is 2.

For the odd prime case, we recall some information about the group of units U_{p^e} of $\mathbb{Z}/p^e\mathbb{Z}$.

First, for p an odd prime and all $e \geq 1$, the group U_{p^e} is a cyclic group, generated by some element b. In other words, there is a primitive root b modulo p^e (Section 19C). Thus every number a coprime to p has the form $a \equiv b^s$ (mod p^e) for some exponent s. Since the order of b modulo p^e is $\phi(p^e) = p^{e-1}(p-1)$, the exponent s is uniquely determined modulo $p^{e-1}(p-1)$. That is, if

$$a \equiv b^s \equiv b^t \quad (\text{mod } p^e)$$

then

$$s \equiv t \quad (\text{mod } p^{e-1}(p-1)).$$

Since $p - 1$ is even, this implies that

$$s \equiv t \quad (\text{mod } 2).$$

Thus every number a coprime to p is either an odd power of b or an even power of b modulo p^e, never both.

With that observation, we can prove

Theorem 2. *Let p be an odd prime, and $(a, p) = 1$. Then there is a solution of $x^2 \equiv a$ (mod p^e), $e > 1$, if and only if there is a solution of $x^2 \equiv a$ (mod p).*

Proof. If $c^2 \equiv a$ (mod p^e), then $c^2 \equiv a$ (mod p).

Conversely, suppose a is a square modulo p, so that for some c, $a \equiv c^2$ (mod p).

Let b be a primitive root modulo p^e. Then $a \equiv b^r$ (mod p^e) for some r. We will show that r is even.

We know that b is also a primitive root modulo p by Exercise 27 of Section 19C. If $a \equiv b^r$ (mod p^e), then $a \equiv b^r$ (mod p). Let $c \equiv b^t$ (mod p). Since $a \equiv c^2$ (mod p), therefore $a \equiv b^{2t}$ (mod p). Hence r is even. Thus $r = 2s$ for some s, and so $a \equiv b^r = b^{2s} = (b^s)^2$ (mod p^e). Hence a is a square modulo p^e.   □

Example 1. We can see that 61 is a square modulo 125, because 61 is obviously a square modulo 5:

$$61 \equiv 1 \quad (\text{mod } 5)$$

and $1 = 1^2$ is a square modulo 5.

On the other hand, 62 is not a square modulo $343 = 7^3$ because modulo 7, $62 \equiv 5$, and 5 is not a square modulo 7 (the squares modulo 7 are 1, 2 and 4).

The case where the modulus is a power of 2 is somewhat different.

Theorem 3. *Suppose a is odd. Then:*
(i) a is always a square modulo 2.
(ii) a is a square modulo 4 iff $a \equiv 1$ (mod 4).
(iii) a is a square modulo 2^e, $e \geq 3$, iff $a \equiv 1$ (mod 8).

Proof. (i) and (ii) are easy. For (iii), suppose first that $b^2 \equiv a$ (mod 2^e) for some $e \geq 3$. Then $b^2 \equiv a$ (mod 8), so a must be $\equiv 1$ (mod 8), since 1 is the only odd square modulo 8.

Conversely, suppose $a = 1 + 8n$ for some fixed integer n. We show that for any $e \geq 3$ there is some x with

$$x^2 \equiv a \quad (\text{mod } 2^e)$$

by induction on e.

If $a = 1 + 8n$, then a is obviously a square modulo 8.

Assume $e > 3$ and suppose

$$y^2 \equiv a \quad (\text{mod } 2^{e-1})$$

for some (necessarily odd) number y. Then $y^2 = a + 2^{e-1}u$ for some integer u. Set

$$w = y + 2^{e-2}u.$$

Then, since $e \geq 4$, $2(e-2) > e$, and so

$$w^2 \equiv y^2 + 2^{e-1}yu + 2^{2(e-2)}u^2$$
$$\equiv a + 2^{e-1}u + 2^{e-1}yu$$
$$\equiv a + 2^{e-1}u(1+y) \quad (\text{mod } 2^e).$$

Since y is odd, $w^2 \equiv a$ (mod 2^e). Thus if $a \equiv 1$ (mod 8), then for any $e \geq 3$ we can find some number w so that $w^2 \equiv a$ (mod 2^e).   □

Example 2. Since $17 \equiv 1 \pmod 8$, 17 is a square modulo $2^{10} = 1024$. In fact, we can determine a number w with $w^2 \equiv 17 \pmod{1024}$ by following the induction step in the proof, as follows:

$$17 \equiv 1 = 1^2 \pmod{16}$$

Hence

$$1^2 = 17 + 16(-1): \quad y = 1, u = -1.$$

Then for a number whose square is 17 modulo 32, we set

$$w = y + 8u = 1 + 8(-1) = -7.$$

Then

$$(-7)^2 = 49 \equiv 17 \pmod{32}.$$

We can replace -7 by 7 since squaring each yields the same result. Then

$$7^2 = 17 + 32: \quad y = 7, u = 1.$$

For a number whose square is 17 modulo 64, we set

$$w = y + 16u = 7 + 16 = 23.$$

Then it turns out that

$$w^2 = 23^2 = 529 = 2^9 + 17 \equiv 17 \pmod{2^9}.$$

For a number whose square is 17 modulo $2^{10} = 1024$, we set

$$y = 23 + 2^8 = 273.$$

Then

$$273^2 = 17 + 2^{10} \cdot 63.$$

So

$$273^2 \equiv 17 \pmod{1024}.$$

We can continue this process indefinitely to find a square root of 17 modulo any power of 2.

Exercises.

1. Find some w so that $w^2 \equiv 33 \pmod{128}$.

2. Find some number w so that $w^2 \equiv 65 \pmod{512}$.

3. In Example 1, find some w so that $61 \equiv w^2 \pmod{125}$, following the approach of Example 2.

4. Let $t_n = \frac{n(n+1)}{2}$ be the nth triangular number, for $n \geq 0$. Show that the square of every odd integer has the form $8t_n + 1$ for some n.

B. The Legendre Symbol

Theorems 2 and 3 of the last section imply that to decide when $x^2 \equiv a \pmod{m}$ is solvable, it is enough to find criteria to decide when $x^2 \equiv a \pmod{p}$ is solvable when p is an odd prime.

To help decide whether or not a given number is a square, that is, a quadratic residue, modulo p, we introduce the Legendre symbol.

Definition. Let p be an odd prime, and a any number not divisible by p. Then the Legendre symbol (a/p) is defined by

$$\left(\frac{a}{p}\right) = 1 \text{ if } a \text{ is a quadratic residue mod } p,$$

$$\left(\frac{a}{p}\right) = -1 \text{ if } a \text{ is quadratic nonresidue mod } p.$$

To decide whether a is a square modulo p, we can manipulate Legendre symbols. Here are the rules.

Theorem 4. *Assume p, q are distinct odd primes and a and b are integers coprime to p. Then*

(1) $\left(\dfrac{a^2}{p}\right) = 1$;

(2) *if $a \equiv b \pmod{p}$, then* $\left(\dfrac{a}{p}\right) = \left(\dfrac{b}{p}\right)$;

(3) $\left(\dfrac{ab}{p}\right) = \left(\dfrac{a}{p}\right)\left(\dfrac{b}{p}\right)$;

(4) $\left(\dfrac{-1}{p}\right) = (-1)^{\frac{p-1}{2}}$;

(5) $\left(\dfrac{2}{p}\right) = (-1)^{\frac{p^2-1}{8}}$;

(6) $\left(\dfrac{p}{q}\right)\left(\dfrac{q}{p}\right) = (-1)^{(\frac{p-1}{2})(\frac{q-1}{2})}$.

These are the condensed versions of the formulas. They become easier to understand when we translate the exponents of -1 in Rules (4), (5) and (6) into congruence conditions. As before, p and q are odd primes:

(4) $\left(\dfrac{-1}{p}\right) = 1$ if $p \equiv 1 \pmod{4}$;

$\qquad\qquad = -1$ if $p \equiv 3 \pmod{4}$

(5) $\left(\dfrac{2}{p}\right) = 1$ if $p \equiv 1$ or $7 \pmod{8}$;

$\qquad\qquad = -1$ if $p \equiv 3$ or $5 \pmod{8}$

(6) $\left(\dfrac{q}{p}\right) = \left(\dfrac{p}{q}\right)$ if p or q is $\equiv 1 \pmod 4$;

$\qquad = -\left(\dfrac{p}{q}\right)$ if both p and q are $\equiv 3 \pmod 4$.

Rule (6) is the famous Law of Quadratic Reciprocity.

Using rules (1)-(6), and especially quadratic reciprocity (rule (6)), we can rather easily determine for a prime p whether a number a is a square modulo p.

Example 3. Is -42 a square mod 103? We ask, is $(-42/103) = 1$ or -1? Using the six rules, we manipulate as follows:

$$\left(\frac{-42}{103}\right) = \left(\frac{-1}{103}\right)\left(\frac{2}{103}\right)\left(\frac{3}{103}\right)\left(\frac{7}{103}\right) \text{ by rule (3)}.$$

Rule (6) gives

$$\left(\frac{3}{103}\right) = -\left(\frac{103}{3}\right)$$

and Rule (2) gives

$$\left(\frac{103}{3}\right) = \left(\frac{1}{3}\right) = 1, \text{ so } \left(\frac{103}{3}\right) = -1.$$

Similarly, Rule (6) gives

$$\left(\frac{7}{103}\right) = -\left(\frac{103}{7}\right).$$

and Rule (2) and a quick review of squares modulo 7 give

$$\left(\frac{103}{7}\right) = \left(\frac{5}{7}\right) = -1,$$

hence

$$\left(\frac{7}{103}\right) = 1.$$

Also

$$\left(\frac{-1}{103}\right) = -1$$

by Rule (4) and

$$\left(\frac{2}{103}\right) = 1$$

by Rule (5). So

$$\left(\frac{-42}{103}\right) = \left(\frac{-1}{103}\right)\left(\frac{2}{103}\right)\left(\frac{3}{103}\right)\left(\frac{7}{103}\right)$$
$$= (-1)(1)(1)(-1) = 1.$$

Thus -42 is a square modulo 103. (In fact, $-42 \equiv (24)^2 \pmod{103}$.)

It is apparent that the law of quadratic reciprocity, rule (6), is the rule that gives the most striking results. It permits us, for example, to decide whether 3 is a square mod 103 by seeing whether 103 is a square mod 3. The former question is not so easy to decide directly, but the latter is very easy.

The rest of this section and the next are devoted to proofs of the various properties of the Legendre symbol collected as Theorem 1. The proofs range from trivial to very clever. The section following the proofs has several applications of these ideas.

Proof of Rule (1). Rule (1) restates the definition of the Legendre symbol.

Proof of Rule (2). $(a/p) = 1$ if and only if there is some number c with $c^2 \equiv a$ $(\mathrm{mod}\ p)$. If $a \equiv b$ $(\mathrm{mod}\ p)$ then $c^2 \equiv a$ $(\mathrm{mod}\ p)$ iff $c^2 \equiv b$ $(\mathrm{mod}\ p)$. So $(a/p) = 1$ if and only if $(b/p) = 1$.

The proofs of Rules (3)-(6) are facilitated by

Theorem 5 (Euler's Lemma). *Let p be an odd prime. If $(a, p) = 1$, then*

$$\left(\frac{a}{p}\right) \equiv a^{(p-1)/2} \quad (\mathrm{mod}\ p).$$

Proof. Let b be a primitive root modulo p. Then b has order $p - 1$ modulo p. If $a \equiv b^r$ $(\mathrm{mod}\ p)$, then the congruence class of r $(\mathrm{mod}\ 2)$ is uniquely determined by a, as noted above Theorem 2. So every a coprime to p is either congruent to an even power of b mod p, and hence is a square mod p, or is congruent to an odd power of b, hence cannot be congruent to an even power of b, hence cannot be a square mod p.

Thus the squares modulo p are the numbers congruent to even powers of b.

Now $b^{\frac{p-1}{2}} \equiv -1$ $(\mathrm{mod}\ p)$ since $(b^{\frac{p-1}{2}})^2 \equiv 1$ $(\mathrm{mod}\ p)$ and b has order $p - 1$ mod p. Let $a \equiv b^r$ $(\mathrm{mod}\ p)$. Then

$$a^{\frac{p-1}{2}} \equiv b^{r\frac{p-1}{2}} \equiv (b^{\frac{p-1}{2}})^r \equiv (-1)^r \quad (\mathrm{mod}\ p).$$

If $\left(\frac{a}{p}\right) = 1$ then r is even, so

$$a^{\frac{p-1}{2}} \equiv (-1)^r = 1 \quad (\mathrm{mod}\ p),$$

while if $\left(\frac{a}{p}\right) = -1$ then r is odd, so

$$a^{\frac{p-1}{2}} \equiv (-1)^r = -1 \quad (\mathrm{mod}\ p).$$

Thus in either case,

$$\left(\frac{a}{p}\right) \equiv a^{\frac{p-1}{2}} \quad (\mathrm{mod}\ p).$$

$\square$

Returning to the proof of the rules for the Legendre symbol collected in Theorem 1:

Proof of Rule (4). This is Euler's lemma with $a = -1$.

Proof of Rule (3). This follows from the identity $(ab)^{(p-1)/2} = a^{(p-1)/2} \cdot b^{(p-1)2}$ and Euler's lemma.

The proofs of Rules (5) and (6) are more difficult. We'll do Rule (5) in this section and do Rule (6) in the next.

Proof of Rule (5). Our proof of Rule (5) starts from the same idea used for the proof in Section 9B of Fermat's Theorem: if p is an odd prime and a is coprime to p, then $a^{p-1} \equiv 1 \pmod{p}$.

To prove Fermat's Theorem, we observed that the set

$$\{a, 2a, 3a, \ldots, (p-1)a\}$$

and the set

$$\{1, 2, 3, \ldots, p-1\}$$

are both complete sets of representatives for the non-zero elements of $\mathbb{Z}/p\mathbb{Z}$. Hence multiplying the elements of each of the sets together gives the congruence

$$a \cdot 2a \cdot 3a \cdots (p-1)a \equiv 1 \cdot 2 \cdot 3 \cdots (p-1) \pmod{p}.$$

Cancelling $1, 2, \ldots, p-1$ from both sides yields $a^{p-1} \equiv 1 \pmod{p}$.

For Rule (5), we consider the product D modulo p of the even numbers between 1 and p:

$$2 \cdot 1, 2 \cdot 2, 2 \cdot 3, \ldots, 2 \cdot \left(\frac{p-1}{2}\right)).$$

We look at D modulo p in two ways.

Before doing so in general, we first illustrate the idea with an example.

Example 4. We determine $\left(\frac{2}{19}\right)$, by looking at

$$D = 2 \cdot 4 \cdot 6 \cdot 8 \cdot 10 \cdot 12 \cdot 14 \cdot 16 \cdot 18.$$

On the one hand, the product is

$$D = 2^9 \cdot (1 \cdot 2 \cdot 3 \cdot 4 \cdot 5 \cdot 6 \cdot 7 \cdot 8 \cdot 9)$$
$$= 2^9 \cdot 9!$$
$$\equiv \left(\frac{2}{19}\right) \cdot 9! \pmod{19},$$

the last congruence by Euler's Lemma. On the other hand, if we replace all the even numbers $> \frac{19}{2}$ by their residues closest to zero by subtracting 19 from each, we obtain

$$D = 2 \cdot 4 \cdot 6 \cdot 8 \cdot 10 \cdot 12 \cdot 14 \cdot 16 \cdot 18$$
$$\equiv 2 \cdot 4 \cdot 6 \cdot 8 \cdot (-9) \cdot (-7) \cdot (-5) \cdot (-3) \cdot (-1)$$
$$\equiv (-1)^5 \cdot (1 \cdot 2 \cdot 3 \cdot 4 \cdot 5 \cdot 6 \cdot 7 \cdot 8 \cdot 9)$$
$$\equiv (-1)^5 \cdot 9! \quad (\text{mod } 19).$$

Canceling 9! gives

$$\left(\frac{2}{19}\right) \equiv (-1)^5 \quad (\text{mod } 19),$$

hence

$$\left(\frac{2}{19}\right) = (-1)^5,$$

where the exponent 5 on the right hand side is the number of odd numbers $< 19/2$.

We follow the same idea as in the example, to prove:

Proposition 6. *For p an odd prime,*

$$\left(\frac{2}{p}\right) = (-1)^t,$$

where t is the number of odd numbers $< \frac{p}{2}$.

Proof. Let D be the product of the $(p-1)/2$ even numbers between 1 and p.
On the one hand, the product

$$D = 2 \cdot 4 \cdot 6 \cdot \ldots \cdot p - 1$$
$$= 2^{\frac{p-1}{2}} \left(\frac{p-1}{2}\right)!,$$

and so by Euler's Lemma,

$$D = \left(\frac{2}{p}\right)\left(\frac{p-1}{2}\right)!.$$

On the other hand, if we subtract p from all the even numbers between $p/2$ and p, we get the negatives of the odd numbers $< p/2$. Thus D is congruent modulo p to the product of the even numbers $< p/2$ and the negatives of the odd numbers $< p/2$. Hence

$$D \equiv (1 \cdot 2 \cdot \ldots \cdot \frac{p-1}{2}) \cdot (-1)^t \quad (\text{mod } p) \equiv (\frac{p-1}{2})!(-1)^t \quad (\text{mod } p)$$

where t is the number of odd numbers $< p/2$.
Putting together the two descriptions of D gives

$$\left(\frac{2}{p}\right)\left(\frac{p-1}{2}\right)! \equiv \left(\frac{p-1}{2}\right)!(-1)^t \quad (\text{mod } p).$$

Canceling $(\frac{p-1}{2})!$ gives

$$\left(\frac{2}{p}\right) \equiv (-1)^t \quad (\text{mod } p).$$

Since both sides equal 1 or -1, we have

$$\left(\frac{2}{p}\right) = (-1)^t.$$

□

It is easy to check that

$$t = \frac{p-1}{4} \text{ iff } \frac{p-1}{2} \text{ is even, iff } p \equiv 1 \quad (\text{mod } 4),$$

$$t = \frac{p+1}{4} \text{ iff } \frac{p-1}{2} \text{ is odd, iff } p \equiv 3 \quad (\text{mod } 4).$$

To obtain Rule 5, we have four cases. If $p \equiv 1 \pmod 4$, then

$$\left(\frac{2}{p}\right) = (-1)^{\frac{p-1}{4}}.$$

Thus if $p \equiv 1 \pmod 8$, then

$$\left(\frac{2}{p}\right) = 1$$

while if $p \equiv 5 \pmod 8$, then

$$\left(\frac{2}{p}\right) = -1$$

If $p \equiv 3 \pmod 4$, then

$$\left(\frac{2}{p}\right) = (-1)^{\frac{p+1}{4}}.$$

Thus if $p \equiv 3 \pmod 8$, then

$$\left(\frac{2}{p}\right) = -1$$

while if $p \equiv 7 \pmod 8$, then

$$\left(\frac{2}{p}\right) = 1.$$

That proves Rule (5).

We'll prove Rule (6), the Law of Quadratic Reciprocity, in the next section.

Exercises.

5. Is 45 a square mod 47?

6. Is -13 a square mod 37?

7. Is 48 a square mod 37?

8. Is 14 a square mod 65? (Hint: Use Theorem 1 of Section A).

9. Is 31 a square mod 200?

10. Is 311 a square mod 1001?

11. Show that if p is an odd prime and a,b,c are integers with $(a,p) = 1$, then $ax^2 + bx + c$ factors modulo p iff

$$\left(\frac{b^2 - 4ac}{p} \right) = 1.$$

12. Prove that if p is an odd prime, then

$$\sum_{r=1}^{p-1} \left(\frac{r}{p} \right) = 0.$$

13. (i) Suppose $p \equiv 1 \pmod 6$. Using the strategy of proof of Rule (5), prove that

$$\left(\frac{3}{p} \right) \equiv (-1)^{(p-1)/6}.$$

(ii) Find an analogous rule when $p = 5 \pmod 6$. Check your rule using Rule (6).

C. Proof of Quadratic Reciprocity

In this section we give a proof of Rule (6), the Law of Quadratic Reciprocity:

Theorem 7. *Let p and q be odd primes. Then*

$$\left(\frac{p}{q} \right) = \left(\frac{q}{p} \right) \cdot (-1)^{\frac{p-1}{2}\frac{q-1}{2}}.$$

In words, p is a square modulo q if and only if q is a square modulo p, unless p and q are both congruent to 3 modulo 4, in which case p is a square modulo q if and only if q is not a square modulo p.

The Law of Quadratic Reciprocity is one of the most subtle results in all of mathematics. Legendre first formulated the law in 1788, but was unable to prove it. Gauss was the first to prove the law–he found a proof in April of 1796. But he evidently decided that he had not captured the essence of the result, because he gave three other proofs that same year, and four additional proofs during the next 12 years. After Gauss, many of the giants of 19th century mathematics studied the result and published their own proofs, including Jacobi, Eisenstein (five proofs), Liouville, Kummer, Dedekind, Dirichlet, Kronecker (eight proofs), de la Vallee Poussin, and

Hilbert. By 1900 some 93 proofs had been published. According to Franz Lemmer-meyer, by 2005 there were 221 published proofs of Quadratic Reciprocity. See his website, http://www.rzuser.uni-heidelberg.de/~hb3/rchrono.html for a list.

The proof we give here is proof No. 193, published by G. Rousseau (1991), which is based on the fifth of Gauss's eight proofs. It uses the Chinese Remainder Theorem, Euler's Lemma and multiplication on cosets.

The proof consists of comparing the products of two different sets of representatives of the same quotient group (Section 11G).

Before beginning the proof of Rule (6), we need the following formula:

Lemma 8. *If q is an odd prime, then*

$$(q-1)!(-1)^{\frac{q-1}{2}} \equiv \left(\frac{q-1}{2}\right)!^2 \pmod{q}.$$

Proof. As in the proof of Rule (5), we take the product $(q-1)!$ modulo q and subtract q from each of the factors $>q/2$: That is, we replace the numbers

$$\frac{q+1}{2}, \frac{q+3}{2}, \ldots, q-1$$

by the negative numbers

$$\left(\frac{q+1}{2}-q\right), \left(\frac{q+3}{2}-q\right), \ldots, (q-1)-q.$$

Then, modulo q, we have

$$(q-1)! \equiv 1 \cdot 2 \cdots \frac{q-1}{2} \cdot \frac{q+1}{2} \cdot \frac{q+3}{2} \cdots q-1$$

$$\equiv 1 \cdot 2 \cdots \frac{q-1}{2} \cdot \left(\frac{q+1}{2}-q\right) \cdot \left(\frac{q+3}{2}-q\right) \cdots (q-1-q)$$

$$\equiv 1 \cdot 2 \cdots \frac{q-1}{2} \cdot \left(-\frac{q-1}{2}\right) \cdot \left(-\frac{q-3}{2}\right) \cdots (-1)$$

$$\equiv \left(\frac{q-1}{2}\right)!^2 (-1)^{\left(\frac{q-1}{2}\right)} \pmod{q}.$$

$\square$

Multiplying both sides by $(-1)^{\left(\frac{q-1}{2}\right)}$ and raising both sides to the $\frac{p-1}{2}$ power for any odd prime p yields

Corollary 9.

$$\left(\frac{q-1}{2}\right)!^{(p-1)} \equiv (q-1)!^{\left(\frac{p-1}{2}\right)}(-1)^{\left(\frac{q-1}{2}\right)\left(\frac{p-1}{2}\right)} \pmod{q}.$$

(Compare the exponent on the -1 with the exponent on -1 in the Law of Quadratic Reciprocity.)

With that, we proceed to the proof of Rule (6), the Law of Quadratic Reciprocity.

Proof. Recall that U_m is the group of units of $\mathbb{Z}/m\mathbb{Z}$. For ease of notation, we will identify congruence classes $[k]_m$ of integers k relatively prime to m with the integers k themselves, as long as the modulus is clear.

Let p and q be distinct odd primes. Related to the Chinese Remainder Theorem, we showed in Chapter 12 that U_{pq}, the group of units of $\mathbb{Z}/pq\mathbb{Z}$, has order $\phi(pq) = \phi(p)\phi(q) = (p-1)(q-1)$. We proved this by showing that the map:

$$\psi : U_{pq} \to U_p \times U_q$$

defined by $\psi(k) = (k,k)$, or more precisely, $\psi([k]_{pq}) = ([k]_p, [k]_q)$, is an isomorphism between U_{pq} and $U_p \times U_q$. (Hence there are as many elements in U_{pq}, namely $\phi(pq)$, as there are in $U_p \times U_q$, namely, $\phi(p)\phi(q)$.)

Observe that the subset $M = \{-1, 1\}$ of U_{pq} is a subgroup of U_{pq}, and, under ψ, M is mapped to the subgroup $N = \{(-1,-1), (1,1)\}$ of $U_p \times U_q$. Thus ψ induces an isomorphism from the quotient group U_{pq}/M to the quotient group $(U_p \times U_q)/N$ (c.f. Section 11G):

$$U_{pq}/M \cong (U_p \times U_q)/N$$

given by sending the coset kM to $(k,k)N$, for k any unit modulo pq.

Note that $(U_p \times U_q)/N$ has order $\frac{(p-1)(q-1)}{2}$ by Lagrange's Theorem.

Our proof of quadratic reciprocity consists of finding, then multiplying together, two different sets of elements of $U_p \times U_q$ that represent the $\frac{(p-1)(q-1)}{2}$ cosets of N in $U_p \times U_q$. Comparing the two products will give us the result.

The first set of representatives. For the first set, observe that for each element (i,j) of $U_p \times U_q$, the coset $(i,j)N$ contains also the element $(i,j)(-1,-1) \equiv (p-i, q-j)$ (and no other element). Thus to find a set of elements in $U_p \times U_q$ that represent all the cosets of N, we choose in each coset the element (i,j) for which $1 \leq j \leq \frac{q-1}{2}$. Doing so, we may describe $(U_p \times U_q)/N$ as

$$(U_p \times U_q)/N = \left\{ (i,j)N : 1 \leq i \leq p-1, 1 \leq j \leq \frac{q-1}{2} \right\}.$$

Thus the set

$$A = \left\{ (i,j) : 1 \leq i \leq p-1, 1 \leq j \leq \frac{q-1}{2} \right\}$$

is a set of representatives for the cosets of N in $U_p \times U_q$.

The second set of representatives. For the second set of representatives of the cosets of N in $U_p \times U_q$, we first look at U_{pq}/M.

Since M is the subset $\{-1, 1\}$, then for each k coprime to pq, the coset kM consists of the two numbers k and $-k$ modulo pq. Then

$$kM = \{k, pq - k\}.$$

We choose to represent each coset of M in U_{pq} by the smaller of k and $pq - k$, that is, by a number in the range $1 \leq k \leq \frac{pq-1}{2}$. That is,

$$U_{pq}/M = \left\{ kM : 1 \leq k \leq \frac{pq - 1}{2} \text{ and } (k, pq) = 1 \right\}.$$

Under the isomorphism from U_{pq}/M to $(U_p \times U_q)/N$ induced by ψ, the image in $(U_p \times U_q)/N$ of that set of coset representatives gives a second set of representatives for the cosets of N in $U_p \times U_q$. Thus,

$$(U_p \times U_q)/N = \left\{ (k,k)N : 1 \leq k \leq \frac{pq - 1}{2} \text{ and } (k, pq) = 1 \right\}.$$

The sets

$$A = \left\{ (i,j) : 1 \leq i \leq p - 1, 1 \leq j \leq \frac{q - 1}{2} \right\}$$

and

$$B = \left\{ (k,k) : 1 \leq k \leq \frac{pq - 1}{2} \text{ and } (k, pq) = 1 \right\}$$

then each represent all the cosets of N in $U_p \times U_q$.

Now multiplication of cosets using representatives of the cosets is well-defined, independent of the choice of representatives (see Section 11G). This means here that the product in $U_p \times U_q$ of the elements in the set A will be in the same coset modulo N as the product of the elements in the set B. Computing and comparing those two products will yield the Law of Quadratic Reciprocity.

The product of the elements in A. Computing the product of the elements in A is fairly routine.

We first compute the product of the pairs (i, j), where i is fixed and j varies between 1 and $\frac{q-1}{2}$. We get

$$\left(i^{\frac{q-1}{2}}, \left(\frac{q - 1}{2} \right)! \right).$$

Then taking the product of these as i goes from 1 to $p - 1$, we get the product of the elements in A:

$$\left((p - 1)!^{\frac{q-1}{2}}, \left(\left(\frac{q - 1}{2} \right)! \right)^{p-1} \right).$$

Now we apply Corollary 9 above to the second term. We find that the product of the elements in the set A is

$$\left((p-1)!^{\frac{q-1}{2}}, ((q - 1)!)^{\frac{p-1}{2}} (-1)^{\left(\frac{p-1}{2} \right)\left(\frac{q-1}{2} \right)} \right).$$

The product of the elements in B. Computing the product of the elements in B is a bit more subtle. We do it one component at a time.

The first component of the product of the elements in B is the product modulo p of the numbers k with $1 \leq k \leq \frac{pq-1}{2}$ that are coprime to p and q. Note that

$$\frac{pq-1}{2} = p\left(\frac{q-1}{2}\right) + \frac{p-1}{2} = q\left(\frac{p-1}{2}\right) + \frac{q-1}{2}.$$

We first write down the product of all the k's from 1 to $\frac{pq-1}{2}$ that are relatively prime to p, laying out the products in rows:

$$1 \cdot 2 \cdots (p-1)$$
$$\cdot (p+1) \cdot (p+2) \cdots (p+(p-1))$$
$$\cdot (2p+1) \cdot (2p+2) \cdots (2p+(p-1))$$

$$\vdots$$

$$\cdot \left(\left(\frac{q-1}{2}-1\right)p+1\right) \cdot \left(\left(\frac{q-1}{2}-1\right)p+2\right) \cdots \left(\left(\frac{q-1}{2}-1\right)p+(p-1)\right)$$
$$\cdot \left(\left(\frac{q-1}{2}\right)p+1\right) \cdot \left(\left(\frac{q-1}{2}\right)p+2\right) \cdots \left(\left(\frac{q-1}{2}\right)p+\left(\frac{p-1}{2}\right)\right).$$

Each line but the last is congruent modulo p to $(p-1)!$. So the entire product is congruent modulo p to

$$((p-1)!)^{\left(\frac{q-1}{2}\right)} \cdot \left(\frac{p-1}{2}\right)!,$$

where the last factor comes from the last line of the product above.

To get the product of all k with $1 \leq k \leq \frac{pq-1}{2}$ that are coprime to p and also coprime to q, we must divide the above product by all the multiples of q in the interval between 1 and $\frac{pq-1}{2} = (\frac{p-1}{2})q + \frac{q-1}{2}$. That set of multiples of q multiply together to give

$$q \cdot 2q \cdots \left(\frac{p-1}{2}\right)q = q^{\left(\frac{p-1}{2}\right)}\left(\frac{p-1}{2}\right)!,$$

a unit modulo p. So multiplying the entire product modulo p by the inverse of this last product, and then canceling the common factor $(\frac{p-1}{2})!$, we see that the first component of the product of the elements of B is congruent modulo p to

$$\frac{(p-1)!^{\left(\frac{q-1}{2}\right)}}{q^{\left(\frac{p-1}{2}\right)}}.$$

Now recall that Euler's Lemma gives

$$\left(\frac{q}{p}\right) \equiv q^{\frac{p-1}{2}} \pmod{p},$$

Substituting the Legendre symbol $\left(\frac{q}{p}\right)$ into this last product, and noting that the Legendre symbol is either 1 or -1, hence can be placed either in the denominator or the numerator, we see that the first component of the product of the elements in B is congruent modulo p to

$$(p-1)!^{(\frac{q-1}{2})}\left(\frac{q}{p}\right).$$

The same computation with the roles of p and q interchanged yields that the second component of the product B is congruent modulo q to

$$(q-1)!^{(\frac{p-1}{2})}\left(\frac{p}{q}\right).$$

Thus the product of the elements of the set B in $U_p \times U_q$ is the ordered pair

$$\left((p-1)!^{(\frac{q-1}{2})}\left(\frac{q}{p}\right),(q-1)!^{(\frac{p-1}{2})}\left(\frac{p}{q}\right)\right).$$

Equating the two descriptions of the product of the elements of $(U_p \times U_q)/N$.
We've seen that the product of all the elements of $(U_p \times U_q)/N$ may be represented by the product of the elements of A and also by the product of the elements of B. Thus the two products represent the same coset:

$$\left((p-1)!^{(\frac{q-1}{2})}\left(\frac{q}{p}\right),(q-1)!^{(\frac{p-1}{2})}\left(\frac{p}{q}\right)\right)N$$
$$= ((p-1)!^{(\frac{q-1}{2})},(q-1)!^{(\frac{p-1}{2})}(-1)^{(\frac{p-1}{2})(\frac{q-1}{2})})N.$$

Since $N = \{(1,1),(-1,-1)\}$, this means that in $U_p \times U_q$, either

$$\left((p-1)!^{(\frac{q-1}{2})}\left(\frac{q}{p}\right),(q-1)!^{(\frac{p-1}{2})}\left(\frac{p}{q}\right)\right)$$
$$= ((p-1)!^{(\frac{q-1}{2})},(q-1)!^{(\frac{p-1}{2})}(-1)^{(\frac{p-1}{2})(\frac{q-1}{2})})(1,1),$$

or

$$\left((p-1)!^{(\frac{q-1}{2})}\left(\frac{q}{p}\right),(q-1)!^{(\frac{p-1}{2})}\left(\frac{p}{q}\right)\right)$$
$$= ((p-1)!^{(\frac{q-1}{2})},(q-1)!^{(\frac{p-1}{2})}(-1)^{(\frac{p-1}{2})(\frac{q-1}{2})})(-1,-1).$$

The two possibilities give two cases.
    In the first case, we have

$$((p-1)!^{(\frac{q-1}{2})}\left(\frac{q}{p}\right) \equiv (p-1)!^{(\frac{q-1}{2})} \pmod{p}$$

and

$$(q-1)!^{(\frac{p-1}{2})}\left(\frac{p}{q}\right) \equiv (q-1)!^{(\frac{p-1}{2})}(-1)^{(\frac{p-1}{2})(\frac{q-1}{2})} \pmod{q}.$$

Simplifying these congruences gives

$$\left(\frac{q}{p}\right) \equiv 1 \pmod{p}$$

and

$$\left(\frac{p}{q}\right) \equiv (-1)^{(\frac{p-1}{2})(\frac{q-1}{2})} \pmod{q}.$$

Putting these together gives

$$\left(\frac{p}{q}\right) = \left(\frac{q}{p}\right) \cdot (-1)^{(\frac{p-1}{2})(\frac{q-1}{2})},$$

which is the law of quadratic reciprocity if $\left(\frac{q}{p}\right) = 1$.

In the second case, we have

$$(p-1)!^{(\frac{q-1}{2})}\left(\frac{q}{p}\right) \equiv (-1)(p-1)!^{(\frac{q-1}{2})} \pmod{p}$$

and

$$(q-1)!^{(\frac{p-1}{2})}\left(\frac{p}{q}\right) \equiv (-1)(q-1)!^{(\frac{p-1}{2})}(-1)^{(\frac{p-1}{2})(\frac{q-1}{2})} \pmod{q}.$$

Simplifying these congruences gives

$$\left(\frac{q}{p}\right) = -1$$

and

$$\left(\frac{p}{q}\right) = (-1)(-1)^{(\frac{p-1}{2})(\frac{q-1}{2})}.$$

Putting these together gives

$$\left(\frac{p}{q}\right) = \left(\frac{q}{p}\right) \cdot (-1)^{(\frac{p-1}{2})(\frac{q-1}{2})},$$

which is the law of quadratic reciprocity if $\left(\frac{q}{p}\right) = -1$.
The two cases complete the proof of Theorem 7. □

Exercises.

14. Let $p = 5, q = 7$ and consider the group $(U_5 \times U_7)/\{(1,1),(-1,-1)\}$. Explicitly write down the sets A and B for this group. Show that the product of the elements in the set A is

$$(4!^3, 3!^4) = (4!^3, 6!^4(-1)^{2\cdot3}),$$

while the product of the elements in the set B is

$$\left(\frac{(4!)^3 \cdot 2!}{7 \cdot 14}, \frac{(6!)^2 \cdot 3!}{5 \cdot 10 \cdot 15}\right) = \left(\frac{(4!)^3}{7^2}, \frac{(6!)^2}{5^3}\right)$$

where the first components are modulo 5, and the second components are mod 7. Use this and Euler's Lemma to verify that

$$\left(\frac{5}{7}\right) = \left(\frac{7}{5}\right).$$

15. Determine whether or not there is a solution of

$$x^4 \equiv 36 \pmod{1009}$$

(1009 is prime.)

16. Decide whether or not there is a solution of

$$x^2 - 43x + 55 \equiv 0 \pmod{73}$$

17. Decide whether or not there is a solution of

$$x^2 - 21x - 17 \equiv 0 \pmod{37}$$

18. Using Rules (1)–(6) as appropriate, prove that for all odd primes p,

$$\left(\frac{-3}{p}\right) = \left(\frac{p}{3}\right).$$

19. Prove that if p and q are odd primes with $q \equiv 3 \pmod 4$, then

$$\left(\frac{-q}{p}\right) = \left(\frac{p}{q}\right).$$

20. For which odd primes p is there a number m coprime to p so that both m and $-m$ are squares modulo p?

21. Use Wilson's Theorem (Section 14A, Exercises) that for p an odd prime, $(p-1)! \equiv -1 \pmod p$, to show (c.f Lemma 8) that

$$\left(\frac{p-1}{2}\right)^2! \equiv (-1)^{\frac{p+1}{2}} \pmod p.$$

D. The Jacobi Symbol

It is important for both computations and applications to extend the Legendre symbol to cover denominators that are not prime.

Definition. For an odd number $m > 1$ and a number a coprime to m, the Jacobi symbol (a/m) is defined in terms of the Legendre symbol as follows. If $m = p_1^{e_1} p_2^{e_2} \cdots p_g^{e_g}$ is the factorization of m into a product of primes, then

$$\left(\frac{a}{m}\right) = \left(\frac{a}{p_1}\right)^{e_1} \left(\frac{a}{p_2}\right)^{e_2} \left(\frac{a}{p_g}\right)^{e_g}.$$

Thus $(a/m) = 1$ or -1 for any a coprime to m.

If $(a/m) = 1$, it does not follow that a is a square (mod m). For example, 2 is not a square mod 15 because 2 is not a square mod 3 or mod 5. But

$$\left(\frac{2}{15}\right) = \left(\frac{2}{5}\right) \left(\frac{2}{3}\right) = (-1)(-1) = 1.$$

On the other hand, if $(a/m) = -1$, then necessarily $(a/p) = -1$ for some prime p dividing m. So a is not a square modulo p, from which it follows that a cannot be a square modulo m.

From the definition of the Jacobi symbol it follows immediately that if $m = rs$ is odd and a is coprime to m, then $(a/m) = (a/r)(a/s)$. From that fact it is fairly easy to show that the Jacobi symbol satisfies Rules (1)–(6) that we proved for the Legendre symbol. For example:

Rule (3).

$$\left(\frac{ab}{m}\right) = \left(\frac{a}{m}\right) \left(\frac{b}{m}\right).$$

Proof. This is true if m is prime, since we proved Rule (3) for the Legendre symbol.

Suppose $m = rs$, with $r, s < m$. Assume by induction that Rule (3) is true for r and s. Then

$$\left(\frac{ab}{r}\right) = \left(\frac{a}{r}\right) \left(\frac{b}{r}\right)$$

and

$$\left(\frac{ab}{s}\right) = \left(\frac{a}{s}\right) \left(\frac{b}{s}\right).$$

So

$$\left(\frac{ab}{m}\right) = \left(\frac{ab}{r}\right) \left(\frac{ab}{s}\right) = \left(\frac{a}{r}\right) \left(\frac{b}{r}\right) \left(\frac{a}{s}\right) \left(\frac{b}{s}\right)$$
$$= \left(\frac{a}{r}\right) \left(\frac{a}{s}\right) \left(\frac{b}{r}\right) \left(\frac{b}{s}\right) = \left(\frac{a}{m}\right) \left(\frac{b}{m}\right).$$

$\square$

For Rules (4) and (6) we set $m' = \frac{m-1}{2}$ and observe:

Lemma 10. *If m is odd and $m = rs$, then $m' \equiv r' + s' \pmod{2}$.*

Proof.

$$m' = \frac{m-1}{2} = \frac{rs-1}{2}$$

$$= \frac{rs-s+s-1}{2} = sr' + s' \equiv r' + s' \pmod 2.$$

□

Rule (4).

$$\left(\frac{-1}{m}\right) = (-1)^{m'}.$$

Proof. We prove Rule (4) by induction as with Rule (3):

Rule (4) was proved for the Legendre symbol, so if m is prime then Rule (4) is true. Assuming $m = rs$, we have by induction and Lemma 10:

$$\left(\frac{-1}{m}\right) = \left(\frac{-1}{r}\right)\left(\frac{-1}{s}\right) = (-1)^{r'}(-1)^{s'} = (-1)^{r's+s'} = (-1)^{m'}.$$

□

Rule (6).

For m, n positive odd numbers,

$$\left(\frac{n}{m}\right) = (-1)^{m'n'}\left(\frac{m}{n}\right).$$

Proof. For a, r, s odd with a and rs coprime, we have

$$\left(\frac{a}{rs}\right) = \left(\frac{a}{r}\right)\left(\frac{a}{s}\right)$$

by definition, and

$$\left(\frac{rs}{a}\right) = \left(\frac{r}{a}\right)\left(\frac{s}{a}\right)$$

by Rule (3). Using Lemma 10 we can prove Rule (6), first for $\left(\frac{p}{m}\right)$ with p prime, by induction on the number of prime divisors of m, and then for $\left(\frac{n}{m}\right)$ by induction on the number of prime divisors of n. We prove the first and leave the second as an exercise.

We have

$$\left(\frac{p}{q}\right) = (-1)^{q'p'}\left(\frac{q}{p}\right)$$

if q and p are distinct primes, by Quadratic Reciprocity. Suppose

$$\left(\frac{p}{m_1}\right) = (-1)^{m_1'p'}\left(\frac{m_1}{p}\right)$$

if m is a product of r primes. Let $m = m_1 q$ with q a prime $\neq p$. Then

$$\left(\frac{p}{m}\right) = \left(\frac{p}{m_1}\right)\left(\frac{p}{q}\right)$$

and by Quadratic Reciprocity and induction,

$$\left(\frac{p}{m_1}\right)\left(\frac{p}{q}\right) = \left(\frac{m_1}{p}\right)\left(\frac{q}{p}\right)(-1)^{m'_1 p'}(-1)^{q'p'}.$$

Since

$$m'_1 p' + q'p' = p'(m'_1 + q') \equiv p'm' \pmod{2}$$

by Lemma 10, we have

$$\left(\frac{p}{m}\right) = \left(\frac{m}{p}\right)(-1)^{p'm'}.$$

This completes rule (6).   □

These rules allow us to compute the Jacobi symbol (a/m) for every a coprime to m in exactly the same way we compute the Legendre symbol.

Example 5. Since 217 and 41 are $\equiv 1 \pmod 4$, we have

$$\left(\frac{217}{475}\right) = \left(\frac{475}{217}\right) = \left(\frac{41}{217}\right)$$

$$= \left(\frac{217}{41}\right) = \left(\frac{12}{41}\right) = \left(\frac{4}{41}\right)\left(\frac{3}{41}\right)$$

$$= \left(\frac{3}{41}\right) = \left(\frac{41}{3}\right) = \left(\frac{2}{3}\right) = -1.$$

Thus 217 is not a square modulo 475. (If we found that $\left(\frac{217}{475}\right) = 1$, we wouldn't know whether or not 217 is a square modulo 475.)

For m prime, the symbol $\left(\frac{n}{m}\right)$ is the Legendre symbol. If we compute $\left(\frac{n}{m}\right)$ either as a Jacobi symbol, or as a Legendre symbol, the outcome will be the same. The significance of this is that we can manipulate Jacobi symbols without knowing how to factor either m or n.

Example 6. Consider $\left(\frac{1501}{3739}\right)$. The denominator 3739 is prime, but not the numerator. As a Jacobi symbol, we compute as follows, noting that 1501 and 737 are congruent to 1 modulo 4:

$$\left(\frac{1501}{3739}\right) = \left(\frac{3739}{1501}\right) = \left(\frac{737}{1501}\right)$$

$$= \left(\frac{1501}{737}\right) = \left(\frac{27}{737}\right) = \left(\frac{3}{737}\right)\left(\frac{9}{737}\right)$$

$$= \left(\frac{3}{737}\right) = \left(\frac{737}{3}\right) = \left(\frac{2}{3}\right) = -1,$$

and so 1501 is not a square modulo 3739.

If we did this as a Legendre symbol, we would need to know how to factor 1501 as $1501 = 19 \cdot 79$. Then

$$\left(\frac{1501}{3739}\right) = \left(\frac{19}{3739}\right)\left(\frac{79}{3739}\right) = (-1)\left(\frac{3739}{19}\right)(-1)\left(\frac{3739}{79}\right)$$

$$= \left(\frac{-4}{19}\right)\left(\frac{26}{79}\right) = \left(\frac{-1}{19}\right)\left(\frac{4}{19}\right)\left(\frac{2}{79}\right)\left(\frac{13}{79}\right)$$

$$= (-1)(1)(1)\left(\frac{79}{13}\right) = (-1)\left(\frac{1}{13}\right) = -1$$

Using Rule (6), the amount of time to compute the Jacobi symbol is similar to the time needed to compute Euclid's algorithm. In the example, Euclid's algorithm for 1501 and 3739 is

$$3739 = 1501 \cdot 2 + 737$$
$$1501 = 737 \cdot 2 + 27$$
$$737 = 27 \cdot 27 + 8$$
$$27 = 8 \cdot 3 + 3$$
$$8 = 3 \cdot 2 + 2$$
$$3 = 2 \cdot 1 + 1.$$

The number of steps needed to compute the Jacobi symbol is approximately twice the number of steps in Euclid's algorithm–one reduction of the top number in the Jacobi symbol plus one use of reciprocity for each step of Euclid's algorithm, plus computing the symbol for any power of 2 that may arise in the numerator. Since Euclid's Algorithm is a fast algorithm, while factoring numbers into products of primes is slow and hard, there is great benefit from being able to determine a Legendre symble by viewing it as a Jacobi symbol.

E. Euler a-Pseudoprimes

Now that the Jacobi symbol is defined, we can use Euler's Lemma as a compositeness test, much as we used Fermat's Theorem.

We know that if m is prime (and $m' = (m-1)/2$), then for all a coprime to m, by Euler's Lemma,

$$\left(\frac{a}{m}\right) \equiv a^{m'} \pmod{m}.$$

So if $\left(\frac{a}{m}\right) \not\equiv a^{m'} \pmod{m}$ for some $a < m$, then m is composite.

What about a converse? For an odd number m that we suspect may be prime, we may ask:

If $a > 1$ is coprime to m and $(a/m') \equiv a^{m'} \pmod{m}$, is m necessarily prime? The answer is "no", but examples are not so common. So we give them a name:

Definition. An odd number $m > 1$ is an Euler a-pseudoprime if m is composite and

$$\left(\frac{a}{m}\right) \equiv a^{m'} \pmod{m}.$$

If m is an Euler a-pseudoprime, then squaring this last congruence gives

$$1 = \left(\frac{a}{m}\right)^2 \equiv (a^{m'})^2 \equiv a^{m-1} \pmod{m}.$$

so m is an a-pseudoprime.

On the other hand, m may be an a-pseudoprime but not an Euler a-pseudoprime.

Example 7. The Carmichael number 1729 is an 11-pseudoprime but not an Euler 11-pseudoprime. To see the latter, we have

$$\left(\frac{11}{1729}\right) = \left(\frac{1729}{11}\right) = \left(\frac{2}{11}\right) = -1$$

since $11 \equiv 3 \pmod 8$; on the other hand, observing that $1729 = 7 \cdot 13 \cdot 19$, we have

$$11^{864} \equiv (11^6)^{144} \equiv 1 \pmod 7$$
$$11^{864} \equiv (11^{12})^{72} \equiv 1 \pmod{13}$$
$$11^{864} \equiv (11^{18})^{48} \equiv 1 \pmod{19}$$

by Fermat's Theorem. Thus

$$11^{864} \equiv 1 \pmod{1729}.$$

We now prove that no composite number m is resistant to trial a-Euler pseudoprime testing, in the same way that no composite number m is resistant to trial strong a-pseudoprime testing.

For m an odd integer > 1, let

$$E_m = \{[a]_m \text{ in } U_m | \left(\frac{a}{m}\right) \equiv a^{m'} \pmod{m}\}.$$

Since (a/m) only depends on the congruence class of a in $\mathbb{Z}/m\mathbb{Z}$ (Rule (2)) and $(ab/m) = (a/m)(b/m)$ (Rule (3)), it is easy to see that E_m is a subgroup of U_m. Using this fact, we have:

Proposition 11. *If m is an composite odd number >1, then m is an Euler a-pseudoprime for less than half of all numbers $a < m$.*

Proof. Since E_m is a subgroup of U_m, by Lagrange's theorem, the number of elements of E_m divides U_m. So it is enough to show that $E_m \neq U_m$, that is, to show that m is not an Euler a-pseudoprime for some a coprime to m.

If m is not a Carmichael number, then for some a, m is not an a-pseudoprime, and therefore not an Euler a-pseudoprime. So we can assume that m is a Carmichael number. This implies that m is a product of distinct odd prime numbers.

We consider two cases.

Case 1, $a^{m'} \equiv 1 \pmod{m}$ for all a coprime to m. We will find some a so that $(a/m) = -1$.

Let $m = pq$ where p is an odd prime. Then $U_m \cong U_p \times U_q$. Let b be a primitive root modulo p. Let $[a]_m$ in U_m correspond to the pair $([b]_p, [1]_q)$ in $U_p \times U_q$. In other words, let $x = a$ be a solution of the congruences

$$x \equiv b \pmod{p},$$
$$x \equiv 1 \pmod{q}.$$

Since b is a primitive root modulo p, b is not a square modulo p, and so $(b/p) = -1$. However, $(1/q) = 1$, and so

$$\left(\frac{a}{m}\right) = \left(\frac{a}{p}\right)\left(\frac{a}{q}\right) = \left(\frac{b}{p}\right)\left(\frac{1}{q}\right) = -1$$

Hence m is not an Euler a-pseudoprime.

Case 2, $a^{m'} \not\equiv 1 \pmod{m}$ for some a coprime to m. Since m is a product of distinct primes, by the Chinese remainder theorem, there is some prime p dividing m so that $a^{m'} \not\equiv 1 \pmod{p}$. Let $m = pq$ with p and q coprime, then $U_m \cong U_p \times U_q$. Let $[c]_m$ in U_m correspond to the pair $([a]_p, [1]_q)$ in $U_p \times U_q$. Then $[c]^{m'}$ corresponds to $([a]^{m'}, [1]^{m'}) = ([a]^{m'}, [1])$. Since $[a]_p^{m'} \neq [1]$, it follows that $[c]^{m'}$ can be neither $[1]_m$ nor $[-1]_m$. Thus m cannot be an Euler c-pseudoprime. $\square$

The Euler a-pseudoprime test as a probabilistic primality test has been replaced by the strong a-pseudoprime test. It can be shown that if m is a strong a-pseudoprime, then m is an Euler a-pseudoprime. (see Crandall and Pomerance (2005), p. 166, Exercise 3.21). Moreover, while it is possible for a Carmichael number m to be an Euler a-pseudoprime for half of the elements of U_m (see Exercise 24), Rabin's theorem shows that m can be a strong a-pseudoprime for at most $1/4$ of the elements of U_m. Thus if m is composite, finding an a for which m fails the strong a-pseudoprime test is never harder, and often easier, than finding an a for which m fails the Euler a-pseudoprime test. Since the computation times for the Euler and strong a-pseudoprime tests are similar, the strong a-pseudoprime test is the test of choice.

Exercises.

22. Prove that if m is any odd number >1, then the other rules of Theorem 1 of Section B hold for the Jacobi symbol, namely:

(1) $(a^2/m) = 1$ for any integer a;
(2) if $a \equiv b \pmod{m}$, then $(a/m) = (b/m)$;
(5) $(2/m) \equiv (-1)^{(m^2-1)/8}$; and

23. Complete the proof of Rule (6) for the Jacobi symbol.

24. (i) Show that for every odd number m, there exists an odd number a so that $(a/m) = -1$.

   (ii) Show that the set $\{[a] \text{ in } U_m | (a/m) = 1\}$ is a subgroup of U_m of index 2 (that is, of order $\phi(m)/2$).

   (iii) Show that if $a^{m'} \equiv 1 \pmod{m}$ for all a coprime to m, then E_m has index 2 in U_m. (Hence this is true for $m = 1729$–see Example 7.)

25. For $m = 1729$, how many elements of E_m are false witnesses for m? (see Section 20B).

26. Show that if $m = c_k = (6k+1)(12k+1)(18k+1)$ is a Carmichael number (thus the three factors are simultaneously prime) and k is odd, then E_m is a subgroup of U_m of index 2.

27. Find the index of E_m in U_m for $m =$
   (i) 697;
   (ii) 1333;
   (iii) 65;
   (iv)) 1001.

Chapter 22
Quadratic Applications

In this chapter we consider several ideas related to quadratic residues, random number generation and factoring.

A. Safeprimes

Safeprimes are prime numbers p of the form $p = 2q + 1$ with q prime. They are so called because composite numbers whose prime factors are safeprimes are resistant to some factoring algorithms, such as the Pollard $p - 1$ algorithm, and hence are considered particularly appropriate as factors of moduli in RSA cryptosystems. It is conjectured but not proven that there are infinitely many safeprimes. The first few safeprimes are 5, 7, 11, 23, 47, 59, 83, 107, 167, ... The On-Line Encyclopedia of Integer Sequences [Sloane] has a much longer list of safeprimes. MAPLE has a command, "safeprime(m)" that will return the smallest safeprime $\geq m$.

In this section we collect some results related to safeprimes and primitive roots.

We begin with

Proposition 1. *If $p = 2p' + 1$ is a safeprime, then for every number a coprime to p, a is a primitive root modulo p if and only if $a \not\equiv 1$ or $-1 \pmod{p}$ and a is not a quadratic residue modulo p.*

Proof. A primitive root is not a quadratic residue modulo p. So if $\left(\frac{a}{p}\right) = 1$, then a cannot be a primitive root.

Suppose p is a safeprime, $p = 2p' + 1$, where p' is also prime. Let β be a primitive root in $\mathbb{Z}/p\mathbb{Z}$, that is, a generator of the group U_p of units of $\mathbb{Z}/p\mathbb{Z}$. Then β has order $2p'$. For every exponent e,

$$e \text{ is odd, if and only if } \beta^e \text{ is not a square,}$$

$$e \text{ is even, if and only if } \beta^e \text{ is a square.}$$

L.N. Childs, *A Concrete Introduction to Higher Algebra*, Undergraduate Texts in Mathematics, © Springer Science+Business Media LLC 2009

Now the order of β^e is $2p'/(e, 2p')$. Thus for any e with $1 \le e \le 2p'$ other than $e = p'$ and $e = 2p'$ ($\beta^{p'} = [-1]$ and $\beta^{2p'} = [1]$), β^e is not a square if and only if the order of β^e is $2p'$, that is, β^e is a primitive root.   □

To apply this proposition we recall from an exercise in Chapter 10 some congruence properties of safeprimes.

Proposition 2. *If p is a safeprime > 11, then*

$$p \equiv 2 \quad (\bmod\ 3),$$
$$p \equiv 3 \quad (\bmod\ 4),$$
$$p \equiv 2, 3, \text{ or } 4 \quad (\bmod\ 5).$$

Hence $p \equiv 23, 47$ or 59 (mod 60)

Using these congruences and properties of the Legendre symbol, we can check various numbers to see if they are primitive roots modulo safeprimes. For example,

Proposition 3. *If $p = 2p' + 1$ is a safeprime > 11, then $(3/p) = 1$, and hence 3 is not a primitive root modulo p.*

Proof. By quadratic reciprocity, $(3/p) = (p/3)(-1)^{p'}$. Now p' is odd, so $(-1)^{p'} = -1$; and $p \equiv 2$ (mod 3), so $(p/3) = (2/3) = -1$. So $(3/p) = 1$.   □

This result has a partial converse, due to McCurley (1989):

Proposition 4. *Let $q > 3$ be prime. Then $2q + 1 = m$ is prime if and only if $3^q \equiv 1$ (mod m).*

Proof. If m is prime, then by Proposition 3 and Euler's Lemma,

$$1 = \left(\frac{3}{m}\right) \equiv 3^q \quad (\bmod\ m).$$

Conversely, suppose $q > 3$ is prime, $m = 2q + 1$ and $3^q \equiv 1$ (mod m). Since q is prime, the order of 3 modulo m is q. So q divides $\phi(m)$ by Euler's Theorem.

Let p be the largest prime dividing m. Then $\phi(m)$ is divisible only by primes $\le p$. If m is not prime, then $m = pt$ with $t \ge 3$, so

$$p \le \frac{m}{3} < \frac{m-1}{2} = q.$$

Hence q cannot divide $\phi(m)$, a contradiction. So m must be prime.   □

Gauss's conjecture is that 10 is a primitive root for infinitely many primes. The next result relates Gauss's conjecture to the question of whether there are infinitely many safeprimes.

Proposition 5. *10 is a primitive root modulo a safeprime p if $p \equiv 23, 47$ or 59 (mod 120). 10 is not a primitive root if $p \equiv 83, 107$ or 119 (mod 120).*

Proof. Again, we compute $(10/p) = (5/p)(2/p)$. Now

$$\left(\frac{2}{p}\right) = \begin{cases} 1 \text{ if } p \equiv 1 \text{ or } 7 \pmod 8, \\ -1 \text{ if } p \equiv 3 \text{ or } 5 \pmod 8, \end{cases}$$

and

$$\left(\frac{5}{p}\right) = \left(\frac{p}{5}\right) = \begin{cases} 1 \text{ if } p \equiv 1 \text{ or } 4 \pmod 5, \\ -1 \text{ if } p \equiv 2 \text{ or } 3 \pmod 5. \end{cases}$$

So 10 is a primitive root modulo p iff $(10/p) = -1$, which will be true if $p \equiv 3$ or 5 (mod 8) and $p \equiv 1$ or 4 (mod 5), or if $p \equiv 1$ or 7 (mod 8) and $p \equiv 2$ or 3 (mod 5). Since a safeprime p cannot be congruent to 1 modulo 5, and cannot be congruent to 1 or 5 modulo 8, we are left with three possible cases:

$$p \equiv 3 \pmod 8, \quad p \equiv 4 \pmod 5, \text{ or}$$
$$p \equiv 7 \pmod 8, \quad p \equiv 2 \pmod 5, \text{ or}$$
$$p \equiv 7 \pmod 8, \quad p \equiv 3 \pmod 5.$$

Since $p \equiv 2 \pmod 3$, we have three congruence classes modulo 120 (here, 120 is the least common multiple of 8, 5, and 3) where $(10/p) = -1$: they are easily seen to be the classes of 59, 47, and 23, respectively. $\qquad\qquad\square$

For general results of this kind, see [Ribenboim, (1996), pp. 379ff]

Sophie Germain primes. If p is a safeprime, so that $p = 2q + 1$ and q is prime, then q is called a Sophie Germain prime, after the early 19th century French mathematician Sophie Germain. Her name is attached to those primes because of the following theorem, related to Fermat's Last Theorem:

Theorem 6. *(Germain). If q is a Sophie Germain prime, then Fermat's equation $x^q + y^q = z^q$ has no nontrivial solutions modulo q.*

This shows that the "first case" of Fermat's last theorem holds for primes $q = (p-1)/2$ where p is a safeprime. For a proof, see [Ribenboim (1979)].

Here is a nice factorization result for Mersenne numbers involving Sophie Germain primes, first proved by Euler.

Proposition 7. *Let q be a Sophie Germain prime $\equiv 3$ (mod 4) with $q > 3$. Let $p = 2q + 1$. Then p divides the Mersenne number $M_q = 2^q - 1$.*

Thus M_q is composite for the Sophie Germain primes

$$q = 11, 23, 83, 131, 179, 191, 239, 251, \dots.$$

Proof. Since $q \equiv 3 \pmod 4$, we have $p \equiv 7 \pmod 8$, and so

$$\left(\frac{2}{p}\right) = 1,$$

which means that there is a number m so that $2 \equiv m^2 \pmod{p}$. Then

$$2^q = 2^{\frac{p-1}{2}} \equiv m^{p-1} \equiv 1 \pmod{p},$$

by Fermat's Theorem. Thus

$$M_q = 2^q - 1 \equiv 0 \pmod{p}.$$

Thus p divides M_q. Since $p < M_q$ for $q > 3$, this means that M_q is composite. □

Exercises.

1. Let p be a Mersenne prime, $p = 2^q - 1$ where q is prime. Show that 10 is not a primitive root modulo p if $q = 1 \pmod 4$.

2. Let $p \geq 5$ be a Fermat prime, $p = 2^{2^n} + 1$ for some $n \geq 1$.
   (i) Show that b is a primitive root modulo p iff b is not a square modulo p.
   (ii) Show that 7 is a primitive root modulo every Fermat prime ≥ 5.

3. Show that -10 is a primitive root for every safeprime congruent to 83, 107 or 119 (modulo 120). Conclude that if there are infinitely many safeprimes, then 10 is a primitive root for infinitely many primes, or -10 is a primitive root for infinitely many primes.

B. Primes in Congruence Classes

In Section 4C we gave Euclid's proof that there are infinitely many primes. Using properties of the Legendre symbol we can obtain several results, generalizing Euclid's method, that show that there are infinitely many primes in a particular congruence class. Here is one example.

Proposition 8. *There are infinitely many primes congruent to 4 modulo 5.*

Proof. Let $p_1, p_2, \ldots, p_r$ be primes congruent to 4 modulo 5. We will produce another one. Suppose $p_1, p_2, \ldots, p_r$ are each $\leq n$. Let

$$A = 5(n!)^2 - 1.$$

Then A is a product of primes $> n$. Let $p > n$ be a prime dividing A. Then

$$5(n!)^2 \equiv 1 \pmod{p}.$$

Since $n!$ is a unit modulo p, 5 is then a square modulo p, and so

$$1 = \left(\frac{5}{p}\right) = \left(\frac{p}{5}\right).$$

Thus $p \equiv 1$ or $4 \pmod 5$. This is true for all primes dividing A. If all the prime dividing A were congruent to 1 modulo 5, then A would be congrent to 1 modulo 5. Hence at least one prime dividing A must be congruent to 4 modulo 5.

Thus there cannot be a finite number of primes congruent to 4 modulo 5. $\square$

The exercises have more examples of proofs of this kind. Keith Conrad, www. math.uconn.edu/ kconrad/blurbs/gradnumthy/dirichleteuclid.pdf , has described all congruence classes for which proofs like Euclid's may be obtained.

Exercises.

4. Show that there are infinitely many primes congruent to 1 modulo 4, as follows: suppose $p_1, \ldots, p_n$ are all such primes. Let $M = (p_1 \cdots p_n)^2 + 1$. If a prime p divides M then -1 is a square modulo p, so $p = 1 \pmod 4 \ldots$.

5. Show that there are infinitely many primes congruent to 7 $\pmod{12}$, as follows. Suppose to the contrary that $p_1, \ldots, p_n$ are all such primes. Let

$$t = (2p_1 \cdots p_n)^2 + 3.$$

(i) Show that t is divisible by at least one prime $\equiv 3 \pmod 4$ with $p \neq p_1, \ldots, p_n$.
(ii) Show that if p divides t, then $(-3/p) = 1$.
(iii) Show that if $(-3/p) = 1$, then $p = 1 \pmod 3$.
(iv) Use (i) and (iii) to conclude that there is a prime p dividing t with $p \equiv 7 \pmod{12}$.

C. Blum, Blum and Shub's Pseudorandom Number Generator

In Section 19H we presented Lehmer's multiplicative congruential method for generating sequences of numbers that "look" random. The sequence $\{x_0, x_1, x_2, \ldots\}$ was defined by $x_{i+1} = Ax_i \bmod M$, where A is some multiplier coprime to M. For appropriate choices of M (large) and A (of large order modulo M) the resulting sequences (usually) pass various statistical tests for randomness. However, a Lehmer sequence has the property that with knowledge of M and A, then starting at any number x_k one can go either forward or backward equally easily–to go backward, simply multiply x_k by the inverse of A modulo M, which is easily found by the extended Euclidean algorithm (Bezout's identity).

Of course, for really random sequences, such as a sequence of numbers arising from tossing a fair die, knowing x_k, the result of the kth toss, would give no idea of what x_{k-1} or x_{k+1} is.

To generate a sequence of pseudorandom numbers, the next number in the sequence is necessarily a known function of the previous numbers in the sequence, as with Lehmer's method: $x_{k+1} = Ax_k \bmod M$. But L. Blum, M. Blum, and M. Shub

(1986) proposed a pseudorandom number generator that has the following intriguing property: given a number x_k in the sequence, it is easy to find the next number x_{k+1} in the sequence, but essentially impossible to find the previous number x_{k-1} in the sequence without special knowledge. As they put it, their sequence is polynomial-time unpredictable to the left, in the sense that given a piece $x_k, x_{k+1}, \ldots, x_l$ of the sequence (with $k > 0$), one can do no better in polynomial time at guessing x_{k-1} than by a truly random guess.

Work of A. Yao implies that such sequences will pass every statistical test for randomness that can run in polynomial time.

The Blum, Blum, Shub (BBS) generator works as follows: pick a modulus M and a starting number y, let $x_0 = y^2 \bmod M$, and generate the sequence $x_1, x_2, \ldots$ by $x_{k+1} = x_k^2 \bmod M$. BBS then yields a sequence of bits (0's and 1's) by defining $b_k = x_k \bmod 2$, that is, $b_k = 0$ if x_k is even, $b_k = 1$ if x_k is odd.

Here is an example:

Let $M = 143$ and let $y = 74$. Then $42 = 74^2 \bmod 143$, so is a quadratic residue. The sequence $x_0, x_1, x_2, \ldots$ is then

$$42, 48, 16, 113, 42, 48, 16, 113, 42, \ldots.$$

The corresponding sequence of bits b_i is

$$0, 0, 0, 1, 0, 0, 0, 1, 0, \ldots.$$

Of course this sequence has a very short period. Suitable sequences should be much longer.

We first discuss the unpredictability to the left, and then describe how to obtain sequences with long periods.

To begin, we restrict the modulus M to be the product of two primes each $\equiv 3$ (mod 4). The reason for doing so is based on the following considerations.

For p a prime, let U_p be the group of units modulo p, and let QR_p denote the set of quadratic residues modulo p. Then QR_p is a subgroup of U_p because it is the image of the squaring map θ from U_p to U_p defined by $\theta(a) = a^2$, a group homomorphism. Evidently θ restricts to a group homomorphism on QR_p. Notice that the squaring map $\theta : U_p \to U_p$ is not one-to-one; in fact, the kernel consists of 1 and $-1 \bmod p$, so θ is a 2 to 1 map on U_p: if x is in U_p and $x = y^2$, then also $x = (-y)^2$. However, we have

Proposition 9. *If p is a prime $\equiv 3$ (mod 4), then the squaring map $\theta : QR_p \to QR_p$ is an isomorphism (that is, one-to-one and onto).*

In less technical language, every quadratic residue modulo p has a unique square root that is also a quadratic residue.

Proof. Since the image of the squaring map θ on U_p is QR_p, θ induces a homomorphism from QR_p to QR_p. We will show that the inverse of $\theta : QR_p \to QR_p$ is the map ω defined on U_p by

$$\omega(x) = x^{\left(\frac{p+1}{4}\right)}.$$

For x in QR_p,

$$\theta(\omega(x)) = (x^{(\frac{p+1}{4})})^2 = x^{(\frac{p+1}{2})} = x^{(\frac{p-1}{2})} \cdot x.$$

By Euler's Lemma, since x is a quadratic residue,

$$x^{(\frac{p-1}{2})} \equiv \left(\frac{x}{p}\right) = 1 \pmod{p}.$$

Hence $\theta(\omega(x)) = x$. Similarly, $\omega(\theta(y)) = y$ for y in QR_p, and so θ and ω are inverse isomorphisms. $\qquad\square$

Now let $M = pq$ where p and q are primes congruent to 3 modulo 4. Then the squaring map $\theta : U_M \to U_M$ is a homomorphism. The map $j : U_m \to U_p \times U_q$ defined by

$$j(x \bmod M) = (x \bmod p, x \bmod q),$$

is an isomorphism of groups and induces by restriction an isomorphism from QR_m onto $QR_p \times QR_q$. Since the squaring map is one-to-one and onto on both QR_p and QR_q, then the squaring map is one-to-one and onto QR_M. (On the other hand, the squaring map is a 4 to 1 map on U_M, because it is 2 to 1 on each of U_p and U_q.)

Expressed more concretely, if x is a quadratic residue modulo M, then among the four square roots of x modulo M, exactly one is itself a quadratic residue. That quadratic residue is easy to find if we know the factorization of $M = pq$. In fact:

Lemma 10. *If $M = pq$ with $p = 2p_1 + 1, q = 2q_1 + 1$ and p_1, q_1 odd, then the inverse of the squaring function $\theta : U_m \to U_m$ is the function $\omega : U_m \to U_m$ defined by*

$$\omega(z) \equiv z^{\frac{p_1 q_1 + 1}{2}}.$$

Proof. To show that $\theta\omega$ and $\omega\theta$ are the identity maps on U_m, it suffices to show that

$$(z^2)^{\frac{p_1 q_1 + 1}{2}} \equiv z^{p_1 q_1 + 1} \equiv z \pmod{m}.$$

But z is in QR_m, so $z \bmod p$ is in QR_p and $z \bmod q$ is in QR_q, so $z^{p_1} \equiv 1 \pmod{p}$ and $z^{q_1} \equiv 1 \pmod{q}$ by Euler's Lemma. Thus $z^{p_1 q_1 + 1} \equiv z \pmod{p}$ and also modulo q, hence modulo m. $\qquad\square$

A modulus $M = pq$ with $p, q \equiv 3 \pmod 4$ then has the following remarkable property:

Theorem 11. *Let $M = pq$ with $p, q \equiv 3 \pmod 4$, and let $\theta : QR_M \to QR_M$ be the squaring function. Then computing the inverse function of θ is equivalent in difficulty to factoring M.*

Proof. If we know the factorization of M as $M = pq$, we just showed how to take a quadratic residue x modulo M and find its unique square root that is a quadratic residue modulo M.

Conversely, suppose we know how to take any quadratic residue modulo M and find its unique square root that is a quadratic residue modulo M. Let z be any number whose Jacobi symbol $\left(\frac{z}{M}\right) = -1$. Since half of all the numbers modulo M have Jacobi symbol $= -1$, such a number z is easy to find. Then also $\left(\frac{-z}{M}\right) = -1$ because $\left(\frac{-1}{M}\right) = \left(\frac{-1}{p}\right)\left(\frac{-1}{q}\right) = (-1)(-1) = 1$. Let $x \equiv z^2 \pmod{M}$. Let y be the unique square root of x that is a quadratic residue modulo M. Then $y^2 \equiv x \equiv z^2 \pmod{M}$. Also, $\left(\frac{y}{M}\right) = 1$, so $y \not\equiv z$ or $-z \pmod{M}$. So M divides $y^2 - z^2 = (y-z)(y+z)$ and M doesn't divide $(y-z)$ or $(y+z)$. Thus $\gcd(M, y-z)$ is a non-trivial factor of M, hence must be p or q, and M has been factored.   □

Since factoring a number M that is a product $M = pq$ of two large primes is known to be a hard problem, it follows that finding the unique quadratic residue y that is the square root of a given quadratic residue x modulo M is then an equally hard problem. In particular, if x_k is a number in the BBS sequence $x_0, x_1, \ldots$ obtained by starting from a quadratic residue x_0 and successively squaring modulo M, then unless we can factor M, we do not know how to find the number x_{k-1} that immediately precedes x_k.

Thus, under the assumption that factoring M is impractically difficult, the BBS sequence is then unpredictable to the left. This means that given the bit b_k (which is the parity of x_k), we can do no better than toss a coin to guess what the bit b_{k-1} is.

In that sense, the BBS sequence behaves just like a random sequence.

The other property that the BBS sequence should have is a long period.

To see how to achieve a long period, we need to understand how to find the period.

The BBS sequence is obtained by successive squaring modulo M. So starting from an arbitrary number y coprime to M, the sequence is

$$y, y^2, y^{2^2}, y^{2^3}, y^{2^4}, y^{2^5}, \ldots \pmod{M}.$$

If y has order d modulo M, then the exponents $2, 2^2, 2^3, \ldots$ of y are all modulo d. We find:

Proposition 12. *Let M be a modulus, y a number coprime to M, and $d = 2^k f$ be the order of y modulo M, with f odd. Then the period of the sequence*

$$y, y^2, y^{2^2}, y^{2^3}, y^{2^4}, y^{2^5}, \ldots \pmod{M}$$

is equal to the order of 2 modulo f.

Proof. To see when the sequence

$$y, y^2, y^{2^2}, y^{2^3}, y^{2^4}, y^{2^5}, \ldots,$$

modulo d starts to repeat, we look for r and $e > 0$ so that

$$2^{r+e} \equiv 2^r \pmod{d},$$

that is,

$$2^{r+e} \equiv 2^r \pmod{2^k f}.$$

Once this congruence holds for some r, it holds for $r+1, r+2, \ldots$. So we can assume $r \geq k$. Then we can cancel 2^k from both sides and the modulus; we get

$$2^{(r-k)+e} \equiv 2^{r-k} \pmod{f}.$$

Since f is odd, we may cancel 2^{r-k} from both sides, to get

$$2^e \equiv 1 \pmod{f}.$$

The least $e > 0$ for which this holds is the order of 2 modulo f. □

To achieve a large order for 2 modulo d, the strategy of Blum, Blum and Shub is to use a modulus $M = pq$ where p and q are "special" primes.

Definition. A prime p is *special* if $p = 2p_1 + 1$ and $p_1 = 2p_2 + 1$ where all of p, p_1 and p_2 are prime. Thus both p and p_1 are safeprimes.

Suppose p and q are special primes. (Examples: 23, 47, 167.)

Assume y is not ± 1 mod p and y is not ± 1 mod q. Then y has order p_1 or $2p_1$ modulo p, and order q_1 or $2q_1$ modulo q. Hence $x = y^2$ has order p_1 mod p and order q_1 mod q. So the order of x mod M is $d = p_1 q_1$. Thus (Exercise 12) $(p-3)(q-3)$ of the $(p-1)(q-1)$ units y mod M yield $x = y^2$ of order d mod M.

For $x = y^2$ of order $d = p_1 q_1$, the period of the BBS sequence generated by y is the order of 2 modulo f, the odd part of d, that is, the order of 2 modulo $p_1 q_1$. The order of 2 modulo $p_1 q_1$ is the least common multiple of the orders of 2 modulo p_1 and modulo q_1. The order of 2 modulo p_1 divides $p_1 - 1 = 2p_2$, and the order of 2 modulo q_1 divides $q_1 - 1 = 2q_2$. Thus the order of 2 modulo f, and hence the period of the BBS sequence starting from almost every y coprime to M, is at least $p_2 q_2$. For large M with $M = pq$ and p and q special primes, then, the period of the BBS sequence is asymptotic to $M/16$.

The question arises whether or not large special primes can be found. Blum, Blum and Shub cite a heuristic estimate by Shanks that the proportion of numbers p between 2^n and 2^{n+1} that are special primes is asymptotically $1/(n^3 \ln^3 2)$. If we seek a 90-digit special prime, then $n = 300$, so by that estimate, perhaps one of every 10^7 numbers of 90 digits is a special prime. If one is looking for smaller special primes, one can find them quickly using the "safeprime" command in MAPLE. For example, I found

$$(p_2, p_1, p) = (5130431863961, 10260863727923, 20521727455847)$$

and

$$(q_2, q_1, q) = (6553710049871, 13107420099743, 26214840199487)$$

in almost no time. For these p and q, if $M = pq$, then the period of a BBS sequence starting from almost every random y will be at least $p_2 \cdot q_2 \geq 10^{25}$.

Blum-Goldwasser cryptography. A BBS sequence can be used for cryptography. The cryptosystem is known as the Blum-Goldwasser cryptosystem [Blum and Gold-wasser (1985)].

Suppose Manuel wants to send a g-bit message to Shafi. Shafi secretly finds two large special primes p and q, sets $M = pq$ and broadcasts M to Manuel.

Manuel selects a random number y coprime to M and generates the BBS sequence $x_0 \equiv y^2, x_1, \ldots, x_g$ modulo M. Let $b = (b_0, b_1, \ldots, b_{g-1})$ be the vector of bits defined by $b_j = x_j \bmod 2$. If $m = (m_0, m_1, \ldots, m_{g-1})$ is Manuel's message (where each $m_j = 0$ or 1), then Manuel encrypts his message by

$$c = m + b,$$

the addition taking place in the vector space $\mathbb{F}_2^g$ of g-tuples with entries in the field $\mathbb{F}_2$ of two elements, 0 and 1. Then Manuel sends Shafi the pair (c, z) where $z = x_g$.

Shafi takes the number x_g and reconstructs the BBS sequence from x_g by successively computing $x_{g-1}, x_{g-2}, \ldots, x_0$ modulo M. She can do this because she knows the factorization $M = pq$, so can compute $x_{g-1}, x_{g-2}, \ldots, x_0$ modulo M by using the inverse ω of the squaring function θ modulo M, as in Lemma 10, above. From $x_0, \ldots, x_{g-1}$ she obtains the parity vector

$$b = (b_0, b_1, \ldots, b_{g-1})$$

where $b_j = x_j \bmod 2$, and using b she can recover Manuel's message by

$$m = c + b$$

in $\mathbb{F}_2^g$.

By Theorem 11, an eavesdropper, Lenore, would need to know how to factor M into its prime factorization $M = pq$ in order to obtain the sequence b and read the message m.

Thus the Blum-Goldwasser cryptosystem is secure as long as the factorization of M is unknown. Thus it is at least as secure as an RSA cryptosystem using the same modulus M. However, decryption appears to be slower than RSA decryption.

Exercises.

6. Find a prime p and a number x so that the BBS sequence $x^2 = y, y^2, y^4, y^8, \ldots$ modulo p has period between 50 and 100.

7. Let $p = 223$, a prime. Find the period of the BBS sequence starting with $y = 47$.

8. Let $m = 8693$, a prime. Then 132 is a primitive root modulo m, and $m - 1 = 4 \cdot 41 \cdot 53$. The order of 2 mod 41 is 20, and the order of 2 mod 53 is 52.

    (i) What is the period of the BBS sequence starting with $y = 132^2 \equiv 38 \pmod{m}$?

    (ii) What is the period of the BBS sequence if we start with $y = 1444 \equiv 38^2 \pmod{m}$?

(iii) Find a quadratic residue y with $1 < y < m - 1$ so that the period of the BBS sequence starting with y is less than 200.

9. Shafi sends Manuel $M = 253$ $(= 11 \cdot 23)$. Manuel wants to send the message $(1, 0, 1)$ to Shafi. He starts the BBS sequence with $y = 49$. What does Manuel send Shafi?

10. Suppose Shafi knows that $M = 253 = 11 \cdot 23$ and she receives $((1, 1, 0), 141)$ from Manuel. What message did Manuel send?

11. Let $M = 83 \cdot 107 = 8881$, a product of two safeprimes. Let $y = 5183 \equiv 3375^2$ (mod M). Find the unique square root of y modulo 8881 that is a quadratic residue modulo M.

12. Show that if $M = pq$ is the product of two safeprimes, then $(p - 3)(q - 3)$ of the units y modulo M yield $y^2 = x$ of order $p_1 q_1$ modulo M.

13. If $p_2 > 5, p_1 = 2p_2 + 1, p = 2p_1 + 1$ are all primes (so that p is a special prime), show that $p_2 \equiv 11$ or 29 (mod 30).

14. (i) Show that if $m = pq$ with p, q distinct odd primes, then the isomorphism $j : U_m \to U_p \times U_q$ defined by $j(a \bmod m) = (a \bmod p, a \bmod q)$ restricts to a homomorphism j_0 from QR_m to $QR_p \times QR_q$.
(ii) Show that j_0 is one-to-one.
(iii) Show that j_0 is onto.

15. Show that if $p = 2p_1 + 1, q = 2q_1 + 1$ are safeprimes > 5 and $M = pq$, then the map from QR_p to QR_p by $y \mapsto y^{\frac{p_1 q_1 + 1}{2}}$ is the same as the map $y \mapsto y^{\frac{p+1}{4}}$.

D. Factoring by the Pollard Rho Method

Trial division on a composite number N takes at least $p / \ln p$ steps, where p is the smallest prime divisor of N, as we showed in Section 6G. The slowness of trial division has led to a search for more efficient factoring algorithms. One such is the Pollard $p - 1$ algorithm, described in Section 10C.

Among the factoring algorithms proposed since the rise of computers, one of the most elegant is the Pollard rho method, also discovered by J.M. Pollard (1975). It is based on three ideas.

The first is that squaring modulo N can be used to generate sequences of numbers that look random, as we described in the last section.

The second is that Euclid's algorithm for finding the greatest common divisor of two numbers is very efficient, as we showed in Section 3F.

The third is the fact from probability that if you roll an n-sided die until a face previously rolled is observed again, then the expected number of rolls is $< 1.3\sqrt{n}$ for $n \geq 100$. (This fact is sometimes exploited at parties in the following guise: if you have a group of 25 or more people, the chance is better than even that two of them will have the same birthday. Think of people as rolls of a die with 366 faces, one face for each day of the year. Then $1.3 \times \sqrt{366} \sim 25$.)

The Pollard rho method works as follows.

Assume that N is the number to be factored. Start with any number $x_0 < N$. Compute a sequence of numbers as follows:

$$x_{n+1} = x_n^2 + 1 \bmod N \text{ for } n = 0, 1, 2, \ldots.$$

Suppose p is the smallest prime divisor of N. Think of the sequence $\{x_0, x_1, \ldots, x_n, \ldots\}$ as a sequence of pseudorandom numbers modulo p. Then, after generating approximately $1.3\sqrt{p}$ numbers of the sequence, it is likely that two of them will be congruent modulo p. For if the numbers x_n were truly random, then obtaining them mod p would be like throwing a p-sided die.

Suppose $x_n \equiv x_m \pmod{p}$, where $n > m$. Then p divides $x_n - x_m$. Since also p divides N, p will divide the greatest common divisor of $x_n - x_m$ and N, so $(N, x_n - x_m) > 1$. If also $x_n \not\equiv x_m \pmod{N}$, then $(N, x_n - x_m) < N$, hence $(N, x_n - x_m)$ is a nontrivial proper factor of N, which we can find explicitly by Euclid's algorithm.

So within the sequence $\{x_0, x_1, \ldots, x_n, \ldots\}$ we search for two numbers x_n and x_m whose greatest common divisor with N is is greater than 1. There should be two such numbers between x_1 and x_m where m is some small multiple of $\sqrt{p}$. In practice, $m = 2\sqrt{p}$ seems to suffice in almost every case.

The main subtlety of the algorithm is to efficiently find two numbers x_n and x_m whose greatest common divisor $(x_n - x_m, N)$ is greater than 1. For if we were to test the difference $x_n - x_m$ for every $m < n$ between 1 and $2\sqrt{p}$ to see if it has a nontrivial common divisor with N, we would end up doing Euclid's algorithm approximately $2p$ times, and then the Pollard rho method would be less efficient than trial division.

To resolve this difficulty we look only at the greatest common divisor of N and $x_{2t} - x_t$ for $t = 1, 2, \ldots, 2\sqrt{p}$. Doing so is just as effective as looking at $(x_n - x_m, N)$ for all $n, m < 2\sqrt{p}$. To see this, suppose $x_n \equiv x_m \pmod{p}$ for some m, n with $1 < m < n < 2\sqrt{p}$. Let $m + d = n$, so $d \geq 1$ and $x_m \equiv x_{m+d} \pmod{p}$. Then

$$x_{m+1} \equiv x_m^2 + 1 \equiv x_{m+d}^2 + 1 \equiv x_{(m+1)+d} \pmod{p},$$

and in the same way, one sees easily that

$$x_{m+r} \equiv x_{(m+r)+d} \pmod{p}$$

for all $r > 0$. Now among the numbers $m, m+1, m+2, \ldots, m+d-1$, exactly one is a multiple of d. Suppose k is the number with $0 < k < d$ so that $m + k = ed$ is a multiple of d. Then

$$x_{ed} \equiv x_{m+k} \equiv x_{(m+k)+d} \equiv x_{ed+d} \pmod{p},$$

and also

$$x_{ed+d} \equiv x_{(ed+d)+d} \equiv x_{ed+2d}, \pmod{p},$$

$$x_{ed+2d} \equiv x_{(ed+2d)+d} \equiv x_{ed+3d}, \pmod{p},$$

etc. So

$$x_{ed} \equiv x_{ed+ed} = x_{2ed} \pmod{p}.$$

Since $ed = m + k < m + d = n < 2\sqrt{p}$, we conclude that

Proposition 13. *If $x_m \equiv x_n \pmod{p}$ for some m and n with $1 \le m < n < 2\sqrt{p}$, then there is some $t < 2\sqrt{p}$ so that*

$$x_t \equiv x_{2t} \pmod{p}.$$

Thus, if $x_m \equiv x_n \pmod{p}$ for some $m \ne n < 2\sqrt{p}$, then computing $(x_{2t} - x_t, N)$ for $1 \le t < 2\sqrt{p}$ will yield some t so that p divides the greatest common divisor of N and $x_{2t} - x_t$. Unless we are unlucky and N divides $x_{2t} - x_t$, we will have found a factor of N.

If p is the smallest prime factor of N, then to test the greatest common divisor of $x_{2t} - x_t$ and N for $1 \le t \le 2\sqrt{p}$ requires at most $2\sqrt{p}$ applications of Euclid's Algorithm, rather than $2p$ applications with the naive approach. Since Euclid's Algorithm requires at most $5 \log_{10}(N)$ divisions (and on average requires around $2 \log_{10}(N)$ divisions), the Pollard rho algorithm would require at most $10\sqrt{p} \log_{10}(N)$ divisions to find p.

If N is a product of two primes of similar size, then the running time would be at most

$$(2\sqrt{p})(5 \log_{10}(p^2)) = 20\sqrt{p} \log_{10}(p),$$

which for large p is less than $p/\ln p$, the number of trial divisions of N by primes $\le p$.

Of course, we do not know in advance the size of any prime p that divides N, so we don't know how long to run the procedure. So we proceed as follows:

Pollard Rho Procedure. Given N, a number we wish to factor:

- Pick x_0.
- For each $n > 0$, compute $x_n \equiv x_{n-1}^2 + 1 \pmod{N}$, $x_n < N$.
- If n is even, compute $(x_n - x_{n/2}, N)$.
- Repeat until either:

$(x_n - x_{n/2}, N) > 1$ but N does not divide $x_n - x_{n/2}$ (success), or
    we run out of time (failure).

    The procedure will fail if N has no sufficiently small prime factors.

    We illustrate the algorithm with two examples.

Example 1. Let $N = 2771 = 17 \cdot 163$. We let $x_0 = 6$ and for each $n > 0$, set $x_{n+1} \equiv x_n^2 + 1 \pmod{N}$. Thus

$$
\begin{aligned}
x_1 &= 37, \\
x_2 &= 1370, & x_2 - x_1, &= 1333, & (1333, 2771) &= 1, \\
x_3 &= 934, \\
x_4 &= 2263, & x_4 - x_2 &= 893 & (893, 2771) &= 1, \\
x_5 &= 362,
\end{aligned}
$$

$$x_6 = 808, \qquad x_6 - x_3 = -126 \qquad (-126, 2771) = 1,$$
$$x_7 = 1680,$$
$$x_8 = 1523, \qquad x_8 - x_4 = -740 \qquad (-740, 2771) = 1,$$
$$x_9 = 203,$$
$$x_{10} = 2416, \qquad x_{10} - x_5 = 2054 \qquad (2054, 2771) = 1,$$
$$x_{11} = 1331,$$
$$x_{12} = 893, \qquad x_{12} - x_6 = 85 \qquad (85, 2771) = 17.$$

At this point we can stop. If we were to continue, we would reach

$$x_{20} = 1438, \qquad x_{20} - x_{10} = -978, \qquad (-978, 2771) = 163.$$

The probability assumption of the algorithm predicted that we should find the prime 17 dividing $x_{2t} - x_t$ for $t < 2\sqrt{17} < 9$, and the prime 163 for $t < 2\sqrt{163} < 26$.

Example 2. Let $N = 47783 = 71 \cdot 673$, $x_0 = 6$. Then the sequence $\{x_n\}$, $x_{n+1} = x_n^2 + 1 \pmod{N}$, becomes:

37, 1370, 13364, 31426, 14433, 25393, 20648, 19979, 29043, 30334, 42109, 36318, 42976, 28061, 3665, 5203, 26032, 6519, 182751 20239, 21246, 34299,

and we find that

$$x_{20} - x_{10} = 20239 - 30334 = -10095 \text{ and } (-10095, 47783) = 673;$$
$$x_{22} - x_{11} = 34299 - 42109 = -7810 \text{ and } (-7810, 47783) = 71.$$

Here the theory would predict a discovery of 71 for $t \le 2\sqrt{71} < 18$ and a discovery of 673 for $t < 2\sqrt{673} < 52$.

At least for these examples, the Pollard rho algorithm works at least as well as the assumption of randomness of the sequence $\{x_n\}$ would imply.

The reason for the name "rho" in "Pollard rho" is illustrated by reviewing Example 1. If we take the sequence $x_0, x_1, x_2 \ldots$ where $x_n \equiv x_{n-1}^2 + 1 \pmod{2771}$, $x_n < 2771$, and look at the sequence modulo 17, starting with $x_0 = 6$, we get

$$6, 3, 10, 16, 2, 5, 9, 14, 10, 16, 2, 5, \ldots. \qquad (*)$$

This is the same as the sequence we would obtain by starting with $y_0 = 6$ and computing $y_n \equiv y_{n-1} + 1 \pmod{17}, y_n < 17$: then for all n, $y_n \equiv x_n \pmod{17}$. Now if we diagram the sequence (*) appropriately, we get a diagram that has the shape of the Greek letter "rho", ρ:

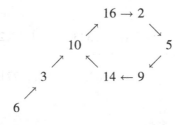

The loop of the "rho" contains six numbers. So the sequence (*) repeats with period six. In particular, $y_8 = y_2 = 10, y_9 = y_3 = 16, \ldots, y_{12} = y_6 = 9$, and so $x_{12} \equiv x_6 \pmod{17}$. Hence $(x_{12} - x_6, 2771) = 17$. In fact, $x_{12} = 893$ and $x_6 = 808$ are both congruent to $y_6 = 9$ modulo 17, and so 17 divides $893 - 808$, as we found in Example 1.

The same analysis is valid for any number N having 17 as a factor, not just $N = 2771$: starting from $x_0 = 6$ and obtaining $x_r + 1 = x_r^2 + 1 \pmod{N}$ we will always find that 17 divides $(x_{12} - x_6, N)$.

Suppose N has 100 digits. How good is the Pollard rho method?

If we assume that the sequence $\{x_0, x_1, \ldots, x_n, \ldots\}$ begins repeating for some $n < 2\sqrt{p}$ modulo p, the smallest prime divisor of N, then the number of steps this algorithm takes to find a nontrivial factor of N is at most $2\sqrt{p}$ times $5\log_{10}(N)$, as observed above. Thus the Pollard rho algorithm on a number N of 100 digits takes at most $1000\sqrt{p}$ steps.

Hence the Pollard rho method is substantially more effective than trial division. Suppose we wish to factor a number N of 100 digits, and we have time for 10^9 divisions. Since each greatest common divisor computation can take 500 divisions, we would have time to do 2×10^6 greatest common divisor tests, that is, compute $(N, x_{2t} - x_t)$ for $t < 2 \times 10^6$. We would be likely to find any prime factor p of N with $2\sqrt{p} < 2 \times 10^6$, that is, any prime factor $p < 10^{12}$, and, if we were fortunate, larger prime factors.

By comparison, trial division 10^9 times using the wheel mod 30300 would find any prime factor under 5×10^9, as noted in Section 6G.

If we have time for 10^{10} divisions, then with Pollard rho we would be likely to find any prime factor $< 10^{14}$, while with trial division we would find any prime factors $< 5 \times 10^{10}$.

The efficiency of the Pollard rho method has made it a successful algorithm in practice. R.P. Brent and J.M. Pollard (1981) used it to find the least prime factor of all the Fermat numbers $F_n = 2^{2^n} + 1$ for $5 \le n \le 13$. The number F_8 has 78 decimal digits, and had not previously been factored; the smallest prime factor of F_8 was found to be the 16-digit prime number 1,238,926,361,552,897, and took 2 hours on a UNIVAC 1100/42. Penk, in 1979, used the Pollard rho algorithm to find a 15 digit factor of the 78 digit Mersenne number $2^{257} - 1$, namely 535,006,138,814,359 (see also Crandall and Penk (1979)).

Exercises.

16. Draw the rho diagram for the sequence (mod p) starting with x_0 where:
   (i) $p = 23, x_0 = 4$;
   (ii) $p = 31, x_0 = 3$;
   (iii) $p = 41, x_0 = 1$;
   (iv) $p = 37, x_0 = 3$;
   (v) $p = 73, x_0 = 6$.

17. Verify (assuming you did (iii) of the last exercise) that starting from $x_0 = 1$ and computing the sequence modulo $1763 = 41 \cdot 43$, that 41 divides $(x_{14} - x_7, N)$.

18. The Pollard rho method can fail if the first t such that $(x_{2t} - x_t, N) > 1$ has $x_{2t} - x_t$ a multiple of N. For example, let $N = 17 \cdot 73 = 1241$, let $x_0 = 6$.
(i) Show that $x_8 - x_2 \equiv 0 \pmod{17}$.
(ii) Show that $x_8 - x_2 \equiv 0 \pmod{73}$.
(iii) Show that N divides $x_{12} - x_6$.
(iv) Find other values of x_0 for which which N divides $x_{12} - x_6$. and values of x_0 for which N doesn't divide $x_{12} - x_6$.

19. Factor, using the Pollard rho algorithm with x_0 of your choosing:
(i) 77;
(ii) 143;
(iii) 341;
(iv) 851;
(v) 2047;
(vi) 3337.

20. Factor, using the Pollard rho algorithm with x_0 of your choosing:
(i) 1194637;
(ii) 8388607;
(iii) 10266001.

To test the accuracy of the estimate $t < 2\sqrt{p}$ for finding t so that p divides $x_{2t} - x_t$, we can look, at least for small p, at the entire diagram of the map $y \equiv x^2 + 1$ (mod p). We write $m \to n$ if $n \equiv m^2 + 1 \pmod{p}$. Here is the diagram for $p = 13$:

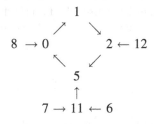

Using the diagram, we can compute a table: in the table, x_0 is the starting value, n and $n + d$ are the least values with $x_n = x_{n+d}$, and t is the least t with $x_{2t} = x_t$:

x_0	n	$n+d$	t
0	0	4	4
1	0	4	4
2	0	4	4
3	1	2	1
4	0	1	1
5	0	4	4
6	2	6	4
7	2	6	4
8	1	5	4
9	1	2	1
10	0	1	1
11	1	5	4
12	1	5	4

From the last column of the table we can determine the maximum value of t, $t_{max} = 4$, and also the expected value of t, $E(t)$, which is defined to be the average value in the last column. With $p = 13$, the last column sums to 40, so $E(t) = 40/13 = 3.1$.

Note that $E(t) < \sqrt{13}$ and $t_{max} = 4 = 1.1\sqrt{13}$, both of which are less than the estimate 2 that we used in analyzing the effectiveness of the Pollard rho algorithm.

21. Draw the diagram and compute t_{max} and $E(t)$ for:
(i) $p = 7$;
(ii) $p = 17$;
(iii) $p = 19$.

22. For $p = 29$, find t_{max} and show that $t_{max} \geq 11 > 2\sqrt{29}$. If you pick x_0 at random, $0 < x_0 < 28$, what is the chance that the resulting t will be ≥ 11?

23. Find the smallest m so that $m/\ln m > 20\sqrt{m}\log_{10} m$.

24. Let $1 + QR_p$ denote the image of the map $\alpha : \mathbb{Z}/p\mathbb{Z} \to \mathbb{Z}/p\mathbb{Z}$ defined by $\alpha(x) = x^2 + 1 \pmod{p}$. Show that α is not one-to-one on $1 + QR_p$. (I found it convenient to know two facts from number theory: the Sum of Two Squares Theorem, that every prime number $\equiv 1 \pmod 4$ is the sum of two squares, and Dirichlet's Theorem, that in every congruence class $a \mod m$ where a and m are coprime, there are infinitely many primes).

Part VI
Finite Fields

Chapter 23
Congruence Classes Modulo a Polynomial

In this chapter we extend the notion of congruence classes from numbers to polynomials. Using congruence classes, we show how to find a root of any polynomial.

A. New Numbers

This section is devoted to several examples, some familiar, one unfamiliar, of "invented" numbers.

Example 1. Our first example is i.

The complex numbers arose because certain polynomials in $\mathbb{R}[x]$ had no roots in $\mathbb{R}$. In particular, the polynomial $x^2 + 1$ has no root in $\mathbb{R}$, so the "imaginary" number $i = \sqrt{-1}$ was invented as a root of $x^2 + 1$. The defining property of i is that $i^2 = -1$.

Having invented i, 16th century algebraists such as Cardano (1545) found that roots of polynomials of degree ≤ 4 could be written in terms of i (see Section 15E). (The most striking example of this was Cardano's "casus irreducibilis", Section 15E) Bombelli, around 1574, showed how to manipulate numbers of the form $a + bi$ (a, b real) and showed, in effect, that the set $\mathbb{C}$ of complex numbers is a field.

We review the basic operations for complex numbers from section 15D. To add or multiply complex numbers, we view a complex number $a + bi$ as the result of evaluating the polynomial $a + bx$ in $\mathbb{R}[x]$ at $x = i$. We add two complex numbers $a + bi$ and $a' + b'i$ by adding the corresponding polynomials:

$$(a + bx) + (a' + b'x) = (a + a') + (b + b')x,$$

and then evaluating at $x = i$:

$$(a + bi) + (a' + b'i) = (a + a') + (b + b')i.$$

To multiply two complex numbers, we multiply the corresponding polynomials:

$$(a + bx)(a' + b'x) = aa' + (ab' + ba')x + bb'x^2;$$

L.N. Childs, *A Concrete Introduction to Higher Algebra*, Undergraduate Texts in Mathematics, © Springer Science+Business Media LLC 2009

evaluate this at i:

$$(a+bi)(a'+b'i) = aa' + (ab'+ba')i + bb'i^2.$$

then use the defining property of i that $i^2 = -1$ to reduce the result:

$$\begin{aligned}(a+bi)(a'+b'i) &= aa' + (ab'+ba')i + bb'i^2 \\ &= aa' + (ab'+ba')i + bb'(-1) \\ &= (aa'-bb') + (ab'+ba')i.\end{aligned}$$

Thus we may think of $\mathbb{C}$ as $\mathbb{R}[i]$, polynomials with real coefficients evaluated at i.

Of course $i = \sqrt{-1}$ is not the only new number invented in the course of trying to solve equations. All kinds of radicals arise as roots of equations.

Example 2. The polynomial $x^3 - 2$, viewed as a polynomial with rational coefficients, has no roots–there is no rational number whose cube is 2. So the notation $\sqrt[3]{2}$ was invented to describe the real root of $x^3 - 2$. Its fundamental algebraic property is that $(\sqrt[3]{2})^3 = 2$.

Analogous to $\mathbb{C} = \mathbb{R}[i]$, we let $\mathbb{Q}[\sqrt[3]{2}]$ denote the set of all real numbers obtained by taking polynomials with rational coefficients and evaluating them at $x = \sqrt[3]{2}$.

Since $\sqrt[3]{2}$ is a root of $x^3 - 2$, we can see that if $f(x)$ is any polynomial with coefficients in $\mathbb{Q}$, and we divide $f(x)$ by $x^3 - 2$:

$$f(x) = (x^3 - 2)q(x) + r(x)$$

with $\deg r(x) \leq 2$, then

$$f(\sqrt[3]{2}) = (\sqrt[3]{2}^3 - 2)q(\sqrt[3]{2}) + r(\sqrt[3]{2}) = r(\sqrt[3]{2}).$$

Thus we may identify $\mathbb{Q}[\sqrt[3]{2}]$ as the set of polynomials of degree ≤ 2 in $\mathbb{Q}[x]$ evaluated at $\sqrt[3]{2}$.

We add and multiply in $\mathbb{Q}[\sqrt[3]{2}]$ as follows: If $f(\sqrt[3]{2})$ and $g(\sqrt[3]{2})$ are polynomials evaluated at $\sqrt[3]{2}$, then $f(\sqrt[3]{2}) + g(\sqrt[3]{2})$ is the polynomial $f(x) + g(x)$ evaluated at $x = \sqrt[3]{2}$, and $f(\sqrt[3]{2}) \cdot g(\sqrt[3]{2})$ is the polynomial $f(x)g(x)$ evaluated at $x = \sqrt[3]{2}$. Thus addition is:

$$(a + b\sqrt[3]{2} + c\sqrt[3]{2}^2) + (a' + b'\sqrt[3]{2} + c'\sqrt[3]{2}^2)$$
$$= ((a+a') + (b+b')\sqrt[3]{2} + (c+c')\sqrt[3]{2}^2$$

and multiplication is:

$$(a + b\sqrt[3]{2} + c\sqrt[3]{2}^2)(a' + b'\sqrt[3]{2} + c'\sqrt[3]{2}^2)$$
$$= aa' + (ab'+ba')\sqrt[3]{2} + (ac'+bb'+ca')\sqrt[3]{2}^2 + (bc'+cb')\sqrt[3]{2}^3 + cc'\sqrt[3]{2}^4.$$

If we want the result to be a polynomial of degree ≤ 2 evaluated at $\sqrt[3]{2}$, then we substitute, using the defining property that $\sqrt[3]{2}^3 = 2$:

$$= aa' + (ab' + ba')\sqrt[3]{2} + (ac' + bb' + ca')\sqrt[3]{2}^2 + (bc' + cb')2 + cc'2\sqrt[3]{2}$$
$$= (aa' + 2bc' + 2cb') + (ab' + ba' + 2cc')\sqrt[3]{2} + (ac' + bb' + ca')\sqrt[3]{2}^2.$$

Thus, as with complex numbers, addition and multiplication of polynomials of degree ≤ 2 evaluated at $\sqrt[3]{2}$ can be viewed as a three-step process:

- first, view the elements of $\mathbb{Q}[\sqrt[3]{2}]$ as polynomials evaluated at $\sqrt[3]{2}$, and add or multiply the polynomials;
- second, replace the resulting polynomial expression by the remainder when divided by $x^3 - 2$.
- third, evaluate the remainder at $x = \sqrt[3]{2}$.

Example 3. Let

$$f(x) = 5x^2 - 3x + 6, \quad g(x) = 3x^2 - x - 1,$$

then

$$f(x)g(x) = 15x^4 - 14x^3 - 16x^2 - 3x - 6.$$

Dividing $f(x)g(x)$ by $x^3 - 2$ gives

$$15x^4 - 14x^3 - 16x^2 - 3x - 6 = (x^3 - 2)(15x - 14) + (-16x^2 + 27x + 34),$$

so $r(x) = -16x^2 + 27x + 34$. Thus

$$(5\sqrt[3]{2}^2 - 3\sqrt[3]{2} + 6)(3\sqrt[3]{2}^2 - \sqrt[3]{2} - 1)$$
$$= r(\sqrt[3]{2}) = -16\sqrt[3]{2}^2 + 27\sqrt[3]{2} + 34.$$

Alternatively, one can multiply the polynomials, evaluate the result at $x = \sqrt[3]{2}^3$, and, starting with the highest power of $\sqrt[3]{2}$ in $f(\sqrt[3]{2})g(\sqrt[3]{2})$, successively replace that highest power by a linear combination of lower powers of $\sqrt[3]{2}$. Thus $\sqrt[3]{2}^3 = 2, \sqrt[3]{2}^4 = 2\sqrt[3]{2}$, so

$$f(\sqrt[3]{2})g(\sqrt[3]{2}) = 15\sqrt[3]{2}^4 - 14\sqrt[3]{2}^3 - 16\sqrt[3]{2}^2 - 3\sqrt[3]{2} - 6$$
$$= 15 \cdot 2\sqrt[3]{2} - 14 \cdot 2 - 16\sqrt[3]{2}^2 - 3\sqrt[3]{2} - 6$$
$$= -16\sqrt[3]{2}^2 + 27\sqrt[3]{2} - 34.$$

We'll verify below that $\mathbb{Q}[\sqrt[3]{2}]$ is a field. Assuming so, then every non-zero element of $\mathbb{Q}[\sqrt[3]{2}]$ has an inverse, and we can find the inverse by either of two methods.

One method involves solving a set of linear equations. For example, to find the inverse of $1 - 2\sqrt[3]{2}$, we look for an inverse of the form $a + b\sqrt[3]{2} + c\sqrt[3]{2}^2$. So we set up the equation

$$(1 - 2\sqrt[3]{2})(a + b\sqrt[3]{2} + c\sqrt[3]{2}^2) = 1.$$

and try to find rational solutions a, b, c. If we multiply the left side out, reduce it to a polynomial of degree ≤ 2 in $\sqrt[3]{2}$ and then collect the coefficients of 1, of $\sqrt[3]{2}$ and of $\sqrt[3]{2}^2$, we get three equations in a, b and c, namely,

$$-4c + a = 1,$$
$$-2b + c = 0,$$
$$-2a + b = 0,$$

which has a solution $a = -\frac{1}{15}, b = -\frac{2}{15}, c = -\frac{4}{15}$.

Another method for finding inverses uses Bezout's Identity. We'll illustrate that method in Section C.

We can construct roots of polynomials in situations where the roots cannot be thought of as complex numbers.

Example 4. Let F be the field $\mathbb{F}_2$ of two elements, denoted 0 and 1. Let $p(x) = x^3 + x + 1$. Then $p(x)$ is irreducible in $\mathbb{F}_2[x]$ (because it has degree 3 and had no roots in $\mathbb{F}_2$).

Let's invent a root of $p(x)$, call it α. For the moment, let's not worry about what α is, just that α is some "number" satisfying the equation $p(\alpha) = \alpha^3 + \alpha + 1 = 0$. Let $\mathbb{F}_2[\alpha]$ be the set of all polynomials in $\mathbb{F}_2[x]$ evaluated at α.

Since $p(x)$ has degree 3, any polynomial $f(x)$ in $\mathbb{F}_2[x]$ satisfies

$$f(x) = p(x)q(x) + r(x)$$

where $r(x)$ has degree ≤ 2, by the Division Theorem. Then since $p(\alpha) = 0$, evaluating this last equation at $x = \alpha$ yields $f(\alpha) = r(\alpha)$, and so every element of $\mathbb{F}_2[\alpha]$ is a polynomial in α of degree ≤ 2. There are then eight elements of $\mathbb{F}_2[\alpha]$:

$$\mathbb{F}_2[\alpha] = \{0, 1, \alpha, \alpha + 1, \alpha^2, \alpha^2 + 1, \alpha^2 + \alpha, \alpha^2 + \alpha + 1\}.$$

Using the defining property, $\alpha^3 + \alpha + 1 = 0$, or $\alpha^3 = \alpha + 1$ (recall that the field of coefficients is $\mathbb{F}_2$), here are some examples of multiplication of elements of $\mathbb{F}_2[\alpha]$:

$$\alpha^2 \cdot \alpha^2 = \alpha^4 = \alpha^3 \alpha = (\alpha + 1)\alpha = \alpha^2 + \alpha;$$
$$(\alpha + \alpha^2)(1 + \alpha) = \alpha + \alpha^2 + \alpha^2 + \alpha^3 = \alpha + 2\alpha^2 + \alpha + 1 = 1;$$
$$(\alpha + \alpha^2)\alpha^2 = \alpha^3 + \alpha^4 = \alpha + 1 + \alpha^2 + \alpha = \alpha^2 + 1.$$

Table 23.1 is the multiplication table for $\mathbb{F}_2[\alpha]$. Note that in the table, every row but the first contains 1, hence every non-zero element of $\mathbb{F}_2[\alpha]$ has an inverse. Thus $\mathbb{F}_2[\alpha]$ is a field. As we'll see, $\mathbb{F}_2[\alpha]$ is a field because the polynomial $p(x) = x^3 + x + 1$ is irreducible in $\mathbb{F}_2[x]$.

Table 23.1

	0	1	α	$\alpha+1$	α^2	α^2+1	$\alpha^2+\alpha$	$\alpha^2+\alpha+1$
0	0	0	0	0	0	0	0	0
1	0	1	α	$\alpha+1$	α^2	α^2+1	$\alpha^2+\alpha$	$\alpha^2+\alpha+1$
α	0	α	α^2	$\alpha^2+\alpha$	$\alpha+1$	1	$\alpha^2+\alpha+1$	α^2+1
$\alpha+1$	0	$\alpha+1$	$\alpha^2+\alpha$	α^2+1	$\alpha^2+\alpha+1$	α^2	1	α
α^2	0	α^2	$\alpha+1$	$\alpha^2+\alpha+1$	$\alpha^2+\alpha$	α	α^2+1	1
α^2+1	0	α^2+1	1	α^2	α	$\alpha^2+\alpha+1$	$\alpha+1$	$\alpha^2+\alpha$
$\alpha^2+\alpha$	0	$\alpha^2+\alpha$	$\alpha^2+\alpha+1$	1	α^2+1	$\alpha+1$	α	α^2
$\alpha^2+\alpha+1$	0	$\alpha^2+\alpha+1$	α^2+1	α	1	$\alpha^2+\alpha$	α^2	$\alpha+1$

Exercises.

1. In Table 23.1, verify that the column headed by α^2+1 is correct.

2. Let i be a root of x^2+1 in $\mathbb{F}_3[x]$. Write down the multiplication table for $\mathbb{F}_3[i]$.

3. In Example 4, find α^n as a polynomial in α of degree ≤ 2 for every $n \ge 3$.

4. Let α be a root of $p(x) = x^2+x+2$ in $\mathbb{F}_3[x]$. Find:
(i) $(\alpha+1)(\alpha+2)$;
(ii) The inverse of $2\alpha+1$;
(iii) α^6 as a polynomial in α of degree ≤ 1.
(iv) The inverse of α^7.

5. Let α be a root of $p(x) = x^2+2x+2$ in $\mathbb{F}_3[x]$. Find the units of $\mathbb{F}_3[\alpha]$ and for each unit, determine its inverse.

6. In Example 4, show that x^3+x+1 has roots α, α^2 and α^4.

7. Let $F = \mathbb{F}_3[\alpha]$, where $\alpha^3+2\alpha+2 = 0$. Using the Root Theorem (Chapter 15), factor x^3+2x+2 in $F[x]$.

B. Congruence Classes and $F[x]/(m(x))$

In Section 17A we extended the idea of congruence from numbers to polynomials. In this section we use the idea of congruence for polynomials to construct new rings and fields made up of congruence classes, just as we did with numbers in Section 6B. Doing so will give meaning to the new roots we worked with in the last section.

Let F be a field. Let $m(x)$ be a polynomial (of degree ≥ 1) with coefficients in F. Recall that two polynomials $a(x)$ and $b(x)$ in $F[x]$ are congruent modulo $m(x)$:

$$a(x) \equiv b(x) \quad (\text{mod } m(x))$$

if $a(x) = b(x) + m(x)q(x)$ for some polynomial $q(x)$ in $F[x]$.

As we observed in Section 17A, congruence modulo $m(x)$ is an equivalence relation:

Proposition 1. *For $m(x) \neq 0$ in $F[x]$, congruence modulo $m(x)$ is:*
Reflexive: for all $a(x)$ in $F[x]$, $a(x) \equiv a(x) \pmod{m(x)}$;
Symmetric: for all $a(x), b(x)$ in $F[x]$, if $a(x) \equiv b(x) \pmod{m(x)}$, then $b(x) \equiv a(x)$
$\pmod{m(x)}$
Transitive: for all $a(x), b(x), c(x)$ in $F[x]$, if $a(x) \equiv b(x) \pmod{m(x)}$ and $b(x) \equiv c(x)$
$\pmod{m(x)}$ then $a(x) \equiv c(x) \pmod{m(x)}$

(The proofs are very easy, using the definition of congruence.)

Since congruence modulo $m(x)$ is an equivalence relation, the relation $a(x) \equiv b(x) \pmod{m(x)}$ partitions the set $F[x]$ into equivalence classes, called *congruence classes*. The congruence class of $a(x)$ modulo $m(x)$, written $[a(x)]_{m(x)}$, is the set of all polynomials $b(x)$ in $F[x]$ that are congruent to $a(x)$ modulo $m(x)$. In symbols,

$$[a(x)]_{m(x)} = \{b(x) \text{ in } F[x] | b(x) \equiv a(x) \pmod{m(x)}\}.$$

Often we will write $[a(x)]$ rather than $[a(x)]_{m(x)}$ if the modulus polynomial $m(x)$ is clear from the context, as in:

Proposition 2. *Two congruence classes $[a(x)]$ and $[b(x)]$ modulo $m(x)$ are equal, $[a(x)] = [b(x)]$, if and only if $a(x) \equiv b(x) \pmod{m(x)}$.*

In words, $a(x)$ is congruent to $b(x)$ modulo $m(x)$, if and only if the set of polynomials which are congruent to $a(x)$ is the same as the set of polynomials which are congruent to $b(x)$.

The proof of this follows easily from symmetry and transitivity of congruence.

Definition. The set of congruence classes of polynomials in $F[x]$ modulo $m(x)$ is denoted by $F[x]/(m(x))$.

Any polynomial $b(x)$ in $[a(x)]_{m(x)}$ is a *representative* of the congruence class $[a(x)]_{m(x)}$. Thus $a(x)$ is a representative of $[a(x)]_{m(x)}$, but so is any polynomial $b(x)$ congruent to $a(x)$ modulo $m(x)$. Then $b(x)$ is a representative of $[a(x)]_{m(x)}$ if and only if $[b(x)]_{m(x)} = [a(x)]_{m(x)}$.

By the Division Theorem, any polynomial $f(x)$ in $F[x]$ may be divided by $m(x)$:

$$f(x) = m(x)q(x) + r(x),$$

where the remainder $r(x)$ has degree $< d = \deg(m(x))$, and is unique. Then $f(x) \equiv r(x) \pmod{m(x)}$, and so $[f(x)]_{m(x)} = [r(x)]_{m(x)}$. Since the remainder $r(x)$ is unique, every congruence class modulo $m(x)$ is represented by a unique polynomial $r(x)$ of degree $< \deg(m(x))$. In other terminology, the set of polynomials $r(x)$ of degree $< \deg(m(x))$ is a *complete set of representatives for* $F[x]/(m(x))$. This property is analogous to the property that the numbers $0, 1, 2, \ldots, m-1$ form a complete set of representatives for $\mathbb{Z}/m\mathbb{Z}$.

We can therefore describe the set $F[x]/(m(x))$ of all congruence classes modulo $m(x)$ as the the the set of classes $[r(x)]_{m(x)}$ of the polynomials $r(x)$ in $F[x]$ of degree $< \deg(m(x))$. Thus

Proposition 3. *If F has q elements and $m(x)$ has degree d, then the set $F[x]/(m(x))$ has q^d elements.*

Proof. The set of polynomials

$$r(x) = r_0 + r_1 x + \ldots + r_{d-1} x^{d-1}$$

of degree $< d$ is a complete set of representatives for $F[x]/(m(x))$. Since there are q choices for each of the d coefficients $r_0, \ldots, r_{d-1}$, the total number of such polynomials is q^d. $\qquad\square$

Example 5. Let $F = \mathbb{Z}/3\mathbb{Z} = \mathbb{F}_3 = \{0, 1, 2 \pmod 3\}$, and let $m(x) = x^3 + 1$. Then the polynomials of degree ≤ 2 form a complete set of representatives for $\mathbb{F}_3[x]/(m(x))$. To find a representative of degree ≤ 2 for the congruence class of $[f(x)]$ we take the remainder on dividing $f(x)$ by $x^3 + 1$. For example, dividing $x^5 + x^4 + 2x$ by $x^3 + 1$ gives

$$x^5 + x^4 + 2x = (x^3 + 1)(x^2 + x) + 2x^2 + x,$$

and so

$$[x^5 + x^4 + 2x] = [2x^2 + x].$$

The set $\mathbb{F}_3[x]/(x^3 + 1)$ has 27 elements, since there are $27 = 3^3$ polynomials of degree ≤ 2 with coefficients in $\mathbb{F}_3$.

Example 6. In $\mathbb{Q}[x]$ consider congruence modulo the polynomial $x^2 - 5$. Each element of $\mathbb{Q}[x]/(x^2 - 5)$ has a representative of degree ≤ 1. In particular, the congruence class of x^n has a representative of degree ≤ 1 for each n: since

$$x^2 \equiv 5 \pmod{x^2 - 5},$$

we have $[x^2] = [5]$; similarly, since

$$x^3 \equiv 5x \pmod{x^2 - 5},$$

we have $[x^3] = [5x]$. Continuing, we have

$$[x^4] = [5x^2] = [25], [x^5] = [25x]$$

etc. It's easy to see that

$$[x^{2n}] = [5^n], \text{ and } [x^{2n+1}] = [5^n x]$$

for every $n > 1$.

Using this information allows us to take the congruence class of any polynomial and find its representative of degree ≤ 1 by substitution. For example,

$$
\begin{aligned}
[x^6 - 3x^5 + 5x^3 - 2x + 1] &= [x^6] - [3x^5] + [5x^3] - [2x] + [1] \\
&= [x^6] - [3][x^5] + [5][x^3] - [2][x] + [1] \\
&= [5^3] - [3][5^2 x] + [5][5x] - [2x] + [1] \\
&= [(-75 + 25 - 2)x] + [125 + 1] \\
&= [-52x + 126].
\end{aligned}
$$

Example 7. Let $F = \mathbb{F}_2$, let $m(x) = x^4 + x + 1$. Then the congruence classes modulo $x^4 + x + 1$ are the classes of all polynomials in $F[x]$ of degree ≤ 3, namely, the $2^4 = 16$ classes

$$
\begin{aligned}
&[0], [1], [x], [x+1], \\
&[x^2], [x^2 + 1], [x^2 + x], [x^2 + x + 1], \\
&[x^3], [x^3 + 1], [x^3 + x], [x^3 + x + 1], \\
&[x^3 + x^2], [x^3 + x^2 + 1], [x^3 + x^2 + x], [x^3 + x^2 + x + 1].
\end{aligned}
$$

Example 8. Let $\mathbb{R}[x]/(x^2 + 1)$ denote the set of congruence classes of polynomials with real coefficients modulo $x^2 + 1$.

As with Example 6, every polynomial $f(x)$ is congruent modulo $x^2 + 1$ to a polynomial of degree ≤ 1, so a typical element of $\mathbb{R}[x]/(x^2 + 1)$ has the form $[a + bx]_{x^2 + 1}$ for a, b in $\mathbb{R}$; that is, every congruence class is represented by a polynomial $a + bx$ in $\mathbb{R}[x]$ of degree ≤ 1. In particular, $[x^2] = [-1]$. Thus the elements of $\mathbb{R}[x]/(x^2 + 1)$ are in one-to-one correspondence (by $[a + bx] \mapsto (a, b)$) with vectors in the two-dimensional vector space $\mathbb{R}^2$, or with the complex numbers $\mathbb{C}$ by the one-to-one correspondence $[a + bx] \mapsto a + bi$.

Exercises.

8. Show that if $[a(x)]$ and $b[x]$ are two congruence classes modulo $m(x)$, then either $[a(x)] = [b(x)]$ or $[a(x)] \cap [b(x)]$ is the empty set.

C. Algebraic Operations

We define addition of congruence classes by

$$
[a(x)]_{m(x)} + [b(x)]_{m(x)} = [a(x) + b(x)]_{m(x)},
$$

Thus we add congruence classes modulo $m(x)$ by adding representatives. This operation on congruence classes is *well-defined*, that is, doesn't depend on the choice of representatives, because of the result:

if $a(x) \equiv a'(x)$ (mod $m(x)$), and $b(x) \equiv b'(x)$ (mod $m(x)$), then $a(x) + b(x) \equiv a'(x) + b'(x)$ (mod $m(x)$),

whose proof is easy and just like the proof for congruence of numbers. Similarly, we multiply congruence classes using representatives:

$$[a(x)]_{m(x)}[b(x)]_{m(x)} = [a(x)b(x)]_{m(x)},$$

which is a well-defined operation because of the result:

if $a(x) \equiv a'(x)$ (mod $m(x)$), and $b(x) \equiv b'(x)$ (mod $m(x)$), then $a(x)b(x) \equiv a'(x)b'(x)$ (mod $m(x)$),

which we proved in Proposition 1 of Section 17A.

Since $F[x]$ is a commutative ring, and addition and multiplication on $F[x]/(m(x))$ are defined by using representatives, that is, by using the addition and multiplication on $F[x]$, all of the properties of a commutative ring hold for $F[x]/(m(x))$, because they hold for $F[x]$. For example, commutativity of multiplication is proved as follows:

$$[a(x)][b(x)] = [a(x)b(x)] = [b(x)a(x)] = [b(x)][a(x)],$$

where the second equality follows from commutativity of multiplication in $F[x]$. Or distributivity:

$$[a(x)][b(x) + c(x)] = [a(x)(b(x) + c(x))] = [a(x)b(x) + a(x)c(x)]$$
$$= [a(x)b(x)] + [a(x)c(x)]$$
$$= [a(x)][b(x)] + [a(x)][c(x)],$$

where the second equality follows from distributivity in $F[x]$.

The zero element of $F[x]/(m(x))$ is $0 = [0]_{m(x)}$, the negative of an element $[a(x)]$ is $[-a(x)]$, and the multiplicative identity element is $1 = [1]_{m(x)}$.

With integers, we found that $\mathbb{Z}/m\mathbb{Z}$ is a field if and only if m is prime. Here is the corresponding result for polynomials:

Proposition 4. *Let F be a field, $m(x)$ a polynomial of degree ≥ 1 with coefficients in F. Then $F[x]/(m(x))$ is a field if and only if $m(x)$ is irreducible.*

Proof. Suppose $m(x)$ is not irreducible, say, $m(x) = r(x)s(x)$ where $r(x)$ and $s(x)$ are polynomials of degree $< \deg(m(x))$. Then $r(x)$ and $s(x)$ represent nonzero congruence classes in $F[x]/(m(x))$, but their product, $m(x)$, represents the zero class in $F[x]/(m(x))$. So $F[x]/(m(x))$ has zero divisors, and therefore cannot be a field. (A zero divisor cannot have an inverse.)

Conversely, suppose $a(x)$ is any nonzero polynomial coprime to $m(x)$. Then by Bezout's Identity, there are polynomials $r(x), s(x)$ in $F[x]$ so that $a(x)r(x) + m(x)s(x) = 1$. But then

$$1 = [1] = [a(x)r(x) + m(x)s(x)]$$
$$= [a(x)r(x)]$$
$$= [a(x)][r(x)].$$

Thus $[a(x)]$ is invertible, and $[r(x)]$ is its inverse. If $m(x)$ is irreducible, then every nonzero polynomial of degree $< \deg(m(x))$ is coprime to $m(x)$, so every non-zero congruence class in $F[x]/(m(x))$ has an inverse. So $F[x]/(m(x))$ is a field.   □

Proposition 4 shows how to construct many new fields. Start with a field F and find an irreducible polynomial $m(x)$ of degree > 1 in $F[x]$. Then $F[x]/(m(x))$ is a field.

Example 9. In $\mathbb{Q}[x], x^2 - 5$ is irreducible. So $\mathbb{Q}[x]/(x^2 - 5)$ is a field. If $[a + bx]$ is not zero, then the inverse of $[a + bx]$ is $[(a/d) - (b/d)x]$, where $d = a^2 - 5b^2$. (You should check that if a or b is not zero, then d is not zero.)

Example 10. $\mathbb{F}_3[x]/(x^3 + 1)$ is not a field because $x^3 + 1$ is not irreducible: $x^3 + 1 = (x + 1)^3$ in $\mathbb{F}_3[x]$. Because $x^3 + 1$ factors, $\mathbb{F}_3[x]/(x^3 + 1)$ has zero divisors:

$$0 = [x^3 + 1] = [x + 1][x^2 - x + 1].$$

Example 11. In $\mathbb{F}_5[x], x^3 + 3x + 3$ is irreducible (because otherwise it would have to have a factor of degree 1, hence a root in $\mathbb{F}_5$). So $E = \mathbb{F}_5[x]/(x^3 + 3x + 3)$ is a field. Since every element of E is represented by a unique polynomial in $\mathbb{F}_5[x]$ of degree ≤ 2, E has $5^3 = 125$ elements.

Generalizing this last example, if $F = \mathbb{F}_p$, the field of p elements, and $m(x)$ in $\mathbb{F}_p[x]$ has degree d, then $\mathbb{F}_p[x]/(m(x))$ has p^d elements. Thus whenever we can find an *irreducible* polynomial of degree d with coefficients in $\mathbb{F}_p$, we can construct a field containing p^d elements.

Definition. A field of the form $F[x]/(m(x))$, where $m(x)$ is an irreducible polynomial with coefficients in F, is called a *simple field extension of F*.

For example, $\mathbb{Q}[x]/(x^2 - 5)$ is a simple field extension of $\mathbb{Q}$, as are the fields of Examples 7 and 11.

Proposition 4 described the Bezout's identity method for finding the inverse of a unit that we promised earlier.

Example 12. In $\mathbb{F}_3[x]/(x^3 + 2x + 1)$, find the inverse of x^5:
First do Euclid's algorithm with x^5 and the modulus $x^3 + 2x + 1$:

$$x^5 = (x^3 + 2x + 1)(x^2 + 1) + (2x^2 + x + 2)$$
$$x^3 + 2x + 1 = (2x^2 + x + 2)(2x + 2) + 2x$$
$$2x^2 + x + 2 = 2x(x + 2) + 2.$$

Then
$$x^5 \equiv 2x^2 + x + 2 \pmod{x^3 + 2x + 1}$$
and
$$2x \equiv -(2x^2 + x + 2)(2x + 2) \equiv (2x^2 + x + 2)(x + 1) \pmod{x^3 + 2x + 1}$$

So
$$1 \equiv 2x(x + 2) - (2x^2 + x + 2)$$
$$\equiv [(2x^2 + x + 2)(x + 1)](x + 2) - (2x^2 + x + 2)$$
$$\equiv (2x^2 + x + 2)(x^2 + 1) \equiv x^5(x^2 + 1),$$

and hence
$$x^5(x^2 + 1) \equiv 1 \pmod{x^3 + 2x + 1}.$$

Example 13. In $\mathbb{F}_2[x]/(x^5 + x^2 + 1)$, we find the inverse of $[x^4 + x^3 + 1]$. To do so, we do Euclid's algorithm in $\mathbb{F}_2[x]$ with $x^4 + x^3 + 1$ and the modulus $x^5 + x^2 + 1$:

$$x^5 + x^2 + 1 = (x^4 + x^3 + 1)(x + 1) + (x^3 + x^2 + 1)$$
$$x^4 + x^3 + 1 = (x^3 + x^2 + 1)(x) + (x^2 + 1)$$
$$x^3 + x^2 + 1 = (x^2 + 1)(x + 1) + 1.$$

Then we use these to obtain Bezout's identity; after substituting for $x^2 + 1$, then $x^3 + x^2 + 1$ in the last equation, we obtain

$$1 = (x^5 + x^2 + 1)(x^2 + x + 1) + (x^4 + x^3 + 1)(x^3 + x).$$

Modulo $x^5 + x^2 + 1$ this becomes

$$1 \equiv (x^4 + x^3 + 1)(x^3 + x) \pmod{x^5 + x^2 + 1},$$

or in terms of congruence classes,

$$[1] = [x^4 + x^3 + 1][x^3 + x)]$$

in $\mathbb{F}_2[x]/(x^5 + x^2 + 1)$.

Exercises. We proved in Chapter 8 that if R is a commutative ring with a finite number of elements, then every non-zero element of R is either a unit or a zero divisor.

9. In $\mathbb{F}_3[x]/(x^3 + 1)$, find the units and the zero divisors.

10. In $\mathbb{F}_2[x]/(x^3 + x^2 + x + 1)$, find the units and the zero divisors.

11. In $\mathbb{F}_5[x]/(x^2 + 1)$, find the units and the zero divisors.

12. Find a simple field extension of $\mathbb{F}_2$ with
   (i) 4 elements;
   (ii) 8 elements;
   (iii) 16 elements;
   (iv) 32 elements;
   (v) 64 elements.

13. Find a simple field extension of $\mathbb{F}_7$ with $7^3 = 343$ elements.

14. In $\mathbb{F}_3[x]/(x^3 + 2x + 1)$, find the inverses of the congruence classes of
   (i) $x^2 + x + 1$;
   (ii) $x^4 + 2x^3$;
   (iii) $x^6 + x^4 + 2x$.

15. Find the inverse of $[1 - 2x]$ in $\mathbb{Q}[x]/(x^3 - 2)$ using Euclid's algorithm.

16. In $\mathbb{F}_2[x]/(x^5 + x^2 + 1)$, find the inverse of $[x^3]$; of $[x^4 + 1]$; of $[x^2]$.

D. Finding a Root of $m(x)$ in $F[x]/(m(x))$

Let F be a field and $m(x)$ a polynomial of degree $d \geq 1$. Elements of $F[x]/(m(x))$ are congruence classes $[f(x)]_{m(x)}$ for polynomials $f(x)$ with coefficients in F.

We may use the algebraic operations in $F[x]/(m(x))$ to simplify the description of elements of $F[x]/(m(x))$.

If $f(x) = a_0 + a_1 x + \ldots + a_n x^n$ with $a_0, a_1, \ldots a_n$ in F, then, denoting $[\ ]_{m(x)}$ by $[\ ]$, we have

$$= [a_0 + a_1 x + \ldots + a_n x^n]$$
$$= [a_0] + [a_1 x] + \ldots + [a_n x^n]$$
$$= [a_0] + [a_1][x] + \ldots + [a_n][x]^n$$

by the definition of addition and multiplication of congruence classes. Now the map from F to $F[x]$, defined by viewing an element of F as a polynomial of degree ≤ 0 in $F[x]$, induces a homomorphism $\iota : F \to F[x]/(m(x))$ given by $\iota(r) = [r]_{m(x)}$ for r in F. In words, ι is the function: take the element r of F, view it as a polynomial, and take its congruence class modulo $m(x)$.

Since the modulus polynomial $m(x)$ has degree 1 or greater, ι is a one-to-one function. So we may identify F with its image in $F[x]/(m(x))$ under the map ι, and for r in F, write the congruence class $[r]_{m(x)}$ as r.

After that identification, then a typical element of $F[x]/(m(x))$ is

$$[f(x)]_{m(x)} = a_0 + a_1[x] + \ldots + a_n[x]^n$$

where $[x] = [x]_{m(x)}$ is the congruence class of x. This description has the interpretation: for $f(x)$ in $F[x]$, the congruence class of $f(x)$ in $F[x]/(m(x))$ is the polynomial $f([x]_{m(x)})$, that is, the polynomial f evaluated at $[x]_{m(x)}$.

Example 14. A typical element of $\mathbb{F}_3[x]/(x^3+2x+1)$ is represented by $f(x) = ax^2 + bx + c$ for some a, b, c in $\mathbb{F}_3$. Thus identifying elements r of $\mathbb{F}_3$ with the congruence classes $[r] = [r]_{x^3+2x+1}$, we can write the congruence class $[f(x)]$ as

$$\begin{aligned}
&= [ax^2 + bx + c] \\
&= [a][x]^2 + [b][x] + [c], \\
&= a[x]^2 + b[x] + c \\
&= f([x]).
\end{aligned}$$

Finally, let us abbreviate $[x]_{m(x)}$ by α. Then $[f(x)]_{m(x)} = f([x]_{m(x)}) = f(\alpha)$: the congruence class of the polynomial $f(x)$ is $f(x)$ evaluated at the congruence class $\alpha = [x]_{m(x)}$.

In this way, we may think of $F[x]/(m(x))$ as $F[\alpha]$, the set of polynomials with coefficients in F evaluated at the congruence class $\alpha = [x]_{m(x)}$. Thus when $\alpha = [x]_{m(x)}$, the following statements are all equivalent:

$$\begin{aligned}
f(\alpha) &= g(\alpha) \\
[f(x)]_{m(x)} &= [g(x)]_{m(x)} \\
f(x) &\equiv g(x) \quad (\bmod\ m(x)) \\
f(x) &= g(x) + m(x)q(x) \text{ for some } q(x) \text{ in } F[x].
\end{aligned}$$

If $m(x)$ is irreducible, then $F[x]/(m(x))$ is a field by Proposition 4.

Having identified $[f(x)]_{m(x)} = f(\alpha)$ where $\alpha = [x]_{m(x)}$, a key observation is that

$$0 = [m(x)]_{m(x)} = m([x]_{m(x)}) = m(\alpha),$$

so that $\alpha = [x]_{m(x)}$ is a root in $F[x]/(m(x))$ of the polynomial $m(x)$ in $F[x]$.

The construction of the field $F[x]/(m(x))$ and the observation that $[x]_{m(x)} = \alpha$ is a root of $m(x)$ in $F[x]/(m(x))$ yields an important theorem.

Theorem 5 (Cauchy-Kronecker-Steinitz Theorem). *Let F be a field, and $m(x)$ an irreducible polynomial with coefficients in F. Then there is a field containing F in which $m(x)$ has a root.*

We can now resolve the mystery of Section A. The roots of polynomials we invented in Section A now have a concrete meaning as congruence classes. To find a root of an irreducible polynomial $m(x)$, simply take the congruence class of x in $F[x]/(m(x))$.

Thus the "imaginary" number i can be identified as $i = [x]_{x^2+1}$ in the field $\mathbb{R}[x]/(x^2+1)$, and with that definition, the fields $\mathbb{R}[x]/(x^2+1)$ and $\mathbb{C}$ look identical. We'll be more precise about this identification in the next chapter.

In Example 4, above, the invented root α of x^3+x+1 in $\mathbb{F}_2[x]$ is just the congruence class $[x]_{x^3+x+1}$ in $\mathbb{F}_2[x]/(x^3+x+1)$.

The Cauchy-Kronecker-Steinitz Theorem has the following important consequence.

Corollary 6 (Splitting Field Theorem). *Let F be a field, $f(x)$ a polynomial of degree ≥ 1 in $F[x]$. Then there exists a field K containing F such that $f(x)$ factors into a product of linear factors in $K[x]$.*

Proof. We proceed by induction on d, the degree of $f(x)$. If $\deg(f(x)) = 1$ the result is trivial.

Let $f(x)$ have degree d. In $F[x]$, suppose $f(x) = p_1(x) \cdot \ldots \cdot p_s(x)$, a product of irreducible polynomials. If $\deg(p_i(x)) = 1$ for all $i = 1, \ldots, s$, then the field K we seek is F itself. Otherwise, renumbering if necessary, suppose that $p_1(x)$ has degree > 1. Let $L = F[y]/(p_1(y))$, and let $\alpha = [y]_{p_1(y)}$. Then L is a field containing F (where we identify F as the congruence classes of polynomials of degree ≤ 0 in L), and α is a root in L of $p_1(x)$. So in $L[x]$, $p_1(x)$ factors as $(x - \alpha)q_1(x)$ by the Root Theorem (Chapter 14). Thus in $L[x]$,

$$f(x) = (x - \alpha)q_1(x)p_2(x) \cdot \ldots \cdot p_s(x).$$

Let

$$g(x) = q_1(x)p_2(x) \cdot \ldots \cdot p_s(x)$$

in $L[x]$. Then $\deg(g(x)) = \deg(f(x)) - 1$. By induction, there is a field K containing L so that in $K[x]$, $g(x)$ factors into a product of linear factors. But then so does $f(x) = (x - \alpha)g(x)$. Since K contains F, we're done. $\qquad\square$

A field K is called a *splitting field* for $f(x)$ if $f(x)$ factors into linear factors in $K[x]$.

Example 15. In $\mathbb{Q}[x], f(x) = x^3 - 2$ is irreducible. It has a root in $\mathbb{R}$, namely $2^{1/3}$, but $\mathbb{R}$ is not a splitting field because $f(x)$ has only one real root. If we let ω be a cube root of unity in $\mathbb{C}$, then the other two roots of $f(x)$ are $\omega \cdot 2^{1/3}$ and $\omega^2 \cdot 2^{1/3}$, so $x^3 - 2$ splits into a product of three linear factors in $\mathbb{C}[x]$. Thus $\mathbb{C}$ is a splitting field for $x^3 - 2$.

By the Fundamental Theorem of Algebra, $\mathbb{C}$ is a splitting field for every polynomial in $\mathbb{Q}[x]$.

The Cauchy-Kronecker-Steinitz Theorem and its corollary, the Splitting Field Theorem, are basic ingredients in several proofs of the Fundamental Theorem of Algebra, for example, Lagrange's 1772 proof [Suzuki (2006)] and Gauss's second proof [Dobbs and Hanks (1992)].

Exercises.

17. The real number $\alpha = \cos 20°$ is a root of the irreducible polynomial

$$f(x) = 4x^3 - 3x - \frac{1}{2}$$

in $\mathbb{Q}[x]$. Let $E = \mathbb{Q}[\cos 20°]$. Show that $f(x)$ splits in E.

18. Show that if $E = \mathbb{F}_2[x]/(x^4 + x + 1)$, then E is a splitting field for $x^4 + x + 1$ and also for $x^2 + x + 1$.

19. Show that for p prime, $\mathbb{F}_p$ is a splitting field for $f(x) = x^p - x$ in $\mathbb{F}_p[x]$.

20. Show that $\mathbb{R}$ is a splitting field for the polynomial $f(x) = x^n - 5$ if and only if $n = 2$.

21. Show that if F is a field and $m(x)$ in $F[x]$ has degree ≥ 1, then the map

$$\iota : F \rightarrow F[x]/(m(x))$$

given by $\iota(r) = [r]_{m(x)}$ is a one-to-one homomorphism.

Chapter 24
Homomorphisms and Finite Fields

In this chapter we describe all finite fields.

A. Homomorphisms and Kernels

The aim of this section is to prove the Isomorphism Theorem, a fundamental result in algebra.

One of the applications of the Isomorphism Theorem will be to resolve the following question.

Consider the polynomial $x^3 - 2$ over the rational numbers $\mathbb{Q}$. On the one hand, there is a well-known real number, $\sqrt[3]{2}$, whose cube is 2 and thus is a root of $x^3 - 2$. On the other hand, from the Cauchy-Kronecker-Steinitz Theorem, we know that the congruence class $[x]_{(x^3-2)}$ is a root of $x^3 - 2$. Is there a relation between the two roots of $x^3 - 2$?

Recall that we completely characterized ring homomorphisms with domain the ring of integers in Chapter 8. Here we recall from Section 13D how to define ring homomorphisms whose domain is a ring of polynomials.

Proposition 1. *Let F be a field, R a commutative ring. A homomorphism $\varphi : F[x] \to R$ is completely determined by the set $\{\varphi(a) | a \text{ in } F\}$ of values of φ on F, and by $\varphi(x)$.*

Proof. If R and S are rings, a function $\varphi : R \to S$ is a ring homomorphism if for all r_1, r_2 in R, $\varphi(r_1 + r_2) = \varphi(r_1) + \varphi(r_2)$, $\varphi(r_1 \cdot r_2) = \varphi(r_1) \cdot \varphi(r_2)$, and $\varphi(1) = 1$. Let $R = F[x]$ and let $p(x) = a_0 + a_1x + \ldots + a_nx^n$ be an element of $F[x]$. Then

$$\varphi(p(x)) = \varphi(a_0 + a_1x + \ldots + a_nx^n)$$
$$= \varphi(a_0) + \varphi(a_1)\varphi(x) + \ldots + \varphi(a_n)\varphi(x)^n.$$

So $\varphi(p(x))$ is determined by the values of φ on the coefficients and by $\varphi(x)$. $\qquad\square$

L.N. Childs, *A Concrete Introduction to Higher Algebra*, Undergraduate Texts in Mathematics, © Springer Science+Business Media LLC 2009

A useful special case of Proposition 1 occurs when $F \subset R$ and φ acts as the identity map on F, i.e., $\varphi(a) = a$ for a in F. In that case, if $\varphi(x) = \alpha$ in R, then $\varphi = \varphi_\alpha$, the evaluation map, defined by $\varphi(p(x)) = p(\alpha)$, or in words, evaluate the polynomial $p(x)$ at $x = \alpha$ in R.

For example, the evaluation map

$$\phi_{\sqrt{2}} : \mathbb{Q}[x] \to \mathbb{R}$$

takes the polynomial $3x^2 + 4x + 1$ to

$$\phi_{\sqrt{2}}(3x^2 + 4x + 1) = 3\sqrt{2}^2 + 4\sqrt{2} + 1 = 7 + 4\sqrt{2}.$$

If $f : R \to S$ is a ring homomorphism, then the kernel of f is the set of elements r in R so that $f(r) = 0$:

$$\ker f = \{r \text{ in } R | f(r) = 0\}$$

We showed in Chapter 8 that f is one-to-one if and only if the kernel of f consists of only the zero element of R. The next result describes the kernel of the evaluation homomorphism φ_a when φ_a is not one-to-one.

Theorem 2 (Minimal Polynomial Theorem). *Let $F \subset E$ be fields and let α be an element of E. Let $\varphi_\alpha : F[x] \to E$ be the evaluation map. Suppose φ_α is not one-to-one, so that there is some polynomial $f(x)$ of degree ≥ 1 such that $\varphi_\alpha(f(x)) = f(\alpha) = 0$. Then*
(i) there exists a unique monic polynomial $p(x)$ of minimal degree > 0 in $F[x]$ with $\varphi_\alpha(p(x)) = p(a) = 0$;
(ii) $p(x)$ is irreducible; and
(iii) $\ker \varphi_\alpha$ consists of all multiples of $p(x)$.

If α is in E and φ_α is not one-to-one, so that $f(\alpha) = 0$ for some polynomial $f(x)$ with coefficients in F, then α is called *algebraic* over F. Some examples of numbers in $\mathbb{C}$ that are algebraic over $\mathbb{Q}$ are elements of $\mathbb{Q}$ itself (if a is in $\mathbb{Q}$, then a is a root of the polynomial $x - a$), $\sqrt{a}$ for any a in $\mathbb{Q}$, and the nth root of any element of $\mathbb{Q}$. There are many more.

If α in E is algebraic over F, then the unique monic polynomial $p(x)$ with $p(\alpha) = 0$ is called the *minimal polynomial* of α over F.

Here is a proof of the Minimal Polynomial Theorem.

Proof. The kernel of φ_α is the set of polynomials $f(x)$ in $F[x]$ so that $\varphi_\alpha(f(x)) = 0$, that is, $f(\alpha) = 0$. If $\ker(\varphi_\alpha)$ contains a nonzero polynomial, then by well-ordering, $\ker(\varphi_\alpha)$ contains a polynomial $p(x)$ of minimal degree > 0. If $p(x)$ is in $\ker(\varphi_\alpha)$, so is any associate of $p(x)$, so we can assume that $p(x)$ is a monic polynomial of minimal degree in $\ker(\varphi_\alpha)$.

We first prove (ii). Suppose $f(x)$ is a polynomial in $\ker(\varphi_\alpha)$. Dividing $f(x)$ by $p(x)$ gives

$$f(x) = p(x)s(x) + r(x),$$

where $r(x)$ has degree $< \deg(p(x))$. Applying φ_α to that equation yields $\varphi_\alpha(r(x)) = 0$, hence $r(x)$ is in $\ker(\varphi_\alpha)$. Unless $r(x) = 0$, this contradicts the assumption that $p(x)$ had minimal degree. So $r(x) = 0$ and $p(x)$ divides $f(x)$. This proves (iii).

To show (i), that $p(x)$ is unique, suppose $q(x)$ is also a monic polynomial of minimal degree in $\ker(\varphi_\alpha)$, then, since $p(x)$ divides $q(x)$, the two polynomials must be associates, and since they are both monic, they are equal.

To prove (ii), let $p(x)$ be the monic polynomial of minimal degree in $\ker(\varphi_\alpha)$. If $p(x)$ factors, $p(x) = f(x)g(x)$ with $\deg f(x) < \deg p(x)$ and $\deg g(x) < \deg p(x)$, then applying the homomorphism φ_α to the factorization of $p(x)$ gives $0 = f(\alpha)g(\alpha)$. This is an equation in E, a field. Since E has no zero divisors, either $f(\alpha) = 0$ or $g(\alpha) = 0$, hence either $f(x)$ or $g(x)$ is in $\ker(\varphi_\alpha)$, contradicting the minimality of the degree of $p(x)$. So $p(x)$ is irreducible. $\qquad\qquad\square$

Here is the fundamental theorem about homomorphisms.

Recall that a ring homomorphism $\varphi : R \to S$ is an isomorphism if φ is one-to-one and maps onto S. That is, the kernel of φ is $\{0\}$ and the image of φ is S.

Theorem 3 (Isomorphism Theorem). *Let $F \subset E$ be fields and let α be a non-zero element of E that is algebraic over F. Let $\varphi_\alpha : F[x] \to E$ be the "evaluation at α" map and let $p(x)$ be the minimal polynomial of α over F. Then φ_α induces a one-to-one homomorphism $\bar\varphi_\alpha : F[x]/(p(x)) \to E$. Let $F[\alpha]$ be the image of φ_α, then $\bar\varphi_\alpha : F[x]/(p(x)) \to F[\alpha]$ is an isomorphism, hence $F[\alpha]$ is a field.*

We write $R \cong S$ if there is an isomorphism between the two rings R and S. Thus in the theorem,

$$F[x]/p(x) \cong F[\alpha].$$

Before proceeding to the proof of this theorem, here are some examples.

Example 1. Let $\varphi_i : \mathbb{R}[x] \to \mathbb{C}$ be the "evaluation at i" homomorphism, where $i^2 = -1$. Then φ_i maps onto $\mathbb{C}$ because any complex number $a + bi = \varphi_i(a + bx)$. The minimal polynomial of i is $p(x) = x^2 + 1$, so φ_i induces a one-to-one homomorphism

$$\bar\varphi_i : \mathbb{R}[x]/(x^2 + 1) \to \mathbb{C},$$

which is an isomorphism because φ_i, hence $\bar\varphi_i$ maps onto $\mathbb{C}$. So

$$\mathbb{R}[x]/(x^2 + 1) \cong \mathbb{C}.$$

Example 2. Let $\varphi : \mathbb{R}[x] \to \mathbb{C}$ be the "evaluation at $\sqrt{-2}$" homomorphism. Then φ is onto and induces an isomorphism, $\mathbb{R}[x]/(x^2 + 2) \cong \mathbb{C}$.

Here is the proof of the Isomorphism Theorem.

Proof. Define $\bar\varphi_\alpha : F[x]/(p(x)) \to E$ by $\bar\varphi_\alpha([f(x)]) = \varphi_\alpha(f(x)) = f(\alpha)$. We must show that $\bar\varphi_\alpha$ is well defined, in the sense that if $[f(x)] = [g(x)]$ in $F[x]/(p(x))$, then $f(\alpha) = g(\alpha)$. (That is, the value of $\bar\varphi_\alpha$ on a congruence class does not depend on the choice of representative of the congruence class.) But if $[f(x)] = [g(x)]$, then $f(x) = g(x) + t(x)p(x)$ for some polynomial $t(x)$. Since $p(\alpha) = 0$, evaluating at $x = \alpha$ yields $f(\alpha) = g(\alpha) + t(\alpha)p(\alpha) = g(\alpha)$.

Since φ is a homomorphism and $\bar{\varphi}$ is defined via φ, it's easy to see that $\bar{\varphi}$ is a homomorphism. The image of $\bar{\varphi}_\alpha$ is the set of all elements of E of the form $f(\alpha)$ for f in $F[x]$. So the image of $\bar{\varphi}_\alpha$ is $F[\alpha]$. To show that $\bar{\varphi}_\alpha$ is one-to-one follows from the fact that since $p(x)$ is irreducible, $F[x]/p(x)$ is a field, and any nonzero ring homomorphism from a field must be one-to-one (Section 7D, Proposition 17). (Note that $\bar{\varphi}_\alpha$ acts like the inclusion function from F to E on polynomials of degree ≤ 0, so φ_α is not the zero homomorphism.)   □

Example 3. Let $\varphi_{-i} : \mathbb{R}[x] \to \mathbb{C}$ be the "evaluation at $-i$" homomorphism. Then the monic irreducible polynomial which generates the kernel of φ_{-i} is $p(x) = x^2 + 1$, and so φ_{-i} induces a homomorphism

$$\bar{\varphi}_{-i} : \mathbb{R}[x]/(x^2 + 1) \to \mathbb{C},$$

which is one-to-one and onto, hence an isomorphism between $\mathbb{R}[x]/(x^2 + 1)$ and $\mathbb{C}$. This isomorphism is different from that defined in Example 1. The composite function $\bar{\varphi}_{-i} \circ \bar{\varphi}_i^{-1}$ takes $a + bi$ to $a - bi$, so is the complex conjugation homomorphism.

Example 4. Let $\mathbb{Q}[\zeta]$ be the set of all complex numbers of the form $f(\zeta)$ where $\zeta = e^{\frac{2\pi i}{p}}$ is a pth root of 1 in $\mathbb{C}$, where p is a prime number. Let φ_ζ be the evaluation homomorphism

$$\varphi_\zeta : \mathbb{Q}[x] \to \mathbb{C}.$$

Then the image of φ_ζ is $\mathbb{Q}[\zeta]$. Since $\zeta^p = 1$, ζ is a root of the polynomial

$$x^p - 1 = (x - 1)(x^{p-1} + x^{p-2} + \ldots + x + 1)$$

and is not a root of $x - 1$. Hence ζ is a root of $x^{p-1} + x^{p-2} + \ldots + x + 1$, which is irreducible (see Section 18B). Thus $x^{p-1} + x^{p-2} + \ldots + x + 1$ is the minimal polynomial over $\mathbb{Q}$ of ζ, and φ_ζ induces an isomorphism

$$\mathbb{Q}[x]/(x^{p-1} + x^{p-2} + \ldots + x + 1) \cong \mathbb{Q}[\zeta].$$

For example, with $p = 3$, $\zeta = \frac{-1 + \sqrt{-3}}{2}$, and this last isomorphism becomes

$$\mathbb{Q}[x]/(x^2 + x + 1) \cong \mathbb{Q}[\frac{-1 + \sqrt{-3}}{2}].$$

Exercises.

1. Find the minimal polynomial of $1 + i$ over $\mathbb{Q}$.

2. Find the minimal polynomial of $1 + \sqrt{2}$ over $\mathbb{Q}$.

3. Find the minimal polynomial of $a + b\sqrt{2}$ over $\mathbb{Q}$ for all a, b in $\mathbb{Q}$.

4. Find the minimal polynomial of $\sqrt{2} + \sqrt{3}$ over $\mathbb{Q}$.

5. Find the minimal polynomial of $\cos 15°$ over $\mathbb{Q}$. (Use that $\cos 60° = 1/2$.)

6. Show that if $p(x)$ is an irreducible polynomial in $\mathbb{Q}[x]$, then the only one-to-one homomorphisms from $\mathbb{Q}[x]/(p(x))$ into $\mathbb{C}$ are evaluation maps ϕ_α where α is a root of $p(x)$ in $\mathbb{C}$.

7. Let $F \subset E$ be fields and let φ_α be the "evaluation at α" map. Show that if $p(x)$ is a monic irreducible polynomial in $\ker \varphi_\alpha$, then $p(x)$ is the minimal polynomial of α over F.

B. Finite Fields Are Simple

In Section 23C we constructed a collection of *simple field extensions*, that is, fields of the form $K = F[x]/(p(x))$, by starting with a field F and an irreducible polynomial $p(x)$ in $F[x]$, and letting K be the set of congruence classes of polynomials modulo $p(x)$. We observed that we could also think of K as $K = F[\alpha]$, polynomials with coefficients in F evaluated at α, where $\alpha = [x]_{p(x)}$.

In particular, we can construct such extensions when $F = \mathbb{F}_p$, the field of p elements, p a prime number. If $q(x)$ is an irreducible polynomial of degree d with coefficients in $\mathbb{F}_p$, then $\mathbb{F}_p[x]/q(x)$ is a field with p^d elements. We can construct in this way many finite fields (i.e., fields with finitely many elements) as simple field extensions of $\mathbb{F}_p$.

Are there finite fields which are not simple field extensions of $\mathbb{F}_p$ for some p?

Perhaps surprisingly, the answer is no.

Theorem 4. *Every finite field is isomorphic to a simple field extension of $\mathbb{F}_p$ for some prime p.*

Proof. We need to show that if K is a finite field with m elements, then there is a prime p, an irreducible polynomial $q(x)$ in $\mathbb{F}_p[x]$ and an isomorphism ϕ : $\mathbb{F}_p[x]/q(x) \to K$.

Since K is a finite field of characteristic p for some prime p, K contains a subfield isomorphic to $\mathbb{F}_p$, namely $\{n \cdot 1 \mid n \text{ in } \mathbb{Z}\}$. (Section 7D).

Also, K has a primitive root, namely, an element α such that every nonzero element of K is a power of α. There are $m - 1$ non-zero elements in K, and so α has order $m - 1$. In particular, $\alpha^{m-1} = 1$ (Section 19A).

Define $\varphi : \mathbb{F}_p[x] \to K$ as follows: Let $\varphi([n]_p) = n \cdot 1$ in K for any n in $\mathbb{Z}$. Let φ send the indeterminate x to the primitive root α. Thus for any $f(x)$ in $\mathbb{F}_p[x]$, $\varphi(f(x)) = f(\alpha)$, so φ is evaluation at α once the coefficients of $f(x)$ in $\mathbb{F}_p$ are replaced by their images in K.

The homomorphism φ is clearly onto, since every nonzero element of K is a power of α. The kernel of φ is the set of polynomials in $\mathbb{F}_p[x]$ with α as a root. Then $\ker(\varphi)$ is nonzero; in fact, as we observed, $x^{m-1} - 1$ is in $\ker(\varphi)$. Therefore the set of polynomials in $\mathbb{F}_p[x]$ with α as a root contains a nonzero monic polynomial $q(x)$ of minimal degree ≥ 1, namely, the minimal polynomial of α over $\mathbb{F}_p[x]$. By Proposition 2 of Section A, the homomorphism $\varphi : \mathbb{F}_p[x] \to K$ induces a one-to-one homomorphism $\bar{\varphi} : \mathbb{F}_p[x]/q(x) \to K$ defined by $\bar{\varphi}([f(x)]) = f(\alpha)$.

Since φ maps onto K, so does $\bar{\varphi}$. Thus $\bar{\varphi}$ is an isomorphism, and hence K is simple. $\qquad\square$

One important consequence of Theorem 4 is that there are severe restrictions on the cardinality of a finite field:

Theorem 5. *If K is a finite field, then K has p^d elements for some prime p. Thus if n is not a prime power, there is no field with n elements.*

Proof. We can prove this in two ways. The first proof uses Theorem 4:

If K is a finite field, then K is isomorphic to $\mathbb{F}_p[x]/q(x)$ for some prime p and some irreducible polynomial $q(x)$ in $\mathbb{F}_p[x]$. If $q(x)$ has degree d, then $\mathbb{F}_p[x]/q(x)$ has p^d elements, hence so does K.

The second proof uses Cauchy's Theorem (Section 11G, Theorem 18):

Suppose K has m elements, and let q be a prime dividing m. View K as an abelian group under addition. By Cauchy's Theorem, K has an element α of order q. But K has characteristic p, so $p\alpha = 0$. So the order q of α divides p. Since p is a prime, therefore $q = p$. So no prime other than p divides m, hence $m = p^d$ for some d. $\quad\square$

Table 24.1 gives a list of fields with n elements for small n.

In the next section we shall prove the converse of the corollary, namely, if $n = p^d$, p prime, is a prime power, then there is a field with n elements.

Table 24.1. A list of fields with n elements for small n.

$n =$	Fields
2	$\mathbb{F}_2$
3	$\mathbb{F}_3$
4	$\mathbb{F}_2[x]/(x^2+x+1)$
5	$\mathbb{F}_5$
6	none
7	$\mathbb{F}_7$
8	$\mathbb{F}_2[x]/(x^3+x+1)$ and $\mathbb{F}_2[x]/(x^3+x^2+1)$ (which are isomorphic)
9	$\mathbb{F}_3[x]/(x^2+1)$ (are there others?)
10	none
11	$\mathbb{F}_{11}$
12	none
13	$\mathbb{F}_{13}$
14	none
15	none
16	$\mathbb{F}_2[x]/(x^4+x+1)$ (others??)

Exercises.

8. (i) Find a primitive root β of $\mathbb{F}_2[x]/(x^4+x^3+x^2+x+1)$.
    (ii) Find the minimal polynomial $q(x)$ in $\mathbb{F}_2[x]$ of β.
    (iii) Show that $\mathbb{F}_2[x]/(x^4+x^3+x^2+x+1)$ is isomorphic to $\mathbb{F}_2[x]/(q(x))$.

9. (i) Find a primitive root β of $\mathbb{F}_3[x]/(x^2+1)$.
    (ii) Find the minimal polynomial $q(x)$ in $\mathbb{F}_3[x]$ of β.
    (iii) Show that $\mathbb{F}_3[x]/(x^2+1)$ is isomorphic to $\mathbb{F}_3[x]/q(x)$.

10. (i) Find a root in $K = \mathbb{F}_2[x]/(x^4+x+1)$ of x^2+x+1.
    (ii) Describe a homomorphism from $\mathbb{F}_2[x]/(x^2+x+1)$ into K.

11. Let $K = \mathbb{F}_2[x]/(x^3+x+1)$ and $L = \mathbb{F}_2[x]/(x^3+x^2+1)$.
    (i) Find a root β of x^3+x+1 in L.
    (ii) Show that the evaluation map $\varphi_\beta : \mathbb{F}_2[x] \to L$ yields an isomorphism (a one-to-one homomorphism) $\bar{\varphi}_\beta$ from K onto L.

C. Constructing and Classifying Finite Fields

We showed in the last section that if K is a field with q elements, then q must be a power of a prime. In this section we begin by showing the converse:

Theorem 6. *Given any prime p and any $n > 0$, there is a field with exactly p^n elements.*

Proof. Consider $f(x) = x^{p^n} - x$ in $\mathbb{F}_p[x]$. By the Splitting Field Theorem (Section 23D), there is a splitting field K for $f(x)$, that is, some field K containing $\mathbb{F}_p[x]$ such that in $K[x]$, $f(x)$ factors into a product of linear factors.

Let F be the subset of K consisting of all the roots of $x^{p^n} - x$ in K. We shall show F is a field with p^n elements. First we show:

F contains p^n distinct elements of K.

To prove this claim, recall (Section 15G) that the derivative $f'(x)$ of a polynomial $f(x)$ has the property that if $f(x)$ and $f'(x)$ are coprime in $K[x]$, then $f(x)$ has no multiple roots in K. Computing the derivative of $x^{p^n} - x$, we get $(d/dx)(x^{p^n} - x) = p^n x^{p^n-1} - 1 = -1$ since $p = 0$ in K. Thus $x^{p^n} - x$ has no multiple roots. That means that when $x^{p^n} - x$ factors in $K[x]$ into a product of linear factors, there are p^n distinct linear factors. So $x^{p^n} - x$ has p^n distinct roots in K, as claimed.

Now we show:

F is a field.

Recall that F is the set of elements α of K that satisfy $\alpha^{p^n} = \alpha$. Thus, if α, β are in F, then:

(i) so is $\alpha + \beta$: for $(\alpha + \beta)^{p^n} = \alpha^{p^n} + \beta^{p^n} = \alpha + \beta$ (the first equality is by Corollary 17 of Section 9E);

(ii) so is $\alpha\beta$: for $(\alpha\beta)^{p^n} = \alpha^{p^n}\beta^{p^n} = \alpha\beta$;

(iii) so is $-\alpha$: for $(-\alpha)^{p^n} = (-1)^{p^n} \alpha^{p^n} = -\alpha$; and
(iv) so is α^{-1}: for $(\alpha^{-1})^{p^n} = (\alpha^{p^n})^{-1} = \alpha^{-1}$.

Since 0 and 1 are in F, and addition and multiplication in F is the same as that in K, therefore F is a field. That completes the proof.   □

Corollary 7. *There is an irreducible polynomial in $\mathbb{F}_p[x]$ of degree n for each n.*

Proof. Let F be a field with p^n elements. By the theorem of the last section, F is a simple field extension of $\mathbb{F}_p$, that is, F is isomorphic to $\mathbb{F}_p[x]/(q(x))$ for some irreducible polynomial $q(x)$ in $\mathbb{F}_p[x]$. Since F has p^n elements, $\mathbb{F}_p[x]/(q(x))$ must have p^n elements, so $q(x)$ must have degree n, and is the desired polynomial.   □

Now we show that, up to isomorphism, there is only one field with p^d elements.

We know that every field with p^n elements is a simple field extension of $\mathbb{F}_p$ for some irreducible polynomial $q(x)$ of degree n.

Suppose we have two different irreducible polynomials of degree n in $\mathbb{F}_p[x]$. Each defines a simple field extension of $\mathbb{F}_p[x]$ with p^n elements.

For example, in $\mathbb{F}_2[x]$ there are three different polynomials of degree 4: $x^4 + x + 1$, $x^4 + x^3 + 1$, and $x^4 + x^3 + x^2 + x + 1$. Each defines a simple field extension of $\mathbb{F}_2$. Each has 16 elements. Are they really different?

The remarkable fact is that if F_1 and F_2 are any two fields with p^n elements, then F_1 and F_2 are isomorphic. Thus, rather than a different field for each irreducible polynomial of degree n over $\mathbb{F}_p$, there is, up to isomorphism, only one, which can be presented as a simple field extension of $\mathbb{F}_p$ in different ways.

The following theorem was proved by E.H. Moore in 1893 (see Dickson (1901)).

Theorem 8. *Any two fields with p^n elements are isomorphic.*

Proof. We prove this theorem by showing that any field with p^n elements is isomorphic to the field F consisting of all roots of $x^{p^n} - x$ that we constructed in Theorem 6.

Let K be a field with p^n elements. Then K is a simple field extension of $\mathbb{F}_p$, so $K = \mathbb{F}_p[x]/(q(x))$ for some irreducible polynomial $q(x)$ of degree n. Then $\alpha = [x]_{q(x)}$ is a root of $q(x)$ in $\mathbb{F}_p[x]/(q(x))$.

Since K is a field with p^n elements, the group of units of K has $p^n - 1$ elements (all the elements of K except 0). So the element α of K satisfies $\alpha^{p^n-1} = 1$, and hence is a root of $x^{p^n} - x$ in $\mathbb{F}_p[x]$.

By the Division Theorem in $\mathbb{F}_p[x]$,

$$x^{p^n} - x = q(x)g(x) + h(x)$$

with quotient $g(x)$ and remainder $h(x)$ of degree $< \deg q(x)$. Evaluating this equation at α in K, we find that $h(\alpha) = 0$. But since $\alpha = [x]_{q(x)}$, the polynomial $q(x)$ is the polynomial of smallest degree ≥ 0 in $\mathbb{F}_p[x]$ with α as a root. Thus $h(x) = 0$ and $q(x)$ divides $x^{p^n} - x$:

$$x^{p^n} - x = q(x)g(x).$$

Since $x^{p^n} - x$ has p^n roots in F, at least one of them must also be a root of $q(x)$ (for otherwise all roots would be roots of $g(x)$, and $g(x)$ would have degree $< p^n$ with p^n roots in F, contradicting D'Alembert's Theorem). Let β be a root of $q(x)$ in F.

Let φ_β be the evaluation homomorphism from $\mathbb{F}_p[x]$ to F, defined by $\varphi_\beta(f(x)) = f(\beta)$. Since $q(x)$ is irreducible and $q(\beta) = 0$, φ_β induces a one-to-one homomorphism $\bar{\varphi}_\beta$ from $\mathbb{F}_p[x]/q(x)$ to F, by the Isomorphism Theorem (Theorem 3 of Section A). Since both $\mathbb{F}_p[x]/(q(x))$ and F have p^n elements, $\bar{\varphi}_\beta$ is an isomorphism from $\mathbb{F}_p[x]/(q(x))$ onto F.

Such an isomorphism to F may be found for every simple field extension of $\mathbb{F}_p$ defined by a polynomial $q(x)$ of degree n. But since two fields that are isomorphic to F must be isomorphic to each other, the proof is complete. $\square$

Since there is essentially only one field with $\ell = p^n$ elements, it is customary to denote it by a special symbol, namely $\mathbb{F}_\ell$.

Corollary 9. *Every irreducible polynomial of degree n in $\mathbb{F}_p[x]$ has a root in every field with p^n elements.*

Proof. If $q(x)$ is the polynomial, and $\mathbb{F}_\ell$ is the field, where $\ell = p^n$, then $\mathbb{F}_\ell \cong \mathbb{F}_p[x]/(q(x))$ by Theorem 8. Since $q(x)$ has a root in $\mathbb{F}_p[x]/(q(x))$, namely $[x]_{q(x)}$, it has a root in $\mathbb{F}_\ell$. $\square$

The last theorem of this section completes a development which is quite remarkable. Starting from nothing but the natural numbers, we have given a complete description of all finite fields, up to isomorphism. For a mathematician who studies "algebra" this is a very satisfying outcome. To ask analogous questions, such as, "Describe all commutative rings with unity, up to isomorphism," or, "Describe all finite groups, up to isomorphism" is to raise unsolved questions which have motivated the mathematical research of hundreds of mathematicians over the past century or more.

Exercises.

12. Let $F = \mathbb{F}_2[x]/(x^4 + x + 1)$. In F:
   (i) find a root of $x^4 + x^3 + 1$;
   (ii) find a root of $x^4 + x^3 + x^2 + x + 1$.

13. Show that every irreducible polynomial of degree 4 in $\mathbb{F}_2[x]$ divides $x^{16} - x$.

14. Show that every irreducible polynomial of degree 1 or 2 in $\mathbb{F}_3[x]$ divides $x^9 - x$.

15. (i) Find an element of $\mathbb{F}_3[x]/(x^2 + 1)$ that is a root of $x^2 + x + 2$.
   (ii) Construct an isomorphism from $\mathbb{F}_3[x]/(x^2 + x + 2)$ to $\mathbb{F}_3[x]/(x^2 + 1)$.

16. Find an isomorphism from $\mathbb{F}_3[x]/(x^2 + 2x + 2)$ to $\mathbb{F}_3[x]/(x^2 + 1)$.

17. Construct an isomorphism from $\mathbb{F}_7[x]/(x^2 - x + 3)$ to $\mathbb{F}_7[x]/(x^2 + 1)$.

18. For every field F and every a in F, find an isomorphism from $F[x]/(x - a)$ to F.

D. Latin Squares

In this section we give an application of the classification of finite fields to a combinatorial question that arises in the design of experiments in statistics.

An $n \times n$ Latin square is a square matrix in which each of the numbers from 1 to n occurs once in each row and once in each column. Here is an example:

$$
\begin{matrix}
4 & 1 & 2 & 3 \\
1 & 2 & 3 & 4 \\
2 & 3 & 4 & 1 \\
3 & 4 & 1 & 2
\end{matrix}
$$

You may recognize this example as the table for addition in $\mathbb{Z}/4\mathbb{Z} = \{1,2,3,4\}$. Similarly, the addition table for $\mathbb{Z}/n\mathbb{Z}$ is an $n \times n$ Latin square for any $n \geq 2$. More generally, if G is any group with operation $*$ and elements $a_1, a_2, \ldots, a_n$, then the multiplication table for G is a table whose subscripts form a Latin square. For example, if we let G be the set of invertible elements of $\mathbb{Z}/8\mathbb{Z}$, namely $G = \{1,3,5,7\}$ under multiplication, then the multiplication table is

$$
\begin{array}{c|cccc}
 & 1 & 3 & 5 & 7 \\
\hline
1 & 1 & 3 & 5 & 7 \\
3 & 3 & 1 & 7 & 5 \\
5 & 5 & 7 & 1 & 3 \\
7 & 7 & 5 & 3 & 1
\end{array}
$$

if we now replace 1, 3, 5, 7 by 1, 2, 3, 4 we get the Latin square:

$$
\begin{matrix}
1 & 2 & 3 & 4 \\
2 & 1 & 4 & 3 \\
3 & 4 & 1 & 2 \\
4 & 3 & 2 & 1
\end{matrix}
$$

Latin squares are of interest in agricultural experiments (see Fisher (1935)). Here are two examples.

Example 5. Suppose five strains of wheat are to be tested for yield on a field = rectangular plot of land. The yield depends not only on the strain of wheat but also on the fertility of the soil, which may vary around the field. Suppose, for example, that the north side of the field happens to be more fertile than the south side. Suppose the experimenters did not know how the fertility varied around the field, and planted the five strains of wheat (labeled 1 – 5) as follows.

North side of field

```
1 1 1 1 1
2 2 2 2 2
3 3 3 3 3
4 4 4 4 4
5 5 5 5 5
```

South side of field

If the yield of strain 1 were higher than that of strain 5 the experimenters would not know whether the result was caused by the fertility of the soil or the difference in the strains.

Fertility tends to be more uniform along strips parallel to the edges of the field, because of the mixing effect of plowing parallel to the edges. So in doing the wheat yield experiment, the problem is to plant the wheat in such a way that variations in fertility of the soil along strips parallel to the edges can be neglected. A useful way to do this is to plant the strains of wheat in a Latin square arrangement, like so:

```
1 2 3 4 5
2 4 5 3 1
4 3 1 5 2
5 1 4 2 3
3 5 2 1 4
```

Example 6. Three diets–all hay, half hay and half corn, all corn–are to be tested on three dairy cows, to see the effect of diet on milk yield. Different cows have different milk yields, and the same cow's milk yield varies over time. To try to test diet independent of these variations, a Latin square is a useful design.

Week\ Cow	1	2	3
1	Corn	1/2	Hay
2	1/2	Hay	Corn
3	Hay	Corn	1/2

Returning to Example 1, suppose in addition to testing five strains of wheat, five kinds of fertilizer are also to be tested. We would like to use a Latin square arrangement for the fertilizer in such a way that each kind of fertilizer is used with each strain of wheat. What is needed, therefore, are two *orthogonal* Latin squares, that is, two 5×5 Latin squares such that each ordered pair (r, s) of (wheat, fertilizer) occurs exactly once on a plot. Here is such a pair.

$$
I: \begin{array}{ccccc}
1 & 2 & 3 & 4 & 5 \\
2 & 4 & 5 & 3 & 1 \\
4 & 3 & 1 & 5 & 2 \\
5 & 1 & 4 & 2 & 3 \\
3 & 5 & 2 & 1 & 4
\end{array}
$$

(*wheat*)

and

$$II : \begin{array}{ccccc} 1 & 2 & 3 & 4 & 5 \\ 4 & 3 & 1 & 5 & 2 \\ 5 & 1 & 4 & 2 & 3 \\ 3 & 5 & 2 & 1 & 4 \\ 2 & 4 & 5 & 3 & 1 \end{array}$$

(*fertilizer*)

Suppose in addition we wish simultaneously to test the effect on yield of five kinds of fungicides. For that we would like to test each fungicide with each fertilizer, and with each strain of wheat, so we need to find another Latin square orthogonal to each of the two above. Here is one.

$$III : \begin{array}{ccccc} 1 & 2 & 3 & 4 & 5 \\ 5 & 1 & 4 & 2 & 3 \\ 3 & 5 & 2 & 1 & 4 \\ 2 & 4 & 5 & 3 & 1 \\ 4 & 3 & 1 & 5 & 2 \end{array}$$

(*fungicide*)

Suppose we wish also to test five kinds of herbicides; we would like yet another Latin square orthogonal to the previous three. Here is one.

$$IV : \begin{array}{ccccc} 1 & 2 & 3 & 4 & 5 \\ 3 & 5 & 2 & 1 & 4 \\ 2 & 4 & 5 & 3 & 1 \\ 4 & 3 & 1 & 5 & 2 \\ 5 & 1 & 4 & 2 & 3 \end{array}$$

(*herbicide*)

Suppose we wished also to test five levels of soil acidity; we would like one more Latin square orthogonal to the previous four. But there is none. For if we had such a square, we could number the five levels of acidity appearing on the top row by 1 2 3 4 5 and then the new square would start

$$V : \begin{array}{ccccc} 1 & 2 & 3 & 4 & 5 \\ a & & & & \end{array}$$

But if V is to be orthogonal to all of the other squares, then the number $a \neq 1$ cannot coincide with the corresponding number in any other square. For example, $a \neq 2$, for otherwise the pair $(2,2)$ would occur twice in the pair of squares (V,I), once at the second entry of the top row, and once in the first entry of the second row, so V and I would not be orthogonal. The same argument prevents a from being 3 or 4 or 5; $a \neq 1$, since 1 already occurs in the first column of V.

This leads to the following problem: Given m, how many pairwise orthogonal $m \times m$ Latin squares can be constructed?

Here are two facts:

Theorem 10. *(a) There cannot be more than $m-1$ pairwise orthogonal $m \times m$ Latin squares.*

(b) If there is a field with m elements then there are $m-1$ pairwise orthogonal $m \times m$ Latin squares.

Proof. Given the soil acidity example above, we leave the proof of the first statement of the theorem as an exercise, below.

To prove part (b), suppose we have a field F with m elements. Let α be a primitive root of F. Then $\alpha^{m-1} = 1$ and every nonzero element of F is a power of α. Consider the addition table for F set up as follows:

$+$	α	$\cdots$	α^s	$\cdots$	α^{m-1}	0
0	α	$\cdots$	α^s	$\cdots$	α^{m-1}	0
α^i	$\alpha^i + \alpha$	$\cdots$	$\alpha^i + \alpha^s$	$\cdots$	$\alpha^i + \alpha^{m-1}$	α^i
α^{i+1}	$\alpha^{i+1} + \alpha$	$\cdots$	$\alpha^{i+1} + \alpha^s$	$\cdots$	$\alpha^{i+1} + \alpha^{m-1}$	α^{i+1}
$\vdots$	$\vdots$					$\vdots$
α^{i+r}	$\alpha^{i+r} + \alpha$	$\cdots$	$\alpha^{i+r} + \alpha^s$	$\cdots$	$\alpha^{i+r} + \alpha^{m-1}$	α^{i+r}
$\vdots$	$\vdots$					$\vdots$
$\alpha^{i+(m-2)}$	$\alpha^{i+(m-2)} + \alpha$	$\cdots$	$\alpha^{i+(m-2)} + \alpha^s$	$\cdots$	$\alpha^{i+(m-2)} + \alpha^{m-1}$	$\alpha^{i+(m-2)}$

Examining the entries of the table, we see that each element of F occurs once in each row and once in each column (Exercise 22). If we write the nonzero entries of the table as powers of α (possible because α is a primitive root of F) and then replace the elements of F by the numbers 1 to m, using the correspondence

$$\alpha \quad \alpha^2 \quad \alpha^3 \quad \cdots \quad \alpha^{m-1} \quad 0$$
$$1 \quad 2 \quad 3 \quad \cdots \quad m-1 \quad m$$

we get a Latin square; call it L_i.

If i, j are two different integers between 1 and $m-1$, then L_i and L_j are orthogonal Latin squares. For example, with $m = 5, \alpha = 2, i = 1$, we get L_1:

$+$	2	2^2	2^3	2^4	0
0	2	2^2	2^3	2^4	0
2	2^2	2^4	0	2^3	2
2^2	2^4	2^3	2	0	2^2
2^3	0	2	2^4	2^2	2^3
2^4	2^3	0	2^2	2	2^4

or

$$
\begin{array}{ccccc}
1 & 2 & 3 & 4 & 5 \\
2 & 4 & 5 & 3 & 1 \\
4 & 3 & 1 & 5 & 2 \\
5 & 1 & 4 & 2 & 3 \\
3 & 5 & 2 & 1 & 4
\end{array}
$$

This is Latin Square I we gave in the wheat example above. With $i = 3$ we get L_3:

$+$	2	2^2	2^3	2^4	0
0	2	2^2	2^3	2^4	0
2^3	0	2	2^4	2^2	2^3
2^4	2^3	0	2^2	2	2^4
2	2^2	2^4	0	2^3	2
2^2	2^4	2^3	2	0	2^2

or

$$
\begin{array}{ccccc}
1 & 2 & 3 & 4 & 5 \\
5 & 1 & 4 & 2 & 3 \\
3 & 5 & 2 & 1 & 4 \\
2 & 4 & 5 & 3 & 1 \\
4 & 3 & 1 & 5 & 2
\end{array}
$$

This was Square III above.

This construction gives $m - 1$ pairwise orthogonal Latin squares. For the pair of entries in L_i and L_j at the (r,s)th position is $(\alpha^{i+r} + \alpha^s, \alpha^{j+r} + \alpha^s)$. Suppose $i \neq j$. If the pair of entries at the (r,s)th position is equal to the pair of entries at the (p,q)th position, then the pairs

$$
(\alpha^{i+r} + \alpha^s, \alpha^{j+r} + \alpha^s) \text{ and } (\alpha^{i+p} + \alpha^q, \alpha^{j+p} + \alpha^q)
$$

are the same, so

$$
\alpha^{i+r} + \alpha^s = \alpha^{i+p} + \alpha^q
$$

and

$$
\alpha^{j+r} + \alpha^s = \alpha^{j+p} + \alpha^q.
$$

Hence

$$
\alpha^{i+r} - \alpha^{i+p} = \alpha^q - \alpha^s = \alpha^{j+r} - \alpha^{j+p}.
$$

So

$$
\alpha^i(\alpha^r - \alpha^p) = \alpha^j(\alpha^r - \alpha^p).
$$

Since $i \neq j$, we must have $\alpha^r - \alpha^p = 0$, so $r = p$, hence $\alpha^q = \alpha^s$ and $q = s$. Thus L_i and L_j are orthogonal if $i \neq j$. That completes the proof. $\square$

We know that if n is any number which is a power of a prime, $n = p^e$, then there is a field with n elements. For such numbers n the theorem says that there are $n - 1$ pairwise orthogonal $n \times n$ Latin squares, but not n pairwise orthogonal $n \times n$ Latin squares.

When n is not a prime power, the question remains, how many pairwise orthogonal $n \times n$ Latin squares can there be? For many n the answer is unknown.

The smallest nonprime power is $n = 6$, and that question was the content of a famous problem of Euler, called the problem of 36 officers. It goes as follows. 36 officers are to be placed in review in a square, 6 rows deep with 6 men in each row. The officers come from 6 different regiments, and each regiment is represented by 6 officers, each of different ranks. For reasons of protocol it is desired that each row and column is to have one officer from each regiment and one officer of each rank. Can this be done? If it could be done, then one would have a pair of orthogonal 6×6 Latin squares.

Euler believed that it could not be done, but it was not proved impossible for well over 100 years after Euler–a proof was finally achieved by M.G. Tarry in 1901.

Thus the situation for nonprime powers is apparently much different than for prime powers. There are 4 pairwise orthogonal 5×5 Latin squares and 6 pairwise orthogonal 7×7 Latin squares, but no two 6×6 Latin squares are orthogonal.

Bose, Shrikhande, and Parker (1960) proved that for every $n > 6$, there are at least two orthogonal $n \times n$ Latin squares.

The construction of orthogonal Latin squares described here is due to R.C. Bose in 1938. See Mann (1949).

Exercises.

19. Solve the 16 officers problem. Take the aces, kings, queens and jacks out of an ordinary deck of playing cards, and lay them in a 4×4 square array so that each row and each column has all four suits and all four ranks.

20. Find three pairwise orthogonal 4×4 Latin squares.

21. Use the construction in the proof of part (2) of Theorem 10 with $\alpha = 3$, to find 4 pairwise orthogonal 5×5 Latin squares.

22. In the proof of part (2) of Theorem 10, verify that each L_i is a Latin square.

23. Prove part (1) of Theorem 10: that is, show that there cannot be m pairwise orthogonal $m \times m$ Latin squares.

24. Find three pairwise orthogonal 8×8 Latin squares.

25. Show that if G is a group under multiplication, with n elements, then the multiplication table for G yields a Latin square.

26. Find a Latin square which cannot be viewed as the multiplication table for a group.

27. Prove that there is no field with 6 elements.

28. Let F be a field with q elements,

$$F = \{a_1, a_2, \ldots, a_{q-1}, a_q = 0\}.$$

Let A_p be the Latin square whose entry in the ith row, jth, column, is $a_p a_i + a_j$. Show that A_q is not a Latin square, but $A_1, \ldots, A_{q-1}$ are pairwise orthogonal Latin squares.

29. (i) If $A = (a_{i,j})$ is an $m \times m$ Latin square, and $B = (b_{k,l})$ is an $n \times n$ Latin square, define $A \times B$ to be the $mn \times mn$ square which consists of an $m \times m$ array of $n \times n$ squares, such that the $(i - j)$-th square is $(a_{i,j}, B)$. Show that $A \times B$ is an $mn \times mn$ Latin square.

(ii) Show that if A and A' are orthogonal, and B and B' are orthogonal, then $A \times B$ and $A' \times B'$ are orthogonal.

(iii) Show that if n is odd, or a multiple of 4, then there are at least two orthogonal $n \times n$ Latin squares.

Chapter 25
BCH Codes

This chapter describes a collection of multiple error correcting codes. The codes are constructed using simple field extensions of $\mathbb{F}_2$.

A. Error Correcting Codes

In Section 8E we looked at ways of encoding messages so that if an error occurs in the transmission of the message, we can correct the error. Those codes, called Hamming codes, were based on describing encoded words as vectors of solutions in $\mathbb{F}_2$ to sets of linear equations.

In this chapter we use finite fields to describe codes that correct multiple errors. These codes were discovered in 1960 by Bose, Chaudhuri, and Hocquenghem, hence are called BCH codes.

The encoded words of BCH codes are vectors of coefficients of polynomials in $\mathbb{F}_2[x]$. The polynomials have as roots certain powers of a primitive root of some appropriate field extension of $\mathbb{F}_2$.

To illustrate the idea we start with a single-error correcting example. We begin by describing the field that we need for the code.

Let $m(x) = x^3 + x + 1$ in $\mathbb{F}_2[x]$. Then $m(x)$ is irreducible in $\mathbb{F}_2[x]$, and so $\mathbb{F}_2[x]/m(x)$ is a field with 8 elements. Denote the congruence class of x, $[x]_{m(x)}$, by α. Then $\mathbb{F}_2[x]/m(x)$ can be viewed as polynomials in α, where $\alpha^3 + \alpha + 1 = 0$, and so we shall denote $\mathbb{F}_2[x]/m(x)$ by $\mathbb{F}_2[\alpha]$. Then α is a primitive root of $\mathbb{F}_2[\alpha]$, so the elements of $\mathbb{F}_2[\alpha]$ may be described as powers of α as in Table 25.1.

The first code, like one in Section 8E, sends out coded words of length 7 with 4 information digits.

Code III. (Codes I and II were the Hamming codes of Section 8E).

Encoding. We want to send the word $w = (a, b, c, d)$, where a, b, c, d are in $\mathbb{F}_2 = \{0, 1\}$. To encode w, we form the polynomial

$$w(x) = ax^6 + bx^5 + cx^4 + dx^3,$$

L.N. Childs, *A Concrete Introduction to Higher Algebra*, Undergraduate Texts in Mathematics, © Springer Science+Business Media LLC 2009

Table 25.1

$\mathbb{F}_8 = \mathbb{F}_2[x]/(x^3+x+1) = \mathbb{F}_2[\alpha]$	
0	0
1	1
α	α
α^2	$\alpha+1 = \alpha^3$
$\alpha^3 = \alpha+1$	α^2
$\alpha^4 = \alpha^2+\alpha$	$\alpha^2+1 = \alpha^6$
$\alpha^5 = \alpha^2+\alpha+1$	$\alpha^2+\alpha = \alpha^4$
$\alpha^6 = \alpha^2+1$	$\alpha^2+\alpha+1 = \alpha^5$
$\alpha^7 = 1$	

and divide $w(x)$ by $m(x) = x^3+x+1$ in $\mathbb{F}_2[x]$:

$$w(x) = m(x)q(x)+z(x),$$

where the remainder $z(x)$ has degree $<3 = \deg(m(x))$. Then

$$z(x) = rx^2+sx+t$$

for some r,s,t in $\mathbb{F}_2$. We set $C(x) = w(x)+z(x)$. Since $-1 = 1$ in $\mathbb{F}_2$, we get

$$
\begin{aligned}
C(x) &= w(x)+z(x) \\
&= ax^6+bx^5+cx^4+dx^3+rx^2+sx+t \\
&= m(x)q(x),
\end{aligned}
$$

and so the polynomial $C(x)$ has the important property that when evaluated at the root α of $m(x)$,

$$C(\alpha) = m(\alpha)q(\alpha) = 0.$$

The encoded word is $C = (a,b,c,d,r,s,t)$, the vector of coefficients of the polynomial $C(x)$. Then C is characterized by the property that it corresponds to the unique polynomial of degree 6 with given top degree coefficients a,b,c,d and having α as a root.

We transmit C.

Decoding. Suppose we receive (a',b',c',d',r',s',t'). We form the polynomial

$$R(x) = a'x^6+b'x^5+c'x^4+d'x^3+r'x^2+s'x+t',$$

and assume that at most one error occurred in the transmission. Under that assumption, $C(x) - R(x) = E(x)$ is either the zero polynomial or consists of a single term, x^e, whose coefficient in $R(x)$ was erroneous. To decide, we look at $R(\alpha)$:

Case 0. If $R(\alpha) = 0$, then, since $C(\alpha) = 0, E(\alpha) = 0$ and we decide that no errors occurred.

Case 1. If $R(\alpha) = \alpha^e$, then, since $C(\alpha) = 0, E(\alpha) = \alpha^e$ and we decide that one error occurred, at the coefficient of x^e.

Thus by evaluating $R(x)$ at $x = \alpha$, we can decide whether an error occurred and, if so, where, so that the error can be corrected.

If two or more errors occurred, and we think at most one error occurred, we would be misled. But if more than one error is very unlikely to occur, this is an effective code.

Example 1. To encode $(1,1,0,1)$, we take $w(x) = x^6 + x^5 + x^3$ and divide it by $x^3 + x + 1$. The remainder is $z(x) = 1$. So $C(x) = w(x) + z(x) = x^6 + x^5 + x^3 + 1$.

(Using Table I, we can confirm that $C(\alpha) = \alpha^6 + \alpha^5 + \alpha^3 + 1 = (\alpha^2 + 1) + (\alpha^2 + \alpha + 1) - (\alpha + 1) + 1 = 0$.)

We send $C = (1,1,0,1,0,0,1)$.

Suppose we receive $R = (1,0,0,1,0,0,1)$. We form the polynomial

$$R(x) = x^6 + x^3 + 1.$$

We use Table 25.1 to evaluate $R(x)$ at α: We find that $R(\alpha)$ is the sum of the following terms:

$$\alpha^6 = \alpha^2 \quad +1$$
$$\alpha^3 = \quad \alpha+1$$
$$1 = \quad 1.$$

Thus

$$R(\alpha) = \alpha^6 + \alpha^3 + 1 = \alpha^2 + \alpha + 1 = \alpha^5.$$

So we change the coefficient of x^5 in $R(x)$, and correct R to $(1,1,0,1,0,0,1)$. Since only one error occurred in the transmission, we correctly determined C.

The next code corrects two errors, using a field of 16 elements. For decoding it is convenient to use matrices.

First, we describe the field.

Let $m(x) = x^4 + x + 1$ in $\mathbb{F}_2[x]$. Then $m(x)$ is irreducible, and so $\mathbb{F}_2[x]/(x^4 + x + 1)$ is a field. Setting $[x]_{m(x)} = \alpha$, we may describe $\mathbb{F}_2[x]/(x^4 + x + 1)$ as $\mathbb{F}_2[\alpha]$, polynomials in α with coefficients in $\mathbb{F}_2$, where $\alpha^4 + \alpha + 1 = 0$. Since $\mathbb{F}_2[\alpha]$ has 16 elements, and every field with 16 elements is isomorphic to $\mathbb{F}_2[\alpha]$, we can also refer to the field $\mathbb{F}_2[\alpha]$ as $\mathbb{F}_{16}$.

It turns out that α is a primitive root of $\mathbb{F}_2[\alpha]$. Thus every nonzero element of $\mathbb{F}_2[\alpha]$ is a power of α. This is exhibited in Table 25.2, which will be convenient for computing in $\mathbb{F}_2[\alpha]$.

We use $\mathbb{F}_{16}$ to construct a code that corrects two errors. The idea is that code words are vectors of coefficients of polynomials of degree 14 in $\mathbb{F}_2[x]$ having α and α^3 as roots.

We know that the minimal polynomial of α, that is, the polynomial of smallest degree in $\mathbb{F}_2[x]$ with α as a root, is $m(x) = x^4 + x + 1$.

Table 25.2

$$\mathbb{F}_{16} = \mathbb{F}_2[x]/(x^4 + x + 1) = \mathbb{F}_2[\alpha]$$

0	0
1	1
α	α
α^2	$\alpha + 1 = \alpha^3$
α^3	α^2
$\alpha^4 = \alpha + 1$	$\alpha^2 + 1 = \alpha^6$
$\alpha^5 = \alpha^2 + \alpha$	$\alpha^2 + \alpha = \alpha^8$
$\alpha^6 = \alpha^3 + \alpha^2$	$\alpha^2 + \alpha + 1 = \alpha^{10}$
$\alpha^7 = \alpha^3 + \alpha + 1$	α^3
$\alpha^8 = \alpha^2 + 1$	$\alpha^3 + 1 = \alpha^{14}$
$\alpha^9 = \alpha^3 + \alpha$	$\alpha^3 + \alpha = \alpha^9$
$\alpha^{10} = \alpha^2 + \alpha + 1$	$\alpha^3 + \alpha + 1 = \alpha^7$
$\alpha^{11} = \alpha^3 + \alpha^2 + \alpha$	$\alpha^3 + \alpha^2 = \alpha^6$
$\alpha^{12} = \alpha^3 + \alpha^2 + \alpha + 1$	$\alpha^3 + \alpha^2 + 1 = \alpha^{13}$
$\alpha^{13} = \alpha^3 + \alpha^2 + 1$	$\alpha^3 + \alpha^2 + \alpha = \alpha^{11}$
$\alpha^{14} = \alpha^3 + 1$	$\alpha^3 + \alpha^2 + \alpha + 1 = \alpha^{12}$
$\alpha^{15} = 1$	

Lemma 1. *The minimal polynomial over* $\mathbb{F}_2$ *of* α^3 *is*

$$m_3(x) = x^4 + x^3 + x^2 + x + 1.$$

Proof. If we evaluate the polynomial $m_3(x)$ at $x = \alpha^3$ we get

$$m_3(\alpha^3) = \alpha^{12} + \alpha^9 + \alpha^6 + \alpha^3 + 1.$$

Using Table 25.2 to replace the powers of α by polynomials in α of degree < 3, we find that $m_3(\alpha^3) = 0$.

We can verify that $m_3(x)$ is irreducible in $\mathbb{F}_2[x]$ by observing that since 0 and 1 are not roots of $m_3(x)$, $m_3(x)$ has no irreducible factors of degree 1; and also $m_3(x)$ is not divisible by $x^2 + x + 1$, the only irreducible polynomial in $\mathbb{F}_2[x]$ of degree 2.

Since $m_3(x)$ has α^3 as a root and is irreducible, $m_3(x)$ must be the minimal polynomial of α^3. $\qquad\qquad\square$

The polynomial of smallest degree with both α and α^3 as roots is the least common multiple of $m(x)$ and $m_3(x)$; but since they are both irreducible in $\mathbb{F}_2[x]$, their least common multiple is their product

$$m(x)m_3(x) = x^8 + x^7 + x^6 + x^4 + 1.$$

We recall the useful fact (see Theorem 15 of Section 9E) that if $p(x)$ is a polynomial in $\mathbb{F}_2[x]$, then $p(x^2) = (p(x))^2$. Thus since α is a root of $m(x)$, so are α^2 and α^4. So we set

$$m_4(x) = m(x)m_3(x) = x^8 + x^7 + x^6 + x^4 + 1,$$

since it is the polynomial of smallest degree in $\mathbb{F}_2[x]$ with with $\alpha, \alpha^2, \alpha^3$ and α^4 as roots.

Code IV.

Encoding. Since $m_4(x)$ has degree 8, encoded words will have length 15 with 7 information digits.

We encode as follows. Let

$$w = (a_{14}, a_{13}, \ldots, a_8)$$

be the seven-bit information word, with corresponding polynomial

$$w(x) = a_{14}x^{14} + a_{13}x^{13} + \ldots + a_8x^8.$$

Divide $w(x)$ by $m_4(x)$:

$$w(x) = m_4(x)q(x) + z(x),$$

where $z(x)$, the remainder, has degree ≤ 7,

$$z(x) = a_7x^7 + \ldots + a_1x + a_0.$$

Then

$$C(x) = a_{14}x^{14} + \ldots + a_1x + a_0 = w(x) + z(x) = m_4(x)q(x).$$

Since the remainder $z(x)$ in the division algorithm is unique, $C(x)$ is the unique polynomial of degree ≤ 14 with given coefficients $a_{14}, \ldots, a_8$ and with $\alpha, \alpha^2, \alpha^3$ and α^4 as roots.

The encoded word $C = (a_{14}, a_{13}, \ldots, a_0)$ is the vector of coefficients of $C(x)$. We transmit C.

Decoding. Suppose we receive R. Set $R = C + E$, where E is the error vector. We view R, C, E as coefficient vectors of polynomials $R(x), C(x), E(x)$, all of degree ≤ 14. Since $m_4(x)$ divides $C(x)$, $C(\alpha) = C(\alpha^2) = C(\alpha^3) = C(\alpha^4) = 0$, and so $R(\alpha) = E(\alpha), R(\alpha^2) = E(\alpha^2), R(\alpha^3) = E(\alpha^3)$ and $R(\alpha^4) = E(\alpha^4)$.

Let $R(\alpha^i) = S_i$ for $i = 1, 2, 3, 4$, and let

$$\mathbf{S} = \begin{pmatrix} S_1 & S_2 \\ S_2 & S_3 \end{pmatrix}.$$

Using this 2×2 matrix with entries in $\mathbb{F}_{16}$ we can correct 0, 1, or 2 errors as follows:

Case 0. No errors. Then $E(x) = 0$, so $\mathbf{S} = 0$ (and in particular, the row rank of $\mathbf{S}$ is 0).

Case 1. One error. Then $E(x) = x^i$ for some i, $0 \leq i \leq 14$; thus the matrix $\mathbf{S}$ is

$$\mathbf{S} = \begin{pmatrix} \alpha^i & \alpha^{2i} \\ \alpha^{2i} & \alpha^{3i} \end{pmatrix}.$$

The second row of $\mathbf{S}$ is α^i times the first row. So $\mathbf{S}$ has (row) rank 1. (We can also verify that $\mathbf{S}$ has row rank 1 by taking the determinant of $\mathbf{S}$. If $\mathbf{S}$ is non-zero but has determinant 0, then it must have rank 1.)

Once we know that $\mathbf{S}$ has rank 1, we can correct the received vector R as follows. We know that $R(\alpha) = \alpha^i$ for some exponent i. We find α^i using Table 25.2, and then change the coefficient of x^i in $R(x)$ to find $C(x)$.

Case 2. Two errors. Then $E(x) = x^i + x^j$ where i and j are the locations of the errors. So

$$\mathbf{S} = \begin{pmatrix} S_1 & S_2 \\ S_2 & S_3 \end{pmatrix} = \begin{pmatrix} \alpha^i + \alpha^j & \alpha^{2i} + \alpha^{2j} \\ \alpha^{2i} + \alpha^{2j} & \alpha^{3i} + \alpha^{3j} \end{pmatrix}.$$

Then

$$\mathbf{S} = \begin{pmatrix} 1 & 1 \\ \alpha^i & \alpha^j \end{pmatrix} \begin{pmatrix} \alpha^i & 0 \\ 0 & \alpha^j \end{pmatrix} \begin{pmatrix} 1 & \alpha^i \\ 1 & \alpha^j \end{pmatrix},$$

a product of three invertible matrices. Thus $\det(\mathbf{S}) \neq 0$, and $\mathbf{S}$ has row rank 2.

For this code, we see that *the rank of* $\mathbf{S}$ = *the number of errors*.

We find i and j in two slightly mysterious steps, as follows.

(a) Solve

$$\begin{pmatrix} S_1 & S_2 \\ S_2 & S_3 \end{pmatrix} \begin{pmatrix} \sigma_2 \\ \sigma_1 \end{pmatrix} = \begin{pmatrix} S_3 \\ S_4 \end{pmatrix}.$$

in $\mathbb{F}_{16}$ for σ_1 and σ_2.

(b) Find the roots in $\mathbb{F}_{16}$ of $p_2(x) = x^2 + \sigma_1 x + \sigma_2$.

We can solve (a) because $\mathbf{S}$ is invertible, and we can use the inverse of $\mathbf{S}$, which for 2×2 matrices is easy to write down. There will be a unique solution.

The reason we want to solve (b) is that

Proposition 2. *The roots in* $\mathbb{F}_{16}$ *of* $x^2 + \sigma_1 x + \sigma_2 = 0$ *are* α^i *and* α^j, *the powers of* α *corresponding to where the errors occur in* $R(x)$.

Proof. To see this, we examine the coefficients τ_1 and τ_2 of

$$g(x) = (x - \alpha^i)(x - \alpha^j) = x^2 + \tau_1 x + \tau_2$$

in $\mathbb{F}_{16}$. Since α^i and α^j are the roots of $g(x)$, we have

$$0 = g(\alpha^i) = \alpha^{2i} + \tau_1 \alpha^i + \tau_2,$$
$$0 = g(\alpha^j) = \alpha^{2j} + \tau_1 \alpha^j + \tau_2.$$

Multiplying these equations by α^i and α^j, respectively, gives

$$0 = \alpha^i g(\alpha^i) = \alpha^{3i} + \tau_1 \alpha^{2i} + \tau_2 \alpha^i,$$
$$0 = \alpha^j g(\alpha^j) = \alpha^{3j} + \tau_1 \alpha^{2j} + \tau_2 \alpha^j.$$

Adding, we get

$$0 = \alpha^i g(\alpha^i) + \alpha^j g(\alpha^j) = S_3 + \tau_1 S_2 + \tau_2 S_1.$$

Similarly, multiplying the equations by α^{2i} and α^{2j}, respectively, and adding, we get

$$0 = \alpha^{2i}g(\alpha^i) + \alpha^{2j}g(\alpha^j) = S_4 + \tau_1 S_3 + \tau_2 S_2.$$

Writing these last two equations in matrix form (noting $+ = -$ in $\mathbb{F}_2$) gives

$$\begin{pmatrix} S_1 & S_2 \\ S_2 & S_3 \end{pmatrix} \begin{pmatrix} \tau_2 \\ \tau_1 \end{pmatrix} = \begin{pmatrix} S_3 \\ S_4 \end{pmatrix}.$$

Thus if we solve the equation in Step (a), we will find the coefficients of the polynomial with α^i and α^j as roots. □

The way to find the roots α^i and α^j of $p_2(x) = x^2 + \sigma_1 x + \sigma_2$ is by trial and error, as before. There are two roots and only 15 candidates, so we evaluate $p_2(x)$ at $x = 1, \alpha, \alpha^2, \alpha^3, \ldots, \alpha^{14}$ until one of them, say α^i, is found to give 0. Then $x - \alpha^i$ is a factor of $p_2(x)$. Dividing $p_2(x)$ by $x - \alpha^i$ gives a quotient $x - \alpha^j$ for some j; then the other root of $p_2(x)$ is α^j.

We can set up a table to systematically search for a root of $p(x)$, starting from $x = \alpha^0 = 1$:

x	x^2	$\sigma_1 x$	σ_2	sum
1	1	σ_1	σ_2	
α	α^2	$\sigma_1 \alpha$	σ_2	
α^2	α^4	$\sigma_1 \alpha^2$	σ_2	
		$\vdots$		
α^k	λ_2	λ_1	$\lambda_0 = \sigma_2$	
α^{k+1}	$\lambda_2 \alpha^2$	$\lambda_1 \alpha$	$\lambda_0 = \sigma_2$	
		$\vdots$		

continuing until we find a line where the sum of the $x^2, \sigma_1 x$ and σ_2 entries is 0. This approach to the search for a root is called a *Chien search*.

To sum up: in this code we send words of length 15, of which 7 are information digits. If we receive a word in this code, we can decide whether 0, 1, or 2 errors occurred, and correct them. We will be misled in case three or more errors occur in a word.

Our next example is a code that corrects three errors in words of length 15. This code also uses $\mathbb{F}_{16}$ as in our last example, and some more matrix theory.

We find the polynomial in $\mathbb{F}_2[x]$ of smallest degree with α, α^3, and α^5 as roots. Since $(\alpha^5)^3 = 1$, the minimal polynomial of α^5 is the irreducible polynomial $m_5(x) = x^2 + x + 1$. So the polynomial of smallest degree with α, α^3, and α^5 as roots is $m_6(x) = m_4(x)m_5(x)$, a polynomial of degree 10. So in each word we are allowed five information digits, $w = (a_{14}, \ldots, a_{10})$. Note that $m_6(x)$ has α^2, α^4 and α^6 as roots also, since the square of a root is also a root.

Code V.

Encoding. For encoding, let $w = (a_{14}, a_{13}, a_{12}, a_{11}, a_{10})$. We let w be the coefficients of the polynomial

$$w(x) = a_{14}x^{14} + \ldots + a_{10}x^{10}.$$

Divide $w(x)$ by $m_6(x)$ to get a remainder $z(x)$ of degree ≤ 9. Then $C(x) = w(x) + z(x)$ is a multiple of $m_6(x)$ the unique polynomial of degree ≤ 14 with given $a_{14}, \ldots, a_{10}$ and having $\alpha, \alpha^2, \alpha^3, \alpha^4, \alpha^5$ and α^6 as roots.

We send C, the vector of coefficients of $C(x)$.

Decoding. Suppose we receive the vector R. Let $E(x) = R(x) - C(x)$. We assume $E(x)$ is a sum of at most three powers of x, the powers corresponding to the errors in R.

For $i = 1, \ldots 6$, we compute $R(\alpha^i)$. Since $C(\alpha^i) = 0$, $R(\alpha^i) = E(\alpha^i)$. Set $S_i = R(\alpha^i)$ for $i = 1, \ldots, 6$ and consider the matrix

$$\mathbf{S} = \begin{pmatrix} S_1 & S_2 & S_3 \\ S_2 & S_3 & S_4 \\ S_3 & S_4 & S_5 \end{pmatrix}.$$

We can determine the rank of $\mathbf{S}$ in a couple of ways.

One way is to do row operations on $\mathbf{S}$ to reduce it to echelon form. Then the number of non-zero rows is the rank of $\mathbf{S}$.

Another method, which works because of the special form of the matrix $\mathbf{S}$, is to look at the matrices

$$\mathbf{U}_1 = (S_1), \mathbf{U}_2 = \begin{pmatrix} S_1 & S_2 \\ S_2 & S_3 \end{pmatrix}, \mathbf{U}_3 = \mathbf{S},$$

the square submatrices of $\mathbf{S}$ in the upper left corner of $\mathbf{S}$.

If $\mathbf{S} = 0$, then $\mathbf{S}$ has rank 0;

If $\det(\mathbf{U}_1) \neq 0$ but $\det(\mathbf{U}_2) = 0$, then $\mathbf{S}$ has rank 1;

If $\det(\mathbf{U}_2) \neq 0$ but $\det(\mathbf{U}_3) = \det(\mathbf{S}) = 0$, then $\mathbf{S}$ has rank 2;

If $\det(\mathbf{U}_3) = \det(\mathbf{S}) \neq 0$, then $\mathbf{S}$ has rank 3.

Once the rank is determined, then *the number of errors = the rank of* $\mathbf{S}$, and we may find the error locations as follows.

Case 0. No errors. $E(x) = 0$. Then $\mathbf{S} = 0$.

Case 1. One error. $E(x) = x^k$. We look at $R(\alpha) = \alpha^k$ and decide that the single error is at the coefficient of x^i in $R(x)$. We correct that coefficient to find the original code word C.

Case 2. Two errors. $E(x) = x^j + x^k$. We look at $\mathbf{U}_2 = \begin{pmatrix} S_1 & S_2 \\ S_2 & S_3 \end{pmatrix}$ and correct the received word $R(x)$ as in Code IV.

Case 3. Three errors. $E(x) = x^i + x^j + x^k$. Then

$$\mathbf{S} = \begin{pmatrix} S_1 & S_2 & S_3 \\ S_2 & S_3 & S_4 \\ S_3 & S_4 & S_5 \end{pmatrix} = \begin{pmatrix} \alpha^i + \alpha^j + \alpha^k & \alpha^{2i} + \alpha^{2j} + \alpha^{2k} & \alpha^{3i} + \alpha^{3j} + \alpha^{3k} \\ \alpha^{2i} + \alpha^{2j} + \alpha^{2k} & \alpha^{3i} + \alpha^{3j} + \alpha^{3k} & \alpha^{4i} + \alpha^{4j} + \alpha^{4k} \\ \alpha^{3i} + \alpha^{3j} + \alpha^{3k} & \alpha^{4i} + \alpha^{4j} + \alpha^{4k} & \alpha^{5i} + \alpha^{5j} + \alpha^{5k} \end{pmatrix}$$

has rank 3, so is invertible.

Since $\mathbf{S}$ is invertible, we can solve uniquely the equation

$$\mathbf{S} \begin{pmatrix} \sigma_3 \\ \sigma_2 \\ \sigma_1 \end{pmatrix} = \begin{pmatrix} S_4 \\ S_5 \\ S_6 \end{pmatrix}$$

for $\sigma_1, \sigma_2, \sigma_3$, and then find the roots in $\mathbb{F}_{16}$ of

$$p(x) = x^3 + \sigma_1 x^2 + \sigma_2 x + \sigma_3 = 0.$$

As we showed above for Code IV, the three roots of this equation will be the three powers $\alpha^i, \alpha^j, \alpha^k$ of α corresponding to where the three errors are.

Example 2. Let $p(x) = x^3 + \alpha^6 x^2 + \alpha^8 x + \alpha^7$. We find a root of $p(x)$ by a Chien search, using Table 25.2:

x	x^3	$\alpha^6 x^2$	$\alpha^8 x$	α^7	sum
1	1	α^6	α^8	α^7	
=	1	$\alpha^3 + \alpha^2$	$\alpha^2 + 1$	$\alpha^3 + \alpha + 1$	$= \alpha + 1 \neq 0$
α	α^3	α^8	α^9	α^7	
=	α^3	$\alpha^2 + 1$	$\alpha^3 + \alpha$	$\alpha^3 + \alpha + 1$	$= \alpha^3 + \alpha^2 \neq 0$
α^2	α^6	α^{10}	α^{10}	α^7	
=	$\alpha^3 + \alpha^2$	$\alpha^2 + \alpha + 1$	$\alpha^2 + \alpha + 1$	$\alpha^3 + \alpha + 1$	$= \alpha^2 + \alpha + 1 \neq 0$
α^3	α^9	α^{12}	α^{11}	α^7	
=	$\alpha^3 + \alpha$	$\alpha^3 + \alpha^2 + \alpha + 1$	$\alpha^3 + \alpha^2 + \alpha$	$\alpha^3 + \alpha + 1$	$= 0.$

So α^3 is a root. Dividing $x^3 + \alpha^6 x^2 + \alpha^8 x + \alpha^7$ by $x - \alpha^3$ yields a quotient $q(x) = x^2 + \alpha^2 x + \alpha^4$. Looking for roots of $q(x)$ starting with $x = \alpha^4$ yields α^7 as a root. Then dividing $q(x)$ by $x - \alpha^7$ yields a quotient of $x - \alpha^{12}$. So α^3, α^7 and α^{12} are the three roots of $p(x)$.

To sum up, we can decode in Code V by

(1) finding the row rank of the matrix $\mathbf{S}$, which tells the number of errors (up to 3); and

(2) finding where the errors are by the techniques described above.

We will be misled if four or more errors occurred in the transmission of an encoded word C.

In the next section we will generalize these examples.

Exercises.

1. Using Code III, encode the messages:
   (i) $(1,0,0,0)$;
   (ii) $(0,1,1,0)$;
   (iii) $(1,1,1,0)$.

2. Using Code III, decode the received words:
   (i) $(1,1,1,0,0,0,1)$;
   (ii) $(1,0,1,1,0,1,1)$;
   (iii) $(0,1,0,1,0,1,0)$.

3. Prove (using Theorem 15 of Section 9E) that any polynomial $p(x)$ with coefficients in $\mathbb{F}_2$ has the property that $(p(x))^2 = p(x^2)$.

4. Using Code IV, encode
   (i) $(1,1,1,0,0,1,1)$;
   (ii) $(0,0,1,1,0,1,1)$;
   (iii) $(1,0,1,0,1,0,1)$.

5. In Code IV, use the matrix **S** to decode:
   (i) $(110,001,110,010,110)$;
   (ii) $(101,011,110,010,110)$;
   (iii) $(110,010,111,110,110)$.

6. Using Code IV, decode:
   (i) $(011,001,011,101,100)$;
   (ii) $(011,110,101,110,110)$;
   (iii) $(100,100,100,100,100)$.

7. In $\mathbb{F}_2[x]/(x^4+x+1) = \mathbb{F}_2[\alpha] = \mathbb{F}_{16}$, solve

$$\begin{pmatrix} \alpha^7 & \alpha^{14} \\ \alpha^{14} & \alpha^8 \end{pmatrix} \begin{pmatrix} \sigma_2 \\ \sigma_1 \end{pmatrix} = \begin{pmatrix} \alpha^8 \\ \alpha^{13} \end{pmatrix}$$

for σ_1, σ_2. Find the roots in $\mathbb{F}_2[\alpha]$ of $x^2 + \sigma_1 x + \sigma_2 = 0$.

8. In Code IV, show that if $p_2(x) = x^2 + \sigma_1 x + \sigma_2$ is the polynomial whose coefficients come from the matrix equation

$$\mathbf{S} \begin{pmatrix} \sigma_2 \\ \sigma_1 \end{pmatrix} = \begin{pmatrix} S_3 \\ S_4 \end{pmatrix}.$$

then

$$R(\alpha)p_2(x) = R(\alpha)x^2 + R(\alpha^2)x + (R(\alpha^3) + R(\alpha)R\alpha^2)).$$

Thus we may decode in Code IV by finding the roots in $\mathbb{F}_{16}$ of $P(x) = R(\alpha)p_2(x)$.

9. Define, by analogy to Code IV, a double error-correcting code using $\mathbb{F}_8$. How many information digits will it have?

10. In Code IV, you receive $R(x) = x^{14} + x^{11} + x^9 + x^8 + x^4 + x^2 + x + 1$, and assume that at most two errors were made. What was $C(x)$?

11. In Code IV, let
$$R = (011, 011, 110, 010, 000).$$

Show that the corresponding polynomial $p_2(x)$ has no roots in $\mathbb{F}_{16}$. Hence R must have resulted from at least three errors in C.

12. Solve in $\mathbb{F}_{16}$,
$$\begin{pmatrix} \alpha^8 & \alpha & \alpha^6 \\ \alpha & \alpha^6 & \alpha^2 \\ \alpha^6 & \alpha^2 & \alpha^5 \end{pmatrix} \begin{pmatrix} \sigma_3 \\ \sigma_2 \\ \sigma_1 \end{pmatrix} = \begin{pmatrix} \alpha^2 \\ \alpha^5 \\ \alpha^{12} \end{pmatrix}$$

for $\sigma_3, \sigma_2, \sigma_1$. Verify that
$$x^3 + \sigma_1 x^2 + \sigma_2 x + \sigma_3 = (x - \alpha^2)(x - \alpha^5)(x - \alpha^{10}).$$

13. Using Code V:
(i) encode (10111);
(ii) decode (101100110011100);
(iii) decode (101000010011110).

14. In Code V, show that if α^i, α^j and α^k are all distinct then the matrix $\mathbf{S}$ is nonsingular. Generalize to the $n \times n$ case where $a_1^{i_1}, a_2^{i_2}, \ldots, a_n^{i_n}$ are all distinct.

15. i) Show that in $\mathbb{F}_8$, every element other than 0 and 1 is a primitive root.
ii) For which $n > 3$ is it true that every element of $\mathbb{F}_{2^n}$ other than 0 and 1 is a primitive root?

16. Does there exist an example of an irreducible polynomial $p(x)$ in $\mathbb{F}_2[x]$ such that in $\mathbb{Z}_2[x]/p(x) = \mathbb{F}_2[\alpha]$, α is not a primitive element?

B. General BCH Codes

In this section we describe the general strategy for encoding and decoding in t-error-correcting BCH codes.

Encoding. We begin with the field $\mathbb{F} = \mathbb{F}_{2^d}$. Let α be a primitive root of $\mathbb{F}$. Then $\alpha^{2^d - 1} = 1$ and
$$\mathbb{F}_{2^d} = \{0, 1, \alpha, \alpha^2, \ldots, \alpha^{2^d - 2}\}.$$

For a t-error correcting BCH code, we let $m_{2t}(x)$ be the polynomial of smallest degree in $\mathbb{F}_2[x]$ with $\alpha, \alpha^2, \ldots, \alpha^{2t-1}, \alpha^{2t}$ as roots. Suppose $m_{2t}(x)$ has degree e. As long as $e < 2^d - 2$, we will have a non-trivial code.

Message words are vectors w with $(2^d - 1) - e$ components

$$w = (a_{2^d-2}, a_{2^d-3}, \ldots, a_e).$$

To encode, we divide the corresponding polynomial

$$w(x) = a_{2^d-2}x^{2^d-2} + a_{2^d-3}x^{2^d-3} + \ldots + a_e x^e$$

by $m_{2t}(x)$:

$$w(x) = m_{2t}(x)q(x) + z(x)$$

Then $z(x)$ has degree $< e - 1$, so no power of x occurs as a monomial in both $w(x)$ and $z(x)$. We let $C(x) = w(x) + z(x)$. Then C, the vector of coefficients of $C(x)$, is the encoded message.

The polynomial $C(x)$ is a multiple of $m_{2t}(x)$, hence has the property that $C(\alpha^r) = 0$ for $r = 1, 2, \ldots, 2t$.

Decoding. Suppose the encoded message $C(x)$ is sent through a noisy channel and we receive $R(x)$. We assume that $R(x) - C(x) = E(x)$, the error polynomial, is a polynomial

$$E(x) = x^{e_1} + x^{e_2} + \ldots + x^{e_r} \qquad (r \leq t)$$

in x with at most t nonzero coefficients.

To find $m_{2t}(x)$, we need to find

- r, the number of errors; and
- $e_1, e_2, \ldots, e_r$, the locations of the errors.

Once $E(x)$ is found, then $C(x) = R(x) - E(x)$ is the encoded polynomial, and we can recover the original message.

Determining r, the Number of Errors. Let

$$E(x) = x^{e_1} + x^{e_2} + \ldots + x^{e_r}$$

be the error polynomial, with $r \leq t$. Since the code is designed to correct up to t errors, we set

$$S = \begin{pmatrix} S_1 & S_2 & \cdots & S_t \\ S_2 & S_3 & \cdots & S_{t+1} \\ \vdots & & & \vdots \\ S_t & S_{t+1} & \cdots & S_{2t-1} \end{pmatrix}$$

where $S_1 = R(\alpha), S_2 = R(\alpha^2), \ldots, S_{2t-1} = R(\alpha^{2t-1}), S_{2t} = R(\alpha^{2t})$.

Since $E(x) = R(x) - C(x)$ and $C(\alpha^i) = 0$ for $i = 1, \ldots, 2t$, it follows that

$$S_1 = E(\alpha) = \alpha^{e_1} + \alpha^{e_2} + \ldots + \alpha^{e_r},$$
$$S_2 = E(\alpha^2) = \alpha^{2e_1} + \alpha^{2e_2} + \ldots + \alpha^{2e_r},$$
$$\vdots$$
$$S_{2t} = E(\alpha^{2t}) = \alpha^{2te_1} + \alpha^{2te_2} + \ldots + \alpha^{2te_r}.$$

We can determine the number of errors by determining the rank of $\mathbf{S}$:

Proposition 3. *If r is the number of errors, then $\mathbf{S}$ has rank r, and*

$$
\mathbf{U}_r = \begin{pmatrix}
S_1 & S_2 & \cdots & S_r \\
S_2 & S_3 & \cdots & S_{r+1} \\
\vdots & & & \vdots \\
S_r & S_{r+1} & \cdots & S_{2r-1}
\end{pmatrix},
$$

the $r \times r$ matrix in the upper left corner of $\mathbf{S}$ (= the $r \times r$ principal minor of $\mathbf{S}$) is invertible.

Proof. For $k = 1, \ldots, r$, the k-th row of $\mathbf{S}$ is

$$
\begin{pmatrix} S_k & S_{k+1} & \cdots & S_{k+(t-1)} \end{pmatrix}
$$

where

$$
S_l = \alpha^{le_1} + \alpha^{le_2} + \ldots + \alpha^{le_r}.
$$

If we set

$$
\mathbf{v_1} = (\alpha^{e_1}, \alpha^{2e_1}, \ldots, \alpha^{te_1})
$$
$$
\mathbf{v_2} = (\alpha^{e_2}, \alpha^{2e_2}, \ldots, \alpha^{te_2})
$$
$$
\vdots
$$
$$
\mathbf{v_r} = (\alpha^{e_r}, \alpha^{2e_r}, \ldots, \alpha^{te_r}),
$$

then for each l, $1 \le l \le t$, the l-th component of

$$
\alpha^{(k-1)e_1} \mathbf{v_1} + \ldots + \alpha^{(k-1)e_r} \mathbf{v_r}
$$

is

$$
\alpha^{(k-1)e_1} \alpha^{le_1} + \ldots + \alpha^{(k-1)e_r} \alpha^{le_r}
$$
$$
= \alpha^{(k-1+l)e_1} + \ldots + \alpha^{(k-1+l)e_r}
$$
$$
= S_{k+(l-1)}.
$$

Thus the vector

$$
\begin{pmatrix} S_k & S_{k+1} & \cdots & S_{k+(t-1)} \end{pmatrix}
$$
$$
= \alpha^{(k-1)e_1} \mathbf{v_1} + \alpha^{(k-1)e_2} \mathbf{v_2} + \ldots + \alpha^{(k-1)e_r} \mathbf{v_r}.
$$

Hence the row space of $\mathbf{S}$ is spanned by $\mathbf{v_1}, \mathbf{v_2}, \ldots, \mathbf{v_r}$, hence has dimension $< r$. Thus the rank of $\mathbf{S}$ is $< r$.

To show that the rank of $\mathbf{S} = r$, it suffices to show that $\mathbf{U}_r$ is invertible. For that implies that the r rows of $\mathbf{U}_r$ are linearly independent, and hence that the first r rows of $\mathbf{S}$ are linearly independent, which in turn implies that the rank of $\mathbf{S}$ is $\ge r$.

Now $\mathbf{U}_r$ may be written as the product

$$
\mathbf{U}_r = \mathbf{ADA}^t,
$$

where

$$A = \begin{pmatrix} 1 & 1 & \cdots & 1 \\ \alpha^{e_1} & \alpha^{e_2} & \cdots & \alpha^{e_r} \\ \alpha^{2e_1} & \alpha^{2e_2} & \cdots & \alpha^{2e_r} \\ & & \vdots & \\ \alpha^{(r-1)e_1} & \alpha^{(r-1)e_2} & \cdots & \alpha^{(r-1)e_r} \end{pmatrix},$$

is an $r \times r$ matrix, A^t is the transpose of A, and

$$D = \begin{pmatrix} \alpha^{e_1} & \cdots & 0 \\ \vdots & \ddots & \vdots \\ 0 & \cdots & \alpha^{e_r} \end{pmatrix}$$

is an $r \times r$ diagonal matrix with nonzero diagonal entries. We illustrated the 2×2 case in Code IV.

Since $\alpha^{e_1}, \ldots, \alpha^{e_r}$ are distinct elements of $\mathbb{F}$, the matrix A is invertible: in fact, it is a Vandermonde matrix, whose determinant is

$$\det(A) = \pm \prod_{1 \le i < j \le r} (\alpha^{e_j} - \alpha^{e_i}).$$

$\square$

Determining the Error Locations. Suppose r is the number of errors, and

$$E(x) = x^{e_1} + \ldots + x^{e_r}$$

is the error polynomial. For $i = 1, \ldots, 2r$, let $S_i = R(\alpha^i)$, as before; then since $C(\alpha^i) = 0$, $S_i = E(\alpha^i)$. So

$$S_1 = \alpha^{e_1} + \ldots + \alpha^{e_r},$$
$$S_2 = \alpha^{2e_1} + \ldots + \alpha^{2e_r}$$
$$\vdots$$
$$S_{2r} = \alpha^{2re_1} + \ldots + \alpha^{2re_r}.$$

To find the locations $e_1, \ldots, e_r$, of the errors, we solve the matrix equation

$$\begin{pmatrix} S_1 & S_2 & \cdots & S_r \\ S_2 & S_3 & \cdots & S_{r+1} \\ & & \vdots & \\ S_r & S_{r+1} & \cdots & S_{2r-1} \end{pmatrix} \begin{pmatrix} \sigma_r \\ \sigma_{r-1} \\ \vdots \\ \sigma_1 \end{pmatrix} = \begin{pmatrix} S_{r+1} \\ S_{r+2} \\ \vdots \\ S_{2r} \end{pmatrix}. \tag{25.1}$$

Then the polynomial $z^r + \sigma_1 z^{r-1} + \ldots + \sigma_r$ has as its roots, $\alpha^{e_1}, \ldots \alpha^{e_r}$, the powers of α corresponding to the locations of the errors. To see this, consider the polynomial

$$\sigma(z) = (z - \alpha^{e_1}) \cdot \ldots \cdot (z - \alpha^{e_r})$$
$$= z^r + \tau_1 z^{r-1} + \ldots + \tau_r.$$

Evaluate the last line at $z = \alpha^{e_i}$ to get

$$0 = \alpha^{re_i} + \tau_1 \alpha^{(r-1)e_i} + \ldots + \tau_{r-1} \alpha^{e_i} + \tau_r$$

Multiplying by α^{e_i} yields

$$0 = \alpha^{(r+1)e_i} + \tau_1 \alpha^{re_i} + \ldots + \tau_{r-1} \alpha^{2e_i} + \tau_r \alpha^{e_i}$$

Doing this for $i = 1, \ldots, t$ and adding the equations yields

$$0 = S_{r+1} + \tau_1 S_r + \ldots + \tau_{r-1} S_2 + \tau_r S_1.$$

In the same way, multiplying by α^{ke_i} for $k \geq 2$ yields

$$0 = S_{r+k} + \tau_1 S_{r+k-1} + \ldots + \tau_r S_k.$$

These equations, in matrix form, mean that

$$\begin{pmatrix} S_1 & S_2 & \cdots & S_r \\ S_2 & S_3 & \cdots & S_{r+1} \\ & & \vdots & \\ S_r & S_{r+1} & \cdots & S_{2r-1} \end{pmatrix} \begin{pmatrix} \tau_r \\ \tau_{r-1} \\ \vdots \\ \tau_1 \end{pmatrix} = \begin{pmatrix} S_{r+1} \\ S_{r+2} \\ \vdots \\ S_{2r} \end{pmatrix}.$$

Comparing this last equation with equation (25.1) shows that

$$\begin{pmatrix} \sigma_r \\ \sigma_{r-1} \\ \vdots \\ \sigma_1 \end{pmatrix} = \begin{pmatrix} \tau_r \\ \tau_{r-1} \\ \vdots \\ \tau_1 \end{pmatrix}$$

is a solution of the matrix equation above. Since U_r is invertible, we have found the unique solution of (25.1).

Thus solving that matrix equation and then finding the roots of the polynomial

$$\sigma(z) = z^r + \sigma_1 z^{r-1} + \ldots + \sigma_r$$

finds the locations of the errors.

These methods enable the construction of BCH codes that correct any desired number of errors.

Example 3. Suppose we wish to construct a four-error correcting BCH code using $\mathbb{F}_{16}$. Then code words have 15 bits, $c_{14}, \ldots, c_1, c_0)$, and the corresponding polynomial $C(x)$ has $\alpha, \alpha^3, \alpha^5$ and α^7 as roots. The minimal polynomials of these elements are as follows:

The minimal polynomial of α over $\mathbb{F}_2$ is $x^4 + x + 1$;
The minimal polynomial of α^3 over $\mathbb{F}_2$ is $x^4 + x^3 + x^2 + x + 1$;
The minimal polynomial of α^5 over $\mathbb{F}_2$ is $x^2 + x + 1$;
The minimal polynomial of α^7 over $\mathbb{F}_2$ is $x^4 + x^3 + 1$.

Since these four polynomials are irreducible over $\mathbb{F}_2$, the polynomial of smallest degree with $\alpha, \alpha^3, \alpha^5$ and α^7 as roots is the product of these polynomials, namely

$$m(x) = x^{14} + x^{13} + x^{12} + \ldots + x + 1 = \frac{x^{15} - 1}{x - 1}.$$

Thus in encoding, we are allowed only one message bit. If the bit is a 1, then the word polynomial is $w(x) = x^{14}$, the encoded polynomial $C(x)$ is $m(x)$, and the code word is $C = (1,1,1,\ldots,1,1)$, while if the bit is a 0, then the encoded polynomial is $C(x) = 0$ and the code word is $C = (0,0,0,\ldots,0,0)$. This code is just the repetition code described in Section 8E.

Thus for a code that corrects four errors, we need to use a larger field.

Example 4. We set up a 4-error correcting code using $\mathbb{F}_{32}$.

Since $\mathbb{F}_{32}$ has 31 non-zero elements, every element of $\mathbb{F}_{32}$ except 0 and 1 is a primitive element. So if we take any irreducible polynomial $m(x)$ in $\mathbb{F}_2[x]$ of degree 5, then the congruence class $[x]_{m(x)}$ will be a primitive root of $\mathbb{F}_{32}$. Let's use

$$m(x) = x^5 + x^2 + 1,$$

and let $\alpha = [x]_{m(x)}$. Then $\alpha^5 = \alpha^2 + 1$, and $\alpha^{31} = 1$. Using that relation, we can set up a log table for $\mathbb{F}_{32}$.

Log table for $\mathbb{F}_{32}$

1	1	1	1
α	α	α	α
α^2	α^2	α^{18}	$\alpha + 1$
α^3	α^3	α^2	α^2
α^4	α^4	α^5	$\alpha^2 + 1$
α^5	$\alpha^2 + 1$	α^{19}	$\alpha^2 + \alpha$
α^6	$\alpha^3 + \alpha$	α^{11}	$\alpha^2 + \alpha + 1$
α^7	$\alpha^4 + \alpha^2$	α^3	α^3
α^8	$\alpha^3 + \alpha^2 + 1$	α^{29}	$\alpha^3 + 1$
α^9	$\alpha^4 + \alpha^3 + \alpha$	α^6	$\alpha^3 + \alpha$
α^{10}	$\alpha^4 + 1$	α^{27}	$\alpha^3 + \alpha + 1$
α^{11}	$\alpha^2 + \alpha + 1$	α^{20}	$\alpha^3 + \alpha^2$
α^{12}	$\alpha^3 + \alpha^2 + \alpha$	α^8	$\alpha^3 + \alpha^2 + 1$
α^{13}	$\alpha^4 + \alpha^3 + \alpha^2$	α^{12}	$\alpha^3 + \alpha^2 + \alpha$
α^{14}	$\alpha^4 + \alpha^3 + \alpha^2 + 1$	α^{23}	$\alpha^3 + \alpha^2 + \alpha + 1$
α^{15}	$\alpha^4 + \alpha^3 + \alpha^2 + \alpha + 1$	α^4	α^4
α^{16}	$\alpha^4 + \alpha^3 + \alpha + 1$	α^{10}	$\alpha^4 + 1$
α^{17}	$\alpha^4 + \alpha + 1$	α^{30}	$\alpha^4 + \alpha$
α^{18}	$\alpha + 1$	α^{17}	$\alpha^4 + \alpha + 1$
α^{19}	$\alpha^2 + \alpha$	α^7	$\alpha^4 + \alpha^2$
α^{20}	$\alpha^3 + \alpha^2$	α^{22}	$\alpha^4 + \alpha^2 + 1$
α^{21}	$\alpha^4 + \alpha^3$	α^{28}	$\alpha^4 + \alpha^2 + \alpha$
α^{22}	$\alpha^4 + \alpha^2 + 1$	α^{26}	$\alpha^4 + \alpha^2 + \alpha + 1$
α^{23}	$\alpha^3 + \alpha^2 + \alpha + 1$	α^{21}	$\alpha^4 + \alpha^3$
α^{24}	$\alpha^4 + \alpha^3 + \alpha^2 + \alpha$	α^{25}	$\alpha^4 + \alpha^3 + 1$
α^{25}	$\alpha^4 + \alpha^3 + 1$	α^9	$\alpha^4 + \alpha^3 + \alpha$
α^{26}	$\alpha^4 + \alpha^2 + \alpha + 1$	α^{16}	$\alpha^4 + \alpha^3 + \alpha + 1$
α^{27}	$\alpha^3 + \alpha + 1$	α^{13}	$\alpha^4 + \alpha^3 + \alpha^2$
α^{28}	$\alpha^4 + \alpha^2 + \alpha$	α^{14}	$\alpha^4 + \alpha^3 + \alpha^2 + 1$
α^{29}	$\alpha^3 + 1$	α^{24}	$\alpha^4 + \alpha^3 + \alpha^2 + \alpha$
α^{30}	$\alpha^4 + \alpha$	α^{15}	$\alpha^4 + \alpha^3 + \alpha^2 + \alpha + 1$

To set up a four-error correcting code, we need to find the polynomial of smallest degree with $\alpha, \alpha^3, \alpha^5$ and α^7 as roots. Using the table, we find that

the minimal polynomial of α is $x^5 + x^2 + 1$;
the minimal polynomial of α^3 is $x^5 + x^4 + x^3 + x^2 + 1$;
the minimal polynomial of α^5 is $x^5 + x^4 + x^2 + x + 1$;
the minimal polynomial of α^7 is $x^5 + x^3 + x^2 + x + 1$.

The polynomial $m(x)$ of smallest degree in $\mathbb{F}_2[x]$ with $\alpha, \alpha^3, \alpha^5$ and α^7 as roots is the product of these four minimal polynomials, and hence has degree 20. Since encoded words based on $\mathbb{F}_{32}$ have 31 bits, we are allowed 11 bit information words, $w = (c_{30}, c_{29}, \ldots, c_{20})$. We encode by dividing the corresponding polynomial

$$w(x) = c_{30}x^{30} + c_{29}x^{29} + \ldots + c_{20}x^{20}$$

by $m(x)$ to get the remainder polynomial $z(x)$, then the encoded polynomial, $C(x) = w(x) + z(x)$, will have $\alpha, \ldots, \alpha^8$ as roots.

We note that $\alpha^9 = \alpha^{40} = (\alpha^5)^8$, so α^9 is a root of the minimal polynomial of α^5. Thus the polynomial of smallest degree in $\mathbb{F}_2[x]$ with $\alpha, \alpha^3, \alpha^5$ and α^7 as roots also has α^9 as a root. Thus this code, designed to correct four errors, in fact will correct five errors.

In the same way, using $\mathbb{F}_{32}$ we can construct a BCH code with encoded words of length 31 with 6 information bits and with up to 7 errors correctable. Or using $\mathbb{F}_{64}$, BCH codes that send out encoded words of length 63 can contain:

30 information bits with up to 6 errors correctable;
24 information bits with up to 7 errors correctable;
18 information bits with up to 10 errors correctable;
16 information bits with up to 11 errors correctable;
10 information bits with up to 13 errors correctable; and
7 information bits with up to 15 errors correctable.

In each, encoding and decoding is as described in this section. The main variation is the product of the minimal polynomials needed for encoding.

Finally, we note that there is a variant of BCH codes, known as Reed-Solomon codes, that take messages that are elements of a finite field $\mathbb{F}_q$, where q is usually a power of 2, and transform the messages into encoded messages, polynomials with coefficients in $\mathbb{F}_q$ that have powers of a primitive root of $\mathbb{F}_q$ as roots. Decoding is similar to BCH codes except that we need to find not only the number and location of the errors, but also the values of the errors, since the coefficients of the polynomials are not just 0 or 1, but can be any element of the field $\mathbb{F}_q$.

For $q = 2^n$, an element of $\mathbb{F}_q$ corresponds to an n-tuple of elements of $\mathbb{F}_2$, so correcting an error in $\mathbb{F}_q$ is the same as correcting a set of n consecutive binary bits.

This ability to correct errors in consecutive binary bits is particularly useful in situations where most errors tend to arise in "bursts", that is, when errors involve a set of consecutive binary bits. Thus Reed-Solomon codes are commonly used in satellite transmissions and in hard disk drives in computers. For compact disks, an additional step of distributing the data around the disk improves the effectiveness of the Reed-Solomon code against imperfections in the disk. The encoding is called a cross-interleaved Reed-Solomon code (CIRC).

Part VII
Factoring Polynomials

Chapter 26
Factoring in $\mathbb{Z}[x]$

In Section 17D we showed that if $f(x)$ is a polynomial with coefficients in $\mathbb{Z}$, then one can determine in a finite number of steps the complete factorization of $f(x)$ into a product of irreducible polynomials in $\mathbb{Q}[x]$. The method, Lagrange interpolation, is quite slow in practice.

In this chapter, we'll give another proof of the finiteness of the process of factoring, and then describe some refinements of the proof that speed up the process.

A. Factoring Polynomials in $\mathbb{Z}[x]$

We showed in Section 15F that given a polynomial $f(x)$ in $\mathbb{C}[x]$, there is some real number $B > 0$ so that for every complex number z with $|z| > B$, then $|f(z)| > 0$. This implies that if r is a root of $f(x)$, then $|r| \leq B$; that is, there is a bound on the size of the roots of the polynomial $f(x)$.

From a bound on the roots of $f(x)$, we obtain a bound on the coefficients of any polynomial in $\mathbb{C}[x]$ that divides $f(x)$:

Proposition 1. *Let $f(x)$ in $\mathbb{C}[x]$ be a monic polynomial of degree n, and suppose $B > 0$ is some real number so that every root r in $\mathbb{C}$ of $f(x)$ satisfies $|r| \leq B$. Let*

$$g(x) = x^d + b_1 x^{d-1} + \ldots + b_{d-1} x + b_d$$

be a degree d factor of $f(x)$. Then for each k with $1 \leq k \leq d$,

$$|b_k| \leq \binom{d}{k} B^k.$$

Proof. If $r_1, \ldots, r_d$ are the complex roots of $g(x)$, then

$$g(x) = (x - r_1)(x - r_2) \cdots (x - r_n).$$

L.N. Childs, *A Concrete Introduction to Higher Algebra*, Undergraduate Texts in Mathematics, © Springer Science+Business Media LLC 2009

When we multiply out the right hand side and collect coefficients of the powers of x, we see that up to a sign, the coefficient b_k is the sum of all products of k of the roots $r_1, \ldots, r_d$. For example,

$$-b_1 = r_1 + r_2 + \ldots + r_d,$$
$$b_2 = r_1 r_2 + r_1 r_3 + r_2 r_3 + r_1 r_4 + r_2 r_4 + \cdots + r_{d-1} r_d$$
$$\vdots$$
$$(-1)^d b_d = r_1 r_2 r_3 \cdot \ldots \cdot r_d.$$

In general, for each k with $1 \leq k \leq d$,

$$(-1)^k b_k = \sum_{1 \leq i_1 < i_2 < \ldots < i_k \leq d} r_{i_1} r_{i_2} \cdot \ldots \cdot r_{i_k}.$$

Taking absolute values of both sides and applying the Triangle Inequality,

$$|b_k| \leq \sum_{1 \leq i_1 < i_2 < \ldots < i_k \leq d} |r_{i_1} r_{i_2} \cdot \ldots \cdot r_{i_k}|.$$

Each root has absolute value $\leq B$, hence each product in this last sum is $\leq B^k$. There are $\binom{d}{k}$ products in the sum, since there are $\binom{d}{k}$ ways to choose k roots out of the set of d roots $r_1, r_2, \ldots, r_d$. Thus

$$|b_k| \leq \binom{d}{k} B^k$$

for all k.   □

Since there is a bound on the coefficients of any factor of degree d of $f(x)$, we have

Theorem 2. *Finding a factorization in $\mathbb{Q}[x]$ of a monic polynomial $f(x)$ of degree n in $\mathbb{Z}[x]$, or showing that $f(x)$ is irreducible in $\mathbb{Q}[x]$, takes finitely many steps.*

Proof. Let $f(x)$ be a monic polynomial of degree n in $\mathbb{Z}[x]$, and let B be a bound on the roots of $f(x)$.

If $f(x)$ factors in $\mathbb{Q}[x]$, then $f(x)$ has as a factor a monic irreducible polynomial $g(x)$ of degree d for some $d \leq n/2$ in $\mathbb{Z}[x]$ (an easy exercise in Chapter 14). Now for each d with $1 \leq d \leq n/2$ and each k with $1 \leq k \leq d$, there is a finite number of integers b_k that satisfy

$$|b_k| \leq \binom{d}{k} B^k,$$

hence for each d, there is a finite number of monic polynomials

$$g(x) = x^d + b_1 x^{d-1} + \ldots + b_{d-1} x + b_d$$

in $\mathbb{Z}[x]$ such that for $k = 0, 1, \ldots, d-1$,

$$|b_k| \leq \binom{d}{k} B^k.$$

Thus for each d with $1 \leq d \leq n/2$, only finitely many monic polynomials of degree d are eligible to be factors of $f(x)$. If any one of them, call it $g(x)$, divides $f(x)$, then $f(x) = g(x)h(x)$ and we have factored $f(x)$. If none of the eligible polynomials of degree d divides $f(x)$ for every $d \leq n/2$, then $f(x)$ is irreducible, and we've found that out by finitely many trial divisions. That completes the proof. $\square$

Note: we assumed that our polynomial was a monic polynomial with integer coefficients. We observed in Section 16A that given an arbitrary polynomial in $\mathbb{Q}[x]$, we can multiply the polynomial by some rational number to make it a polynomial in $\mathbb{Z}[x]$ that is primitive (the greatest common divisor of its coefficients is 1), but it need not be monic. But if

$$h(x) = a_0 x^n + a_1 x^{n-1} + \ldots + a_{n-1}x + a_n$$

is a polynomial with integer coefficients, then, as noted in Section 16A, the polynomial

$$g(y) = a_0^{n-1} h(\frac{y}{a_0}) = y^n + a_1 y^{n-1} + a_2 a_0 y^{n-2} + \ldots + a_0^{n-2} a_{n-1} y + a_0^{n-1} a_n$$

is a monic polynomial with integer coefficients, and $h(x)$ will factor exactly the same way $g(y)$ will. So if we can factor every monic polynomial in $\mathbb{Z}[x]$ in finitely many steps, the same will be true for every polynomial in $\mathbb{Q}[x]$.

Theorem 2 shows that factoring in $\mathbb{Z}[x]$ is a *finite* process. That does not mean it is a *feasible* process. And in fact, the argument in Theorem 2 gives a process that in its naive application is almost impossibly long, because there will be so many polynomials eligible to be factors.

In the remainder of this chapter we present results that refine these ideas to make the factoring process more feasible. The refinements involve two ideas.

One is to find a better bound for the coefficients of polynomial divisors of a given polynomial.

The other is to find an efficient way to factor a polynomial modulo M for M a suitably large modulus.

In particular, suppose we are trying to find an irreducible factor $g(x)$ of degree d of a monic polynomial $f(x)$ of degree n. If we know that the coefficients of $g(x)$ are all bounded in absolute value by B, and if $h_1(x), h_2(x), \ldots, h_k(x)$ are all the irreducible degree d factors of $f(x)$ modulo M where $M \geq 2B$, then the number of eligible polynomials $g(x)$ of degree d that might divide $f(x)$ is k, rather than some power of B.

Example 1. To see how these ideas might apply, suppose we consider

$$f(x) = x^6 - 7x^5 + 16x^4 + 143x^3 - 939x^2 + 786x - 144,$$

and we find that for this polynomial, $B = 339$ is a bound on the coefficients of any degree 3 factor of $f(x)$.

It turns out that

$$f(x) \equiv (x^3 + 2x + 1)(x^3 + 3x^2 + 4x + 1) \quad (\text{mod } 5),$$

a product of two polynomials that are irreducible modulo 5. Thus we know that if $f(x)$ factors in $\mathbb{Z}[x]$, it must factor into the product of two irreducible polynomials of degree 3.

If $g(x) = x^3 + ax^2 + bx + c$ is a factor of $f(x)$ in $\mathbb{Z}[x]$, and $g(x) \equiv x^3 + 2x + 1$ (mod 5), then we know only that $|a| \leq 339, |b| \leq 339$ and $|c| \leq 339$. Since $335/5 = 67$, there are 135 numbers a with $a \equiv 0$ (mod 5) and $-339 \leq a \leq 339$, and a similar number of possibilities for b and c, and so there are approximately $135^3 = 2460375$ eligible polynomial factors of $f(x)$ in $\mathbb{Z}[x]$. So finding a factor of $f(x)$, or showing that $f(x)$ has no factor, would be rather time-consuming.

On the other hand, suppose we find that

$$f(x) \equiv (x^3 + 660x + 6)(x^3 + 676x^2 + 399x + 659) \quad (\text{mod } 683),$$

a product of irreducible polynomials mod 683, a prime number larger than $2 \cdot 339$. If $f(x)$ factors in $\mathbb{Z}[x]$, then we know that $f(x)$ must be the product of two irreducible polynomials of degree 3, and one of them, call it $g(x)$, must be congruent modulo 683 to the polynomial $x^3 + 660x + 6$. But if $g(x) = x^3 + ax^2 + bx + c$, we also know that $|a| \leq 339, |b| \leq 339$ and $|c| \leq 339$. The only polynomial $g(x)$ in $\mathbb{Z}[x]$ that satisfies those bounds on a, b and c and is congruent modulo 683 to $x^3 + 660x + 6$ is the polynomial

$$g(x) = x^3 + (660 - 683)x + 6 = x^3 - 23x + 6.$$

A single trial division shows that $g(x)$ does divide $f(x)$:

$$f(x) = (x^3 - 23x + 6)(x^3 - 7x^2 + 39x - 24).$$

Example 2. Now consider

$$f(x) = x^6 - 31x^5 - 105x^4 + 757x^3 + 790x^2 - 176x + 97.$$

Now $f(x)$ factors as

$$f(x) \equiv (x^2 + x + 1)(x^4 + x + 1) \quad (\text{mod } 2),$$

a product of polynomials that are irreducible modulo 2. (It factors in a similar way modulo 3, 5, 7, 11 and 13.) Thus if $f(x)$ factors in $\mathbb{Z}[x]$, it must factor into the product of an irreducible polynomial $g(x)$ of degree 2 times an irreducible polynomial of degree 4. It happens that a bound on the coefficients of a degree 2 factor of $f(x)$ is $B = 2236$.

We factor $f(x)$ modulo 4513, a prime number $> 2 \cdot 2236$, and find that

$$f(x) \equiv (x^4 + 671x^3 + 559x^2 + 516x + 706)(x^2 + 1608x + 1285) \quad (\text{mod } 4513),$$

a product of polynomials that are irreducible modulo 4513. Thus if $g(x)$ is an irreducible factor in $\mathbb{Z}[x]$ of $f(x)$ of degree 2, then $g(x)$ must be congruent modulo 4513 to

$$x^2 + 1608x + 1285.$$

Since the coefficients of $g(x)$ are bounded by 2236, the only possibility for $g(x)$ in $\mathbb{Z}[x]$ is $g(x) = x^2 + 1608x + 1285$.

But a single trial division shows that $g(x)$ doesn't divide $f(x)$.

Thus $f(x)$ has no irreducible factor of degree 2, and therefore must be irreducible in $\mathbb{Z}[x]$.

(It turns out that $f(x)$ is irreducible modulo 29, hence had to be irreducible in $\mathbb{Z}[x]$.)

Exercises.

1. Show that a monic polynomial in $\mathbb{Z}[x]$ can be factored completely into irreducible polynomials in $\mathbb{Q}[x]$ in finitely many steps.

B. Bounding Roots and Coefficients of Factors

Let

$$f(x) = x^n + a_1 x^{n-1} + \ldots + a_{n-1}x + a_n$$

be a monic polynomial with integer coefficients. In this section we show how to obtain an upper bound on the coefficients of any monic factor $g(x)$ of $f(x)$ in $\mathbb{Z}[x]$.

Given Proposition 1 in Section A, one way to find such an upper bound is to find an upper bound on the norms of the complex roots of $f(x)$ (where for a complex number $\alpha = a + bi$, the norm $|\alpha| = \sqrt{a^2 + b^2}$).

In Section 15F we found that

$$B_0 = 1 + |a_1| + \ldots + |a_n|$$

is an upper bound on the roots of f. In this section, following Polya and Szego (1972), p, 106, we will find a bound on the roots that is almost always smaller than B_0. We will also state (but not prove) a bound on the coefficients, due to Mignotte (1974), and compare Mignotte's bound with the coefficient bound obtained from the root bound.

We begin with an auxiliary result.

Proposition 3. *Let*

$$p(x) = x^n - (p_1 x^{n-1} + \ldots + p_{n-1}x + p_n)$$

in $\mathbb{R}[x]$ *where* $p_1, p_2, \ldots, p_n \geq 0$ *with at least one* $p_j > 0$. *Then* $p(x)$ *has a unique positive real root* r. *Moreover,* $p(x) < 0$ *for* $0 < x < r$ *and* $p(x) > 0$ *for* $x > r$.

Proof. Write

$$h(x) = \frac{p(x)}{x^n} = 1 - q(x)$$

where

$$q(x) = \frac{p_1}{x} + \frac{p_2}{x^2} + \ldots + \frac{p_{n-1}}{x^{n-1}} + \frac{p_n}{x^n}.$$

Since all of $p_1, \ldots, p_n$ are non-negative and $q(x) \neq 0$, it is clear that for x close to 0, $q(x)$ is close to $+\infty$, while for x sufficiently large, $q(x)$ is close to 0. Moreover, for $x > 0$, the derivative of $q(x)$,

$$q'(x) = -\left(\frac{p_1}{x^2} + 2\frac{p_2}{x^3} + \ldots + (n-1)\frac{p_{n-1}}{x^n} + n\frac{p_n}{x^{n+1}} \right),$$

is the negative of a sum of terms ≥ 0, so is always negative for $x > 0$. Thus $q(x)$ is monotone decreasing as x increases from 0 to $+\infty$.

Then $h(x) = 1 - q(x)$ is monotone increasing, is negative for x near 0, and approaches 1 as x goes to infinity. Hence $h(x)$ has a unique positive real root, c. Then c is the unique positive real root of $p(x) = x^n h(x)$. Moreover, $h(x)$, hence also $p(x)$, is <0 for $0 < x < r$, and is >0 for $x > r$.   □

From this result, we obtain a bound on the roots of any polynomial in $\mathbb{C}[x]$:

Proposition 4. *Let*

$$f(x) = x^n + a_1 x^{n-1} + \ldots + a_{n-1} x + a_n$$

be a polynomial in $\mathbb{C}[x]$. Let $p_1 = |a_1|, p_2 = |a_2|, \ldots, p_n = |a_n|$ and let

$$p(x) = x^n - (p_1 x^{n-1} + \ldots + p_{n-1} x + p_n).$$

If c is the unique positive real root of $p(x)$, then for every complex root α of $f(x)$, $|\alpha| \leq c$.

Proof. Suppose $f(\alpha) = 0$. Then

$$\alpha^n = -(a_1 \alpha^{n-1} + \ldots + a_{n-1}\alpha + a_n),$$

so by the triangle inequality,

$$|\alpha|^n = |\alpha^n| = |a_1 \alpha^{n-1} + \ldots + a_{n-1}\alpha + a_n|,$$
$$\leq |a_1||\alpha^{n-1}| + \ldots + |a_{n-1}||\alpha| + |a_n|,$$
$$= p_1 |\alpha|^{n-1} + \ldots + p_{n-1}|\alpha| + p_n.$$

Setting $|\alpha| = r$, this gives

$$r^n \leq p_1 r^{n-1} + \ldots + p_{n-1} r + p_n,$$

hence $p(r) \leq 0$. Thus $|\alpha| = r \leq c$ by the last proposition.   □

From these last two propositions we can get several bounds on the norms of the complex roots of

$$f(x) = x^n + a_1 x^{n-1} + \ldots + a_{n-1} x + a_n,$$

such as

$$B_1 = \max_{1 \leq k \leq n} \left\{ \left(\frac{2^n - 1}{\binom{n}{k}} |a_k| \right)^{1/k} \right\},$$

or

$$B_2 = \max_{1 \leq k \leq n} \{ 2(|a_k|)^{1/k} \},$$

or

$$B_3 = \max_{1 \leq k \leq n} \{ (n|a_k|)^{1/k} \}.$$

All of these may be shown to be bounds by showing that they are larger than the unique positive root c of the related real polynomial $p(x)$ in Proposition 4. For example,

Proposition 5. *Given $f(x)$, $p(x)$ as in Proposition 4 and B_2 as above, then $B_2 \geq c$ where c is the unique positive root of $p(x)$.*

Proof. Letting $B_2 = b$, we have $b \geq 2|a_k|^{1/k}$ for all k, so

$$|a_k| \leq \frac{b^k}{2^k}.$$

Hence

$$p(b) = b^n - (|a_1| b^{n-1} + |a_2| b^{n-2} + \ldots + |a_k| b^{n-k} + \ldots + |a_n|)$$

$$\geq b^n - (\frac{b}{2} b^{n-1} + \ldots + \frac{b^k}{2^k} b^{n-k} + \ldots + \frac{b^n}{2^n})$$

$$= b^n (1 - (\frac{1}{2} + \frac{1}{2^2} + \ldots + \frac{1}{2^n})) > 0.$$

So $B_2 = b > c$ by Proposition 3. $\qquad\square$

For any particular polynomial $f(x)$, we can use the unique positive root c of $p(x)$ as a bound on the norms of the roots of $f(x)$.

To find c, we could use the method of bisection on $p(x)$ or Newton's method on the function $h(x) = p(x)/x^n$. For bisection, find a_0, b_0 with $p(a_0) < 0, p(b_0) > 0$, then look at $p((a_0 + b_0)/2)$. If $p((a_0 + b_0)/2) > 0$, let $a_1 = a_0$ and $b_1 = (a_0 + b_0)/2$ and repeat. If $p((a_0 + b_0)/2) < 0$, let $a_1 = (a_0 + b_0)/2$ and $b_1 = b_0$ and repeat. Bisection is slow but reliable.

For Newton's method, start with x_0 with $h(x_0) = p(x_0)/x_0^n < 0$, let $x_1 = x_0 - h(x_0)/h'(x_0)$ and repeat. One can show that starting Newton's method at any x_0 with $0 < x_0 \leq c$ will yield a sequence of approximations $\{x_n\}$ that converges to c. (The proof is a calculus exercise that we omit).

Example 3. Let

$$f(x) = x^6 - 7x^5 + 16x^4 + 143x^3 - 939x^2 + 786x - 144.$$

Then

$$h(x) = 1 - \left(\frac{7}{x} + \frac{16}{x^2} + \frac{143}{x^3} + \frac{939}{x^4} + \frac{786}{x^5} + \frac{144}{x^6}\right)$$

and $h(1) < 0$. After 14 iterations of Newton's method, starting with $x_0 = 1$, we find
that

$$c = 10.62080617.$$

By comparison, the coefficients of $f(x)$ satisfy

$$|a_1| = 7,$$
$$|a_2|^{1/2} = |16|^{1/2} = 4,$$
$$|a_3|^{1/3} = |143|^{1/3} = 5.22,$$
$$|a_4|^{1/4} = |939|^{1/4} = 5.54,$$
$$|a_5|^{1/5} = |786|^{1/5} = 3.79,$$
$$|a_6|^{1/6} = |144|^{1/6} = 2.29,$$

so the three bounds B_1, B_2, B_3 are

$$B_1 = \max_{1 \le k \le n} \left\{\left(\frac{2^n - 1}{\binom{n}{k}}|a_k|\right)^{1/k}\right\} = 73.5,$$

$$B_2 = \max_{1 \le k \le n} \{2(|a_k|)^{1/k}\} = 14;$$

$$B_3 = \max_{1 \le k \le n} \{(n|a_k|)^{1/k}\} = 42.$$

The bound B_0 we found in Chapter 15F is

$$B_0 = 1 + 7 + 16 + 143 + 939 + 786 + 144 = 2036.$$

In Example 3, the bound $B_2 = 2\max_{1 \le k \le n}\{|a_k^{1/k}|\}$ was the best of the four bounds
B_0, B_1, B_2, B_3. It is worth observing that B_2 is a reasonably good bound for every
polynomial.

Proposition 6. *Let*

$$f(x) = x^n + a_1 x^{n-1} + \ldots + a_{n-1}x + a_n$$

be a polynomial in $\mathbb{C}[x]$. Let $p_1 = |a_1|, p_2 = |a_2|, \ldots, p_n = |a_n|$ and let

$$p(x) = x^n - (p_1 x^{n-1} + \ldots + p_{n-1}x + p_n).$$

Let c be the unique positive real root of $p(x)$. Then

$$B_2 = 2 \max_{1 \le k \le n} \{|a_k^{1/k}|\} \le 2c.$$

For Example 3, $c = 10.62...$ and $B_2 = 14 < 2c$.

Proof. Let $B_2 = 2|a_l|^{1/l} = 2p_l^{1/l}$. Then $p_l = B_2^l/2^l$. We have

$$0 = p(c) = c^n - p_1 c^{n-1} - p_2 c^{n-2} - \ldots - p_l c^{n-l} - \ldots p_n$$
$$\le c^n - p_l c^{n-l}$$
$$= c^{n-l}(c^l - p_l)$$
$$= c^{n-l}(c^l - \frac{B_2^l}{2^l}).$$

So $2^l c^l - B_2^l \ge 0$, hence $B_2 \le 2c$, as we wished to show. $\qquad\square$

Bounding roots. The bound B_2 or the bound c can complement Descartes' Rational Root Theorem to limit the number of possibilities for integer roots of a monic polynomial with integer coefficients.

Example 4. Let

$$f(x) = x^6 - 7x^5 + 16x^4 + 143x^3 - 939x^2 + 786x - 144,$$

the polynomial of Example 3. Descartes' Theorem shows that any integer root of $f(x)$ must divide 144. But every root of $f(x)$ must have norm $\le c$ where $c = 10.62....$ Thus the possible integer roots of $f(x)$ are $1, 2, 3, 4, 6, 8, 9$ and their negatives, a total of 14 possibilities, rather than the 30 possibilities given by Descartes' Theorem.

Example 5. For a more extreme example, let

$$f(x) = x^{18} + 5x^{10} + 2^9 3^9 5^9.$$

For the bound B_2, we have

$$B_2 = 2(2^9 3^9 5^9)^{1/18} = 2\sqrt{30} < 11.$$

Thus among the 2000 divisors of $2^9 3^9 5^9$ that are possible roots by Descartes' Theorem, only 18 of them $(1, 2, 3, 4, 5, 6, 8, 9, 10$ and their negatives) have absolute value < 11.

Bounds on coefficients. From a bound B on the roots of $f(x)$, we obtained a bound on the coefficients of any polynomial that divides $f(x)$ in $\mathbb{C}[x]$ in Proposition 1, section A, namely, if

$$g(x) = x^d + b_1 x^{d-1} + \ldots + b_{d-1}x + b_d$$

is a monic degree d factor of $f(x)$, then for each k with $1 \leq k \leq d$,

$$|b_k| \leq \binom{d}{k} B^k.$$

Mignotte (1974) obtained a bound on the coefficients of a polynomial factor of $f(x)$ that does not use a root bound as in Proposition 1.

For a polynomial

$$f(x) = a_0 x^n + a_1 x^{n-1} + \ldots + a_{n-1}x + a_n$$

with integer coefficients, define the norm of f to be

$$\|f\| = (|a_0|^2 + |a_1|^2 + \ldots + |a_n|^2)^{1/2},$$

the length of the vector of coefficients $(a_0, a_1, \ldots, a_n)$ in $\mathbb{C}^{d+1}$. Mignotte shows that if

$$g(x) = b_0 x^d + b_1 x^{d-1} + \ldots + b_{d-1}x + b_d$$

in $\mathbb{Z}[x]$ is a factor of $f(x)$, so that $g(x)g_1(x) = f(x)$ for $g_1(x)$ in $\mathbb{Z}[x]$, then for each k with $0 \leq k \leq e$,

$$|b_k| \leq \binom{d}{k} \|f\|.$$

Example 6. Let $f(x) = x^{10} + 4x^7 - 2x^3 + 5x - 1$. Then

$$\|f\| = (1 + 16 + 4 + 25 + 1)^{1/2} = \sqrt{47}.$$

If $f(x) = g(x)g_1(x)$ in $\mathbb{Z}[x]$ where

$$g(x) = b_0 x^5 + b_1 x^4 + b_2 x^3 + b_3 x^2 + b_4 x + b_5$$

in $\mathbb{Z}[x]$, then we can assume $g(x), g_1(x)$ are monic (so that $b_0 = 1$), and $b_5 = 1$ or -1. Then Mignotte's bound implies that the other coefficients of $g(x)$ must satisfy

$$|b_1| \leq \binom{5}{1} \sqrt{47} < 35;$$

$$|b_2| \leq \binom{5}{2} \sqrt{47} < 69;$$

$$|b_3| \leq \binom{5}{3} \sqrt{47} < 69;$$

$$|b_4| \leq \binom{5}{4} \sqrt{47} < 35$$

By comparison, the unique positive root c of the corresponding polynomial $h(x)$ is $c = 1.7966$. So for a monic factor of degree 5 of $f(x)$ we have

$$|b_1| \leq \binom{5}{1}c < 9;$$

$$|b_2| \leq \binom{5}{2}c^2 < 33;$$

$$|b_3| \leq \binom{5}{3}c^3 < 58;$$

$$|b_4| \leq \binom{5}{4}c^4 < 52.$$

Example 7. Let

$$f(x) = x^6 - 7x^5 + 16x^4 + 143x^3 - 939x^2 + 786x - 144.$$

Then

$$\|f\| = (1 + 7^2 + 16^2 + 143^2 + 939^2 + 786^2 + 144^2)^{1/2} = \sqrt{15410087} = 1241.4.$$

In Example 1 we found that a bound on the absolute values of the roots of f is $c = 10.62$.

Suppose we seek a factor of $f(x)$ of the form

$$g(x) = x^3 + b_1x^2 + b_2x + b_3.$$

Then comparing the coefficient bounds from the root bound and from the Mignotte norm, we have (rounding up to the next integer),

$|b_k| <$	$\binom{3}{k}\|f\|$	$\binom{3}{k}c^k$
$|b_1| <$	3725	32
$|b_2| <$	3725	339
$|b_3| <$	1242	1198

(Note that in fact, $|b_3| \leq 144$ since b_3 is an integer that divides 144).

These examples suggest that the root bound approach may be better for bounding coefficients of high powers of x in a factor $g(x)$ of a polynomial $f(x)$, while the Mignotte norm bound may be better for bounding coefficients of low powers of x.

The Mignotte bound gives a quickly computed uniform bound on all of the coefficients of a factor $g(x)$ of degree d. The largest binomial coefficient $\binom{d}{k}$ occurs with $k = d/2$ or $k = (d-1)/2$. Then for $g(x)$ of even degree, Mignotte's bound yields a uniform bound on all the coefficients of $g(x)$, namely,

$$|b_k| < \binom{d}{d/2}\|f\|$$

for all $k = 1 \cdots d$. Using Stirling's formula from calculus, one can replace $\binom{d}{d/2}$ by the slightly larger approximation $(2^d)/\sqrt{(d/2)\pi}$. Thus we may use

$$B_u = \frac{2^d}{\sqrt{(d/2)\pi}}\|f\|$$

as a uniform bound on the coefficients of a degree d factor of the polynomial $f(x)$ in $\mathbb{Z}[x]$ when d is even. The same bound also works for d odd.

Exercises.

2. For
$$f(x) = x^6 - 7x^5 + 16x^4 + 143x^3 - 939x^2 + 786x - 144,$$
find the uniform bound B_u on coefficients of a possible factor of degree 3, and compare it to the coefficient bound in Example 7.

3. Let
$$f(x) = x^6 + 3342x^5 + 1.$$
Suppose we look for a factor $g(x)$ of $f(x)$ of degree 3,
$$g(x) = x^3 + b_1x^2 + b_2x + b_3.$$

(i) Estimate the root bound c on f, and find the bound B_2.
(ii) What is the Mignotte bound $\|f\|$?
(iii) Compare the bounds on b_1, b_2 and b_3 given by the root bound c, the root bound B_2 and the Mignotte bound.

4. Let
$$f(x) = x^6 + x^5 + 5x + 3342.$$
Suppose we look for a factor $g(x)$ of $f(x)$ of degree 3,
$$g(x) = x^3 + b_1x^2 + b_2x + b_3.$$

(i) Estimate the root bound c on f and the bound B_2.
(ii) What is the Mignotte bound $\|f\|$?
(iii) Compare the bounds on b_1, b_2 and b_3 given by the root bounds c and B_2 and by the Mignotte bound.

5. If we seek a monic degree 1 factor $x - b$ of a monic polynomial $f(x)$ in $\mathbb{Z}[x]$, then the Mignotte bound gives a bound on $|b|$, but we also have a bound on $|b|$ given by the constant term $|f(0)|$ of f. How do the two bounds compare?

6. Let $n = 2d$ and $f(x) = x^n + a_1x^{n-1} + \ldots + a + n$, and suppose
$$\max\{|a_k|^{1/k}\} = |a_1| \geq 2.$$

Then $|a_k| \leq |a_1|^k$.
(i) Show that $\|f\| \leq |a_1|^{d+1}$.

(ii) Show that if $B = 2\max\{|a_k|^{1/k}\} = 2|a_1|$, then

$$\binom{d}{k} B^k \geq \binom{d}{k} \|f\|$$

for all $k \leq d$ for which

$$|a_1| \leq 2^{\frac{k}{d+1-k}}.$$

(iii) Show that if $|a_1| = 2$ and $|a_k| \leq 2^k$ for $k > 1$, then for $B = 2\max\{|a_k|^{1/k}\}$,

$$\binom{d}{k} B^k \geq \binom{d}{k} \|f\|$$

for $\frac{d+1}{2} \leq k \leq d$, and hence the Mignotte bound is better than the bound obtained from B for almost half of the coefficients of a degree d factor of $f(x)$.

C. Berlekamp's Factoring Algorithm

We showed in Section A that we can factor any monic polynomial f in $\mathbb{Z}[x]$, or show that f is irreducible, by choosing a large enough modulus M and factoring f modulo M. In section B we found a bound B on the coefficients of a factor of f, so that if we factor f modulo M where $M > 2B$, then a factorization modulo M yields at most one possible factorization of f in $\mathbb{Z}[x]$. So the factoring problem is reduced to factoring the polynomial f modulo M.

In this section we show how to factor a polynomial modulo p for p a prime. The clever algorithm was published by E. R. Berlekamp in 1967. It combines Fermat's Theorem and elementary linear algebra.

First, a word on notation. Let p be a prime number, and denote the field of p elements, that is, the set of congruence classes of integers modulo p, by $\mathbb{F}_p$. As usual, we'll use the integers $0, 1, 2, \ldots, p-1$ to denote the elements of $\mathbb{F}_p$–that is, we'll denote the elements of $\mathbb{F}_p$ by integers that represent the congruence classes of $\mathbb{F}_p$, rather using the congruence classes themselves. Thus the polynomial $x^2 + 2x + 2$ in $\mathbb{F}_3[x]$ really means $[1]_3 x^2 + [2]_3 x + [2]_3$. As long as we recall that all coefficients of polynomials are meant "mod p", there should be no confusion.

Suppose we want to factor a polynomial $f(x)$ of degree d in $\mathbb{F}_p[x]$. Just as with numbers, we can factor $f(x)$ by trial division, because there are only finitely many polynomials of degree $\leq d/2$ in $\mathbb{F}_p[x]$, and we can simply check them all as possible factors of $f(x)$, using the division theorem.

However, as with trial division of natural numbers, trial division as a factoring method for polynomials in $\mathbb{F}_p[x]$ is impractical except when p is small and the polynomial to be factored has small degree. To get an idea of practicality, if we have a possibly irreducible polynomial f of degree 16 in $\mathbb{F}_7[x]$, then we would need to divide f by all of the 861580 irreducible polynomials of degree ≤ 8 in $\mathbb{F}_7[x]$ to be sure that f is irreducible.

Thus we need a better idea. Berlekamp's algorithm answers that need.

The idea behind Berlekamp's algorithm for factoring $f(x)$ of degree d in $\mathbb{F}_p[x]$ is that if we can find some nonconstant polynomial $h(x)$ of degree $\leq d$ so that $f(x)$ divides $h(x)^p - h(x)$, then we will obtain a factorization of $f(x)$. More precisely,

Theorem 7. *Given* $f(x)$ *in* $\mathbb{F}_p[x]$ *of degree* $d > 1$, *let* $h(x)$ *in* $\mathbb{F}_p[x]$ *be a polynomial of degree* ≥ 1 *and* $\leq d$ *such that* $f(x)$ *divides* $h(x)^p - h(x)$. *Then*

$$f(x) = gcd(f(x), h(x)) \cdot gcd(f(x), h(x) - 1) \cdots gcd(f(x), h(x) - (p-1))$$

is a nontrivial factorization of $f(x)$ *in* $\mathbb{F}_p[x]$.

Proof. Suppose $f(x)$ divides $h(x)^p - h(x)$. We use two facts.

First, by Fermat's Theorem, the polynomial $u^p - u$ has p roots in $\mathbb{F}_p$, namely $u = 0, 1, 2, \ldots, p-1$. Thus by the Root Theorem, $u^p - u$ factors in $\mathbb{F}_p$ into

$$u^p - u = u(u-1)(u-2) \cdots (u-(p-1)).$$

Setting $u = h(x)$ yields

$$h(x)^p - h(x) = h(x)(h(x) - 1)(h(x) - 2) \cdots (h(x) - (p-1))$$

in $\mathbb{F}_p[x]$. This is a factorization of $h(x)^p - h(x)$ into a product of pairwise coprime polynomials in $\mathbb{F}_p[x]$.

Second, if a and b are coprime polynomials in $F[x]$, F a field, then for every polynomial f in $F[x]$,

$$gcd(f, ab) = gcd(f, a) \cdot gcd(f, b).$$

By induction, this fact generalizes to the case of the greatest common divisor of f and more than two pairwise coprime factors. Now since $f(x)$ divides $h(x)^p - h(x)$, we have that

$$f(x) = gcd(f(x), h(x)^p - h(x)).$$

Since $h(x) - r$ and $h(x) - s$ are coprime for $r \neq s$, we have:

$$
\begin{aligned}
f(x) &= gcd(f(x), h(x)^p - h(x)) \\
&= gcd(f(x), (h(x)(h(x) - 1) \cdots (h(x) - (p-1)))) \\
&= gcd(f(x), h(x)) \cdot gcd(f(x), h(x) - 1) \cdots gcd(f(x), h(x) - (p-1)).
\end{aligned}
$$

Since $\deg(h(x) - s) < \deg f(x)$, the greatest common divisor of $f(x)$ and $h(x) - s$ cannot be $f(x)$ for any s. So the factorization

$$f(x) = gcd(f(x), h(x)) \cdot gcd(f(x), h(x) - 1) \cdots gcd(f(x), h(x) - (p-1))$$

must involve only polynomials of degree $\leq d = \deg f(x)$, and hence is a nontrivial factorization of $f(x)$. $\square$

In the factorization of Theorem 7, the greatest common divisors may be found very efficiently by Euclid's algorithm in $\mathbb{F}_p[x]$.

Example 8. In $\mathbb{F}_2[x]$, let $f(x) = x^5 + x + 1$. It turns out, as we shall see below, that for $h(x) = x^4 + x^3 + x$, then $f(x)$ divides $h(x)^2 - h(x) = x^8 + x^6 + x^4 + x^3 + x^2$. (In fact, $x^8 + x^6 + x^4 + x^3 + x^2 = (x^5 + x + 1)(x^3 + x)$.) So $f(x) = gcd(f(x), h(x))gcd(f(x), h(x) - 1)$.

To find the two greatest common divisors, we use Euclid's algorithm:
First, we find that $gcd(f(x), h(x) - 1) = x^2 + x + 1$:

$$x^5 + x + 1 = (x^4 + x^3 + x + 1)(x + 1) + (x^3 + x^2 + x);$$
$$x^4 + x^3 + x + 1 = (x^3 + x^2 + x)x + (x^2 + x + 1);$$
$$x^3 + x^2 + x = (x^2 + x + 1)x.$$

So
$$gcd(x^5 + x + 1, x^4 + x^3 + x + 1) = x^2 + x + 1.$$

Similarly, we find that

$$gcd(f(x), h(x)) = gcd(x^5 + x + 1, x^4 + x^3 + x) = x^3 + x^2 + 1.$$

Then the factorization of $f(x)$ is

$$x^5 + x + 1 = (x^3 + x^2 + 1)(x^2 + x + 1).$$

To factor $f(x)$ in $\mathbb{F}_p[x]$ by the strategy of Theorem 7, then, we seek a polynomial $h(x)$ of degree e, where $1 \le e < d$, such that $f(x)$ divides $h(x)^p - h(x)$. This is done by setting up and solving a set of linear equations for the coefficients of $h(x)$, in the following way.

Let
$$h(x) = b_0 + b_1 x + b_2 x^2 + + b_{d-1} x^{d-1},$$

where $b_0, b_1, \ldots, b_{d-1}$ in $\mathbb{F}_p$ are coefficients to be determined. By Proposition 12 of Section 9E, we have

$$h(x)^p = b_0^p + b_1^p x^p + b_2^p x^{2p} + \ldots + b_{d-1}^p x^{p(d-1)}$$

By Fermat's Theorem, $b^p = b$ for all b in $\mathbb{F}_p$, and so

$$h(x)^p = b_0 + b_1 x^p + \ldots + b_{d-1} x^{(d-1)p} = g(x^p). \tag{26.1}$$

To find the remainder when we divide $h(x)^p$ by $f(x)$, we find $x^{ip} \bmod f(x)$ for $i = 0, 1, \ldots, d-1$:
$$x^{ip} = f(x)q_i(x) + r_i(x),$$

with $\deg r_i(x) < d = \deg f(x)$. Hence

$$x^{ip} = r_i(x) \quad (\bmod f(x)),$$

and so (26.1) yields

$$h(x)^p = b_0 r_0(x) + b_1 r_1(x) + \ldots + b_{d-1} r_{d-1}(x) \quad (\bmod\ f(x)).$$

Then $f(x)$ divides $h(x)^p - h(x)$ if and only if $f(x)$ divides the polynomial

$$b_0 r_0(x) + b_1 r_1(x) + \ldots + b_{d-1} r_{d-1}(x) - [b_0 + b_1 x + \ldots + b_{d-1} x^{d-1}].$$

But this polynomial has degree $\leq d - 1$, and so is divisible by $f(x)$ (which has degree d) if and only if it is the zero polynomial in $\mathbb{F}_p[x]$. This last condition is the condition we will use to determine the coefficients $b_0, b_1, \ldots, b_{d-1}$ of $h(x)$. Namely, $b_0, b_1, \ldots, b_{d-1}$ must satisfy

$$b_0 r_0(x) + b_1 r_1(x) + \ldots + b_{d-1} r_{d-1}(x) - [b_0 + b_1 x + \ldots + b_{d-1} x^{d-1}] = 0. \quad (26.2)$$

If we collect the coefficients of $1, x, x^2, \ldots, x^{d-1}$ in equation (26.2), we get d simultaneous linear equations in the d unknowns $b_0, b_1, \ldots, b_{d-1}$ where the coefficients in $\mathbb{F}_p$ of the b_j's in the equations are the coefficients of the (known) remainder polynomials $r_j(x)$.

Solving this set of equations gives elements $b_0, \ldots, b_{d-1}$ which are coefficients of a polynomial $h(x)$ such that $f(x)$ divides $h(x)^p - h(x)$.

Example 9. Let $f(x) = x^5 + x + 1$ in $\mathbb{F}_2[x]$. We find the remainder polynomials $r_i(x) = x^{2i} \bmod f(x)$, obtained by dividing x^{2i} by $f(x)$, $i = 0, \ldots, 4$, as follows:

$$r_0(x) = 1,$$
$$r_1(x) = x^2,$$
$$r_2(x) = x^4$$

$$\text{For } r_3(x) : x^6 = x f(x) + (x^2 + x), \text{ so } r_3(x) = x^2 + x.$$
$$\text{For } r_4(x) : x^8 = x^3 f(x) + (x^4 + x^3), \text{ so } r_4(x) = x^4 + x^3.$$

The equation (26.2) becomes

$$0 = b_0 + b_1 x^2 + b_2 x^4 + b_3 (x^2 + x) + b_4 (x^4 + x^3) - (b_0 + b_1 x + b_2 x^2 + b_3 x^3 + b_4 x^4).$$

Collecting coefficients of $1, x, x^2, x^3$ and x^4, we have:

power	coefficients
1	$0 = b_0 - b_0$
x	$0 = b_3 - b_1$
x^2	$0 = b_1 + b_3 - b_2$
x^3	$0 = b_4 - b_3$
x^4	$0 = b_2 + b_4 - b_4.$

These reduce to $b_2 = 0, b_1 = b_3 = b_4, b_0$ arbitrary. In order that $h(x)$ have degree ≥ 1, we must choose $b_1 - b_3 = b_4 = 1$. We then have two choices for $h(x)$, corresponding to $b_0 = 0$ and $b_0 = 1$:

$$g_0(x) = x^4 + x^3 + x$$

and

$$g_1(x) = x^4 + x^3 + x + 1 = g_0(x) + 1.$$

Then, as we showed in Example 6,

$$\begin{aligned} f(x) &= gcd(f(x), g_0(x)) \cdot gcd(f(x), g_1(x)) \\ &= (x^3 + x^2 + 1)(x^2 + x + 1). \end{aligned}$$

It is convenient to put equation (26.2) into matrix form. Let $\mathbf{I}$ denote the $d \times d$ identity matrix,

$$\mathbf{I} = \begin{pmatrix} 1 & 0 & \cdots & 0 \\ 0 & 1 & \cdots & 0 \\ \vdots & & & \vdots \\ 0 & 0 & \cdots & 1 \end{pmatrix},$$

let

$$r_i(x) = r_{i,0} + r_{i,1}x + r_{i,2}x^2 + \ldots + r_{i,d-1}x^{d-1}$$

for each i, and let

$$\mathbf{Q} = \begin{pmatrix} r_{0,0} & r_{0,1} & \cdots & r_{0,d-1} \\ r_{1,0} & r_{1,1} & \cdots & r_{1,d-1} \\ \vdots & \vdots & & \vdots \\ r_{d-1,0} & r_{d-1,1} & \cdots & r_{d-1,d-1} \end{pmatrix}$$

be the matrix whose rows are the coefficients of the remainder polynomials $r_0(x), \ldots, r_{d-1}(x)$. Then it is easily verified that the components of the vector $\mathbf{b} = (b_0, b_1, \ldots, b_{d-1})$ give a solution of equation (26.2) if and only if

$$\mathbf{b}(\mathbf{Q} - \mathbf{I}) = \mathbf{0} = (0, \ldots, 0). \tag{26.3}$$

Combining all this with Theorem 1, we get

Theorem 8. *(Berlekamp's Factoring Algorithm). Let $f(x)$ in $\mathbb{F}_p[x]$ have degree d. Let $\mathbf{Q}$ be the $d \times d$ matrix whose ith row is the vector of coefficients of the remainder polynomial $r_i(x) = x^{pi} \bmod f(x)$ for $i = 0, 1, \ldots, d - 1$. Let $\mathbf{b} = (b_0, b_1, \ldots, b_{d-1})$ be a solution of*

$$\mathbf{b}(\mathbf{Q} - \mathbf{I}) = \mathbf{0}$$

(i.e., of equation (26.3)), and let $h(x) = b_0 + b_1 x + \ldots + b_{d-1}x^{d-1}$. If $h(x)$ has degree ≥ 1, then for some s in $\mathbb{F}_p$, $h(x) - s$ and $f(x)$ have a common factor of degree ≥ 1.

Example 10. Let $f(x) = x^6 + x^5 + x^4 + x^3 + x^2 + x + 1$, a polynomial in $\mathbb{F}_2[x]$ of degree 6. To find the matrix $\mathbf{Q}$, we divide $f(x)$ into x^{2i} for $i = 0, \ldots, 5$, to get $r_i(x)$:

$$x^0 = f(x) \cdot 0 + 1, \text{ so } r_0(x) = 1,$$
$$x^2 = f(x) \cdot 0 + x^2, \text{ so } r_1(x) = x^2,$$
$$x^4 = f(x) \cdot 0 + x^4, \text{ so } r_2(x) = x^4,$$
$$x^6 = f(x) \cdot 1 + (x^5 + x^4 + x^3 + x^2 + x + 1),$$
$$\text{so } r_3(x) = 1 + x + x^2 + x^3 + x^4 + x^5,$$
$$x^8 = f(x)(x^2 + x) + x, \text{ so } r_4(x) = x,$$
$$x^{10} = f(x) \cdot (x^4 + x^3) + x^3, \text{ so } r_5(x) = x^3.$$

The coefficients of $r_0(x), \ldots, r_5(x)$ form the rows of the matrix Q:

$$Q = \begin{pmatrix} 1 & 0 & 0 & 0 & 0 & 0 \\ 0 & 0 & 1 & 0 & 0 & 0 \\ 0 & 0 & 0 & 0 & 1 & 0 \\ 1 & 1 & 1 & 1 & 1 & 1 \\ 0 & 1 & 0 & 0 & 0 & 0 \\ 0 & 0 & 0 & 1 & 0 & 0 \end{pmatrix}$$

To find $h(x) = b_0 + b_1 x + b_2 x^2 + b_3 x^3 + b_4 x^4 + b_5 x^5$, we solve

$$\mathbf{b} \cdot (Q - I) = \begin{pmatrix} b_0 & b_1 & b_2 & b_3 & b_4 & b_5 \end{pmatrix} \begin{pmatrix} 0 & 0 & 0 & 0 & 0 & 0 \\ 0 & 1 & 1 & 0 & 0 & 0 \\ 0 & 0 & 1 & 0 & 1 & 0 \\ 1 & 1 & 1 & 0 & 1 & 1 \\ 0 & 1 & 0 & 0 & 1 & 0 \\ 0 & 0 & 0 & 1 & 0 & 1 \end{pmatrix} = \mathbf{0}$$

or

$$b_3 = 0,$$
$$b_1 + b_3 + b_4 = 0,$$
$$b_1 + b_2 + b_3 = 0,$$
$$b_5 = 0,$$
$$b_2 + b_3 + b_4 = 0,$$
$$b_3 + b_5 = 0.$$

This reduces quickly to $b_3 = b_5 = 0$, and $b_1 = b_2 = b_4$. The only solutions with $\deg(h(x)) \geq 1$ are $h(x) = x^4 + x^2 + x + b_0$ with $b_0 = 0$ or 1. For either choice of b_0,

$$h(x)^2 - h(x) = x^8 + x = f(x) \cdot (x^2 + x).$$

Thus
$$f(x) = gcd(f(x), x^4 + x^2 + x) \cdot gcd(f(x), x^4 + x^2 + x + 1).$$

By Euclid's algorithm, the left factor is $x^3 + x + 1$, and the right factor is $x^3 + x^2 + 1$. Both are irreducible polynomials, so the factorization of $f(x)$ in $\mathbb{F}_2[x]$ is

$$x^6 + x^5 + x^4 + x^3 + x^2 + x + 1 = (x^3 + x + 1) \cdot (x^3 + x^2 + 1).$$

Counting irreducible factors. Using some ideas of linear algebra, we can determine the number of distinct irreducible factors of $f(x)$ where $\deg(f) = d$.

Let $V = \mathbb{F}_p^d$ denote the vector space over the field $\mathbb{F}_p$ consisting of d-tuples of elements of $\mathbb{F}_p$ (row vectors). Let N be the set of vectors $\mathbf{b} = (b_0, b_1, \ldots, b_{d-1})$ in $\mathbb{F}_p^d$ with $\mathbf{b} \cdot (\mathbf{Q} - \mathbf{I}) = \mathbf{0}$. Then N is the null space of the matrix $\mathbf{Q} - \mathbf{I}$; N is a subspace of V. Let $\{\mathbf{v_1}, \mathbf{v_2}, \ldots, \mathbf{v_g}\}$ be a basis of N, that is, a set of vectors in N of minimal cardinality such that every vector in N is a linear combination of $\mathbf{v_1}, \mathbf{v_2}, \ldots, \mathbf{v_g}$: that is, every b in N may be written as $b = c_1 \mathbf{v_1} + c_2 \mathbf{v_2} + \ldots + c_g \mathbf{v_g}$, for some $c_1, c_2, \ldots, c_g$, in $\mathbb{F}_p$. The smallest g for which such a set $\{\mathbf{v_1}, \mathbf{v_2}, \ldots, \mathbf{v_g}\}$ exists is called the dimension of the space N.

The space N always contains the vectors $(a, 0, 0, \ldots, 0)$ for any a in $\mathbb{F}_p$, because such a vector corresponds to the constant polynomial $h(x) = a$, and $h(x)^p - h(x) = a^p - a = 0$ for any a in $\mathbb{F}_p$ by Fermat's theorem. Thus the dimension of N is at least one. To factor $f(x)$, we need to find some polynomial $h(x)$ of degree ≥ 1, and that means we need to find a vector $\mathbf{b} = (b_0, b_1, \ldots, b_{d-1})$ in N where at least one of the components $b_1, b_2, \ldots, b_{d-1}$ is not zero. If such a vector $\mathbf{b}$ exists, then there are vectors in N which are not of the form $(a, 0, \ldots, 0)$, and so the dimension of N is at least 2. This suggests the following result, which is slightly easier to describe if we assume that f is squarefree, that is, f is a product of distinct irreducible polynomials. (Recall from Section 15G that f is squarefree iff $gcd(f, f') = 1$, a condition that is easily checked. If f is not squarefree and $f' \neq 0$, then we can replace f by its squarefree part $f/(f, f')$.)

Theorem 9. *Let $f(x)$ in $\mathbb{F}_p[x]$ be squarefree. Then:*

(a) The dimension of the null space of $\mathbf{Q} - \mathbf{I}$ is equal to the number of irreducible factors of $f(x)$.

(b) $f(x)$ is irreducible in $\mathbb{F}_p[x]$ if and only if the null space N of $\mathbf{Q} - \mathbf{I}$ has dimension one.

The dimension of the null space of $\mathbf{Q} - \mathbf{I}$ can be computed in the following way. Since $\mathbf{Q} - \mathbf{I}$ is a $d \times d$ matrix, the dimension of the null space of $\mathbf{Q} - \mathbf{I}$ is equal to d minus the row or column rank of $\mathbf{Q} - \mathbf{I}$. The column rank of $\mathbf{Q} - \mathbf{I}$ is equal to the number of nonzero columns after performing column operations on $\mathbf{Q} - \mathbf{I}$ to reduce it to echelon form. This is the same as doing row operations on the transpose of $\mathbf{Q} - \mathbf{I}$ to reduce it to row echelon form.

To illustrate with $\mathbf{Q} - \mathbf{I}$ as in Example 10, a series of column operations transforms $\mathbf{Q} - \mathbf{I}$ into the echelon form

$$\mathbf{E} = \begin{pmatrix} 0 & 0 & 0 & 0 & 0 & 0 \\ 1 & 0 & 0 & 0 & 0 & 0 \\ 0 & 1 & 0 & 0 & 0 & 0 \\ 0 & 0 & 1 & 0 & 0 & 0 \\ 1 & 1 & 0 & 0 & 0 & 0 \\ 0 & 0 & 0 & 1 & 0 & 0 \end{pmatrix}.$$

Since the matrix $\mathbf{E}$ has four nonzero columns, the null space has dimension $6 - 4 = 2$. In fact, the null space can be obtained by solving $\mathbf{b}\mathbf{E} = \mathbf{0}$, since doing column operations to $\mathbf{Q} - \mathbf{I}$ does not change the space of solutions to $\mathbf{b}(\mathbf{Q} - \mathbf{I}) = \mathbf{0}$. The solutions $\mathbf{b}$ of $\mathbf{b}\mathbf{E} = \mathbf{0}$ are the solutions of the equations

$$b_1 + b_4 = 0,$$
$$b_2 + b_4 = 0,$$
$$b_3 = 0,$$
$$b_5 = 0.$$

Hence a vector $\mathbf{b}$ satisfying $\mathbf{b}\mathbf{E} = \mathbf{0}$ may be written as follows, where b_0 and b_4 may be chosen arbitrarily:

$$\mathbf{b} = (b_0, b_1, b_2, b_3, b_4, b_5) = (b_0, b_4, b_4, 0, b_4, 0)$$
$$= b_0(1,0,0,0,0,0) + b_4(0,1,1,0,1,0).$$

We chose $b_4 = 1$, above.

Proof (Proof of Theorem 9). Suppose $f(x)$ has degree d and factors into the product of g distinct irreducible factors,

$$f(x) = p_1(x)p_2(x) \cdots p_g(x),$$

where each $p_i(x)$ is irreducible.

Here is how to construct polynomials $h(x)$ so that $f(x)$ divides $h(x)^p - h(x)$.

For each vector $\mathbf{s} = (s_1, s_2, \ldots s_g)$ of elements of $\mathbb{F}_p$, use the Interpolation Theorem (Corollary 9 of Section 17B) to construct a unique polynomial $h_{\mathbf{s}}(x)$ of degree $\leq d$ so that

$$h_{\mathbf{s}}(x) \equiv s_i \pmod{p_i(x)}$$

for $i = 1, 2, \ldots, g$.

Since $h_{\mathbf{s}}(x) \equiv s_i \pmod{p_i(x)}$, it follows that $p_i(x)$ divides $h_{\mathbf{s}}(x) - s_i$, and so $p_i(x)$ divides

$$h_{\mathbf{s}}(x) \cdot (h_{\mathbf{s}}(x) - 1) \cdot \ldots \cdot (h_{\mathbf{s}}(x) - (p-1))$$

$$= \prod_{r=1}^{g}(h_{\mathbf{s}}(x) - r) = h_{\mathbf{s}}(x)^p - h_{\mathbf{s}}(x).$$

So $f(x)$ divides $h_{\mathbf{s}}(x)^p - h_{\mathbf{s}}(x)$. The map that sends $\mathbf{s}$ to the interpolation polynomial $h_{\mathbf{s}}(x)$ defines a one-to-one function γ from $\mathbb{F}_p^g$ to the set $\mathscr{P}$ of polynomials $h(x)$ so that $f(x)$ divides $h(x)^p - h(x)$.

Now we show that γ is onto.

Given a polynomial $h(x)$ in $\mathscr{P}$, then $f(x)$ divides $h(x)^p - h(x)$, and so for each i from 1 to g, the irreducible factor $p_i(x)$ of $f(x)$ divides

$$h(x)^p - h(x) = h(x) \cdot (h(x) - 1) \cdot \ldots \cdot (h(x) - (p-1)).$$

Since the polynomials $h(x), h(x) - 1, \ldots, h(x) - (p-1)$ are pairwise coprime, and each $p_i(x)$ is irreducible, each $p_i(x)$ divides $h(x) - s_i$ for a unique s_i, $1 \le s_i \le p-1$. Then the vector

$$\mathbf{s} = (s_1, s_2, \ldots, s_g)$$

is in $\mathbb{F}_p^g$ and for all $i = 1, \ldots, g$,

$$h(x) \equiv s_i \pmod{p_i(x)}.$$

Thus $h(x) = h_s(x)$ and the map γ is onto.

Thus γ is a one-to-one correspondence between $\mathbb{F}_p^g$ and the set $\mathscr{P}$ of polynomials $h(x)$ so that $f(x)$ divides $h(x)^p - h(x)$.

Now if $h(x) = b_0 + b_1 x + \ldots + b_{d-1} x^{d-1}$, then $h(x)$ is in $\mathscr{P}$ if and only if the vector of coefficients $(b_0, b_1, \ldots, b_{d-1})$ is in the null space N of $\mathbf{Q} - \mathbf{I}$. Hence the cardinality of N = the cardinality of $\mathscr{P}$ = the cardinality of $\mathbb{F}_p^g = p^g$, where g is the number of distinct irreducible factors of $f(x)$. Hence the dimension of N is g.

Part (b) follows immediately. From (a), the null space N of $\mathbf{Q} - \mathbf{I}$ is one-dimensional if and only if $f(x) = p(x)$, an irreducible polynomial. $\qquad \square$

Example 11. In $\mathbb{F}_3[x]$, how many irreducible factors divide

$$f(x) = x^5 + 2x^4 + x^3 + x^2 + 2?$$

We compute $\mathbf{Q} - \mathbf{I}$:

$$1 = f(x) \cdot 0 + 1,$$
$$x^3 = f(x) \cdot 0 + x^3,$$
$$x^6 = f(x) \cdot (x+1) + (1 + x + 2x^2 + x^3),$$
$$x^9 = f(x) \cdot (x^4 + x^3 + x) + x,$$
$$x^{12} = f(x) \cdot (x^7 + x^6 + x^4) + x^4.$$

So, noting that the rows of $\mathbf{Q}$ are the coefficients of the remainders arranged, from left to right, by increasing powers of x, we have

$$\mathbf{Q} = \begin{pmatrix} 1 & 0 & 0 & 0 & 0 \\ 0 & 0 & 0 & 1 & 0 \\ 1 & 1 & 2 & 1 & 0 \\ 0 & 1 & 0 & 0 & 0 \\ 0 & 0 & 0 & 0 & 1 \end{pmatrix}, \quad \mathbf{Q} - \mathbf{I} = \begin{pmatrix} 0 & 0 & 0 & 0 & 0 \\ 0 & 2 & 0 & 1 & 0 \\ 1 & 1 & 1 & 1 & 0 \\ 0 & 1 & 0 & 2 & 0 \\ 0 & 0 & 0 & 0 & 0 \end{pmatrix},$$

which, after column operations, becomes the echelon form

$$\mathbf{E} = \begin{pmatrix} 0 & 0 & 0 & 0 & 0 \\ 1 & 0 & 0 & 0 & 0 \\ 0 & 1 & 0 & 0 & 0 \\ 2 & 0 & 0 & 0 & 0 \\ 0 & 0 & 0 & 0 & 0 \end{pmatrix}.$$

So $\mathbf{Q} - \mathbf{I} = \mathbf{E}$ has rank 2, the null space of $\mathbf{E}$ has dimension 3, $f(x)$ has three distinct irreducible factors, and the vectors $\mathbf{b}$ satisfying $\mathbf{bE} = 0$ have the form

$$\mathbf{b} = (b_0, b_1, b_2, b_3, b_4) = (b_0, b_3, 0, b_3, b_4),$$

where b_0, b_3 and b_4 are arbitrary. Thus $f(x)$ divides $h(x)^3 - h(x)$ where we can choose $h(x) = x^4$, or $h(x) = x + x^3$, or $h(x) = 1$, or any $\mathbb{F}_3$-linear combination of those three choices.

Exercises.

7. Find the three irreducible factors of $f(x)$ in Example 11.

8. Factor $x^{10} + x^9 + x^7 + x^3 + x^2 + 1$ in $\mathbb{F}_2[x]$.

9. Factor $x^8 + x^7 + x^6 + x^4 + 1$ in $\mathbb{F}_2[x]$.

10. Show that $x^5 + x^2 + 1$ is irreducible in $\mathbb{F}_2[x]$.

11. Show that $x^7 + x^3 + 1$ is irreducible in $\mathbb{F}_2[x]$.

12. Show that $7x^7 + 6x^6 + 4x^4 + 3x^3 + 2x^2 + 2x + 1$ is irreducible in $\mathbb{Q}[x]$ (use the last exercise).

13. Use Berlekamp's algorithm to factor $x^2 - q$ in $\mathbb{F}_p[x]$, where q and p are coprime, and prove Euler's Lemma (Section 21B) that q is a quadratic residue mod p, that is, $x^2 - q \equiv 0 \pmod{p}$ has a root, iff $q^{(p-1)/2} \equiv 1 \pmod{p}$.

D. The Hensel Factorization Method

Given a bound B on the coefficients of factors of a polynomial $f(x)$ in $\mathbb{Z}[x]$, we can look for factorizations of $f(x)$ modulo M for $M \geq 2B$. Any factor of f modulo M corresponds to at most one possible factor of f in $\mathbb{Z}[x]$, because there will be only one polynomial in $\mathbb{Z}[x]$ that will satisfy the bound on coefficients and reduce to the given factor of f modulo M.

Thus we wish to find factorizations of f modulo M, where M may be large.

There are two choices on how to proceed.

One is to find primes $p > 2B$ and use Berlekamp's algorithm to factor f modulo p. If we're lucky, f will have few irreducible factors modulo p, so there will be few choices for factorizations of f in $\mathbb{Z}[x]$.

An alternative is to find a small prime p so that f factors modulo p into few distinct irreducible factors, and then lift the factorization modulo p to a unique factorization modulo p^{2^e} for e so large so that $p^{2^e} > 2B$.

This method, called the Hensel factorization method [Zassenhaus (1978)], uses an extension of coprimeness to polynomials with coefficients not in a field.

Definition. Let R be a commutative ring, and f, g be polynomials of degrees ≥ 1 with coefficients in R. Then f and g are *coprime* if there exist polynomials r, s with coefficients in R so that

$$rf + sg = 1.$$

If $R = \mathbb{Z}/m\mathbb{Z}$ and f, g are polynomials with integer coefficients, we'll say that f and g are *coprime modulo m*, if the images of f and g in $\mathbb{Z}/m\mathbb{Z}[x]$ are coprime, that is, if there exist polynomials r, s in $\mathbb{Z}[x]$ so that $fr + gs \equiv 1 \pmod{m}$.

In short, we extend the definition of coprime by using the Bezout Identity criterion.

Before presenting the main result, we need an auxiliary result about coprime polynomials.

Proposition 10. *Let g, h be monic and coprime in $R[x]$. Then for all k in $R[x]$ there exist polynomials a, b in $R[x]$ with $ag + bh = k$. If $\deg k < \deg(fg)$, then we can choose a, b with $\deg(a) < \deg(h), \deg(b) < \deg(g)$.*

Proof. Since g and h are coprime, there exist polynomials r, s so that $gr + hs = 1$. It follows that $grk + hsk = k$.

Suppose $\deg(k) < \deg(fg)$ and there exist a, b in $R[x]$ so that $ag + bh = k$ with $\deg(b) \geq \deg(g)$. Then $b = gq + s$ with $\deg(s) < \deg(g)$, and

$$ag + (gq + s)h = k.$$

Hence $(a + qh)g + sh = k$, or, letting $r = a + qh$, then

$$rg + sh = k.$$

Since $\deg(s) < \deg(g)$, we have $\deg(sh) < \deg(gh)$, and also $\deg(k) < \deg(gh)$. So $\deg(rg) < \deg(gh)$. Since g is monic, it follows that $\deg(r) < \deg(h)$. $\square$

Here is the main result.

Theorem 11. *Let f be a monic polynomial in $\mathbb{Z}[x]$. Suppose there are monic polynomials g_1, h_1 in $\mathbb{Z}[x]$ so that g_1 and h_1 are coprime modulo m and $f = g_1 h_1 \pmod{m}$. Then there exist unique monic polynomials g_2 and f_2 so that*

$$g_2 \equiv g_1 \pmod{m}$$
$$h_2 \equiv h_1 \pmod{m},$$

g_2 and h_2 are coprime modulo m^2, and

$$f \equiv g_2 h_2 \pmod{m^2}.$$

Proof. The proof shows how to construct g_2 and h_2.

We write

$$g_2 = g_1 + mb$$
$$h_2 = h_1 + mc$$

for polynomials b, c in $\mathbb{Z}[x]$ with $\deg(b) < \deg(g_1), \deg(c) < \deg(h_1)$ that we need to find. To find them, we note that since $f \equiv g_1 h_1 \pmod{m}$, we have

$$f = g_1 h_1 + mk$$

for some polynomial k in $\mathbb{Z}[x]$. Since f, g_1 and h_1 are monic, $\deg(k) < \deg(g_1 h_1)$. Then

$$g_2 h_2 - f = (g_1 + mb)(h_1 + mc) - (g_1 h_1 + mk)$$
$$= g_1 h_1 + mg_1 c + mh_1 b + m^2 bc - g_1 h_1 - mk.$$

For the left side to be congruent to 0 modulo m^2, we need

$$m(g_1 c + h_1 b - k) \equiv 0 \pmod{m^2},$$

or

$$g_1 c + h_1 b - k \equiv 0 \pmod{m}.$$

But since g_1 and h_1 are coprime modulo m, there exist polynomials c and b so that

$$g_1 c + h_1 b \equiv k \pmod{m},$$

and since $\deg(k) < \deg(g_1 h_1)$, we may choose the polynomials c and b so that $\deg c < \deg h_1$ and $\deg b < \deg g_1$. Then by the way we chose c and b, the polynomials $g_2 = g_1 + mb$ and $h_2 = h_1 + mc$ are monic and satisfy

$$f \equiv g_2 h_2 \pmod{m^2}.$$

To finish the proof we need to show that g_2 and h_2 are coprime modulo m^2. So we seek polynomials r_2 and s_2 so that

$$r_2 g_2 + s_2 h_2 \equiv 1 \pmod{m^2}.$$

Since g_1 and h_1 are coprime, there exist polynomials r_1 and s_1 so that $r_1 g_1 + s_1 h_1 = 1 + mz$ for some polynomial z. We write

$$r_2 = r_1 + mw, \quad s_2 = s_1 + my$$

for unknown polynomials w, y in $\mathbb{Z}[x]$, and substitute for r_2, g_2, s_2 and h_2 in the desired congruence

$$r_2 g_2 + s_2 h_2 \equiv 1 \pmod{m^2}.$$

to obtain

$$(r_1 + mw)(g_1 + mb) + (s_1 + my)(h_1 + mc)$$
$$\equiv r_1 g_1 + mwg_1 + mr_1 b + s_1 h_1 + ms_1 c + myh_1 \pmod{m^2}$$
$$\equiv 1 + mz + m(wg_1 + r_1 b + s_1 c + yh_1) \pmod{m^2}.$$

For this last expression to be congruent to 1 modulo m^2, we need to find polynomials w, y so that

$$wg_1 + yh_1 \equiv -z - r_1 b - s_1 c \pmod{m}.$$

But since g_1 and h_1 are coprime modulo m, it follows that we can find w, y satisfying this last congruence. That means there exist $r_2 = r_1 + mw, s_2 = s_1 + my$ so that

$$r_2 g_2 + s_2 h_2 \equiv 1 \pmod{m^2}.$$

Thus g_2 and h_2 are coprime modulo m^2, and that completes the proof.   □

Example 12. Let $f(x) = x^4 + 23x^3 - 15x^2 + 17x - 7$. We find that

$$f(x) \equiv x^4 + 2x^3 + 3x^2 + 2x + 2 = (x^2 + 1)(x^2 + 2x + 2) \pmod{3},$$

so $f(x)$ factors modulo 3 into the product of two distinct polynomials that are irreducible modulo 3, and hence coprime modulo 3.

Now we want to factor $f(x)$ modulo 9. So let $g_1 = x^2 + 1$, $h_1 = x^2 + 2x + 2$, and let

$$g_2 = g_1 + 3b = (x^2 + 1) + 3b$$
$$h_2 = h_1 + 3c = (x^2 + 2x + 2) + 3c$$

for some polynomials b, c with $\deg c < \deg h_1, \deg b < \deg g_1$. Then

$$g_2 h_2 \equiv (x^2 + 1)(x^2 + 2x + 2) + 3c(x^2 + 1) + 3b(x^2 + 2x + 2) \pmod{9}.$$

To find b, c we set up the congruence

$$f \equiv g_2 h_2 \pmod{9}$$

and substitute:

$$x^4 + 23x^3 - 15x^2 + 17x - 7 \equiv (x^4 + 2x^3 + 3x^2 + 2x + 2)$$
$$+ 3c(x^2 + 1) + 3b(x^2 + 2x + 2) \pmod{9};$$

or

$$21x^3 - 18x^2 + 15x - 9 \equiv 3c(x^2 + 1) + 3b(x^2 + 2x + 2) \pmod{9}.$$

Factoring 3 out of everything yields

$$7x^3 - 6x^2 + 5x - 3 \equiv c(x^2 + 1) + b(x^2 + 2x + 2) \pmod{3},$$

which we know we can solve for polynomials b, c of degree ≤ 2 since $x^2 + 1$ and $x^2 + 2x + 2$ are coprime modulo 3.

To solve the congruence for b and c, we set up some linear equations: write $b = rx + s$, $c = tx + v$, then

$$7x^3 - 6x^2 + 5x - 3 \equiv (tx + v)(x^2 + 1) + (rx + s)(x^2 + 2x + 2) \pmod{3}.$$

Equating the coefficients of $1, x, x^2, x^3$ on both sides yields

$$-3 \equiv v + 2s$$
$$5 \equiv t + 2r + 2s$$
$$-6 \equiv v + 2r + s$$
$$7 \equiv t + r \pmod 3.$$

One sees easily that $r = t = 2, s = v = 1$ is the unique solution, so

$$b = 2x + 1, c = 2x + 1.$$

Thus

$$g_2 = g_1 + 3b \equiv (x^2 + 1) + 3(2x + 1) \equiv x^2 + 6x + 4,$$
$$h_2 = h_1 + 3c = (x^2 + 2x + 2) + 3(2x + 1) \equiv x^2 + 8x + 5,$$

and it is easily checked that

$$(x^2 + 6x + 4)(x^2 + 8x + 5) = x^4 + 14x^3 + 57x^2 + 62x + 20$$
$$\equiv x^4 + 23x^3 - 15x^2 + 17x - 7 = f(x) \pmod 9.$$

In a similar way we can lift the factorization modulo 9 to one modulo $9^2 = 81$, then to $81^2 = 6561$ and beyond, until we get past the bound on the coefficients of any degree 2 factor of $f(x)$, at which point we either find a factorization of f in $\mathbb{Z}[x]$ or show that none exists that reduces to $f = g_1 h_1$ modulo 3. In the latter case, f must be irreducible in $\mathbb{Q}[x]$.

Note that $\|f\| = (1^2 + 23^2 + 15^2 + 17^2 + 7^2)^{1/2} = \sqrt{1093} = 33.06$, so using the Mignotte bound we would need only to look at a factorization of f modulo 81 to either find a factorization of f or show that f is irreducible.

It turns out that $f(x)$ is irreducible modulo 5, so must be irreducible in $\mathbb{Q}[x]$.

Exercises.

14. Factor $x^4 - x^3 - 84x^2 + 125x - 13$ modulo 5, then modulo 25, then in $\mathbb{Z}$.

15. Factor $x^4 + 2x^3 - 38x^2 - 69x - 28$ modulo 3, then modulo 9, then in $\mathbb{Z}$.

16. Factor $x^4 + x^2 + 2$ modulo 2, then modulo 4, then modulo 16, then in $\mathbb{Z}$.

Chapter 27
Irreducible Polynomials

We find a formula for the number of irreducible polynomials of degree n in $\mathbb{F}_p[x]$ for any p and n, and use it to show that in some sense, almost every polynomial in $\mathbb{Z}[x]$ is irreducible in $\mathbb{Q}[x]$.

A. Irreducible Polynomials in $\mathbb{F}_p[x]$

We begin by showing

Theorem 1. $x^{p^n} - x$ *is the product of all monic irreducible polynomials in* $\mathbb{F}_p[x]$ *of degree d, for all d dividing n.*

We prove this in two parts.

Theorem 2. *If $q(x)$ is an irreducible polynomial of degree d and d divides n, then $q(x)$ divides $x^{p^n} - x$.*

Proof. Let $F = \mathbb{F}_p[x]/(q(x)) = \mathbb{F}_p[\alpha]$, where $\alpha = [x]_{q(x)}$. Then $q(x)$ is the minimal polynomial over $\mathbb{F}_p$ of α. Now F is a field with p^d elements. So by Fermat's theorem, $\alpha^{p^d} = \alpha$. Since $de = n$ for some integer e,

$$\alpha^{p^n} = \alpha^{p^{de}} = \alpha,$$

so α is a root of $x^{p^n} - x$.

Now $q(x)$ is irreducible in $\mathbb{F}_p[x]$, so either $q(x)$ divides $x^{p^n} - x$ or (by Bezout's identity),

$$s(x)q(x) + t(x)(x^{p^n} - x) = 1$$

for some polynomials $s(x)$, $t(x)$ in $\mathbb{F}_p[x]$. But if the second condition held, then setting $x = \alpha$ would yield $0 = 1$, impossible. Hence $q(x)$ divides $x^{p^n} - x$, as claimed. $\qquad\square$

L.N. Childs, *A Concrete Introduction to Higher Algebra*, Undergraduate Texts in Mathematics, © Springer Science+Business Media LLC 2009

Theorem 3. *If $q(x)$ is an irreducible factor of $x^{p^n} - x$ and has degree d, then d divides n.*

Proof. This proof uses the Isomorphism Theorem of Section 24A.

Let K be a splitting field over $\mathbb{F}_p$ of $x^{p^n} - x$, and let F be the subfield consisting of all of the p^n roots of $x^{p^n} - x$ described in Theorem 6 of Section 24C. Since $q(x)$ divides $x^{p^n} - x$, there is a root β of $q(x)$ in F. Since $q(x)$ is irreducible, $q(x)$ is the minimal polynomial of β over $\mathbb{F}_p$.

Let $\phi_\beta : \mathbb{F}_p[x] \to F$ be the "evaluation at β" homomorphism. Since $q(x)$ is the minimal polynomial of β, the homomorphism ϕ_β induces a 1-1 homomorphism $\bar{\phi}$ from $E = \mathbb{F}_p[x]/(q(x))$ to F by sending $[x]$ to β.

Let L be the image of E in F; L is then a subfield of F isomorphic to E.

Let α be a primitive element of F. Let $s(x)$ be the minimal polynomial of α over L. Then the evaluation homomorphism ϕ_α from $L[x]$ to F sending x to α induces a 1-1 homomorphism ϕ' from $L[x]/s(x)$ to F, which is onto because every non-zero element of F is a power of α. So ϕ' is an isomorphism from $L[x]/(s(x))$ onto F. So $L[x]/(s(x))$ and F have the same number of elements.

How many elements are in $L[x]/(s(x))$? If $s(x)$ has degree e, and L has q elements, then $L[x]/(s(x))$ has q^e elements. But $q = p^d$ and F has p^n elements. So $(p^d)^e = p^n$. So $de = n$, and d, the degree of $q(x)$, divides n. That completes the proof. $\square$

Let $N_n(p)$ be the number of irreducible polynomials of degree n in $\mathbb{F}_p[x]$. We'll write N_n if the prime p is understood.

Using Theorem 1, we will find an explicit formula for $N_n(p)$.

To obtain such a formula, we use the Mobius function, a classical tool in number theory and combinatorics.

Definition. The Mobius function $\mu(n)$ is defined for $n \geq 1$ by

$$\mu(n) = \begin{cases} 1 \text{ if } n = 1, \\ 0 \text{ if } n \text{ is not squarefree} \\ (-1)^r \text{ if } n \text{ is the product of } r \text{ distinct primes.} \end{cases}$$

The formula we want is

$$N_n = \frac{1}{n} \sum_{d \mid n} \mu\left(\frac{n}{d}\right) p^d.$$

This formula is a special case of the Mobius inversion formula, which we now derive. We begin with two facts about the Mobius function.

Proposition 4. *If $(m,n) = 1$, then $\mu(mn) = \mu(m)\mu(n)$.*

This is easy to verify.

A function such as μ that satisfies Proposition 4 is called *multiplicative*. Another example of a multiplicative function is Euler's ϕ function.

Proposition 5. $\sum_{d \mid n} \mu(d) = 0$ *unless $n = 1$.*

The proof of this is an exercise in manipulating sums. Before doing the proof in general we illustrate with $n = 36 = 2^2 3^2$: then the divisors of n are 1, 2, 4, 3, 6, 12, 9, 18 and 36, and we have

$$\sum_{d|36} \mu(d) = [\mu(1) + \mu(2) + \mu(2^2)]$$
$$+ [\mu(3) + \mu(2 \cdot 3) + \mu(2^2 \cdot 3)]$$
$$+ [\mu(3^2) + \mu(2 \cdot 3^2) + \mu(2^2 \cdot 3^2)]$$
$$= \mu(1)[1 + \mu(2) + \mu(2^2)]$$
$$+ \mu(3)[1 + \mu(2) + \mu(2^2)]$$
$$+ \mu(3^2)[1 + \mu(2) + \mu(2^2)].$$

Now $\mu(d) = 0$ if d is divisible by the square of a prime, and $\mu(1) = 1$, so this sum reduces to

$$= \mu(1)[\mu(1) + \mu(2)] + \mu(3)[\mu(1) + \mu(2)]$$
$$= [\mu(1) + \mu(3)][\mu(1) + \mu(2)].$$

Now $\mu(1) = 1$, $\mu(3) = -1$, so $\mu(1) + \mu(3) = 0$. Hence $\sum_{d|36} \mu(d) = 0$.
The proof in general works in a similar way.

Proof. Write $n = p^e q$ with $(p, q) = 1$. Then

$$\sum_{d|n} \mu(d) = \sum_{r=0}^{e} \sum_{b|q} \mu(p^r b)$$
$$= \sum_{r=0}^{e} \sum_{b|q} \mu(p^r)\mu(b).$$

Since $\mu(p^r) = 0$ for $r \geq 2$, this reduces to

$$= \sum_{b|q} \mu(1)\mu(b) + \sum_{b|q} \mu(p)\mu(b)$$
$$= \mu(1) \sum_{b|q} \mu(b) + \mu(p) \sum_{b|q} \mu(b)$$
$$= [\mu(1) + \mu(p)] \sum_{b|q} \mu(b) = 0$$

since $\mu(1) + \mu(p) = 0$.   □

With Proposition 5 we can prove the useful

Proposition 6 (Mobius Inversion Formula). *Let f be a function defined on the natural numbers. If we set*

$$F(n) = \sum_{d|n} f(d) \text{ for every } n \geq 1,$$

then

$$f(n) = \sum_{d|n} \mu(\frac{n}{d}) F(d) = \sum_{e|n} \mu(e) F(\frac{n}{e}).$$

Proof. If we substitute $e = n/d, d = n/e$, then as d runs through all divisors of n, so does e. Hence the last two sums are equal.

Now by definition of F,

$$\sum_{e|n} \mu(e) F(\frac{n}{e}) = \sum_{e|n} \mu(e) (\sum_{d|(n/e)} f(d)) = \sum_{e|n} \sum_{d|(n/e)} (\mu(e) f(d).$$

Interchanging the order of summation (if $d|(n/e)$, then $de|n$ so $e|(n/d)$), we get

$$\sum_{e|n} \mu(e) F(\frac{n}{e}) = \sum_{d|n} \left(\sum_{e|(n/d)} \mu(e) \right) f(d). \qquad (27.1)$$

Now by Proposition 5, for each $m > 1$,

$$\sum_{e|m} \mu(e) = 0.$$

So the coefficient of $f(d)$ is 0 unless $n/d = 1$, that is, $d = n$. Hence the sum (1) reduces to the single term $f(n)$, as was to be shown.   □

With these generalities out of the way, we can get the desired formula for N_n^p. We shall write N_n^p as N_n if p is understood.

Theorem 7. *Let N_n be the number of irreducible polynomials of degree n in $\mathbb{F}_p[x]$. Then*

$$N_n = \frac{1}{n} \sum_{d|n} \mu \left(\frac{n}{d} \right) p^d.$$

Proof. Theorem 1 describes the complete factorization of $x^{p^n} - x$ in $\mathbb{F}_p$ for any n. Since $x^{p^n} - x$ is the product of all the N_d irreducible polynomials of degree d for all d dividing n, we obtain the formula

$$p^n = \sum_{d|n} d N_d$$

by summing the degrees of all the irreducible factors of $x^{p^n} - x$. Now apply the Mobius inversion formula with $F(n) = p^n$, $f(d) = dN_d$. We get

$$n N_n = \sum_{d|n} \mu \left(\frac{n}{d} \right) p^d.$$

Dividing both sides by n yields the desired formula.   □

With that formula we can give another proof of Corollary 7 of Chapter 24, that for every prime p and every $n > 0$ there is an irreducible polynomial in $\mathbb{F}_p[x]$ of degree n.

Proposition 8. *For every prime p and every $n > 0$, $N_n > 0$.*

Proof. Since $\mu(n/n) = 1$ and $\mu(n/d) \geq -1$ for all $d|n$, $d < n$, we have that

$$N_n = \frac{1}{n}p^n + \frac{1}{n}\sum_{d|n, d<n} \mu\left(\frac{n}{d}\right)p^d$$

$$\geq \frac{1}{n}p^n - \frac{1}{n}\sum_{d|n, d<n} p^d$$

$$\geq \frac{1}{n}\left(p^n - \sum_{d=0}^{n-1} p^d\right)$$

Now

$$\sum_{d=0}^{n-1} p^d = \frac{p^n - 1}{p - 1} < p^n,$$

so

$$\frac{1}{n}\left(p^n - \sum_{d=0}^{n-1} p^d\right) > 0.$$

Hence $N_n > 0$. □

The number $N_n(p)$ of irreducible monic polynomials over $\mathbb{F}_p$ of degree n for $n = 1, \ldots, 10$ is given by the following formulas

n	$N_n(p)$	$N_n(2)$	$N_n(3)$	$N_n(5)$	$N_n(7)$
1	p	2	3	5	7
2	$(p^2 - p)/2$	1	3	10	21
3	$(p^3 - p)/3$	2	8	40	112
4	$(p^4 - p^2)/4$	3	18	150	588
5	$(p^5 - p)/5$	6	48	624	3360
6	$(p^6 - p^2 - p^3 + p)/6$	9	116	2580	19544
7	$(p^7 - p)/7$	18	312	11160	117648
8	$(p^8 - p^4)/8$	30	810	48750	720300
9	$(p^9 - p^3)/9$	56	2184	217000	4483696
10	$(p^{10} - p^5 - p^2 + p)/10$	99	5880	976248	28245840

Every irreducible polynomial in $\mathbb{F}_7[x]$ of degree n gives rise to infinitely many different irreducible polynomials of degree n in $\mathbb{Q}[x]$. So there are many irreducible polynomials in $\mathbb{Q}[x]$. We'll get an idea of how many in the next section.

For more discussion on Mobius inversion, see Bender and Goldman (1975).

Exercises.

1. If F is a function defined on natural numbers and f is defined by

$$f(n) = \sum_{d|n} \mu(d)F(n/d),$$

prove that

$$F(n) = \sum_{d|n} f(d).$$

2. If f is a multiplicative function defined on natural numbers and $F(n) = \sum_{d|n} f(d)$, prove that F is multiplicative.

3. Prove Proposition 4.

4. What are the 8 monic irreducible polynomials of degree 3 in $\mathbb{F}_3[x]$?

5. Find the formula for $N_{12}(p)$. Find $N_{12}(2)$.

6. Find the formula for $N_{30}(p)$.

7. If n is divisible by g distinct primes, how many different powers of p appear in the formula for $N_n(p)$?

8. Show that

$$\left(\frac{p^n}{n}\right)(1 - \varepsilon) < N_n < \frac{p^n}{n}$$

for some quantity $\varepsilon = \varepsilon(n)$ where $\varepsilon \to 0$ as $n \to \infty$. Conclude that for n large, approximately one of every n monic polynomials in $\mathbb{F}_p[x]$ of degree n is irreducible. (Asking about the size of N_n for n large is the analogue in $\mathbb{F}_p[x]$ of the Prime Number Theorem discussed in Section 4C.)

9. If d divides n, prove that every irreducible polynomial of degree d in $\mathbb{F}_p[x]$ has a root in every field F with p^n elements.

10. Show that if $q(x)$ in $\mathbb{F}_p[x]$ is irreducible and has degree d, and F is a field with p^n elements, where $d|n$, then F is a splitting field of $q(x)$.

11. Factor $x^{16} - x$ in $\mathbb{F}_2[x]$.

12. Factor $x^9 - x$ in $\mathbb{F}_3[x]$.

13. Factor $x^{25} - x$ in $\mathbb{F}_5[x]$.

14. Show that if p, q are primes, then $x^{p^q} - x = (x^p - x)h(x)$ in $\mathbb{F}_p[x]$, where $h(x)$ is the product of all monic irreducible polynomials in $\mathbb{F}_p[x]$ of degree q.

15. Show that $\mathbb{F}_{16}$ is a splitting field for $x^4 - x$ in $\mathbb{F}_2[x]$. If $\mathbb{F}_{16} = \mathbb{F}_2[\alpha]$ where $\alpha^4 + \alpha + 1 = 0$ (as in Table 2 of Chapter 25A), what are the roots in $\mathbb{F}_{16}$ of $x^4 - x$?

16. Prove Rabin's irreducibility test [Rabin (1980b)] for polynomials $m(x)$ of degree n in $\mathbb{F}_p[x]$: $m(x)$ is irreducible if
(i) $m(x)$ divides $x^{p^n} - x$; and

(ii) for any prime divisor d of n, the greatest common divisor of $m(x)$ and $x^{p^{n/d}} - x$ is 1.

17. Suppose $m(x)$ in $\mathbb{F}_p[x]$ has degree d. Call $m(x)$ *Carmichael* if $m(x)$ is composite, and for every polynomial $a(x)$ in $\mathbb{F}_p[x]$, coprime to $m(x)$,

$$a(x)^{p^d} = a(x) \quad (\mathrm{mod}\ m(x)).$$

(i) Show that if $m(x)$ is irreducible, then for every $a(x)$ coprime to $m(x)$,

$$a(x)^{p^d} = a(x) \quad (\mathrm{mod}\ m(x)).$$

(ii) Prove that the following are equivalent:
(a) $m(x)$ is Carmichael;
(b) $m(x)$ divides $x^{p^d} - x$;
(c) $m(x) = q_1(x) \cdots q_g(x)$, a product of distinct irreducible polynomials, where for each i, if d_i is the degree of $q_i(x)$, then $p^{d_i} - 1$ divides $p^d - 1$;
(d) $m(x) = q_1(x) \cdots q_g(x)$, a product of distinct irreducible polynomials, where for each i, if d_i is the degree of $q_i(x)$, then d_i divides d.

B. Most Polynomials in $\mathbb{Z}[x]$ are Irreducible

In the last section, we computed the number $N_n(p)$ of monic irreducible polynomials of degree n in $\mathbb{Z}_p[x]$ for any n and p. We showed that

$$N_n(p) = \frac{1}{n} \sum_{d \mid n} \mu\left(\frac{n}{d}\right) p^d,$$

where $\mu(e)$ is the Mobius function. Thus we have

Lemma 9.

$$N_n(p) > \frac{p^n}{2n}.$$

Proof. Since $\mu(n/d)$ is either 1, -1 or 0, and $\mu(1) = 1$, the formula

$$N_n(p) = \frac{1}{n} \sum_{d \mid n} \mu\left(\frac{n}{d}\right) p^d,$$

yields

$$nN_n(p) = p^n - \sum_{d \mid n, d < n} p^d.$$

Since every proper divisor of n is $< n/2$, we have

$$\sum_{d|n, d<n} p^d \leq \sum_{d \leq n/2} p^d < p^{\lfloor n/2 \rfloor + 1}$$

where $\lfloor a \rfloor$ denotes the greatest integer $< a$. Hence

$$n N_n(p) > (p^n - p^{\lfloor n/2 \rfloor + 1}).$$

If $n > 2$, then $\lfloor n/2 \rfloor + 1 \leq n - 1$, so

$$N_n(p) > \frac{1}{n}(p^n - p^{n-1}) = \frac{p^n}{n}\left(1 - \frac{1}{p}\right) \geq \frac{p^n}{n}\left(\frac{1}{2}\right).$$

For $n = 2$,

$$N_2(p) = \frac{1}{2}(p^2 - p) = \frac{p^2}{2}\left(1 - \frac{1}{p}\right) \geq \frac{p^2}{2}\left(\frac{1}{2}\right).$$

$\square$

Using this lower bound for N_n^p we will show that almost all monic polynomials in $\mathbb{Z}[x]$ of degree $n \geq 1$ are irreducible. The main idea of the argument is that if $f(x)$ is a monic polynomial in $\mathbb{Z}[x]$ whose image in $\mathbb{F}_p[x]$ is irreducible for some prime p, then $f(x)$ is irreducible in $\mathbb{Z}[x]$.

What do we mean by "almost all"?

The way we will interpret this is as follows.

Pick a bound M. Consider the set $P_n(M)$ of all monic polynomials $f(x)$ in $\mathbb{Z}[x]$,

$$f(x) = x^n + a_{n-1}x^{n-1} + \ldots + a_2 x^2 + a_1 x + a_0,$$

so that each coefficient a_k satisfies $-M < a_k \leq M$. This is a finite set of polynomials: the number of such polynomials is $(2M)^n$, since there are $2M$ possibilities for each of the n coefficients $a_{n-1}, \ldots, a_0$.

We will find a lower bound on the number of irreducible polynomials in the set $P_n(M)$, and show that for a suitable increasing sequence of numbers M, the proportion of irreducible polynomials goes to 1. More precisely,

Theorem 10. *For every $n \geq 2$ and every $g \geq 1$ let M_g be the product of the first g odd primes. Let*

$$I_n(M_g) = \{f(x) \text{ in } P_n(M_g) | f(x) \text{ is irreducible}\}.$$

Then

$$\lim_{g \to \infty} \frac{|I_n(M_g)|}{|P_n(M_g)|} = 1.$$

Proof. For every $M \geq 2$, if

$$f(x) = x^n + a_{n-1}x^{n-1} + \ldots + a_2 x^2 + a_1 x + a_0,$$

is in $P_n(M)$, then each coefficient a_k satisfies $-M < a_k \leq M$ for $0 \leq k \leq n-1$. Since the integers a with $-M < a \leq M$ is a complete set of representatives for $\mathbb{Z}/(2M)\mathbb{Z}$, we have a one-to-one correspondence between $P_n(M)$ and monic polynomials of degree n with coefficients in the ring $\mathbb{Z}/(2M)\mathbb{Z}$.

Now assume $M = M_g = 3 \cdot 5 \cdots p_g$ is the product of the first g odd primes.

By the Chinese remainder theorem, there is an isomorphism

$$\mathbb{Z}/(2M)\mathbb{Z} \cong \mathbb{Z}/2\mathbb{Z} \times \mathbb{Z}/3\mathbb{Z} \times \cdots \times \mathbb{Z}/p_g\mathbb{Z}$$

given by mapping $[a]_{2M}$ to the $(g+1)$-tuple $([a]_2, [a]_3, \ldots, [a]_{p_g})$. This map induces a one-to-one correspondence between polynomials in $P_n(M)$ and $(g+1)$-tuples $([f(x)]_2, [f(x)]_3, \ldots, [f(x)]_{p_g})$ of monic polynomials of degree n in

$$\mathbb{Z}/2\mathbb{Z}[x] \times \mathbb{Z}/3\mathbb{Z}[x] \times \cdots \times \mathbb{Z}/p_g\mathbb{Z}[x].$$

Here if $f(x)$ is in $P_n(M)$, then $[f(x)]_p$ denotes the image of $f(x)$ in $\mathbb{Z}/p\mathbb{Z}[x]$ obtained by replacing the coefficients of $f(x)$ by their congruence classes modulo p.

Under this correspondence between $P_n(M)$ and

$$\mathbb{Z}/2\mathbb{Z}[x] \times \mathbb{Z}/3\mathbb{Z}[x] \times \cdots \times \mathbb{Z}/p_g\mathbb{Z}[x].$$

a polynomial $f(x)$ is irreducible in $\mathbb{Z}[x]$ if for some prime p among $2, 3, \ldots, p_g$, the image $[f(x)]_p$ of $f(x)$ in $\mathbb{Z}/p\mathbb{Z}[x]$ is irreducible.

Thus $|I_n(M)| \geq$ the number of $(g+1)$-tuples of monic polynomials of degree n, $(h_0(x), h_1(x), \ldots, h_g(x))$, with $h_0(x)$ in $\mathbb{Z}/2\mathbb{Z}[x]$, $h_1(x)$ in $\mathbb{Z}/3\mathbb{Z}[x]$, $\ldots$, $h_g(x)$ in $\mathbb{Z}/p_g\mathbb{Z}[x]$, such that at least one of $h_0(x), \ldots, h_g(x)$ is irreducible.

How many $(g+1)$-tuples of polynomials

$$(h_0(x), h_1(x), \ldots, h_g(x)) \text{ in } \mathbb{Z}/2\mathbb{Z}[x] \times \mathbb{Z}/3\mathbb{Z}[x] \times \cdots \times \mathbb{Z}/p_g\mathbb{Z}[x]$$

have the property that none of them is irreducible?

By Lemma 9 above, the number N_n of monic irreducible polynomials of degree n in $\mathbb{Z}/p\mathbb{Z}[x]$ satisfies $N_n^p > p^n/2n$. Thus the number of monic polynomials of degree n in $\mathbb{Z}/p\mathbb{Z}[x]$ that are not irreducible is less than

$$p^n - \frac{p^n}{2n} = p^n \left(1 - \frac{1}{2n}\right).$$

Hence the number of $(g+1)$-tuples of monic degree n polynomials in $\mathbb{Z}/2\mathbb{Z}[x] \times \mathbb{Z}/3\mathbb{Z}[x] \times \cdots \times \mathbb{Z}/p_g\mathbb{Z}[x]$ such that none of the $(g+1)$-polynomials is irreducible, is at most

$$2^n \left(1 - \frac{1}{2n}\right) 3^n \left(1 - \frac{1}{2n}\right) \cdots p_g^n \left(1 - \frac{1}{2n}\right)$$

$$= (2M)^n \left(1 - \frac{1}{2n}\right)^{g+1}.$$

Thus the number of $(g+1)$-tuples of monic degree n polynomials such that at least one of the $g+1$ polynomials is irreducible is at least

$$(2M)^n - (2M)^n \left(1 - \frac{1}{2n}\right)^{g+1} = (2M)^n \left(1 - \left(1 - \frac{1}{2n}\right)^{g+1}\right).$$

But then, since $|P_n(M)| = (2M)^n$, we have

$$\frac{|I_n(M)|}{|P_n(M)|} \geq 1 - \left(1 - \frac{1}{2n}\right)^{g+1}.$$

Letting the number g of primes $p_1, p_2, \ldots, p_g$ increase (recall that $M = M_g = p_1 p_2 \cdot \ldots \cdot p_g$), we have

$$1 \geq \lim_{g \to \infty} \frac{|I_n(M)|}{|P_n(M)|}$$

$$\geq 1 - \lim_{g \to \infty} \left(1 - \frac{1}{2n}\right)^{g+1}.$$

Since the degree n is fixed while g (hence M) goes off to infinity,

$$\lim_{g \to \infty} \left(1 - \frac{1}{2n}\right)^{g+1} = 0.$$

Hence

$$\lim_{g \to \infty} 1 - \left(1 - \frac{1}{2n}\right)^{g+1} = 1.$$

and so

$$\lim_{g \to \infty} \frac{|I_n(M)|}{|P_n(M)|} = 1,$$

as we wished to show.   □

As a numerical example, if we consider monic polynomials of degree 5 and let M be the product of the first 30 odd primes, then among the $(2M)^5$ such polynomials with coefficients a_k satisfying $-M < a_k \leq M$, at least 95.7% of them are irreducible. Here M is slightly larger than 3×10^{52}.

We noted in Section 16C that there are monic irreducible polynomials in $\mathbb{Z}[x]$ that factor modulo every prime. Thus

$$\frac{|I_n(M)|}{|P_n(M)|}$$

is closer to 1 than the estimate of Theorem 2 indicates.

Theorem 2 is a special case of a theorem of Van der Waerden (1934).

Exercises.

18. Let $M = 3 \cdot 5 = 15$ and $n = 2$. Let $\mathscr{S}$ be the set consisting of the $900 = 30^2$ monic polynomials $x^2 + bx + c$ in $\mathbb{Z}[x]$ with coefficients satisfying $-14 \leq b, c \leq 15$. How many polynomials in $\mathscr{S}$ are irreducible? (A polynomial of degree 2 is irreducible if and only if it has no roots, so count the number of polynomials in $\mathscr{S}$ that have a root in $\mathbb{Z}$.)

19. Same question with $n = 3$.

Answers and Hints to the Exercises

Chapter 1

2 a) transitivity fails; b) the three equivalence classes are $3\mathbb{Z}, 1+3\mathbb{Z}, 2+3\mathbb{Z}$; c) reflexivity fails for 0; d) transitivity fails; e) symmetry fails.

Chapter 2

2. Add $(k+1)^3$ to both sides of $P(k)$, then show that

$$\left(\frac{k(k+1)}{2}\right)^2 + (k+q)^3 = \left(\frac{(k+1)(k+2)}{2}\right)^2.$$

5. Write

$$x^{k+1} - 1 = (x^{k+1} - x) + (x - 1) = x(x^k - 1) + (x - 1)$$

and substitute for $x^k - 1$ using $P(k)$

8.
$$(k+1)! = (k+1)k! > (k+1)2^k > 2 \cdot 2^k = 2^{k+1}.$$

9. First show that for $k \geq 4$,

$$(k+1)^4 = k^4 + 4k^3 + 6k^2 + (4k+1) \leq 4k^4.$$

10. For c) use that

$$\frac{1}{(2n+1)^2} = \frac{1}{8t_n + 1} < \frac{1}{8t_n}.$$

11. Use Exercise 5 with $x = a$.
13. $16^n - 16 = 16(16^{n-1} - 1)$: can use Exercise 5 with $x = 16$.
14. Can use Exercise 5 with $x = 8/3$.
15. Can use Exercise 5 with $x = 3^4$.

16. Write $4^{k+2} = 4^k(5^2 - 9)$.

17. Write $2^{2k+3} = 2^{2k+1} \cdot (3+1)$.

20. First observe $a^1 = a$ (why?). Then for each m, prove $P(n) : a^{m+n} = a^m a^n$ for $n \geq 1$ by induction on n.

21. If $N(n)$ is the number of moves needed to move n disks from one pole to another, show $N(n+1) = N(n) + 1 + N(n)$.

22. For 4 disks, the answer is 80.

23. If $r^2 < n \leq (r+1)^2$, then $(r+1)^2 < 2n$?

24. Check the argument for $n = 1$.

26. See the proof of Theorem 6.

27. Adapt the proof of Proposition 4.

28. Divide the polygon into two polygons by an edge connecting two non-adjacent vertices.

29. Let $P(n)$ be "For some $k \geq 1$, $f^{(k)}(n) = 1$," and prove that $P(n)$ is true for all $n \geq 1$.

30. See Section 3A, Example 1.

31. Show that $1 + \frac{5}{3} < (\frac{5}{3})^2$.

32. Try observing that $c(n+1) = 1 + c(1) + c(2) + \ldots + c(n)$.

33. Let $\mathscr{S}$ be the set of numbers $a > 0$ for which there is a number b with $2a^2 = b^2$.

35. Given a non-empty descending chain of natural numbers, let $\mathscr{S}$ be the set of numbers in the chain.

37. Set $x = y = 1$ in the Binomial Theorem.

38. Fix s and do it by induction on $n \geq s$, using Corollary 13.

39. Write $(x+y)^{2n} = (x+y)^n(x+y)^n$, expand $(x+y)^n$ and $(x+y)^{2n}$ by the Binomial Theorem and collect the coefficients of $x^n y^n$.

Chapter 3

1. a) If a is a least element of S, then $a \leq s$ for all s in S; if b is a least element of S, then $b \leq s$ for all s in S, so $b \leq a$. But since b is in S, $a \leq b$.

   b) r is the least element of $\mathscr{S} = \{b - aq \mid q \text{ in } \mathbb{N}\}$. So r is unique.

2. $q = [\frac{b}{a}]$, $\frac{r}{a} = \{\frac{b}{a}\}$.

3. $a = ds, b = dt$ implies $r = b - aq = dt - qds = d(t - qs)$.

4. Use Exercise 3.

5. Let $d = ay + bz$. Divide d into a: if $a = dq + t$, then $a = (ay + bz)q + t$, so t is in J. If $t > 0$, then d is not the least element of J. Then repeat, dividing d into b.

7. $(11111000011)_2 = (3,7,0,3)_8$.

10. 7855 seconds = 2 hrs, 10 mins, 55 seconds.

11. This problem involves distance in base 1760 and time in base 60.

15. With $n = 4$, adding the five products of "digits" $< 10^4$ can exceed 10^8. So 10^n with $n = 3$ is the largest possible base if you do both addition and multiplication on the calculator.

19. a): 5; c): 1 .

20. 17, but see Exercise 33.

22. See Exercise 35.

23. Use (or redo) Exercise 3.

25. If d divides a and b, then d divides $ar + bs$, so.... (See also Corollary 6 of Section D.)

30. Let the m consecutive integers be $a, a+1, \ldots, a+(m-1)$. If $a \geq 0$, then $a = mq + r$, and if $r = 0$, then m divides q. If $r > 0$ then $a + (m-r) \leq a + (m-1)$ and is divisible by m. If $a < 0$ write $-a = mq + s$ with $0 \leq s < m$. If $s = 0$ then m divides a. If $s > 0$, then $a + s < b$ and $a + s$ is divisible by m. (See also Section 6D.)

31. See Section 5B, Proposition 4.

32. Let $m - n = d$. Then $n, n+1, \ldots, n+d$ are d consecutive integers–apply Exercise 30.

35. e) 1.

37. Show that every common divisor of a and b is a divisor of r, hence a common divisor of a and r. Then show that every common divisor of a and r is a common divisor of a and b.

38. Try induction on the number of divisions in Euclid's Algorithm.

40. One example is: 6 divides $2 \cdot 3$.

41. Generalize the examples in Exercise 40.

44. If $ar + ms = 1$ and $bt + mw = 1$, then $1 = (ar + ms)(bt + mw) = aby + mz$ for $y = rt$ and $z = \ldots$.

46. Factor d from $ra + sb$ and use Corollary 6.

48. Let $d = (a, b)$. Then md divides ma and mb, so $md \leq (ma, mb)$. Also, let $d = ar + bs$, then $md = mar + mbs$, so (ma, mb) divides md.

49. If $e = (ab, m)$, then e divides m and $(m, b) = 1$, so $(e, b) = 1$. Since e divides ab, Corollary 7 yields that e divides a. Thus $e \leq (a, m)$. Also, $(a, m) \leq (ab, m) = e$.

51. Use Corollary 6.

52. Can assume r, s are integers with $(r, s) = 1$. Try to show that $s = 1$.

57. b) $x = 19 + 45k, y = -8 - 19k$.

59. a) $x = 13 + 31k, y = 7 + 17k$ with $k \geq 0$
    b) $x = 18 + 31k, y = 10 + 17k$ with $k \geq 0$.

61. $d = 1, r = 731 + 1894k, s = -1440 - 3731k$ for all k in $\mathbb{Z}$.

63. Find integer solutions of $5f = 9c + 160$.

65. Observe that $5 = 16 \cdot 6 - 13 \cdot 7$.

69. Let $b = aq_1 + r_1, a = r_1 q_2 + r_2$ with $r_1 < a$ and $q_2 \geq 1$, and assume $a < F_n$. If $r_1 < F_{n-1}$, then by induction, $N(r, a) \leq n - 4$. If $r_1 \geq F_{n-1}$, then

$$F_{n-1} + F_{n-2} = F_n > a \geq r_1 + r_2 \geq F_{n-1} + r_2,$$

so $r_2 < F_{n-2}$, hence by induction, $N(r_2, r_1) \leq n - 5$. In either case, $N(a, b) \leq n - 3$.

70. a) Do $n = 0, 1$, then check that if the formula is true for $n = k - 2$ and for $n = k - 1$, then it is true for $n = k$.

Chapter 4

2. For the induction step, let $c = a_1 a_2 \cdot \ldots \cdot a_{n-1}$ and $d = a_n$ and apply Lemma 3 to cd.

4. Since $(n,q) \leq (n,q^r)$ in general, it suffices to show that if $(n,q) = 1$ then $(n,q^r) = 1$, or equivalently, if $(n,q^r) > 1$ then $(n,q) > 1$. If $(n,q^r) > 1$, let p be a prime divisor of (n,q^r). Then p divides q^r, so by Exercise 2, p divides q. Since p also divides n, $p \leq (n,q)$ so $(n,q) \neq 1$.

5. Show that n cannot factor into $n = ab$ where both a and b are $> \sqrt{n}$

6. Use Exercise 5 and note that $2021 < 45^2$.

8. a) m in $3\mathbb{N}$ is irreducible iff $n = 3q$ with q not a multiple of 3.

   c) $30 \cdot 3 = 6 \cdot 15$.

9 a) If $a + b\sqrt{-23} = c + d\sqrt{-23}$ with $b \neq d$, then $\sqrt{-23} = \frac{a-c}{d-b}$ would be a rational number. But then the negative number -23 would be the square of a rational number, impossible.

   d) If $a + b\sqrt{-23}$ divides 3, then there are integers c and d so that

$$(a + b\sqrt{-23})(c + d\sqrt{-23}) = 3.$$

This is true iff $ac + 23bd = 3$ and $ad + bc = 0$ by part a), and this in turn is true iff

$$(a - b\sqrt{-23})(c - d\sqrt{-23}) = 3.$$

Multiplying the two equations together yields

$$(a^2 + 23b^2)(c^2 + 23d^2) = 9.$$

The only solutions of this equation must have $b = d = 0$.

13. If $100^{1/5} = a/b$, then $b^5 \cdot 2^2 \cdot 5^2 = a^5$. If $2^e \| a$ and $2^f \| b$, then $5e + 2 = 5f$, impossible.

14 b) "integers" can be negative.

15. First show that for each prime p dividing c, if $p^e \| c$ then $p^{er} \| c^r$. If $p^g \| a$, then p does not divide b, so $g = er$. So if $a > 0$, then a is an r-th power. Similarly for b. Finally, if $a = -a_0$ where $a_0 > 0$ and $a_0 = d^r$, then $a = (-d)^r$ because r is odd.

17. Be sure to check your answer!

19. Write $a = 2^r f, b = 2^s g$ where f, g are odd and coprime. Then check the various possibilities for $\min\{4r, 5s\}$ given that $\min\{r, s\} = 3$.

21. Use Lemma 3 or Chapter 3, Corollary 6.

27. Write $ar + bs = c$ and review the proof of Chapter 3, Corollary 8.

28. Let q be the product of all primes p dividing c such that $(p,a) = 1$. Show that $(a + bq, c) = 1$.

35. See Section 3E.

37. See Exercise 17.

39. Observe that if a divides bk then bk is a common multiple of a and b.

42. Look for examples where $[a,b,c] < abc/(a,b,c)$.

43. Note that if $p_1 = 4k_1 + 1, p_2 = 4k_2 + 1$, then $p_1 p_2 = 4l + 1$ for some l.

44. Every odd prime has the form $p = 6k + 1$ or $p = 6k - 1$.

50. Show that every prime that divides $n! + 1$ must be $> n$.

Chapter 5

1. For $a > 0$ use Proposition 1. For $a < 0$ show there is a smallest $k > 0$ so that $r = a + km \geq 0$, then show that $0 \leq r < m$.

4. New Years Day 2007 was 365 days after New Years Day 2006. Find 365 (mod 7).

8. If $a \equiv r \bmod m$ with $0 \leq r \leq m - 1$, then $a, a + 1, \ldots, a + (m - 1)$ are congruent modulo m to $r, r + 1, \ldots, m - 1, 0, 1, 2, \ldots, r - 1$, respectively. Then use Exercise 1.

12. No. Try $a = 15$.

13. Write $(15 - c)a = 25k$, then find all a so that if 25 divides $(15 - c)a$, then 25 divides $15 - c$.

14. Use that $4, 4^2, 4^3 \equiv 4, 7, 1 \pmod 9$, respectively.

15. $7^{546} \equiv -2 \pmod{17}$.

16 b) Use $68 \equiv -2 \pmod 7$. d) Use $66 \equiv 9 \pmod{19}$.

18. Use that $6 \equiv -5 \pmod{11}$.

19. a) Use $(a + b)^2 = a^2 + ab + ab + b^2$. b) Use induction on n.

20. Do it for $a = 0$ and 1 first.

21. If $b(b - 1) \equiv 0 \pmod{15}$, then 3 divides b or $b - 1$ and 5 divides b or $b - 1$.

22. First show that for $a > 0$, $a^6 \equiv 1 \pmod 7$.

23. Write the number as $100a + 10a_1 + a_0$.

24. Use that $6 = 2 \cdot 3$.

26. a) $10a + b \equiv 0 \pmod 7$ iff $-2(10a + b) \equiv 0 \pmod 7$ iff $a - 2b \equiv 0 \pmod 7$

29. One idea: 13 divides $10a + b$ iff 13 divides $4(10a + b) \equiv a + 4b$.

31. Yes: $12 \equiv 1 \pmod{11}$.

32. 2 and 7 are like Proposition 11; 3 and 11 are like Proposition 9; 5 and 7 are like Proposition 12.

33. One strategy: show that the divisors of m and y are the same as the divisors of m and x.

35. α should $= 8$ (Hungary) or 14 (Norway), the website, YAHOO, and the fruit, ORANGE.

36. a) change 3 to 4.

38. Observe that $n - p(n) \neq m - p(m)$ unless $n = m$ or $n = 0, m = 9$.

39. Note that $561 = 3 \cdot 11 \cdot 17$ and follow Example 8.

41. $6a \equiv 16 \pmod{20}$ iff $3a \equiv 8 \pmod{10}$ iff $a \equiv 56 \pmod{10}$.

43. For the proof, apply first Proposition 16, then Proposition 17.

45. There is no solution to d). For b), the smallest positive solution is $x = 9$.

47. Cancel 3.

49. Note that $7 \cdot 31 = 217 \equiv -1 \pmod{218}$.

51. Show that for all x, $ax^2 + bx + c \equiv 0 \pmod{m}$ iff $(2ax + b)^2 \equiv b^2 - 4ac$ $\pmod{m}$.

52. If $a = b = c$ the congruence holds, so assume $a \leq b \leq c$ and $a < c$.

53. For one direction, observe that squares are congruent to 0 or 1 modulo 4. For the other, experiment to look for patterns, first with odd numbers, then with multiples of 4.

54. Look at squares modulo 16.

Chapter 6

2. b) (This includes a 24 hour layover in Chicago.) c) (This includes a 7 hour layover in Chicago, a five hour layover in Buffalo and a 13 hour layover in Toronto.) d) (This includes a four hour layover in Winnipeg.)

4. It is the same as to find $((-10)(-4)) \bmod 23$, namely 17.

8. b) To solve $[5]_7[x]_7 = [3]_7$. observe that $[3]_7 = [10]_7$.

9. Multiply both sides by $[4]_{11}$.

10. Multiply both sides by the inverse of 11_{13}.

12. 1 plays 3, 4 plays 9, 5 plays 8, 6 plays 7, 2 plays Bye.

14. Bye plays player a in round r if $a + a \equiv r \pmod 9$. Thus in round 4, Bye plays 2.

16. If we ignore cases where the addition table modulo $2n$ has a player playing herself, then the table will describe a tournament just like the tournament in the text, except that it takes $2n$ rounds and during the even-numbered rounds, two opponents sit out.

17. a) and b) can be done by using Exercises 36 and 37. To do d) and e) is the same as to decide if -3 and -2 are primitive roots modulo 7, respectively.

18. Any set of nine consecutive integers will do.

20. Try casting out nines.

21. It's true for k satisfying $(k, 10) = 1$.

25. Try -3.

31. A special case of Exercise 30.

34. Se Chapter 5, Exercise 8.

35. Use Exercise 34.

36. Use Proposition 4 b).

37. Use Proposition 4 b).

38. Note that $-b \equiv a \pmod{a+b}$.

41. This is a special case of the "Problem of Frobenius".

42. a) The inverse in $\mathbb{Z}/13\mathbb{Z}$ of $[4]$ is $[10]$. b) i): $X = [70] = [5]$.

45. In $\mathbb{Z}/14\mathbb{Z}$ the inverses of $[1], [3], [5], [9], [11], [13]$ are $[1], [5], [3], [11], [9],$ $[13]$, resp.

47. $a' = a + mk$, so $(a, m) = (a', m)$.

48. If $(a, m) = 1$ and $(a, n) = 1$ then $(a, mn) = 1$.

51. c) Observe that $2^5 = 32$.

52. If $\{0, b, b^2, \ldots, b^{p-1}\}$ is a complete set of representatives modulo p, then b is a unit, so p is prime and $\{b, b^2, \ldots, b^{p-1}\}$ represent all the $p-1$ units of $\mathbb{Z}/p\mathbb{Z}$.

53. a) If $[ba] = [ba']$, multiply by the inverse of $[b]$.

   b) Show that ba_i is a unit modulo m for all i, and then use a) and Proposition 4.

54. If $[b]$ is a unit of $\mathbb{Z}/m\mathbb{Z}$ with inverse $[c]$, then f_c is the inverse function to f_b. If $[b]$ is not a unit, then $(b, m) = d > 1$, so $b \cdot (m/d) \equiv 0 \pmod{m}$. Then $f_b([\frac{m}{d}]) = f_b([0])$ and f_b is not one-to-one.

55. Every unit mod m is a unit mod d. Conversely, suppose $[b]_d$ is a unit of $\mathbb{Z}/d\mathbb{Z}$. Then $(b + dk, m) = 1$ for some k (use Chapter 4, Exercise 28 or Chapter 6, Exercise 40), so $[b + dk]_m$ is a unit of $\mathbb{Z}/m\mathbb{Z}$ and $[b + dk]_d = [b]_d$.

56. $X = [4 + 5k]_{20}$ for $k = 0, 1, 2, 3$.

58. $(36, 45) = 9$.

60. b) The inverse of $[8]$ in $\mathbb{Z}/11\mathbb{Z}$ is $[7]$.

62. b) $9x \equiv 48 \pmod{30}$ iff $3x \equiv 16 \pmod{10}$ iff $x \equiv 2 \pmod{10}$; the general solution is $X = [2 + 10k]_{30}$, $k = 0, 1, 2$.

64. One way to do this is to count the number of multiples of 2 that are ≤ 210, then the multiples of 3 that are ≤ 210, then adjust the sum because you counted the common multiples of 2 and 3 twice, etc.

65. Any number a not a unit modulo 30 must satisfy $(a, 30) > 1$, hence must be a multiple of 2, 3 or 5.

69. There are seven primes between 200 and 250.

70. There are 14 primes between 700 and 800.

71. The number is < 120130.

Chapter 7

1. Multiply $1 + (-1) = 0$ by (-1) and use Proposition 4.

2. Show that a is the inverse of $-a$.

3. Add $-a$ to both sides of the equation.

5. Multiply $ax = d$ by a^{-1} to find a solution, then use Exercise 4 to show uniqueness.

6. Multiply $a + (-a) = 0$ by b and use Proposition 4.

7. Start with $1 + 0 = 1$.

8. Multiply on the left by the inverse of a.

9. See the suggestion for Exercise 8.

10. Use Exercises 8 and 9.

11. Use Exercise 8.

12. Neither have negatives.

13. (ii) Try distributivity.

14. (i) No. (ii) Yes.

16. $[(m+1)/2]_m$.

17. There are 12 units.

21. $\mathbb{Z}/11\mathbb{Z}$ has none, $\mathbb{Z}/12\mathbb{Z}$ has seven.

24. If x_0 and x_1 are solutions of $ax = b$, then $t = x_1 - x_0$ is a solution of $ax = 0$.

26. $X = [2+3k]_{18}$ for $k = 0, \ldots, 5$.

27. One way: look at (a, m).

28. The zero divisors are congruence classes of 3, 5, 6, 9, 10 and 12. The complementary zero divisors of 12 are 5 and 10.

31. Note that $(22, 26) > 1$.

32. For $[113]$, do Euclid's Algorithm on 365 and 113.

33. Note that $0 = z^2 - 1 = (z+1)(z-1)$.

34. 6.

37. The inverse of $a + bi$ is $\frac{a}{a^2+b^2} - \frac{b}{a^2+b^2}i$.

38. Use that $f(a)f(a^{-1}) = f(aa^{-1}) = f(1) = 1$.

39. If $f : \mathbb{Q} \to \mathbb{Q}$, then $f(\frac{1}{1}) = \frac{1}{1}$, so $f(\frac{n}{1}) = \frac{n}{1}$ for all n in $\mathbb{Z}$. Then by Exercise 38, $f(\frac{1}{n}) = \frac{1}{f(n)} = \frac{1}{n}$, hence for integers a, b, $f(\frac{a}{b}) = f(a)f(\frac{1}{b}) = a \cdot \frac{1}{b} = \frac{a}{b}$.

41. By Exercise 38, $f(U_m) \subset U_d$. If $[a]_d$ is a unit of $\mathbb{Z}/d\mathbb{Z}$, then there is some unit $[a']_m$ of $\mathbb{Z}/m\mathbb{Z}$ so that $[a']_d = [a]_d$ by Chapter 6, Exercise 55.

43. $a + a + \ldots + a = a(1 + 1 + \ldots + 1)$.

44. (i) $a + a = a(1 + 1) = 0$.

(ii) $(a+b)^2 = a^2 + ab + ab + b^2 = a^2 + ab(1+1) + b^2$.

45. Use the additive analogue of the proof of Theorem 14 to show that $0 = 1 + 1 + \ldots + 1$ (n summands) for some $n > 0$, so if $f : \mathbb{Z} \to F$ is defined by $f(1) = 1$, then $\ker(f) \neq 0$. Then use Proposition 20.

Chapter 8

5. See an example in Section A.

6. See Exercise 5.

7. Try to solve the equation $\mathbf{AX} = \mathbf{0}$. There will be a non-zero solution iff A doesn't have an inverse.

8. There are six units.

9. The first received vector has two errors; the third vector has a single error in the last component.

10. The probability of at most two errors is

$$(1 - .001)^8 + \binom{8}{1}(1 - .001)^7(.001) + \binom{8}{2}(1 - .001)^6(.001)^2.$$

12. The sum of three columns of $\mathbf{H}$ is a column of $\mathbf{H}$.

14. The $\mathbf{G}$ for Code II is the $\mathbf{G}$ for Code I with an additional row $(1\ 1\ 1\ 0)$ corresponding to the equation $w + a + b + c = 0$.

15. A coded word is a sum of columns of the matrix $\mathbf{G}$ of Exercise 14. Some observations: i) there are 16 coded words. ii) The difference of two coded words is the same as the sum of two coded words, hence is a sum of columns of $\mathbf{G}$. Thus the problem comes down to counting the number of 1's in each of the 15 non-zero coded words. iii) Since the sum of all four columns is a column of 1's, to count the

number of 1's in a sum of three columns is the same as to count the number of 0's in a single column.

17. The encrypted message starts with UKDL....

18. The plaintext message starts with CONG....

Chapter 9

1. $[4]$ has order 2, $[2]$ and $[3]$ have order 4, $[1]$ has order 1.

5. $[6]_{13}$ has order 12.

8. Use that if $a^t - 1$ is a common multiple of r and s, then $a^t - 1$ is a multiple of m.

10. Use Exercise 9.

11. Use Proposition 3.

12. If $e/(e,f) = 1$ then e divides f?

14. b): 1; c), 5.

15. Use Proposition 2.

21. The order of 7 modulo 167 must be a divisor of 166.

24. This is easier once you find a primitive root modulo 19 (Section 6D).

25. 8.

26. Consider two cases, p divides a and $(p,a) = 1$.

27. Use Fermat's Theorem.

30. Use Fermat's Theorem.

32. Show it modulo 3, 11 and 17.

33. Try showing that the only units of $\mathbb{Z}/p\mathbb{Z}$ that are their own inverses are $[1]$ and $[-1]$, hence that $[2] \cdot [3] \cdots [p-2] = [1]$.

35. Use Proposition 5.

37. A repunit has the form $u_n = \frac{10^n - 1}{9}$. Use Fermat's Theorem.

38. Generalize Exercise 37.

39. a) Use Proposition 2. b) Use Proposition 5.

42. $\phi(21) = \phi(7)\phi(3) = 12$.

44. Use that $40 \equiv -2 \pmod{21}$.

46. Observe that $7^3 \equiv -1 \pmod{43}$.

47. Use Proposition 2, section A.

51. b) Count the number of multiples of p in the set $\{1, 2, \ldots, p^n\}$.

55. $4 < 8$.

56. Use Proposition 7 b).

60. Use Exercise 59.

62. What is the order of 2 modulo 341? (Note that $341 = 11 \cdot 31$.)

63. The order of 10 modulo 17 must be a divisor of 16.

65. Observe that $10^3 \equiv -3^3 \equiv -1 \pmod{13}$. Or observe that $10^3 \equiv -1 \pmod{1001}$.

66. Find all primes p so that p divides $10^3 - 1$ but not $10 - 1$.

68. Use Exercise 67.

70. If a prime p divides t, then $p-1$ divides $\phi(t)$. So if $\phi(t)$ divides 100 and a prime p divides t, then $p-1$ divides 100. So write down the divisors of 100, then identify which primes p have the property that $p-1$ is a divisor of 100. Any possible t then must be a product of some of those primes....

71. The case $(a_1+a_2)^p \equiv a_1^p + a_2^p \pmod{p}$ is Proposition 12. Use that to show the analogous result for $(a_1+a_2+\ldots+a_n)^p$ by induction on n.

73. $a \bmod 2$ and $b \bmod 2$ are 0 or 1. So there are three cases.

76. $69 = 64 + 4 + 1$.

80. $1728 = 1024 + 512 + 128 + 64$.

82. Use the following property of primes: if p is prime, then $x^2 \equiv 1 \pmod{p}$ has only two solutions, because if p divides $x^2 - 1 = (x-1)(x+1)$, then p must divide $x-1$ or $x+1$.

85. Here, $m = 267, r = 1000$. For precomputations, solve $267m' + 1 = 1000r'$ for m' and find $w = 1000^2 \bmod 267$.

86. Here, $m = m' = 7, w = 2, r = 10$.

Chapter 10

1. With $m = 101$, the words 02, 21, 25 are encrypted as 11, 22, 5.

3. The words 8, 15, 12, 4 are encrypted as 27, 10, 12, 9.

5. $1415^{11} \equiv 2551 \pmod{3337}$.

7. Use that $\phi(m) = \phi(pq) = m - p - q + 1$.

10. Use that $3^3 = 27 \equiv 1 \pmod{13}$ and $3^6 \equiv 1 \pmod{7}$, the latter by Fermat's Theorem.

13. The divisors of $m = 2^{p-1}(2^p - 1)$ are 2^r and $2^r(2^p - 1)$ for $r = 0, 1, 2, \ldots, p-1$. Add these up (omitting m).

14. If $2^{n-1} - 1 = nq$, then

$$2^{2^n-1)-1} - 1 = 2^{2^n-2} - 1 = (2^n)^2 q - 1,$$

which has $2^n - 1$ as a factor.

15. Use Proposition 3.

18. Use that $a^{2r+1} + 1$ has $a + 1$ as a factor for every r. (Proving this is a routine induction on r.)

19. Use that $2^{(2^n)} \equiv -1 \pmod{2^{2^n} - 1}$ and that 2^{2^n} is a multiple of 2^{n+1}.

20. This is done in Section 19C.

21. Note that 2 has order q modulo p. If $2^p \equiv 1 \pmod{p^2}$, then $2^p \equiv 1 \pmod{p}$, so q divides p. Since p and q are both primes, $p = q$. But that is impossible since $p = 2^q - 1$.

23 ii). Try $m = 35$.

24. This example is discussed in Section 11D.

25. (iii) If a, b are integers with $ab \equiv -1 \pmod{3}$, then either $a \equiv 1, b \equiv -1 \bmod 3$ or $a \equiv -1, b \equiv 1 \bmod 3$. Hence $a + b \equiv 0 \bmod 3$.

29. Let $N = 13 \cdot 7 = 91$. Then 13 and 7 are both 3-smooth. Let $k = 2$ and $a = 27$. Then $B = 2^6 = 64$, and

$$27^{64} \equiv (27^2)^{32} \equiv 1^{32} \equiv 1 \pmod{91}.$$

30. i) Observe that

$$1 \equiv 5^{32s} \equiv (5^{16s})^2 \equiv (6498)^2 \pmod{7081}.$$

  ii) $(6498 - 1, 7081) < 7081$ because 7081 doesn't divide $6498 - 1$, and $(6498 - 1, 7081) > 1$ because 7081 doesn't divide $6498 + 1$.

31. The approach of Exercise 30 is essentially the strong pseudoprime test of Section 20B.

Chapter 11

1. For solvability: to solve $ax = b$, multiply both sides by a^{-1} (on the left).

2. To find a left identity for G, pick some a in G and let $x = e$ solve $xa = a$. To show that $eb = b$ for every b in G, find some c so that $ac = b$, then $ea = a$ implies $eac = ac$, so $eb = b$. Solving $xb = e$ then shows that every element b of G has a left inverse.

4. Let u be some product of $a_1, \ldots, a_n$. Somewhere in u is an expression $(a_r a_{r+1})$. Let $b = a_r a_{r+1}$. Then u is a product of $n - 1$ elements of G. Then by induction,

$$u = a_1(\ldots(b(\ldots(a_{n-1}a_n)\ldots)) = uv$$

or, if $r = 1$,

$$u = b(\ldots(a_{n-1}a_n)\ldots) = bv$$

If $r > 1$ then replacing b by $a_r a_{r+1}$ in v and applying the induction assumption to v completes the proof. If $r = 1$,

$$u = (a_1 a_2)v = a_1(a_2 v)$$

so applying induction to $a_2 v$ completes the proof.

5. Call the product u. If $u = a_1 v$, apply induction to v. If $u = ca_1 b$, use commutativity to get $u = a_1 cb$ and apply induction to cb.

8. i) If $[[a, b], c] = m$ and $[a, [b, c]] = n$, show that m divides n and n divides m.

  iii) Only one element of $\mathbb{N}$ has an inverse.

  iv) Only when a divides b.

  v) No.

9. See Proposition 2, Chapter 9.

11. ii) Having done i), note that $[12] = [-7]$.

12. i) $\langle [10]_{21} \rangle$ has six elements.

  iii) Note that $[8] = [2^3]$.

14. H contains 1 if and only if $H = G$. If $(b, n) = 1$ then there is some positive integer r so that $br \equiv 1 \pmod{n}$, hence $b + b + \ldots b(r \text{ summands}) = 1$. If $(b, n) > 1$ then H doesn't contain 1.

15. See Exercise 11.

16. No. Which elements are in U_{21}.

17. For equivalence relations, see Chapter 1.

20. i) Use that $(\phi(m), m - 1) = (6, 8) = 2$.

   iv) By Proposition 11 it suffices to find $U_{75}(2)$. There are four classes $[a]_{75}$ whose square is $[1]$.

22. Use Proposition 11.

24. Show that $(n - 1, \phi(n)) = 1$.

25. Show that $(m - 1, \phi(m)) = 2$.

26. 36.

27. i) The cosets of $1, 6, -1, -6$.

   ii) The cosets of $9, -9, 19, -19$.

29. i) It may help to show that 3 is a primitive root modulo 31.

33. i) $\ker(L_{[2]}) = \{[a]_{10}|[2a] = [0]\} = \{[0], [5]\}$.

34. Use Proposition 11.

37. Need to check that f^{-1} sends the identity to the identity and products to products. For products, let $f(a) = a', f(b) = b'$. Since $f(a * b) = f(a) * f(b)$,

$$f^{-1}(a' * b') = f^{-1}(f(a) * f(b)) = f^{-1}(f(a * b)) = a * b = f^{-1}(a') * f^{-1}(b').$$

40. The four elements of G/H are $[2]H, [11]H, [22]H$ and $[1]H$. All have order 2.

41. $[28]H$ has order 4.

43. If b is a primitive root, then $b^r \equiv 1$ or $-1 \pmod{p}$ if and only if $(p - 1)/2$ divides r.

45. If $N = \{[1], [-1]\}$, the cosets are $N, [2]N, [3]N$.

47. iii) The story is that Gauss's 2nd grade teacher, for busywork, asked the students to add up the numbers from 1 to 100. Gauss wrote the numbers out twice:

$$1, 2, 3, \ldots, 98, 99, 100$$
$$100, 99, 98 \ldots, 3, 2, 1$$

and observed that twice the sum is 100 times 101.

49. Use the function L of the proof of Cayley's Theorem (Theorem 18).

51. 48 elements.

Chapter 12

1. $x = 73$.

2. None.

6. $x \equiv 47 \pmod{120}$.

9. $x \equiv -334 \pmod{2100}$.

10. The system is solvable iff there is an integer solution of $1 = mx + nqy$.

12. There are 136 unwounded survivors.

13. There are 46 participants.

17. Each vessel holds 3193 ko, and the total amount of rice stolen was 9563 ko.

21. $x \equiv 31 \pmod{100}$.

26. This system reduces to $x \equiv 10 \pmod{17}, x \equiv 5 \pmod{22}$.

29. Let $(m,n) = d$. For each a_1 with $0 \le a_1 < m_1$, for a solution we must have $a_2 = a_1 + kd$ for some integer k. Modulo m_2 there are m_2/d choices for a_2.

32. Use Exercise 31.

33. (i) The solutions to the four systems of congruences: $x \equiv \pm 1 \pmod{r}, x \equiv \pm 1$ (mod s) all satisfy $x^2 \equiv 1 \pmod{m}$.

34 (i) $e_1 = 400, e_2 = 126$; b) $x = 192$.

38. If we use $1,113,840 = 5 \cdot 7 \cdot 9 \cdot 13 \cdot 16 \cdot 17$, then the table needs 31 entries.

39. (i) If $p = 2q + 1$ with p, q primes >10, then $q \not\equiv 0 \pmod 3$ because q is prime, and $q \not\equiv 1 \pmod 3$ because p is prime.

41. $p \equiv 11, 23, 47, 59$, or $83 \pmod{84}$.

42. For p_2, consider separately $l \equiv 1 \pmod 4$ and $l \equiv 3 \pmod 4$.

43. $95^{1002} \equiv 4 \pmod{31}$ and $\equiv 1 \pmod 7$, so $\equiv 190 \pmod{217}$.

45. $21^{235} \equiv 30 \pmod{391}$.

48. For example, 7 maps to $(1,3)$ and 10 to $(1,2)$.

50. $(1,0) \cdot (0,1) = (0,0)$.

52. The kernel is $\{[0], [12]\}$. The image has 12 pairs, all pairs (b,c) where $b \equiv c$ (mod 2). See Exercise 28.

53. Exactly one solution.

54. (i) $x_0 = 4$.

55. $30030 = 2 \cdot 3 \cdot 5 \cdot 7 \cdot 11 \cdot 13$, so $\lambda(30030) = [1,2,4,6,10,12] = 60$.

56. Write $(t, p-1) = tr + (p-1)s$, then $a^{(t,p-1)} = (a^t)^r \cdot (a^{p-1})^s \equiv 1 \pmod p$.

57. (i) Use that $m - 1 = pq - 1 = pq - p + p - 1 = p(q-1) + (p-1)$.

   (iii) If $a^{m-1} \equiv 1 \pmod p$, then $a^{p(q-1)}a^{p-1} \equiv 1 \pmod p$, so $a^{q-1} \equiv 1 \pmod p$. Hence $a^{(p-1,q-1)} \equiv 1 \pmod p$. Similarly modulo q.

58. To show that $U_m(2)$ has four elements, observe that $a^2 \equiv 1 \pmod m$ iff $(a+1) \cdot (a-1) \equiv 0 \pmod p$ and also modulo q.

59. Use Exercise 57.

61. Use Exercise 57.

62. To show that $U_{65}(4)$ has order 16, show that $x^4 \equiv 1 \pmod{13}$ has exactly four solutions, and the same is true modulo 5.

Chapter 13

1. For example, x^p and x.

3. Apply Proposition 1.

4. Choose two polynomials whose leading coefficients are complementary zero divisors.

6 (iii) $2R[x]$

(iv) $f(x) = x$ is an example.

9. A generalization is $(x^n + x^{n-1} + \ldots + x + 1)(x-1) = x^{n+1} - 1$.

10. $R[x]$ is an infinite set, while $Func(R,R)$ is a subset of $R \times R$ (think of a function defined as a set of ordered pairs), which is a finite set if R is.

11. Let $h(x) = (x^9 + 8x^3 + x) - (x^6 + 2x^2 + x)$. If $h(x)$ is the zero function, then $h(1) = 0$. But $h(1) = 6$, so p can only be 2 or 3. To check 3, look at $h(2)$ modulo 3.

Chapter 14

1. (i) -7 ; (iv) 0 ; (v) $3x^2 - x - 1$.

2. No.

3. Try dividing $2x^2 + 1$ into $6x^2 = 14x^2$.

4. (i) Use the Remainder Theorem.

(iii) Let $y - x^4$ and use the Remainder Theorem in $\mathbb{Q}[y]$.

5. Use the Remainder Theorem, and in part (ii) see for which m the remainder is 0 modulo m.

7. How do you know that the remainder when you divide $x^2 + 2x + 5$ by $x^2 - 3$ is the same as when you divide $f(x)$ by $x^2 - 3$?

10 (i) $x^2 - 2x \equiv 0 \pmod{15}$ iff $x^2 - 2x \equiv 0 \pmod 5$ and $\pmod 3$, iff $x \equiv 0$ or $2 \pmod 5$, and also $\pmod 3$. So there are four roots of $x^2 - 2x$ modulo 15.

11. See Exercise 10.

14. $d = 2$.

17. Try induction on the length of Euclid's Algorithm for f and g. Observe that if Euclid's Algorithm for g and f starts with $g = fq_1 + r_1$, then the rest of Euclid's algorithm for g and f is Euclid's Algorithm for f and r_1.

18. See the suggestion for Exercise 17.

21. The greatest common divisor is $x^2 - 1$.

22. $1 = x^3 + (x^2 + x + 1)(x + 1)$.

25. (i) $x^2 + 1$. (ii) $(x^5 + 2x^3 + x^2 + x + 1)(x^4 + 2x^3 + x + 1)/(x^2 + 1)$.

27. Use that if $n = mq + r$, then

$$x^n - 1 = x^{mq+r} - x^r + x^r - 1 = x^r(x^{mq} - 1) + (x^r - 1).$$

29. $k(x) = \frac{x^2 - 1}{(x^3 - 1, x^2 - 1)}$.

32. Use Bezout's Identity on (f, g) and (f, h).

33. If $d = tf + wg$, write $t = gq + r$ with $\deg r < \deg g$. Then $d = rf + (qf + w)g$ and $s = qf + w$ must have degree $< \deg f$.

35. If p doesn't divide f, then for some polynomials r, s, $prt + fs = 1$. Multiply by g and observe that then p divides the left side.

38. If p and q are monic, then the leading coefficients of p and aq are 1 and a, respectively. So $a = 1$.

40. (i) Use the Intermediate Value Theorem from calculus.

(ii) and (iii): See Section 15G.

(iv) Prove this by induction on the degree of $f(x)$, using the Root Theorem.

42. Is $x^2 - 3x - 4$ irreducible?

45. Use Proposition 1 of Chapter 13.

48. (i). Try dividing by irreducible polynomials of degree 3.

(iii) Try dividing by irreducible polynomials of degree 4.

49. Use D'Alembert's Theorem.

50. (ii) Try $x - a$ dividing $x(x - (a+b))$.

Chapter 15

1. Let

$$l(x) = \frac{f(x)}{g(x)} - \frac{h(x)}{k(x)} = \frac{m(x)}{g(x)k(x)}$$

where $m(x) = k(x)f(x) - h(x)g(x)$, and assume $l(a) = 0$ for infinitely many elements a of F. Then $m(a) = 0$ for infinitely many elements a of F, so $m(x) = 0$. Hence

$$\frac{f(x)}{g(x)} = \frac{h(x)}{k(x)}.$$

3. $x^3 = (x+1)^3 = 3(x+1)^2 + 3(x+1) - 1$.

5. (i)

$$\frac{t+1}{(t-1)(t+2)} = \frac{2/3}{t-1} + \frac{1/3}{t+2}.$$

(ii):

$$\frac{1}{(t+1)(t^2+2)} = \frac{1/3}{t+1} + \frac{-t/3 + 1/3}{t^2+2}.$$

6. One solution is $\frac{17}{180} = \frac{-5}{9} + \frac{1}{4} + \frac{2}{5}$.

7. Let $p(x) = x - x_0$ and let

$$f(x) = a_0 + a_1 p(x) + a_2 p(x)^2 + \ldots + a_n p(x)^n$$

be the expansion of $f(x)$ in base $p(x)$. By repeatedly differentiating $f(x)$, show that $a_0 = f(x_0)$, $a_1 = f'(x_0)$, $a_2 = \frac{f''(x_0)}{2!}, \ldots, a_n = \frac{f^{(n)}(x_0)}{n!}$, and hence $f(x) = T_f(x)$.

8. (ii) $(1 - i)(\frac{1}{2} + \frac{5}{2}i) = 3 + 2i$.

9. (iii) $\frac{1}{37} - \frac{6}{37}i$.

10. (ii) $4\sqrt{2}e^{-i\pi/4}$.

11. (ii) $|\alpha| = \sqrt{c}$.

13. (iv) Use $x^3 + 1 = (x^6 - 1)(x^3 - 1)$.

17. If j is the complex conjugation map from $\mathbb{C}[x]$ to $\mathbb{C}[x]$ defined by: $j(f(x))$ is the polynomial $\overline{f}(x)$ whose coefficients are the complex conjugates of those of $f(x)$, then j is a ring homomorphism, and, if $f(x)$ has real coefficients, then

$j(f(x)) = f(x)$. If $\alpha = r + is$ is a root of $f(x)$, use the Root Theorem, then apply the homomorphism j.

18. Use Exercise 17.

19. (ii) Write $-2 = 2e^{i\pi} = 2e^{i(\pi+2k\pi)}$, then $x = 2^{1/4}e^{i(\pi+2k\pi)/4}$ for $k = 0,1,2,3$.

20. (i) See the discussion about Exercise 17, then apply complex conjugation to $g(x)$.

   (ii) $\mathbb{C}$ has no zero divisors.

21. Use Exercise 20.

22. $t = \sqrt{7} + 1$.

24. $x = (\frac{5}{2} + \frac{\sqrt{29}}{2})^{1/3} + (\frac{5}{2} - \frac{\sqrt{29}}{2})^{1/3}$.

26. $x = 4, 1 + \sqrt{3}, 1 - \sqrt{3}$.

28. Choose $|a_0| < 1$.

30. $p(z) = (z^3 + 2z^2 + 4z) + i(z^3 - z^2 + 2)$, so

$$|p(z)|^2 = 2z^6 + 2z^5 + 13z^4 + 20z^3 + 12z^2 + 4.$$

31. Apply Proposition 6.

33. Write $f(x)^e = f(x)^{e-1}f(x)$ and use induction on e.

34. $x - 1; x^3 - 2x^2 - x + 2$.

36. (f, f') is the same in $\mathbb{C}[x]$ or in $\mathbb{Q}[x]$.

37. $f' = 1$.

39. (i) $x + 1$ is a multiple factor; (ii) $x^2 + x + 1$ is a multiple factor.

42. $f'(x) \neq 0$, so the proof of Theorem 8 for characteristic zero applies.

44. $(x + 1)(x^2 + x + 1)^2$.

46. There is a solution in the "Hints" in the Second Edition.

Chapter 16

1. (i) Multiply $f(x)$ by 18.

2. The polynomial factors into the product of two polynomials of degree 2 in $\mathbb{Z}[x]$.

3. (ii) Let g_1, h_1 be primitive associates of g, h. Then $g_1 h_1 = af$ for some rational number a. Use Proposition 1 and Lemma 2 to show that $a = 1$ or -1. So the leading coefficients of g_1 and h_1 are 1 or -1. Since f is monic we can multiply g_1 or h_1 by -1 as needed to make $a = 1$ and g_1, h_1 monic.

4. Reduce modulo p for every prime p.

6. (i) Let $g_1 = rg, h_1 = sh$ with g_1, h_1 primitive. Then r, s are in $\mathbb{Z}$ and $rsf = g_1 h_1$. Since f is monic and $g_1 h_1$ is primitive, $rs \not\equiv 0 \pmod{p}$ for every prime p. So $rs = 1$ or -1, hence g, h are in $\mathbb{Z}[x]$.

8. Use Exercise 6 (ii).

9. (ii) $x = -1$ is the only rational root.

10. (i) Any root is $x = r/s$ where r divides 2 and s divides 6. One root is $x = -2/3$.

11. Let $x = y - 7$, then $f(x) = f(y-7) = g(y)$ and $f(-7) = g(0) = -1$. So the only possible roots of $g(x)$ are $y = 1$ or -1, and so the only possible roots of $f(x)$ are $x = 1 - 7$ or $-1 - 7$.

12. The polynomial of Exercise 9 (ii) becomes $y^3 + y^2 - 30y - 72$, which has roots 6, -4 and -3.

13. (i) This follows from the observation that if c is a divisor of ab then c factors uniquely as $c = de$ where d is a factor of a and e is a factor of b.

   (iv) Note that $\frac{d}{-e} = \frac{-d}{e}$, so you can assume the denominator of any root is positive.

17. If $f(x) = 2x^4 - 8x^2 + 3$, then $g(y) = y^4 f(1/y) = 3y^4 - 8y^2 + 2$, which is irreducible by Eisenstein. Show that any factorization of $f(x)$ yields a factorization of $g(y)$ and conversely.

20. Suppose $f(x) = g(x)h(x)$ in $\mathbb{Z}[x]$. We can assume that g, h are monic. Then this factorization would hold modulo 2 and modulo 3. But modulo 3, g and h must have degree 2, hence also in $\mathbb{Z}[x]$, hence also modulo 2. But modulo 2, f is not divisible by a polynomial of degree 2.

21. Use Corollary 12.

24. Use Proposition 11.

26. No.

27. Try factoring $f(x) = x^4 - 15x^2 + 1 = (x^2 + cx + 1)((x^2 - cx + 1)$ as in the proof of Proposition 11. Then we must have $-c^2 + 2 = -15$, or $c^2 = 17$. Modulo 32, $17 \equiv 9^2$, so $f(x)$ factors modulo 32. In fact, 17 is a square, hence $f(x)$ factors, modulo every power of 2: see Section 21A.

Chapter 17

1. (i) $f(x) \equiv x + 2$
   (iii) $x^9 \equiv x$

2. The units modulo $m(x)$ are $1, 2, x + 2, 2x + 1$.

8. In order to have a complete set of representatives, x must have order 24. For (i), x has order 12; for (ii), x has order 24; for (iii) x cannot have order 24 because there are fewer than 24 units modulo $x^2 + x + 4$.

9. $x^5 \equiv 1$.

10. See Table 2, Chapter 25.

12. Try choosing $f(x)$ to be a divisor of $m(x)$.

13. p^d elements.

14. (i) and (iii) are possible; (ii) is not.

15. For part (b), if $(m, f) = 1$ then there are polynomials r, s so that $mr + fs = 1$, then $fs \equiv 1 \pmod{m}$. For part (c), Let $(m, f) = d$, let $ds = m$, then $fs \equiv 0$ but $f, s \not\equiv 0 \pmod{m}$.

16. (i) a must be coprime to m. $\phi(m)$ is the number of units modulo m = the number of numbers $< m$ that are coprime to m.

(ii) $\phi_p(m)$ is the number of polynomials in $\mathbb{F}_p[x]$ of degree $< \deg m$ that are coprime to m.

(iii) $\phi_p(m) = 6$.

17. If $p(x) = b$ is a non-zero element of F, then b divides every $f(x)$, hence every difference $f(x) - g(x)$.

18. If $f(a) = g(a)$ for all a in $\mathbb{F}_p$, then $h(x) = f(x) - g(x)$ is a multiple of $x - a$ for every a, by the Root Theorem. Thus $x(x-1)(x-2) \cdots (x-(p-1)) = x^p - x$ divides $f(x) - g(x)$.

19. (i) $f(x) = x^3$.

21. $q(x) = x^3 + 2x^2 + 1$.

22. (iii) For each c of degree $< \deg mn$, let $a = c \bmod m, b = c \bmod n$. If $(c, mn) = 1$, then $(a, m) = 1$ and $(b, n) = 1$. Conversely, let $(a, m) = 1$ and $\deg a < \deg m$, let $(b, n) = 1$ and $\deg b < \deg n$. Then there is a unique c with $\deg c < \deg mn$ and $c \equiv a$ $(\bmod\ m), c \equiv b \pmod{n}$, and $(c, mn) = 1$. So there is a one-to-one correspondence between c with $\deg c < \deg mn$ and $(c, mn) = 1$, and pairs (a, b) with $(a, m) = 1$ and $\deg a < \deg m$, and $(b, n) = 1$ and $\deg b < \deg n$.

24. $f(x) = (-5/24)x^2 + (4/3)x - (1/8)$

26. $a_{(-1,1,5)}(x) = x^2 + 3x + 1$

28. If $a_s(x)$ divides $f(x)$, then so does $-a_s(x) = a_{-s}(x)$.

29. The polynomial $a_s(x) = 1$ corresponds to $s = \{1, 1, 1, \ldots 1\}$.

30. For each $i, 1 \le i \le d$, s_i has four possibilities. So there are 4^d possible vectors s. By the comment on Exercise 28, there are $4^d/2 = 2^{2d-1}$ possible polynomials $a_s(x)$ up to associates.

Chapter 18

2. The k-th coefficient c_k of $f(x)g(x)$ is $a_0b_k + \ldots a_kb_0$ if $k \le d$, and is $a_rb_d + \ldots + a_db_r$ if $k = d + r$. The result follows from the triangle inequality.

8. The MAPLE command "isprime(m)" is very helpful: I found $7 \cdot 2^{20} + 1$ and $177 \cdot 2^{100} + 1$ in a few minutes.

9. Dirichlet's Theorem says that if $(a, m) = 1$ then there are infinitely many primes in the congruence class $a \bmod m$. To find a prime p so that $p - 1$ is a multiple of 2^r, find a prime p congruent to 1 modulo 2^r.

11. $3^{r-1} > 2^r(3r - 1)$ for $r = 12$: $3^{11} = 177147 > 143360 = 2^{12} \cdot 35$.

Chapter 19

1. (i) 2; (ii) 6, (iii) 4.

7. $\mathbb{Z}/m\mathbb{Z}$ is cyclic, so the exponent is $\lambda = m$.

8. One way to produce examples is to observe that the order of a is the same as the order of a^{-1}.

9. (i) 2, (ii) -2, (iv) 21.

11. See the remark for Exercise 8.

13. (ii) For $i = 1, \ldots, r$ pick a primitive root b_i modulo p_i, and let $a \equiv b_i$ modulo p_i.

15. One generator is $[3]_{10}$.

17. Most classes $[a]_{49}$ where a is a number congruent to 3 or 5 modulo 49 will generate U_{49}.

19. $\phi(28) = 12, \lambda(28) = 6$.

21. All $[a]$ with $(a, n) = 1$ will generate $\mathbb{Z}/n\mathbb{Z}$.

22. Observe that if b is a primitive root, then b^r has order $\frac{(p-1)}{(r, p-1)}$.

23. The exponent is 3.

24. No; no.

28. Let c be any number $<p$. Then $(c, p^e) = 1$, so $c \equiv b^s \pmod{p^e}$, hence $c \equiv b^s \pmod{p}$.

29. Given a number $a < p$, $a^{p-1} = 1 + ps$ for some s. If $b = a + rp$, then

$$b^{p-1} \equiv 1 + p(s - a^{p-2}r) \pmod{p^2},$$

so $b^{p-1} \equiv 1 \pmod{p^2}$ for a unique number r modulo p.

30. (i) Solve the four systems $x \equiv \pm 1 \pmod{r}, x \equiv \pm 1 \pmod{s}$.

31. (i) The product of all elements of G is $(b^m)^{2m+1} = b^m$.

32. Use that there is some unit of order $\lambda(m)$.

33 (i) Observe that if a in U_r has order d and b in U_s has order e, then (a, b) in $U_r \times U_s$ has order $[d, e]$.

34. (ii) $|U_{105}| = (104, 2)(104, 4)(104, 6) = 16$, while $\phi(105) = 48$.

35. I found $m = 1767 = 3 \cdot 19 \cdot 31$.

36. (ii) $p(3) \equiv 2 \pmod{11}$.

39. $(r, s) = (3, 32)$.

40. $w = 19, v = 3^{33 \cdot 19} 49^{31 \cdot 19} \pmod{83}$.

43. Show that since $(A, M/3) = 1$, the order of A modulo $M/3$ will be the period of A modulo M.

Chapter 20

1. $1105 = 5 \cdot 13 \cdot 17$ and $1104 = 16 \cdot 69 = 12 \cdot 92 = 4 \cdot 276$.

3. If $k \equiv 2 \pmod 5$ then $12k + 1$ is a multiple of 5, etc.

4. Use Example 3.

5. (i) If $r \equiv 1 \pmod 7$ then $90r + 1$ is a multiple of 7, etc.

6. (i) Use Korselt's criterion.

(iv) No. Put the right side over the common denominator $d_1 d_2 \cdots d_n$. If a prime p divides d_1, then it must divide $d_2 \cdots d_n$.

7. (i) It suffices to show that G is closed under multiplication.

9. One example is $37 \cdot 73 \cdot 181$.

10. The only relevant primes p not congruent to 1 modulo 5 so that $p-1$ divides 120 are 7 and 13. Consider the four cases where a product contains: neither 7 nor 13; just 7; just 13; both 7 and 13.

12. There is an obvious number a with $(a,m) = 1$ and $a^{m-1} \not\equiv 1 \pmod{m}$.

14. Find a so that $a^{45} \equiv 1 \pmod{91}$ by finding a so that $a^{45} \equiv 1 \pmod{7}$ and mod 13. If $a^{45} \equiv 1 \pmod{7}$, then $a^3 \equiv 1 \pmod{3}$, so $a \equiv 1, 2, 4$. Similarly modulo 13. So there are at least eight non-trivial a with $a^{45} \equiv 1 \pmod{91}$. (Then there are some a with $a^{45} \equiv -1 \pmod{91}$.)

16. 35 is an a-pseudoprime for all a for which $a^2 \equiv 1 \pmod{35}$, namely $a = 1, -1, 6, -6$.

19. Since $\lambda(1729) = 36$ and 36 divides 1728, 1729 is Carmichael. Now $1728 = 2^6 \cdot 27$, and since $a^3 6 \equiv 1$, $a^{27 \cdot 4} \equiv 1$ for all a. Also, $a^{27 \cdot 2} \equiv 1$ modulo 7 and modulo 19, and $\equiv 1$ or -1 modulo 13. So

$$\{a \,|\, a^{54} \equiv 1 \pmod{1729}\}$$

has index 2 in U_{1729}. The other coset is

$$\{a \,|\, a^{54} \equiv 1 \bmod 7 \text{ and } \bmod 19, \text{ and } \equiv -1 \pmod{13}\}$$
$$= \{a \,|\, a^{54} \equiv 1065 \pmod{1729}\}.$$

Thus 1729 is not a strong a-pseudoprime for those a with $a^{54} \equiv 1065$, and that set includes half of the elements of U_{1729}.

21. $(10651, 1729) = 13$.

26. Let $b-1 = c, b+1 = d$. Now (m, cd) divides $(m, c)(m, d)$ (apply Bezout's identity to the two factors). For the other direction, show that (m, c) and (m, d) are coprime. Since (m, c) and (m, d) divide (m, cd), then $(m, c)(m, d)$ divides (m, cd).

28. Use that $2^p \equiv 1 \pmod{M_p}$ and that $2^{p-1} - 1$ is odd and a multiple of p.

29. Try $a = 3$.

30. Show that the exponent $\lambda(m) = (p - 1((q-1)/2)$ satisfies $(m-1, \lambda(m)) = 2$, hence $U_m(m-1) = U_m(2)$.

31. Show that $U_m(m-1) = U_m(2)$, as in Exercise 30.

34. From Exercise 33 you can factor $m = pq$, find $\phi(m)$, and solve $133d \equiv 1 \pmod{\phi(m)}$.

Chapter 21

3. First show that $w \equiv 6$ (or 19) modulo 25. Let $w = 6 + 25k$ and find k so that $w^2 \equiv 61$ modulo 125.

5. No. To compute the Legendre symbol one can start either by observing that $45 \equiv -2 \pmod{47}$ or by noticing that $45 = 3^2 \cdot 5$.

8. 14 is a square modulo 65 iff 14 is a square modulo 5 and modulo 13.

11. Apply the Quadratic Formula.

12. One way to do this is to use the fact that every $r \not\equiv 0$ modulo p is (mod p) the power of a primitive root.

15. There is a solution.

16. See Exercise 11.

20. p must be $\equiv 1$ modulo 4.

24. (i) Let $m = p^r q$ with p prime and $(p,q) = 1$. Let b be a primitive root modulo p^r, and let $a \equiv b \pmod{p^r}, a \equiv 1 \pmod{q}$.

(ii) Use that $\left(\frac{-}{m}\right) : U_m \to \{\pm 1\}$ is a homomorphism of groups that is onto by (i); then apply the Fundamental Homomorphism Theorem (Theorem 17) of Chapter 11 and Lagrange's Theorem.

25. 3/4 of (1/2 of $\phi(1729)$).

26. Show that if k is odd, then $a^{(m-1)/2} \equiv 1 \pmod{m}$ for all a coprime to m, by showing that the exponent $\lambda(m) = 36k$ of U_m divides $\frac{m-1}{2}$.

Chapter 22

1. Compute a Legendre symbol by showing that $p \equiv 1 \pmod 5$ and $p \equiv 7 \pmod 8$.

2. (ii) Show that $p \equiv 1 \pmod 4$ and $p \equiv 3$ or 5 $\pmod 7$, hence $\left(\frac{7}{p}\right) = -1$.

3. Follow the proof of Proposition 5.

6. One approach: find a special prime $p = 2p_1 + 1$ with $p_1 = 2p_2 + 1$ so that modulo p_1, 2 has order p_2 and $50 < p_2 < 100$, or 2 has order $2p_2$ and $50 \leq 2p_2 \leq 100$.

8. (i) $[52, 20] = 260$. (iii) Try $y = 132^{82}$ or $y = 132^{106}$.

9. $(c, z) = (1, 0, 0, 82)$.

10. $m = (1, 0, 1)$.

11. See Lemma 10.

12, Let $p = 2p_1 + 1$ with p, p_1 prime. Observe that if a has order p_1 or $2p_1$ modulo p, then a^2 has order p_1. Thus all but two of the $p - 1$ units y modulo p have y^2 of order p_1. Then apply the Chinese Remainder Theorem.

13. Look at p_2 modulo 2, 3, and 5, remembering that, for example, none of p_2, p_1 and p can be congruent to zero.

15. Write $y = z^2$ modulo p and show that both maps yield z modulo p.

20. (i) 241 is a factor; (ii) 47 is a factor; (iii) 401 is a factor.

21. (i) $t_{max} = 3; E(t) = 1.57$.

22. There are two numbers, 13 and 16, for which $t = 12$. Thus the chance is $2/29$.

24. One way to do this is as follows. Since $\alpha(x) = \alpha(-x)$, it suffices to find a number x so that both x and $-x$ are in $1 + QR_p$. Let $\alpha(a) = x, \alpha(b) = -x$, then $1 + a^2 \equiv -1 - b^2 \pmod p$, or $a^2 + b^2 \equiv -2 \pmod p$. We can find a prime q so that $q \equiv -2 \pmod p$ and $q \equiv 1 \pmod 4$, for by the Chinese Remainder Theorem those two congruences are equivalent to a single congruence for which Dirichlet's Theorem applies. Since that prime $q = a^2 + b^2$ for some a, b, that solves the problem.

Chapter 23

3. See Section 25A, Table 1.

4 iii) $\alpha^6 = \alpha + 2$.

5. The units can all be written as powers of α, where $\alpha^8 = 1$. So the inverse of α^r is α^{8-r}.

6. One way to do this is to observe that

$$0 = (\alpha^3 + \alpha + 1)^2 = \alpha^6 + \alpha^2 + 1 = (\alpha^2)^3 + \alpha^2 + 1.$$

7. The roots of $x^3 + 2x + 2$ in $\mathbb{F}_3[\alpha]$ are α, α^3 and α^9.

8. See Chapter 1.

9. The zero divisors are all the non-zero multiples of $[x+1]$

11. The units are the congruence classes of $1,2,3,4$ and $c(x+1)$ and $d(x-1)$ where $c,d = 1,2,3,4..$

12 iv) One answer is $\mathbb{F}_2[x]/(x^5 + x^2 + 1)$, since $x^5 + x^2 + 1$ is irreducible in $\mathbb{F}_2[x]$.

13. $\mathbb{F}_7[x]/(p(x))$ where $p(x)$ is any of the 112 irreducible polynomials in $\mathbb{F}_7[x]$ of degree 3. (See Section 27A.)

14 i) The inverse of $x^2 + x + 1$ mod $x^3 + 2x + 1$ is $2x^2 + x + 1$.

16. The inverse of x^2 is $x^3 + 1$. The inverse of x^3 is $x^4 + x^2 + x$.

17. Use that $\cos 3x = 4\cos^3 x - 3\cos x$, and that $\cos 60° = \cos(60 \pm 360)°$.

18. In $E[x]$, if $\alpha = [x]$, then $x^4 + x^2 + 1$ has roots α, α^2 and α^4, and Since $(x-1) \cdot (x^2 + x + 1) = x^3 - 1$ has roots the cube roots of $1 = \alpha^{15}$, the roots of $x^2 + x + 1$ are α^5 and α^{10}.

19. Use Fermat's Theorem to find p roots of $f(x)$ in $\mathbb{F}_p$.

20. The roots of $x^n - 5$ in $\mathbb{C}$ are $\zeta^k (5)^{1/n}$ for $k = 1, \ldots, n$, where $\zeta = e^{2i\pi/n}$. These roots are not all real if $n > 2$.

21. If $[r] = [s]$, then $m(x)$ divides $r - s$, impossible if $r \neq s$ and $\deg m(x) \geq 1$.

Chapter 24

3. $(x - (a + b\sqrt{2})((x - (a - b\sqrt{2}) = x^2 - 2ax + (a^2 - 2b^2)$.

4. $(x^2 - 1)^2 - 8x^2 = x^4 - 10x^2 + 1$.

5. $\cos(4\theta) = 2\cos^2(2\theta) - 1 = 2(2\cos^2\theta - 1)^2 - 1$. Let $\theta = 15°$.

6. Let β be in $\mathbb{C}$ and let $\phi([x]) = \beta$. Then $\phi([f(x)]) = \phi(f([x])) = f(\beta)$. In order that ϕ is well-defined on $[f(x)]$, we must have that if $g(x) = f(x) + p(x)$, then $g(\beta) = f(\beta)$. But that's true only when $p(\beta) = 0$. So β must be a root of $p(x)$.

7. Let $m(x)$ be the minimal polynomial of α in $F[x]$. Then $m(x)$ divides $p(x)$. Since $p(x)$ is irreducible and both $m(x)$ and $p(x)$ are monic, they must be equal.

8. i) $[x] = \alpha$ doesn't work because $\alpha^5 = 1$. There are eight choices for β.

    iii) Let $\phi_\beta : \mathbb{F}_2[x] \to \mathbb{F}_2[x]/(x^4 + x^3 + x^2 + x + 1)$ be the evaluation homomorphism, and apply the Isomorphism Theorem.

10. Note that if β is a root of $x^2 + x + 1$, then $\beta^3 = 1$.

12. See Section 25B, Example 3.

13. Given an irreducible polynomial in $\mathbb{F}_2[x]$ of degree 4, construct with it a field of 16 elements using Chapter 23, Proposition 4, then follow the proof of Theorem 8.

14. See the suggestion for 13.

15 i) Try $i+1$.

17. Find a root $\beta = ai+b$ of $x^2 - x + 3$ in $\mathbb{F}_7[x]/(x^2+1)$.

18. Use the Isomorphism Theorem.

20. Use $\mathbb{F}_2[x]/(x^2+x+1)$.

24. Use $\mathbb{F}_8 = \mathbb{F}_2[x]/(x^3+x+1)$–see Section 25A for a description.

27. See Section 24B, Theorem 4.

28. First show that A_p is a Latin square for each p. Then show that if $p \neq p'$ and

$$(a_p a_i + a_j, a_{p'} a_i + a_j) = (a_p a_k + a_l, a_{p'} a_k + a_l),$$

then $i = k$ and $j = l$.

29. For notation, let $A = (a_{i,j})$ for $i, j = 1, \ldots, m$ and $B = (b_{k,l})$ for $k, l = 1, \ldots, n$. Then the $(r-1)n+s$-th row of $A \times B$ is

$$(a_{r,1}, b_{s,1}), (a_{r,1}, b_{s,2}), \ldots, (a_{r,1}, b_{s,n}), (a_{r,2}, b_{s,1}), \ldots$$
$$(a_{r,2}, b_{s,n}), \ldots, (a_{r,m}, b_{s,1}), \ (a_{r,m}, b_{s,2}), \ldots, (a_{r,m}, b_{s,n})$$

Chapter 25

1. (i) $(1,0,0,0,1,0,1)$.

2. (i) $C = (0,1,1,0,0,0,1); w = (0,1,1,0)$.

4. (i) $(111,001,100,000,100)$.

5. (i) The errors correspond to α^6 and α^{14}; $w = (0,1,0,0,0,1,1)$.

8. It is helpful to observe that $S_1^2 = S_2, S_2^2 = S_4$.

9. One digit–$m_4(x)$ has degree 6.

14. See Proposition 3 of Section B.

15. Use that all units of $\mathbb{F}_8$ have order dividing 7.

16. There is one of degree 4.

Chapter 26

1. Do an induction argument on n, the degree of $f(x)$. The case $n = 1$ is obvious, and the induction step is Theorem 2.

5. If $f(x) = x^n + a_1 x^{n-1} + \ldots + a_n$, then Mignotte gives

$$|b| \leq \|f\| = (1 + |a_1|^2 + \ldots + |a_n|^2)^{1/2},$$

while Descartes gives

$$|b| \le |a_n| = \sqrt{|a_n|^2} < \|f\|$$

for every f.

6. i) If $|a_k| \le |a_1|^k$, then

$$\|f\| \le (|a_1|^2 + |a_1|^{2 \cdot 2} + \ldots + |a_1|^{2 \cdot 2d})^{1/2}$$

$$\le |a_1| \left(\frac{|a_1|^{2d} - 1}{|a_1| - 1} \right) \le |a_1|^{d+1}.$$

8. Observe that the squarefree part of f, namely $f/(f, f')$, $= x^7 + 1$, hence divides $x^8 - x$, which in turn is the product of all the irreducible polynomials in $\mathbb{F}_2[x]$ of degree 1 or 3. (Or use Berlekamp!)

9. See Chapter 25, Code IV.

10. Show that $\mathbf{Q} - \mathbf{I}$ has rank 4.

12. Reduce the polynomial modulo 2.

13. Show that

$$\mathbf{Q} = \begin{pmatrix} 1 & 0 \\ 0 & q^{\frac{p-1}{2}} \end{pmatrix}.$$

So $x^2 - q$ factors over (hence has a root in) $\mathbb{F}_p$ iff $\mathbf{Q} - \mathbf{I} = \mathbf{0}$ iff $q^{(p-1)/2} \equiv 1 \pmod{p}$.

15. Modulo 3, the polynomial factors as $(x^2 - 2x + 1)(x^2 + x + 2)$. Modulo 9, it factors as

$$(x^2 - 2x + 4)(x^2 + 4x + 2).$$

Chapter 27

1. The proof is similar to that of Theorem 6, with an summation interchange.

3. Do induction on the number of prime factors of n.

4. There are 18 monic polynomials of degree 2 with a non-zero constant term. Determine which of these has a root of 1 or -1 and toss them out. The remaining polynomials will be irreducible.

5. $N_{12} = \frac{1}{12}(p^{12} - p^6 - p^4 + p^2)$.

7. There are 2^8 terms.

8. For the right hand inequality, observe that if q is the smallest prime that divides n, then

$$nN_n = \sum_{d|n} \mu(d) p^{n/d} = p^n - p^{n/q} + \sum_{d|n, d>q} \mu(d) p^{n/d}$$

and

$$\sum_{d|n, d>q} \mu(d) p^{n/d} < \sum_{k<n/q} p^k < p^{n/q}.$$

For the left inequality,

$$nN_n > p^n - \sum_{d|n,d<n} p^d > p^n - \sum_{k=1}^{n/2} p^k;$$

summing the geometric series, show that

$$nN_n \geq \frac{p^n}{n}(1 - \frac{2}{p^{n/2}})$$

and $\varepsilon = \frac{2}{p^{n/2}}$ goes to 0 as n goes to infinity.

10. If F has p^n elements, then F is a splitting field for $x^{p^n} - x$, and since $q(x)$ divides $x^{p^n} - x$, $q(x)$ splits in F.

12. $x^9 - x = x(x-1)(x+1)(x^2+1)(x^2+x+2)(x^2+2x+2)$.

13. It is the product of the five monic irreducible polynomials of degree 1 and the ten monic irreducible polynomials of degree 2 in $\mathbb{F}_5[x]$.

15. The roots are $0, \alpha^5, \alpha^{10}, 1$.

16. If $m(x)$ divides $x^{p^n} - x$, then $m(x)$ is a product of irreducible polynomials of degrees d dividing n. If $q(x)$ is an irreducible factor of $m(x)$ of degree $d < n$, then $q(x)$ divides $(m(x), x^{p^d} - x)$, so if $(m(x), x^{p^d} - x) = 1$, then $m(x)$ cannot be divisible by an irreducible polynomial of degree $d < n$.

17. The equivalence of iii) and iv) follows from the identity

$$(x^a - 1, x^b - 1) = (x^{(a,b)} - 1),$$

which follows from the identity: if $b = aq + r$, then

$$x^b - 1 = (x^{b-qa})(x^{qa} - 1) + (x^r - 1).$$

The equivalence of iv) and i) is from Theorem 1.

ii) implies i) follows from the congruence

$$x^{p^d} \equiv x \pmod{m(x)}.$$

For i) implies ii): first show that if $m(x)$ is Carmichael, then $m(x)$ is squarefree. To do that, let $m = f^e q$ with f irreducible, $(f, q) = 1$ and $e > 1$. Then defining a by the conditions:

$$a \equiv 1 + f \pmod{f^e}$$
$$a \equiv 1 \pmod{q},$$

Then a is coprime to m and has order a power of p modulo m, contradicting the Carmichael assumption.

Then show that if $m = q_1 q_2 \ldots q_r$ of degrees $d_1, d_2, \ldots, d_r$, then $p^{d_i} - 1$ divides $p^d - 1$, by constructing an element a by

$$a \equiv b_i \pmod{q_i}$$
$$a \equiv 1 \pmod{m/q_i},$$

where b_i is a primitive root of $\mathbb{F}_p[x]/(q_i)$. Then a has order $p^{d_i} - 1$ modulo m. Since a is coprime to m and $a^{p^d-1} \equiv 1 \pmod{m}$, $p^{d_i} - 1$ divides $p^d - 1$. (These results were obtained by H. H. Smith III in his 1991 undergraduate thesis at the Univ. at Albany.)

18. One way to proceed is to count the reducible polynomials: write $x^2 + ax + b = (x+r)(x+s)$ with r, s in $\mathbb{Z}$ and find the resulting pairs (a, b) with $-14 \le a, b \le 15$ so that there exist r, s solving

$$a = r + s, b = rs.$$

There are 119 pairs (a, b) so that $x^2 + ax + b$ is reducible. That leaves $900 - 119 = 781$ irreducible polynomials.

By comparison, there are 3 reducible polynomials of degree 2 modulo 2, 6 reducible polynomials of degree 2 modulo 3 and 15 reducible polynomials of degree 2 modulo 5. Each triple of reducible polynomials modulo 2, 3 and 5 defines a unique polynomial modulo 30 that is not provably irreducible. Thus $900 - 270 = 630$ of the 900 polynomials in $\mathbb{Z}[x]$ with coefficients bounded between -14 and 15 are provably irreducible by looking at them modulo 2, 3 or 5, and 151 of the polynomials are irreducible in $\mathbb{Q}[x]$ but are reducible modulo 2, 3 and 5.

References

Alford, W.R., Granville, A. and Pomerance, C. (1994), There are infinitely many Carmichael numbers, Ann. of Math. 139, 703–722. (Chapter 20)

Bender, E.A. and Goldman, J.R. (1975), On the application of Mobius inversion in combinatorial analysis, Amer. Math. Monthly 82, 789–802. (27)

Berggren, J. L. (2003), Episodes in the Mathematics of Medieval Islam, Springer, New York. (15)

Berlekamp, E.R. (1967), Factoring polynomials over finite fields, Bell System Tech. J. 46, 1853–1859. (22)

Blum, L., Blum, M., and Shub, M. (1986), A simple unpredictable pseudorandom number generator, SIAM J. Comput. 15, 364–383. (22)

Blum, M. and Goldwasser, S. (1985), An efficient probabilistic public key encryption scheme which hides all partial information, Proceedings of Advances in Cryptology -CRYPTO '84, pp. 289–299, Springer Verlag, New York. (22)

Blum, M., and Goldwasser, S. (1985), An Efficient Probabilistic Public Key Encryption Scheme which Hides All Partial Information, Proceedings of Advances in Cryptology -CRYPTO '84, pp. 289–299, Springer Verlag. (22)

Brent, R.P. and Pollard, J.M. (1981), Factorization of the eighth Fermat number, Math. Comp. 36, 627–630. (4, 22, 20)

Caldwell, C., primes.utm.edu (4)

Cardano, G. (1545), Ars Magna, sive de Regulis Algebraicis, Witmer, T.R., transl. and ed. (1968), MIT Press, Cambridge MA. (15)

Chernick, J. (1939), On Fermat's simple theorem, Bull. Amer. Math. Soc. 45, 269–274. (25)

Chrystal, G. (1904), Algebra, An Elementary Textbook, repr. Chelsea, 1980. (15)

Cipra, B. (1993), The ubiquitous Reed-Solomon codes, SIAM News. January, 1993. www.eccpage.com/reed_solomon_codes.html (29)

Cipra, B. (1993). The FFT: Making technology fly, SIAM News, May, 1993. (18)

Cochran, W.T., et al. (1967), What is the fast Fourier transform?, Proc. IEEE 55, 1664–1674. (18)

Conrad, K., Euclidean proofs of Dirichlet's theorem, www.math.uconn.edu/~kconrad/blurbs/gradnumthy/dirichleteuclid.pdf (22)

Crandall, R. E. and Pomerance C. (2005), Prime Numbers, A Computational Perspective, 2nd edn. Springer, New York. (10)

Crandall, R.E. and Penk, M.A. (1979), A search for large twin prime pairs, Math. Comp. 33, 383–388. (22)

Dejon, B. and Henrici, P. (1969), Constructive Aspects of the Fundamental Theorem of Algebra, Wiley, London. (16)

Delaurentis, J.M. (1984), A further weakness in the common modulus protocol for the RSA cryptoalgorithm, Cryptologia 8, 253–259. (26)

Descartes, R. (1637), La Geometrie, Dover, New York. (1, 16)

Dickson, L. E. (1901), Linear Groups, With An Exposition Of The Galois Field Theory, dlxs2.library.cornell.edu/m/math/ (24)

Diffee, W. and Hellman, M. (1976), New directions in cryptography, IEEE Trans. Inform. Theory IT22, 644–654. (10, 19)

Dobbs, D., and Hanks, R. (1992), A Modern Course on the Theory of Equations, 2nd ed., Polygonal Publ. House. (23)

Driver, E., Leonard, P. A. and Williams, K. S. (2005), Irreducible quartic polynomials with factorizations modulo p, American Mathematical Monthly 112, 876–890. (16)

Edwards, C. H. Jr. (1979), The Historical Development of the Calculus, Springer-Verlag, New York (19)

Euclid (300 B.C.), The Elements, Heath, T.L., trans. (1925–36), Dover, New York. (2, 3, 4)

Fefferman, C. (1967), An easy proof of the fundamental theorem of algebra, Amer. Math. Monthly 74, 854–855. (15)

Fisher, R.A. (1935), The Design of Experiments, Oliver and Boyd, Edinburgh. (24)

Fishman, G.S. and Moore, L.R. (1982), A statistical evaluation of multiplicative congruential random number generators with modulus 2^31 -1, J. Amer. Statist. Assoc. 77, 129–136. (19)

Gauss, K.F. (1801), Disquisitiones Arithmeticae, transl. by A.A. Clarke, New Haven, corrected 2nd edn. (1986), Springer-Verlag New York (5)

Gillings, R.J. (1972), Mathematics in the Age of the Pharaohs, Dover, New York. (2)

GIMPS (2008), www.mersenne.org (10)

Goldstine, H.H. (1972), The Computer from Pascal to von Neumann, Princeton University Press, Princeton, NJ. (2)

Graham, W. (1984), Divisibility of polynomial expressions, Math. Mag. 57, 232. (5)

Granville, A. (1992), Primality testing and Carmichael numbers, Notices Amer. Math. Soc. 39, 696–700. (20)

Granville, A. (2004), Smooth numbers: computational number theory and beyond, in J. Buhler and P. Stevenhagen, eds. (2008), Algorithmic Number Theory, Cambridge Univ. Press. (10)

Greathouse, C. and Weisstein, E. W. Odd Perfect Number. From MathWorld–A Wolfram Web Resource. http://mathworld.wolfram.com/OddPerfectNumber.html (10)

Guinness Book of Records (1993), edited by P. Matthews, Bantam Books, New York. (4, 10)

Hamming, R.W. (1950), Error detecting and error correcting codes, Bell System Tech. J. 29, 147–160. (8)

Hammond, A. L. (1978), Mathematics, our invisible culture, in Steen, L. A., Mathematics Today, Springer-Verlag, New York. (intro)

Hardy, G.H. (1940), A Mathematician's Apology, forward by C.P. Snow, Cambridge University ress, Cambridge, UK. 1967. (10)

Hardy, G.H. and Wright, E.M. (1979), An Introduction to the Theory of Numbers, 5th edn, Oxford University Press, New York. (3)

Hartshorne, R., http://math.berkeley.edu/~robin/Viete/construction.html (15)

Hellman, M.E. (1979), The mathematics of public-key cryptography, Scientific American, August 1979, 146–157. (19)

Herstein, I (1975), Topics in Algebra, Wiley. (11)

Hill, L.S. (1931), Concerning certain linear transformation apparatus of cryptography, Amer. Math. Monthly 38, 135–154. (8)

Hodges, A. (1983), Alan Turing: The Enigma, Simon and Schuster, New York. (8)

Holdener, J. A. (2002), A Theorem of Touchard on the Form of Odd Perfect Numbers, American Mathematical Monthly 109, 661–663. (10)

Joseph, G. G. (2000), The Crest of the Peacock, Non-European Roots of Mathematics, Princeton Univ. Press. (12)

Kahn, D. (1967), The Codebreakers, Macmillan, New York. (8)

Karatsuba, A. and Ofman, Yu (1962), Multiplication of Many-Digital Numbers by Automatic Computers. Doklady Akad. Nauk SSSR, Vol. 145 (1962), pp. 293–294. (7)

Keller, W., www.prothsearch.net/fermat.html (4)

Knuth, D.E. (1998), The Art of Computer Programming, 3rd edn, Vol. 2, Addison-Wesley, Reading, MA. (2, 3, 19, 20)

Koblitz, N. (1994), A Course in Number Theory and Cryptography, Springer, New York (19)

Kolata, G., 100 Quadrillion Calculations Later, Eureka!, New York Times, April 27, 1994 (4)

Konheim, A.G. (1981), Cryptography, A Primer, Wiley, New York. (8)

Lemmermeyer, F. (2000), Reciprocity Laws: From Euler to Eisenstein, Springer-Verlag, New York (21)

Lemmermeyer, F., www.rzuser.uni-heidelberg.de/~hb3/rchrono.html (21)

Lenstra, A.K. and Lenstra, H.W., Jr. (1993), The development of the number theory sieve, Springer Lecture Notes in Mathematics, 1554. (4)

Lenstra, A.K., Lenstra, H.W., Jr., Manassse, M.S., Pollard, J.M. (1993), The factorization of the ninth Fermat number, Math. Comp. 61, 319–349. (4)

MacWilliams, F.J. and Sloane, N.J.A. (1983), The Theory of Error-Correcting Codes, North-Holland, Amsterdam. (25)

Mann, H.B. (1949), Analysis and Design of Experiments, Dover, New York. (24)

McCurley, K. (1989), A key distribution system equivalent to factoring, J. Cryptology 1, 95–105. (22)

Mehl, S.: serge.mehl.free.fr/chrono/Bachet.html (3)

Mignotte, M. (1974), An inequality about factors of polynomials, Math. Comp. 28, 1153–1157. (26)

Montgomery, P. L. (1985), Multiplication without trial division, Mathematics of Computation 44, 519–521. (9)

Morrison, M.A. and Brillhart, J. (1975), A method of factoring and the factoring of F_7, Math. Comp. 29, 183–205. (4)

National Institute of Standards and Technology, www.itl.nist.gov/fipspubs/fip186.htm (19)

Park, S.K. and Miller, K.W. (1988), Random number generators: Good ones are hard to find, Comm. ACM 31, 1192–1201. (19)

Pinch, R.G.E. (1993), Some primality testing algorithms, Notices Amer. Math. Soc. 40, 1203–1210. (20)

Pless, V. (1998), Introduction to the Theory of Error-Correcting Codes, 3rd edn., Wiley, New York. (25)

Pollard, J.M. (1974), Theorems of factorization and primality testing, Proc. Cambridge Philos. Soc. 76, 521–538. (10)

Pollard, J.M. (1975), A Monte Carlo method for factorization, BIT 15, 331–334. (22)

Polya, G. and Szego, G. (1972), Problems and Theorems in Analysis. Volume I, Springer, Berlin. (26)

Pomerance, C. (1984), Lecture notes on primality testing and factoring, MAA Notes # 4. (10)

Pomerance, C., Ed. (1990), Cryptology and computational number theory, Proc. Symposia in Applied Math, American Mathematical Society. (10)

Pomerance, C., Selfridge, J.L., and Wagstaff, S.S. (1980), The pseudoprimes to 25 x 109, Math. Comp. 35, 1003–1026. (10)

R. C. Bose, Shrikhande, S. S. and Parker, E. T. (1960), Further Results on the Construction of Mutually Orthogonal Latin Squares and the Falsity of Euler's Conjecture, Canadian Journal of Mathematics, 12, 189–203. (24)

Rabin, M.O. (1980a), Probabilistic algorithm for testing primality, J. Number Theory 12, 128–138. (20)

Rabin, M.O. (1980b), Probabilistic algorithms in finite fields, SIAM J. Comput. 9, 273–280. (27)

Ribenboim, P. (1979), 13 Lectures on Fermat's Last Theorem, Springer-Verlag, New York. (10)

Ribenboim, P. (1980), The Book of Prime Number Records, 2nd edn. Springer-Verlag, New York. (10, 22)

Ribenboim, P. (1996), The New Book of Prime Number Records, Springer-Verlag, New York. (22)

Rivest, R., Shamir, A., and Adleman, L. (1978), A method for obtaining digital signatures and public-key cryptosystems, Comm. ACM 21, 120–126. (10)

Rotman, J. (2000), A First Course in Abstract Algebra (2nd. Ed.), Prentice Hall. (11)

Rousseau, G. (1991), On the quadratic reciprocity law, J. Australian Math. Soc. 51, 423–425. (21)

RSA Laboratories (2004), www.rsa.com/rsalabs/node.asp?id=2004 (10)

Saidak, F. (2006), A New Proof of Euclid's Theorem, American Mathematical Monthly 113, 937–938. (4)

Schmandt-Besserat, D. (1993), Before Writing, University of Texas Press, Houston, TX. (1)

Sigler, L. E.,tr. (2003), Fibonacci's Liber Abaci, Leonardo Pisano's Book of Calculation, Springer, New York (3, 5)

Sloane, N. J. A., The On-Line Encyclopedia of Integer Sequences, research.att.com/~njas/sequences/Seis.html (22)

Suzuki, J. (2006), Lagrange's proof of the fundamental theorem of algebra, American Mathematical Monthly 113 (October), 705–714. (23)

Terkelson, F. (1976), The fundamental theorem of algebra, Amer. Math. Monthly 83, 647. (15)

Van der Waerden, B.L. (1934), Die Seltenheit der Gleichungen mit Affekt, Math. Ann. 109, 13–16. (27)

Van der Waerden, B.L. (1983), Geometry and Algebra in Ancient Civilizations, Springer-Verlag, New York. (12)

Van der Waerden, B.L. (1985), A History of Algebra, Springer-Verlag, New York. (8)

Van Lint, J.H. (1982), Introduction to Coding Theory, Springer-Verlag, New York. (25)

von der Gathan, J. and Shoup, V. (1992), Computing Frobenius maps and factoring polynomials, Comput. Complexity 2, 187–224. (22)

Weil, A. (1984), Number Theory, An Approach Through History, Birkhauser, Boston. (10)

Wicker, S. B. and Bhargava, V. K (eds) (1999), Reed-Solomon Codes and Their Applications, Wiley, New York. media.wiley.com/product_data/excerpt/19/07803539/0780353919–2.pdf (25)

Wunderlich, M. and Selfridge, J. (1974), A Design for a Number Theory Package with an Optimized Trial Division Routine. Commun. ACM 17(5): 272–276 (6)

Young, J. and Buell, D.A. (1988), The twentieth Fermat number is composite, Math. Comp. 50, 261–263. (4)

Zagier, D. (1977), The first 50 million prime numbers, Math. Intelligencer 0, 7–19. (4, 10)

Zassenhaus, H. (1978), A remark on the Hensel factorization method, Math. Comp. 32, 287–292. (26)

Index